Introduction to Complex Analysis and Its Applications

Donald W. Trim
The University of Manitoba

PWS Publishing Company

An International Thomson Publishing Company

Boston • Albany • Bonn • Cincinnati • Detroit • London • Madrid
Melbourne • Mexico City • New York • Paris • San Francisco • Singapore
Tokyo • Toronto • Washington

PWS PUBLISHING COMPANY
20 Park Plaza, Boston, MA 02116-4324

To my wife; things are never as complex to her as they seem to me.

I(T)P™
International Thomson Publishing
The trademark ITP is used under license

For more information, contact:

PWS Publishing Company
20 Park Plaza
Boston, MA 02116

International Thomson Publishing Europe
Berkshire House I68–I73
High Holborn
London WC1V 7AA
England

Thomas Nelson Australia
102 Dodds Street
South Melbourne, 3205
Victoria, Australia

Nelson Canada
1120 Birchmount Road
Scarborough, Ontario
Canada M1K 5G4

International Thomson Editores
Campos Eliseos 385, Piso 7
Col. Polanco
11560 Mexico D.F., Mexico

International Thomson Publishing GmbH
Königswinterer Strasse 418
53227 Bonn, Germany

International Thomson Publishing Asia
221 Henderson Road
#05–10 Henderson Building
Singapore 0315

International Thomson Publishing Japan
Hirakawacho Kyowa Building, 31
2-2-1 Hirakawacho
Chiyoda-ku, Tokyo 102
Japan

Library of Congress Cataloging-in-Publication Data
Trim, Donald W.
Introduction to complex analysis and its applications / by Donald W. Trim.
p. cm.
Includes index.
ISBN 0-534-94410-8 (hardcover)
1. Mathematical analysis. 2. Functions of complex variables.
I. Title.
QA300.T69 1995
515.9—dc20 95-43659
CIP

Editorial Assistant: Anna Aleksandrowicz
Production and Design: Pamela Rockwell
Manufacturing Coordinator: Wendy Kilborn
Marketing Manager: Marianne Rutter
Interior Illustrator: Lotus Art
Compositor: Integre Technical Pub. Co., Inc.
Cover Printer: Henry N. Sawyer Co., Inc.
Text Printer and Binder: Quebecor Martinsburg

Printed and bound in the United States of America.
96 97 98 99 — 10 9 8 7 6 5 4 3 2 1

In the final chapter we discuss Laplace transforms. We develop their elementary properties and show how they can be used to replace ordinary differential equations with algebraic equations. Cauchy's integral formula and residues then lead to a direct formula for inverting Laplace transforms. Finally, the transform is used to solve partial differential equations on bounded and semi-infinite domains.

We take every opportunity to compare properties of complex functions to those of real functions. Only then can the power and elegance of complex variable theory truly be appreciated. Plenty of exercises, with answers to even-numbered ones, are provided. Answers to odd-numbered exercises are available from the publisher. Exercises are graded from routine problems for reinforcement of fundamentals to challenging problems for the more enterprising reader.

To the Student

When the domain and range of a function are sets of complex numbers, we speak of a complex function of a complex variable. This text is an introduction to the properties and applications of such functions; in particular, we develop the *calculus* of complex functions. Your earliest calculus studies involved real-valued functions $f(x)$ of a real variable x. You found their derivatives and integrals, and used these in many geometric and physical applications. Your second exposure to calculus expanded single-variable concepts to functions of many real variables $f(x, y, \ldots)$. Once again you differentiated and integrated these functions, and applied the derivatives and integrals in more complex, but more realistic, applications. It is now time to take the third and, for most, the final step: calculus of functions $f(z)$ of a complex variable z.

You are well aware of the connection between integration and differentiation in real calculus; you will be amazed at the intimate relationships among the three major topics of complex calculus — differentiation, integration, and infinite series. Properties in any one of these topics are reflected in properties in the other two, so much so, that we could commence complex calculus with any one of the three topics. Our preference is to begin with differentiation, it being, perhaps, the easiest of the three topics for most students in real calculus. We follow this with integration and infinite series.

Calculus of complex functions has many similarities to both single-variable (real) calculus and multivariable (real) calculus, and we shall point these out as discussions unfold. But there are also striking differences, and we are even more careful to draw these to your attention. For instance, in most applications of real calculus, we are concerned with points where functions are well-behaved. Contrarily, points where complex functions misbehave are often the most valuable in applications. Existence of the first derivative of a real function implies nothing about existence of higher order derivatives, whereas existence of the first derivative of a complex function implies existence of derivatives of all orders.

Complex calculus provides proofs to many results that seem otherwise intractable in real analysis; it also provides a clearer understanding of some topics in real analysis. For example, the often assumed result that every real polynomial has a zero is verified in Chapter 4; why there is a number called the *radius* of convergence

Preface

To the Instructor

This text is written for undergraduate students in science and engineering. At a brisk pace, theory and a selection of applications can be covered in one semester; for complete coverage at a more leisurely pace, allow two semesters. The presentation assumes a working knowledge of single-variable real calculus, including infinite series, and familiarity with partial differentiation and line integrals. Discussions are rigorous, but the approach is intuitive.

Chapter 1 introduces complex numbers. We add, subtract, multiply, divide, and find roots of complex numbers. We introduce the Cartesian and polar forms for complex numbers (the exponential form coming in Chapter 3), illustrate complex numbers geometrically, and solve polynomial equations. Chapter 2 begins with the algebra of complex functions, discusses limits and continuity, and ends with complex differentiation and analyticity. The fact that real and imaginary parts of analytic functions are harmonic suggests the use of complex functions for problems in electrostatics, steady-state temperature distributions, and fluid flows. An introduction to these applications is given in Chapter 2. The complex counterparts of the exponential, trigonometric, hyperbolic, logarithmic, and inverse trigonometric and hyperbolic functions are introduced in Chapter 3. We concentrate on algebraic properties of these functions and their zeros and singularities but, in preparation for conformal mapping in Chapter 7, we also stress the importance of visualizing a function $w = f(z)$ as a mapping from the z-plane to the w-plane.

In Chapter 4 we focus on properties of analytic functions through contour integration. We verify the Cauchy-Goursat theorem, introduce Cauchy's integral formulas and Poisson's integral formulas, and develop modulus principles for complex and harmonic functions. Power series and Laurent series with the subsequent classification of singularities of complex functions are discussed in Chapter 5. Chapters 4 and 5 culminate in the theory of residues in Chapter 6. The power of residues is brought out in the evaluation of real trigonometric and improper integrals, the principle of the argument, and Rouche's theorem.

Although the first six chapters emphasize algebraic consequences of analyticity, and geometric results in the context of conformal mapping are discussed in Chapter 7, we constantly stress the geometric interpretation of a function as a mapping. Having students repeatedly visualize regional mappings by exponential, trigonometric, and hyperbolic functions and their inverses, we pave the way for the application of conformal mappings to problems in heat flow, fluid flow, and electrostatics. These applications are dealt with in detail in Chapter 7, perhaps with more detail than most texts at this level, but they are introduced in Chapter 2 as illustrations of the fact that families of curves defined by harmonic conjugates are orthogonal trajectories. In this way we give students an early indication that complex functions are indeed useful in physical applications.

of a real power series becomes clear when we study complex series in Chapter 5; and why solutions of Laplace's equation do not have relative extrema is verified in Chapter 4.

We use geometric visualizations to introduce and clarify ideas whenever possible. Visualizations of real functions $f(x)$ as curves and $f(x, y)$ as surfaces are unavailable for complex functions $f(z)$. Instead, we interpret $f(z)$ as a mapping from one complex plane to another. The mapping approach helps us appreciate many of the properties of functions, and in addition, paves the way for the important topic of conformal mapping in Chapter 7 with its applications to problems in electrostatics, heat flow, and fluid flow.

The study of calculus of complex functions can be a very rewarding experience. Topics unfold naturally, and each new topic intimately relates to everything that has gone before. Proofs of most results, even the very profound, are usually quite straightforward, and the material is rich in applications. True understanding cannot be achieved simply by reading the text. You must engross yourself in the subject and we have included numerous exercises for this purpose. Try as many as you can; you will not regret the effort. Answers to even-numbered exercises can be found in Appendix A.

Finally, we wish you every success in your studies, and we hope that you will share in our fascination with the subject.

I would like to thank the staff of PWS Publishing Company for their cooperation and contributions in the publication of this text; their efforts made transition from initial manuscript to project completion as effortless as possible. I would also like to express my gratitude to the following reviewers of early versions of the manuscript:

Bruce A. Barnes
University of Oregon

James W. Carlson
Creighton University

Peter Colwell
Iowa State University

Arlo D. Davis
Indiana University of Pennsylvania

Patrick N. Dowling
Miami University

Bruce H. Hanson
St. Olaf College

William R. Klinger
Taylor University

Charles R. MacCluer
Michigan State University

Daniel C. Weiner
Boston University

Dana Williams
Dartmouth College

The final product was much improved by their suggestions.

CHAPTER 1

Complex Numbers

When the domain and range of a function are sets of complex numbers, we speak of a complex function of a complex variable. In this chapter we study complex numbers in preparation for complex functions. We add, subtract, multiply, and divide complex numbers and find their roots. We introduce the Cartesian and polar forms of complex numbers (a third form is introduced in Chapter 3), illustrate complex numbers in the complex plane, and solve polynomial equations.

1.1 Introduction

Perhaps the most natural way to introduce complex numbers is through real quadratic equations, or equations of the form

$$ax^2 + bx + c = 0, \tag{1.1}$$

where a, b, and c are real numbers. It is well known that when the discriminant $b^2 - 4ac$ is positive, this equation has two real distinct roots,

$$x = \frac{-b \pm \sqrt{b^2 - 4ac}}{2a}, \tag{1.2}$$

and when $b^2 - 4ac = 0$, there is a repeated real root, $x = -b/(2a)$. But when $b^2 - 4ac < 0$, the quadratic equation has no real roots.

Complex numbers enable us to solve quadratic equations with negative discriminants. The most fundamental complex number is denoted by i; it is defined as a number whose square is -1:

$$i^2 = -1. \tag{1.3}$$

In other words, i is a solution of the quadratic equation

$$z^2 + 1 = 0 \tag{1.4}$$

(with discriminant equal to -4).

The introduction of i is often met with apprehension because it does not conform to our past experience with real numbers. But the transition from real numbers to complex numbers is really not that difficult if we consider the analogy with the transition from rationals to reals. All rational numbers satisfy linear equations

$$ax + b = 0 \qquad (a \neq 0),$$

with integer coefficients a and b. For example, $3/2$ satisfies $2x-3=0$. The quadratic equation $x^2-3=0$, on the other hand, does not have rational solutions. We denote the solutions of $x^2-3=0$ by $\pm\sqrt{3}$. But what is $\sqrt{3}$? It is not a rational number: It cannot be expressed as an integer divided by an integer, and it does not have a repeating decimal representation. We usually say that $\sqrt{3}$ is a number that gives 3 when multiplied by itself. Over the years we have accepted this explanation and now regard $\sqrt{3}$ with no trepidation whatsoever. The same is true when we extend the real number system to the complex number system. Equation 1.4 has no real solutions. We denote a solution of the equation by the letter i. But what is i? It is a number that, when multiplied by itself, gives -1. As we see then, the definition of i is much the same as the definition of $\sqrt{3}$, and with a little experience, we shall regard i with no more apprehension than $\sqrt{3}$.

Exercises 1.1

In these exercises we discuss operations on real numbers of the form $a+b\sqrt{3}$, where a and b are rational numbers. Each operation is directly analogous to an operation on complex numbers in Section 1.2.

1. Show that $a+b\sqrt{3}=c+d\sqrt{3}$ if and only if $a=c$ and $b=d$.
2. How do we add $a+b\sqrt{3}$ and $c+d\sqrt{3}$?
3. How do we subtract $c+d\sqrt{3}$ from $a+b\sqrt{3}$?
4. How do we multiply $a+b\sqrt{3}$ and $c+d\sqrt{3}$?
5. How do we divide $a+b\sqrt{3}$ by $c+d\sqrt{3}$?

1.2 Complex Numbers

The fundamental complex number is i, a number whose square is -1. The complete set of complex numbers is introduced in the following definition.

Definition 1.1
The **complex number system** C consists of all numbers of the form

$$z=x+yi, \tag{1.5}$$

where x and y are real and i satisfies $i^2=-1$.

Examples are $4+2i$, $3-2i$, $-3+\pi i$, and $-6-3i$. We liken $x+yi$ to numbers of the form $a+b\sqrt{3}$ in Exercises 1.1; we could have $4+2\sqrt{3}$, $3-2\sqrt{3}$, $-3+\pi\sqrt{3}$, and $-6-3\sqrt{3}$. The number $3-2\sqrt{3}$ cannot be written exactly in a simpler form; the 3 and -2 cannot in any way be combined. Likewise, the 3 and -2 in the complex number $3-2i$ are distinct parts, so much so that we give them special names. The number x in 1.5 is called the **real part** of the complex number z; it is denoted by $x=\operatorname{Re} z$. The number y is called the **imaginary part** of z, denoted by $y=\operatorname{Im} z$. For example, $\operatorname{Re}(3-2i)=3$ and $\operatorname{Im}(3-2i)=-2$. Both the real and imaginary parts of a complex number are themselves real numbers. The real number system is a subset of $\mathcal{C}$ obtained when $y = 0$.

To appreciate many results involving complex numbers, it is helpful to have geometric visualizations. Because a complex number has two distinct parts, its real and imaginary parts, and they are independent, a two-dimensional model is required to specify $\mathcal{C}$. We choose a plane, called the *complex* (Argand) *plane*. In particular, some fixed point O is chosen to represent the complex number $0 + 0i$. Through O are drawn two mutually perpendicular axes (Figure 1.1), one called the *real axis*, and the other called the *imaginary axis*. The complex number $x + yi$ is then represented by the point x units in the real direction and y units in the imaginary direction. For example, the complex numbers $1 + 2i$, $-1 - i$, $4 - 3i$, and $-2 + 2i$ are shown in Figure 1.2. With this geometric representation, there is a one-to-one correspondence between numbers in $\mathcal{C}$ and points in the complex plane. In addition, the real number system (which is a subset of $\mathcal{C}$) is represented by points on the real axis.

The point in the complex plane representing the number $x+yi$ is the point (x, y) in the Cartesian plane, and we therefore call $x+yi$ the *Cartesian form* of a complex number.

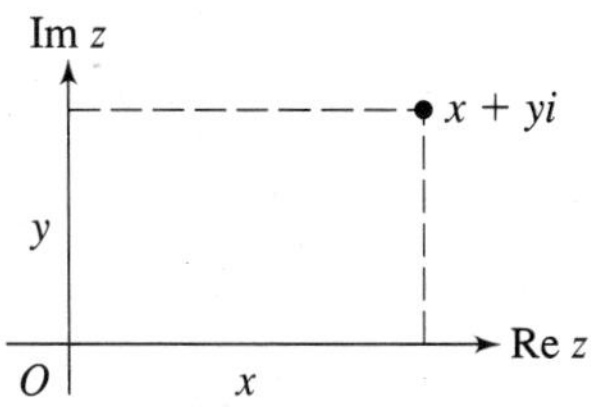

Figure 1.1

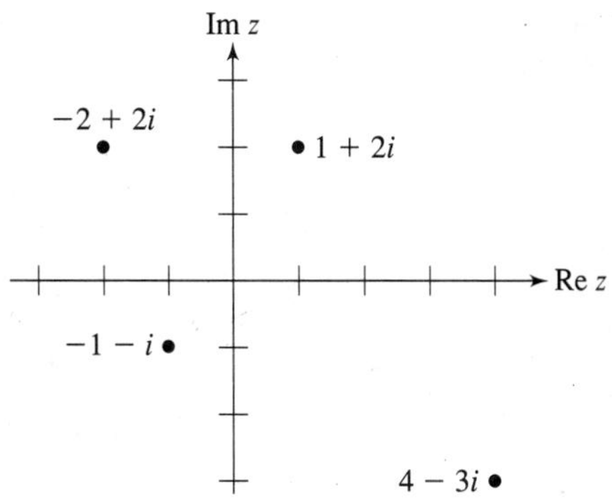

Figure 1.2

In order to add, subtract, multiply, and divide complex numbers, we first define what it means for two complex numbers to be equal, and we do so by analogy with Exercise 1 in Section 1.1.

Definition 1.2
Two complex numbers $z_1 = x + yi$ and $z_2 = a + bi$ are said to be equal if their real and imaginary parts are identical; that is,

$$x + yi = a + bi \quad \Leftrightarrow \quad x = a \text{ and } y = b. \tag{1.6}$$

Geometrically, two complex numbers are equal if they correspond to the same point in the complex plane. Do not pass this definition off too lightly; we use it many times in our work. The first instance is later in this section when we find square roots of complex numbers. But first we learn how to add, subtract, multiply, and divide complex numbers. Addition and subtraction of complex numbers are defined as follows.

Definition 1.3
If $z_1 = x + yi$ and $z_2 = a + bi$, then

$$z_1 + z_2 = (x + a) + (y + b)i, \tag{1.7}$$

$$z_1 - z_2 = (x - a) + (y - b)i. \tag{1.8}$$

In words, complex numbers are added and subtracted by adding and subtracting their real and imaginary parts. (Compare this with the addition and subtraction of numbers of the form $a + b\sqrt{3}$ in Exercises 2 and 3 of Section 1.1.) For example,

$$(3 - 2i) + (6 + i) = (3 + 6) + (-2 + 1)i = 9 - i,$$
$$(3 - 2i) - (6 + i) = (3 - 6) + (-2 - 1)i = -3 - 3i.$$

It is sometimes convenient to consider a complex number z to be represented geometrically by the vector joining O to the point representing z (Figure 1.3) rather than the point itself. In this case, it is clear that equation 1.7 for addition of complex numbers is precisely the law of addition of vectors. Thus, to add two complex numbers z_1 and z_2 geometrically, we perform vector addition by means of either a triangle (Figure 1.4(a)) or a parallelogram (Figure 1.4(b)).

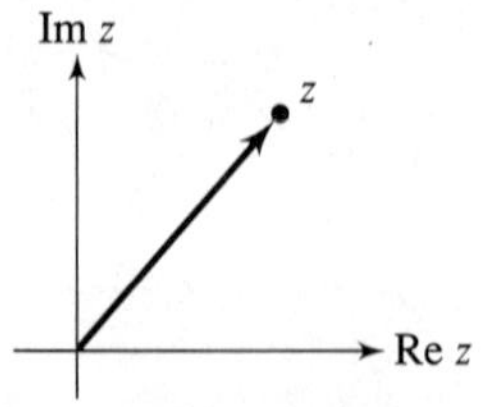

Figure 1.3

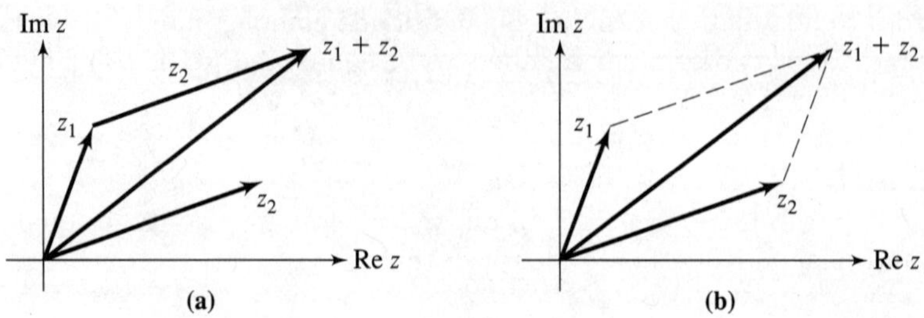

Figure 1.4

Complex numbers can be subtracted vectorially also. Subtraction by triangles is shown in Figure 1.5(a), and by parallelograms in Figure 1.5(b).

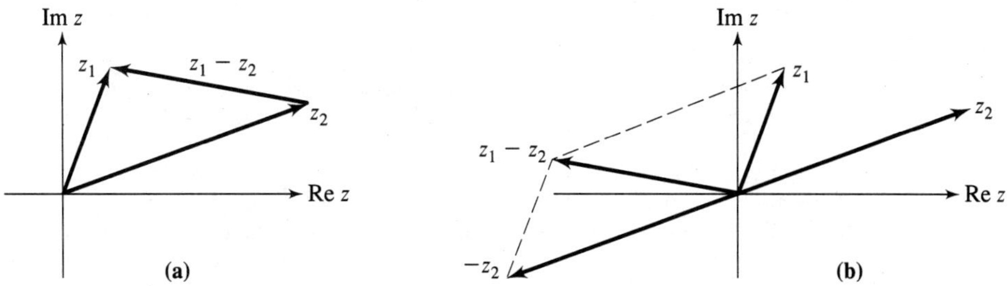

Figure 1.5

Complex numbers are multiplied according to the following definition.

Definition 1.4
If $z_1 = x + yi$ and $z_2 = a + bi$, then

$$z_1 z_2 = (x + yi)(a + bi) = (xa - yb) + (xb + ya)i. \tag{1.9}$$

For example,

$$\begin{aligned}(3 - 2i)(6 + i) &= [(3)(6) - (-2)(1)] + [(3)(1) + (-2)(6)]i \\ &= (18 + 2) + (3 - 12)i = 20 - 9i.\end{aligned}$$

It is not necessary to memorize 1.9 if we note that this definition is precisely what we would expect if the usual laws for multiplying binomials were applied, together with the fact that $i^2 = -1$:

$$\begin{aligned}(3 - 2i)(6 + i) &= (3)(6) + (3)(i) + (-2i)(6) + (-2i)(i) \\ &= 18 + 3i - 12i - 2i^2 \\ &= 18 - 9i - 2(-1) \\ &= 20 - 9i.\end{aligned}$$

With addition, subtraction, and multiplication taken care of, it is natural to turn to division of complex numbers. If we accept that division of any complex number by itself should be equal to 1 and that ordinary rules of algebra should prevail, a definition of division of complex numbers is not necessary. It follows from 1.9. When $z_1 = x + yi$ and $z_2 = a + bi$, we calculate

$$\frac{z_1}{z_2} = \frac{x + yi}{a + bi}$$

by multiplying numerator and denominator by $a - bi$. This action results in

$$\begin{aligned}\frac{z_1}{z_2} = \frac{x + yi}{a + bi} = \frac{(x + yi)(a - bi)}{(a + bi)(a - bi)} &= \frac{(xa + yb) + (-xb + ya)i}{a^2 + b^2} \qquad \text{(using 1.9)} \\ = \left(\frac{xa + yb}{a^2 + b^2}\right) + \left(\frac{ya - xb}{a^2 + b^2}\right) i. \tag{1.10}\end{aligned}$$

For example,

$$\frac{3-2i}{6+i} = \frac{(3-2i)(6-i)}{(6+i)(6-i)} = \frac{16-15i}{37} = \frac{16}{37} - \frac{15}{37}i.$$

In summary, addition, subtraction, multiplication, and division of complex numbers are performed using ordinary rules of algebra with the extra condition that i^2 is always replaced by -1.

EXAMPLE 1.1 ....

Write the following complex numbers in Cartesian form:

(a) $(3+i)(2-i)^2 - i$ **(b)** $\dfrac{i^3}{2+i}$ **(c)** $\dfrac{4-3i^2+2i}{(2-2i^3)^2}$

Solution **(a)** $(3+i)(2-i)^2 - i = (3+i)(3-4i) - i = (13-9i) - i = 13 - 10i$

(b) $\dfrac{i^3}{2+i} = \dfrac{-i(2-i)}{(2+i)(2-i)} = \dfrac{-1-2i}{5} = -\dfrac{1}{5} - \dfrac{2}{5}i$

(c) $\dfrac{4-3i^2+2i}{(2-2i^3)^2} = \dfrac{4+3+2i}{(2+2i)^2} = \dfrac{7+2i}{8i} = \dfrac{(7+2i)(-i)}{(8i)(-i)} = \dfrac{2-7i}{8} = \dfrac{1}{4} - \dfrac{7}{8}i$

Notice in part (c) of this example that we multiplied numerator and denominator by $-i$ rather than $-8i$; the result is the same in either case. Both lead to a real denominator. ....

Definition 1.5
The **complex conjugate** $\overline{z}$ of a complex number $z = x + yi$ is

$$\overline{z} = x - yi. \tag{1.11}$$

For example, the complex conjugates of $1+2i$ and $3-4i$ are $\overline{1+2i} = 1-2i$ and $\overline{3-4i} = 3+4i$. Geometrically, $\overline{z}$ is the reflection of z in the real axis (Figure 1.6).

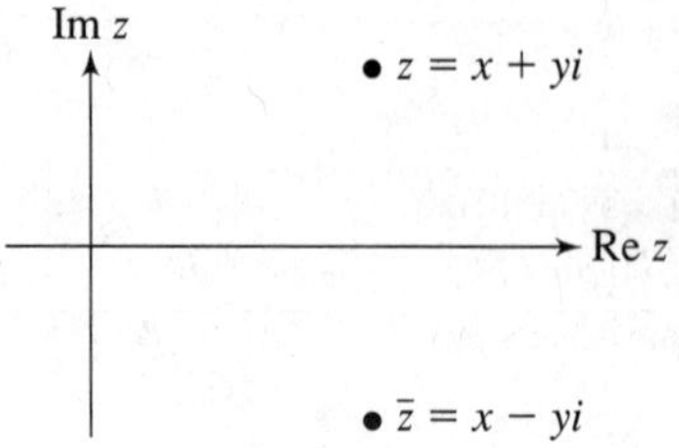

Figure 1.6

The procedure by which two complex numbers are divided (in 1.10) can be stated as follows. To divide z_1 by z_2, multiply z_1 and z_2 by $\overline{z_2}$:

$$\frac{z_1}{z_2} = \frac{z_1\overline{z_2}}{z_2\overline{z_2}}.$$

The denominator will be real, and the Cartesian form is immediate.

Let us now return to the discussion of quadratic equation 1.1. When the discriminant is positive, the equation has two real solutions, and when the discriminant is zero, we regard the quadratic as having two real solutions that are identical. A solution of

$$x^2 + 1 = 0$$

(which has a negative discriminant) is the complex number i, but so also is $-i$, because $(-i)^2 + 1 = -1 + 1 = 0$. The quadratic equation

$$x^2 + 16 = 0$$

has two solutions, $x = \pm 4i$. If we apply quadratic formula 1.2 to the equation

$$x^2 + 2x + 5 = 0,$$

the result is

$$x = \frac{-2 \pm \sqrt{4 - 20}}{2} = \frac{-2 \pm \sqrt{-16}}{2}.$$

By $\sqrt{-16}$ we would seem to mean the number which multiplied by itself is -16. But there are two such complex numbers, namely, $\pm 4i$. We make the agreement that $\sqrt{-16}$ shall denote that complex number whose square is -16 and that has a positive imaginary part. By this agreement,

$$\sqrt{-16} = 4i \quad \text{and} \quad -\sqrt{-16} = -4i.$$

The quadratic formula applied to $x^2 + 2x + 5 = 0$ therefore gives the two complex numbers

$$x = \frac{-2 \pm 4i}{2} = -1 \pm 2i.$$

It is straightforward to verify that these two complex conjugates do indeed satisfy $x^2 + 2x + 5 = 0$.

The agreement made in this last example is worth reiterating as a general principle: When $a > 0$ (is a real number),

$$\sqrt{-a} = \sqrt{a}i. \qquad \textbf{(1.12)}$$

We call $\sqrt{a}i$ the **principal square root** of $-a$; the other square root is $-\sqrt{a}i$.

These examples lead us to the following result.

■ **THEOREM 1.1** Every real quadratic equation

$$ax^2 + bx + c = 0 \tag{1.13a}$$

has two solutions. They are given by the quadratic formula

$$x = \frac{-b \pm \sqrt{b^2 - 4ac}}{2a}. \tag{1.13b}$$

When $b^2 - 4ac > 0$, the roots are real and distinct; when $b^2 - 4ac = 0$, the roots are real and equal; and when $b^2 - 4ac < 0$, the roots are complex conjugates. ■

That 1.13(b) defines two complex numbers that satisfy 1.13(a) when $b^2 - 4ac < 0$ is easily verified by substituting from 1.13(b) into 1.13(a).

EXAMPLE 1.2 ....

Find all solutions of the following equations:

(a) $x^2 + x + 3 = 0$ **(b)** $x^2 - 6x + 9 = 0$

(c) $2x^2 + 17x - 2 = 0$ **(d)** $x^4 + 5x^2 + 4 = 0$

Solution **(a)** By quadratic formula 1.13(b),

$$x = \frac{-1 \pm \sqrt{1 - 12}}{2} = \frac{-1 \pm \sqrt{-11}}{2} = -\frac{1}{2} \pm \frac{\sqrt{11}}{2}i.$$

(b) This quadratic can be factored as $0 = x^2 - 6x + 9 = (x - 3)^2$ and therefore has a double root, $x = 3$.

(c) By 1.13(b),

$$x = \frac{-17 \pm \sqrt{289 + 16}}{4} = \frac{-17 \pm \sqrt{305}}{4},$$

two real roots.

(d) If we set $y = x^2$, then

$$0 = x^4 + 5x^2 + 4 = y^2 + 5y + 4 = (y + 4)(y + 1).$$

Consequently, y is equal to -1 or -4. Because $y = x^2$, we set $x^2 = -1$ and $x^2 = -4$. These equations have roots $x = \pm i$ and $x = \pm 2i$.

....

Theorem 1.1 is valid even when coefficients a, b, and c in equation 1.13(a) are complex, in which case we speak of a complex quadratic equation. However, formula 1.13(b) leads to square roots of complex numbers when a, b, and c are complex. This is not a problem; using Definition 1.2 for equality of complex numbers we can find the square roots of any complex number. (Other methods are developed in Sections 1.4 and 3.6.)

Finding the square roots of a complex number Z is equivalent to solving the following equation for z:

$$z^2 = Z. \tag{1.14}$$

If X and Y are the real and imaginary parts of Z ($Z = X + Yi$), and we let x and y be the real and imaginary parts of z, then

$$X + Yi = (x + yi)^2 = x^2 - y^2 + 2xyi.$$

But according to Definition 1.2, two complex numbers are equal if and only if their real parts and their imaginary parts are equal; that is, we may set

$$x^2 - y^2 = X, \qquad 2xy = Y.$$

We now have two real equations in two real unknowns x and y, the real and imaginary parts of z. Once they are solved, and there should be two solutions, we have the two square roots of Z. We illustrate by finding the square roots of i.

If we set $z = x + yi$ in the equation $z^2 = i$, the result is

$$i = (x + yi)^2 = (x^2 - y^2) + 2xyi.$$

Equating real and imaginary parts of the complex numbers on each side of this equation gives

$$x^2 - y^2 = 0, \qquad 2xy = 1.$$

From the second of these, $y = 1/(2x)$, which we substitute into the first:

$$0 = x^2 - \frac{1}{(2x)^2} = \frac{4x^4 - 1}{4x^2}.$$

This implies that

$$0 = 4x^4 - 1 = (2x^2 + 1)(2x^2 - 1).$$

Because x must be real, we have two solutions, $x = \pm 1/\sqrt{2}$; correspondingly, $y = \pm 1/\sqrt{2}$. Thus, the square roots of i are

$$z = \pm\frac{1}{\sqrt{2}} \pm \frac{1}{\sqrt{2}}i = \pm\left(\frac{1}{\sqrt{2}} + \frac{1}{\sqrt{2}}i\right).$$

One of these is denoted by $\sqrt{i}$ and the other, by $-\sqrt{i}$. The following agreement extends 1.12 to all complex numbers:

Definition 1.6

The **principal square root** of $a + bi$, where $b \neq 0$, is denoted by $\sqrt{a + bi}$. It is that square root of $a + bi$ with positive real part.

With this agreement,

$$\sqrt{i} = \frac{1}{\sqrt{2}} + \frac{1}{\sqrt{2}}i \quad \text{and} \quad -\sqrt{i} = -\frac{1}{\sqrt{2}} - \frac{1}{\sqrt{2}}i.$$

We have uncovered here a most important principle. A complex equation can be replaced by two real equations, the equations obtained by equating real and imaginary parts of each side of the complex equation. Watch for this principle; we use it often.

As we suggested earlier, with the ability to find square roots of complex numbers, we can use formula 1.13(b) to find roots of complex quadratic equations. For example, 1.13(b) applied to

$$iz^2 + 2z + 3 = 0$$

gives

$$z = \frac{-2 \pm \sqrt{4 - 12i}}{2} = -1 \pm \sqrt{1 - 3i}.$$

To find the square roots of $1 - 3i$, we set $w = x + yi$ in $w^2 = 1 - 3i$:

$$1 - 3i = (x + yi)^2 = (x^2 - y^2) + 2xyi.$$

Equating real and imaginary parts gives

$$x^2 - y^2 = 1, \qquad 2xy = -3.$$

From the second of these, $y = -3/(2x)$, which we substitute into the first:

$$x^2 - \left(\frac{-3}{2x}\right)^2 = 1 \Longrightarrow 4x^4 - 4x^2 - 9 = 0.$$

This is essentially a quadratic equation in x^2, namely, $4(x^2)^2 - 4(x^2) - 9 = 0$, and therefore

$$x^2 = \frac{4 \pm \sqrt{16 + 144}}{8} = \frac{1 \pm \sqrt{10}}{2}.$$

Because x^2 must be positive, it follows that

$$x^2 = \frac{1 + \sqrt{10}}{2} \Longrightarrow x = \pm\sqrt{\frac{1 + \sqrt{10}}{2}}.$$

Correspondingly,

$$y = \frac{\mp 3}{\sqrt{2 + 2\sqrt{10}}}.$$

Hence, the square roots of $1 - 3i$ are

$$\sqrt{1 - 3i} = \sqrt{\frac{1 + \sqrt{10}}{2}} - \frac{3}{\sqrt{2 + 2\sqrt{10}}}i, \qquad -\sqrt{1 - 3i} = -\sqrt{\frac{1 + \sqrt{10}}{2}} + \frac{3}{\sqrt{2 + 2\sqrt{10}}}i,$$

and the solutions of the quadratic equation $iz^2 + 2z + 3 = 0$ are

$$z = \frac{1}{i}\left[-1 \pm \left(\sqrt{\frac{1+\sqrt{10}}{2}} - \frac{3}{\sqrt{2+2\sqrt{10}}}i\right)\right] = \frac{\pm 3}{\sqrt{2+2\sqrt{10}}} + \left(1 \pm \sqrt{\frac{1+\sqrt{10}}{2}}\right)i.$$

Exercises 1.2

1. Show each of the following complex numbers in the complex plane: (a) $2-i$, (b) $3+4i$, (c) $-1-5i$, (d) $-3+2i$, (e) $5i$, and (f) $2(1+i)$.

In Exercises 2–15 write the complex expression in Cartesian form.

2. $(2+4i)-(3-2i)$
3. $(1+2i)^2$
4. $(-2+i)(3-4i)$
5. $3i(4i-1)^2$
6. $i^3 - 3i^2 + 2i + 4$
7. $(1+i)^6$
8. $\dfrac{1-i}{3+2i}$
9. $\dfrac{(3+i)^2}{2-i}$
10. $i^{24} - 3i^{13} + 4$
11. $(i-2)[(2+i)(1-i)+3i-2]$
12. $6i\left(\dfrac{1+i}{2-i}\right) + 3\left(\dfrac{i-4}{2i+1}\right)$
13. $\overline{2+i} - (3+\overline{4i})$
14. $\overline{1+i}^2 + \overline{(1+i)^2}$
15. $\left(\dfrac{1}{2} - \dfrac{\sqrt{3}}{2}i\right)^3$
16. Verify the following properties for the complex conjugation operation:
    (a) $\overline{z_1 + z_2} = \overline{z_1} + \overline{z_2}$
    (b) $\overline{z_1 - z_2} = \overline{z_1} - \overline{z_2}$
    (c) $\overline{z_1 z_2} = \overline{z_1}\,\overline{z_2}$
    (d) $\overline{\left(\dfrac{z_1}{z_2}\right)} = \dfrac{\overline{z_1}}{\overline{z_2}}$
    (e) $\overline{z^n} = \overline{z}^{\,n}$, n a positive integer
17. (a) If $P(z)$ is a polynomial with real coefficients, use the results of Exercise 16 to verify that $\overline{P(z)} = P(\overline{z})$.
    (b) Verify that if z is a root of a polynomial equation $P(z) = 0$, with real coefficients, then so also is $\overline{z}$.
18. Verify that all complex numbers z satisfying the equation $z\overline{z} = r^2$, $r > 0$ a real constant, lie on a circle. What are its center and radius?
19. Prove that if $z_1 z_2 = 0$, then at least one of z_1 and z_2 must be zero.
20. We have made the agreement that when $a > 0$ is a real number, $\sqrt{-a}$ denotes a complex number with positive imaginary part (see equation 1.12). Show that with this agreement,

    $$\sqrt{z_1 z_2} \text{ is not always equal to } \sqrt{z_1}\sqrt{z_2}.$$

    Hint: Try $z_1 = z_2 = -1$.

21. Explain the fallacy in

$$-1 = \sqrt{-1}\sqrt{-1} = \sqrt{(-1)(-1)} = \sqrt{1} = 1.$$

In Exercises 22–31 find all solutions of the equation.

22. $x^2 + 5x + 3 = 0$

23. $x^2 + 3x + 5 = 0$

24. $x^2 + 8x + 16 = 0$

25. $x^2 + 2x - 7 = 0$

26. $x^2 + 2x + 7 = 0$

27. $4x^2 - 2x + 5 = 0$

28. $\sqrt{3}x^2 + 5x + \sqrt{15} = 0$

29. $x^4 + 4x^2 - 5 = 0$

30. $x^4 + 4x^2 + 3 = 0$

31. $x^4 + 6x^2 + 3 = 0$

32. Find two numbers whose sum is 6 and whose product is 10.

In Exercises 33–36 find square roots of the complex number.

33. $-7 - 24i$

34. $21 - 20i$

35. $2 + i$

36. $-1 + i$

In Exercises 37–42 find both solutions of the complex quadratic equation.

37. $z^2 + (2 - i)z + 1 - i = 0$

38. $2iz^2 + 4z + 3 - 2i = 0$

39. $z^2 + 2z + i = 0$

40. $iz^2 + 2iz + 3 = 0$

41. $z^2 + 4iz - 1 + i = 0$

42. $z^2 + (2 - i)z + 1 + i = 0$

43. Show that the solutions of the equation in Exercise 37 can be obtained by letting $z = x + yi$, substituting into the equation, and equating real and imaginary parts.

44. Repeat the procedure of Exercise 43 on the equation in Exercise 38.

45. Show that the line joining the complex numbers z_1 and z_2 is perpendicular to the line joining z_3 and z_4 if and only if $\mathrm{Re}[(z_1 - z_2)(\overline{z_3} - \overline{z_4})] = 0$.

46. Show that the line joining the complex numbers z_1 and z_2 is parallel to the line joining z_3 and z_4 if and only if $\mathrm{Im}[(z_1 - z_2)(\overline{z_3} - \overline{z_4})] = 0$.

1.3 Polar Representation of Complex Numbers

In Section 1.2 we worked with complex numbers in Cartesian form, and this proved sufficient to solve quadratic equations. For other applications, it is often advantageous to have what is called the *polar representation* of a complex number. To find this representation, we define the modulus and argument of a complex number.

Definition 1.7
The **modulus** of a complex number $z = x + yi$ is

$$r = |z| = \sqrt{x^2 + y^2}. \tag{1.15}$$

The modulus is the length of the line segment joining the points in the complex plane representing the complex numbers $z = 0$ and $z = x + yi$ (Figure 1.7). It is unique; a complex number has exactly one modulus. For example,

$$|2 + 3i| = \sqrt{2^2 + 3^2} = \sqrt{13}, \quad \text{and} \quad |-3 + 4i| = \sqrt{(-3)^2 + 4^2} = 5.$$

Notice that when z is real, say $z = x$, then $|z| = \sqrt{x^2} = |x|$, where $|x|$ is the absolute value of x. In other words, when z is real, the modulus bars are absolute value bars.

Definition 1.8

The **argument** of a nonzero complex number $z = x + yi$, usually denoted by θ or $\arg z$, is the angle of rotation of the positive real axis in a counterclockwise direction to the line segment joining $z = 0$ and $z = x + yi$. Clockwise rotations are regarded as negative.

Arguments are not unique. For the complex number $z = 1 - i$ (Figure 1.8), one possible value for θ is $7\pi/4$, but another is $-\pi/4$. In fact, there are infinitely many possibilities for θ, namely, $\theta = 2k\pi - \pi/4$, where k is any integer. We call that value of the argument of z that satisfies the restriction $-\pi < \theta \leq \pi$ the **principal value** of the argument and denote it with a capital A:

$$-\pi < \operatorname{Arg} z \leq \pi. \tag{1.16}$$

For $z = 1 - i$, the principal value of the argument is $-\pi/4$.

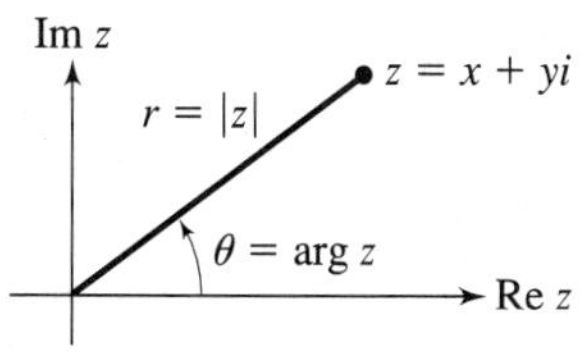

Figure 1.7

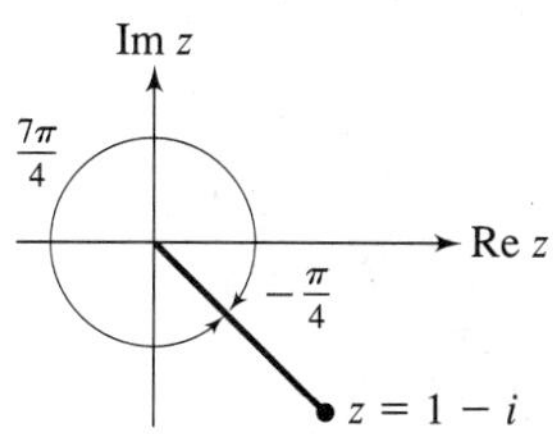

Figure 1.8

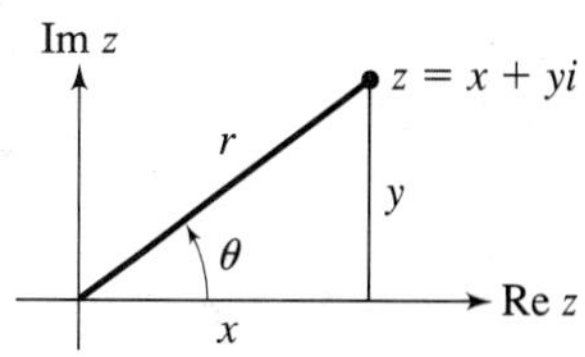

Figure 1.9

The real and imaginary parts of a complex number can be expressed in terms of its modulus and argument. The triangle in Figure 1.9 indicates that

$$x = r\cos\theta, \qquad y = r\sin\theta. \tag{1.17}$$

It is easy to reverse these equations and express r in terms of x and y; equation 1.15 does this. But the case for θ is more complicated. Certainly we can write that

$$\cos\theta = \frac{x}{\sqrt{x^2 + y^2}} \quad \text{and} \quad \sin\theta = \frac{y}{\sqrt{x^2 + y^2}}, \tag{1.18}$$

and these equations do indeed define all arguments of $z = x + yi$. In an attempt to find a single equation defining θ in terms of x and y, there is a tendency to divide the second equation by the first, giving

$$\tan\theta = \frac{y}{x},$$

and take this one step further to

$$\theta = \operatorname{Tan}^{-1}\left(\frac{y}{x}\right). \tag{1.19}$$

Unfortunately, principal values* of the inverse tangent function are between $\pm\pi/2$, and therefore this equation gives a correct argument only for complex numbers in the first and fourth quadrants.

How then are we to find arguments for given $z = x + yi$? Certainly we can find all angles satisfying 1.18. Alternatively, we can find the angle θ defined by 1.19. If z is in the first or fourth quadrants ($x > 0$), then all possible arguments are $\theta + 2k\pi$, where k is an integer. When z is in the second quadrant, arguments are $(\theta + \pi) + 2k\pi = \theta + (2k + 1)\pi$; and when z is in the third quadrant, arguments are $(\theta - \pi) + 2k\pi = \theta + (2k - 1)\pi$.

EXAMPLE 1.3 ....

Find all arguments and the principal value of the argument for the following complex numbers:

(a) $1 + \sqrt{3}i$ **(b)** $2 - 2i$ **(c)** $-4 + 3i$ **(d)** $-3 - 2i$

Solution We have shown all four complex numbers in Figure 1.10.

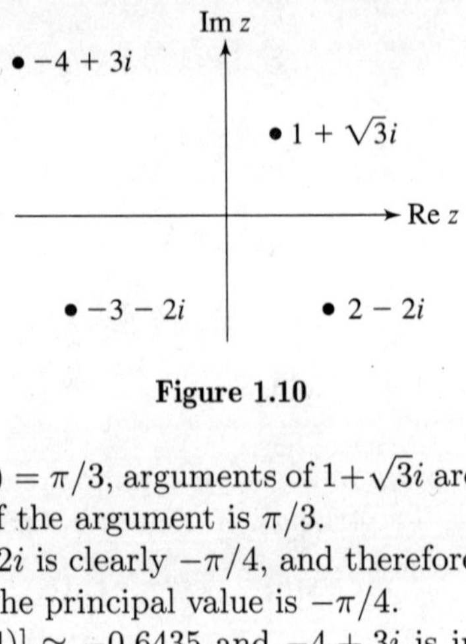

Figure 1.10

(a) Because $\operatorname{Tan}^{-1}(\sqrt{3}/1) = \pi/3$, arguments of $1+\sqrt{3}i$ are $\pi/3+2k\pi = (6k+1)\pi/3$. The principal value of the argument is $\pi/3$.

(b) One argument of $2 - 2i$ is clearly $-\pi/4$, and therefore all arguments are $2k\pi - \pi/4 = (8k - 1)\pi/4$. The principal value is $-\pi/4$.

(c) Because $\operatorname{Tan}^{-1}[3/(-4)] \approx -0.6435$ and $-4 + 3i$ is in the second quadrant, it follows that arguments are $(\pi - 0.6435) + 2k\pi = (2k+1)\pi - 0.6435$. The principal value is $\pi - 0.6435$.

(d) With $\operatorname{Tan}^{-1}[-2/(-3)] \approx 0.5880$ and $-3 - 2i$ in the third quadrant, arguments are $(0.5880 - \pi) + 2k\pi = (2k - 1)\pi + 0.5880$. The principal value is $0.5880 - \pi$.

....

*A capital letter on the inverse tangent function denotes the principal values $-\pi/2 < \operatorname{Tan}^{-1} a < \pi/2$. For the inverse sine and cosine functions, capital letters denote the principal values $-\pi/2 \le \operatorname{Sin}^{-1} a \le \pi/2$ and $0 \le \operatorname{Cos}^{-1} a \le \pi$.

If expressions 1.17 are substituted into the Cartesian form $z = x + yi$ for a complex number,

$$z = x + yi = r\cos\theta + (r\sin\theta)i.$$

When r is factored from both terms,

$$z = r(\cos\theta + \sin\theta\, i). \tag{1.20}$$

This expression is called the **polar representation** of a complex number. Realize again that it is not unique. If 1.20 is a polar representation of a complex number, then

$$r[\cos(\theta + 2k\pi) + \sin(\theta + 2k\pi)\, i],$$

for any integer k, is a polar representation for the same complex number.

EXAMPLE 1.4 ....

Find polar representations for $-1 + i$ and $2 - 3i$.

Solution Because the modulus of $-1 + i$ is $\sqrt{2}$ and an argument is $3\pi/4$ (Figure 1.11), a polar representation is

$$-1 + i = \sqrt{2}[\cos(3\pi/4) + \sin(3\pi/4)\, i].$$

The modulus of $2 - 3i$ is $\sqrt{13}$, and an argument is $\mathrm{Tan}^{-1}(-3/2) \approx -0.983$. Consequently,

$$\begin{aligned} 2 - 3i &= \sqrt{13}\{\cos[\mathrm{Tan}^{-1}(-3/2)] + \sin[\mathrm{Tan}^{-1}(-3/2)]\, i\} \\ &\approx \sqrt{13}[\cos(-0.983) + \sin(-0.983)\, i]. \end{aligned}$$

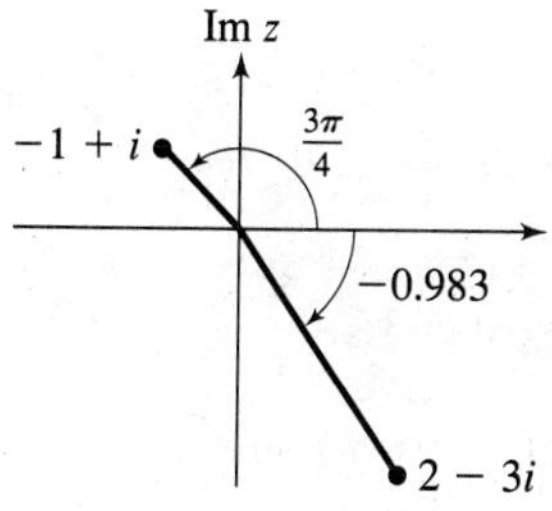

Figure 1.11

....

EXAMPLE 1.5 ....

Is $-2[\cos(\pi/4) + \sin(\pi/4)\, i]$ the polar form for a complex number?

Solution No, because -2 cannot be the modulus of a complex number (moduli must be nonnegative). To express this complex number in polar form we write

$$-2[\cos(\pi/4) + \sin(\pi/4)\, i] = 2[-\cos(\pi/4) - \sin(\pi/4)\, i] = 2[\cos(5\pi/4) + \sin(5\pi/4)\, i].$$

Two complex numbers are equal if and only if they have the same real parts and the same imaginary parts. How do we phrase equality in terms of moduli and arguments? Certainly complex numbers can be equal only if they have the same moduli. Their arguments need not be the same, however, but they must differ by a multiple of 2π. In other words, we have the following: Two nonzero complex numbers z_1 and z_2 are equal if and only if

$$|z_1| = |z_2| \tag{1.21a}$$

and

$$\arg z_1 = \arg z_2 + 2k\pi, \qquad k \text{ an integer.} \tag{1.21b}$$

Because principal values of arguments of complex numbers are unique, an alternative to 1.21 is that complex numbers z_1 and z_2 are equal if and only if

$$|z_1| = |z_2| \tag{1.22a}$$

and

$$\operatorname{Arg} z_1 = \operatorname{Arg} z_2. \tag{1.22b}$$

Complex numbers are easily added and subtracted in Cartesian form. Even multiplication and division of two complex numbers is relatively easy in Cartesian form. On the other hand, multiplication and division in polar form lead to properties that are not evident in Cartesian form. For example, if $z_1 = r_1(\cos\theta_1 + \sin\theta_1\, i)$ and $z_2 = r_2(\cos\theta_2 + \sin\theta_2\, i)$ are any two complex numbers, their product is

$$\begin{aligned} z_1 z_2 &= [r_1(\cos\theta_1 + \sin\theta_1\, i)][r_2(\cos\theta_2 + \sin\theta_2\, i)] \\ &= r_1 r_2[(\cos\theta_1\, \cos\theta_2 - \sin\theta_1\, \sin\theta_2) + (\sin\theta_1\, \cos\theta_2 + \sin\theta_2\, \cos\theta_1)i] \\ &= r_1 r_2[\cos(\theta_1 + \theta_2) + \sin(\theta_1 + \theta_2)\, i]. \end{aligned} \tag{1.23}$$

Because the last expression is the polar representation of $z_1 z_2$, we have proved the following facts:

1. The modulus of the product of two complex numbers is the product of their moduli; i.e.,

$$|z_1 z_2| = |z_1||z_2|. \tag{1.24a}$$

2. An argument of the product of two complex numbers is the sum of their arguments; i.e.,

$$\arg(z_1 z_2) = \arg z_1 + \arg z_2. \tag{1.24b}$$

Be wary of how you interpret 1.24(b); it is not an equation in the normal sense of equation. Each of the three terms has infinitely many possible values (because the argument of a complex number is determined only to multiples of 2π). What this "equation" says is that given any two terms in 1.24(b), there is a value of the third that makes 1.24(b) an equation. For example, positive arguments of $z_1 = 1+i$ and $z_2 = 1-i$ are $\pi/4$ and $7\pi/4$. The product of these complex numbers is $z_1 z_2 = (1+i)(1-i) = 2$. An argument for the complex number 2 is 0, but certainly the arguments $\arg z_1 = \pi/4$, $\arg z_2 = 7\pi/4$, and $\arg z_1 z_2 = 0$ do not satisfy 1.24(b). On the other hand, the arguments $\arg z_1 = \pi/4$, $\arg z_2 = 7\pi/4$, and $\arg z_1 z_2 = 2\pi$ do satisfy 1.24(b).

Equation 1.24(b) cannot be replaced by principal values, because, in general, $\operatorname{Arg} z_1 z_2$ may not be equal to $\operatorname{Arg} z_1 + \operatorname{Arg} z_2$. This is the situation when $z_1 = z_2 = -1$. In this case, $\operatorname{Arg} z_1 z_2 = \operatorname{Arg} 1 = 0$, and $\operatorname{Arg} z_1 + \operatorname{Arg} z_2 = \pi + \pi = 2\pi$.

A similar analysis shows that

$$\frac{z_1}{z_2} = \frac{r_1}{r_2}[\cos(\theta_1 - \theta_2) + \sin(\theta_1 - \theta_2)\, i]. \tag{1.25}$$

In other words, we have the following properties corresponding to 1.24:

$$\left|\frac{z_1}{z_2}\right| = \frac{|z_1|}{|z_2|}, \tag{1.26a}$$

$$\arg\left(\frac{z_1}{z_2}\right) = \arg z_1 - \arg z_2. \tag{1.26b}$$

EXAMPLE 1.6 ....

Find $z_1 z_2$ and z_1/z_2 if $z_1 = 6(\cos 1.1 + \sin 1.1\, i)$ and $z_2 = 3(\cos 2.4 + \sin 2.4\, i)$.

Solution According to 1.23 and 1.25,

$$z_1 z_2 = 18(\cos 3.5 + \sin 3.5\, i) \quad \text{and} \quad \frac{z_1}{z_2} = 2[\cos(-1.3) + \sin(-1.3)\, i].$$

....

The polar form for a complex number is very useful when we raise a complex number to a power, say, $(1+\sqrt{3}i)^8$. To do this we derive a simple formula for z^n when n is a positive integer. With $z = r(\cos\theta + \sin\theta\, i)$,

$$\begin{aligned} z^2 &= [r(\cos\theta + \sin\theta\, i)][r(\cos\theta + \sin\theta\, i)] \\ &= r^2(\cos 2\theta + \sin 2\theta\, i); \\ z^3 &= zz^2 = [r(\cos\theta + \sin\theta\, i)][r^2(\cos 2\theta + \sin 2\theta\, i)] \\ &= r^3(\cos 3\theta + \sin 3\theta\, i). \end{aligned}$$

The pattern emerging (which can be proved by mathematical induction) is

$$z^n = r^n[\cos(n\theta) + \sin(n\theta)\, i]. \tag{1.27}$$

This result is often called **DeMoivre's theorem**.

EXAMPLE 1.7

Evaluate $(1+\sqrt{3}i)^8$.

Solution Because $|1+\sqrt{3}i| = \sqrt{(1)^2+(\sqrt{3})^2} = 2$ and $\arg(1+\sqrt{3}i) = \mathrm{Tan}^{-1}(\sqrt{3}) = \pi/3$, it follows that

$$\begin{aligned}(1+\sqrt{3}i)^8 &= \{2[\cos(\pi/3)+\sin(\pi/3)\,i]\}^8 = 2^8[\cos(8\pi/3)+\sin(8\pi/3)\,i]\\ &= 256[\cos(2\pi/3)+\sin(2\pi/3)\,i] = 256\left(-\frac{1}{2}+\frac{\sqrt{3}}{2}i\right) = 128(-1+\sqrt{3}i).\end{aligned}$$

EXAMPLE 1.8

Describe the position of the complex number $(\cos\phi+\sin\phi\, i)z$ in the complex plane relative to the position of z.

Solution If $z = r(\cos\theta+\sin\theta\, i)$ is the polar representation of z, then equation 1.23 gives

$$(\cos\phi+\sin\phi\, i)z = r[\cos(\theta+\phi)+\sin(\theta+\phi)\, i].$$

The modulus of z does not change, but its argument increases by ϕ. In other words, multiplying a complex number by $\cos\phi+\sin\phi\, i$ rotates it through angle ϕ about the origin.

EXAMPLE 1.9

Use DeMoivre's theorem to verify that

$$\cos 3\theta = \cos^3\theta - 3\cos\theta\sin^2\theta.$$

Solution When the complex number in DeMoivre's theorem 1.27 has modulus $r = 1$,

$$z^n = (\cos\theta+\sin\theta\, i)^n = \cos(n\theta)+\sin(n\theta)\, i.$$

For $n = 3$, this becomes

$$(\cos\theta+\sin\theta\, i)^3 = \cos 3\theta+\sin 3\theta\, i.$$

When the left side is expanded (with the binomial theorem), the result is

$$\begin{aligned}\cos 3\theta + \sin 3\theta\, i &= \cos^3\theta + 3\cos^2\theta(\sin\theta\, i) + 3\cos\theta(\sin\theta\, i)^2 + (\sin\theta\, i)^3\\ &= (\cos^3\theta - 3\cos\theta\,\sin^2\theta) + (3\cos^2\theta\,\sin\theta - \sin^3\theta)i.\end{aligned}$$

On each side of this equation we have a complex number, and two complex numbers are equal if and only if their real and imaginary parts are the same. When we equate real parts we obtain

$$\cos 3\theta = \cos^3\theta - 3\cos\theta\,\sin^2\theta.$$

Imaginary parts give the additional trigonometric identity

$$\sin 3\theta = 3\cos^2\theta\,\sin\theta - \sin^3\theta.$$

Exercises 1.3

In Exercises 1–6 express the complex number in polar form.

1. $-1+i$
2. $-2i$
3. $\sqrt{3}-i$
4. $3+4i$
5. $-1-2i$
6. $-2[\cos(\pi/3) - \sin(\pi/3)\, i]$

In Exercises 7–10 express the complex number in Cartesian form.

7. $3[\cos(\pi/6) + \sin(\pi/6)\, i]$
8. $\cos(1.4) + \sin(1.4)\, i$
9. $2[\cos(\pi/4) + \sin(\pi/4)\, i]^2$
10. $\dfrac{4[\cos(2\pi/3) + \sin(2\pi/3)\, i]}{\cos(\pi/6) + \sin(\pi/6)\, i}$

In Exercises 11–16 simplify the expression as much as possible.

11. $(1-i)^4$
12. $[\cos(\pi/5) + \sin(\pi/5)\, i]^{10}$
13. $\dfrac{1}{(\sqrt{3}+i)^6}$
14. $\dfrac{-2[\cos(\pi/3) - \sin(\pi/3)\, i]^3}{3+i}$
15. $\left(\dfrac{1+i}{2-2i}\right)^5$
16. $\dfrac{(2+i)^6}{(4-i)^3}$

Use DeMoivre's theorem with $r = 1$ to verify the identity in Exercises 17–20.

17. $\cos 4\theta = \cos^4\theta - 6\cos^2\theta\,\sin^2\theta + \sin^4\theta$
18. $\sin 4\theta = 4\cos^3\theta\,\sin\theta - 4\cos\theta\,\sin^3\theta$
19. $\cos 5\theta = \cos^5\theta - 10\cos^3\theta\,\sin^2\theta + 5\cos\theta\,\sin^4\theta$
20. $\sin 5\theta = 5\cos^4\theta\,\sin\theta - 10\cos^2\theta\,\sin^3\theta + \sin^5\theta$

1.4 Applications of the Polar Form for Complex Numbers

Roots of Complex Numbers

An equation of the form

$$a_n z^n + a_{n-1} z^{n-1} + \cdots + a_1 z + a_0 = 0, \tag{1.28}$$

where $n \geq 1$ is an integer and $a_0, a_1, \ldots, a_n$ are constants, is called a **polynomial equation** of degree n. When the a_j are real, we speak of a real polynomial equation, and when the a_j are complex, we have a complex polynomial equation. For instance, the quadratic equation $3z^2 + 2z + 5 = 0$ is a real polynomial equation of degree 2; the equation $(4 - i)z^3 - 2z^2 + 3iz + 6 = 0$ is a complex polynomial equation of degree 3 (in short, a complex cubic equation). Early in your calculus studies or even before, someone quoted to you a theorem called the fundamental theorem of algebra. It was probably stated as follows: Every real polynomial equation of degree $n \geq 1$ has exactly n solutions (counting multiplicities). Recall the meaning of multiplicity. Each root $z = 1$ and $z = 2$ of the quadratic equation $z^2 - 3z + 2 = 0$ has multiplicity 1, but the root $z = 1$ of $z^2 - 2z + 1 = 0$ has multiplicity 2. The sixth-degree equation $(z - 1)^2(z - 2)^3(z + 4) = 0$ has three distinct roots, but it has six roots counting multiplicities — $z = 1$ is a root of multiplicity 2, $z = 2$ has multiplicity 3, and $z = -4$ has multiplicity 1. The fundamental theorem indicates that when multiplicities are taken into account, the number of solutions is equal to the degree of the equation. The theorem is true for complex polynomial equations as well as real ones. You could not prove the real version when it was first given to you, and you cannot prove either version now, but you will be able to in Chapter 4. This presents no problem, because the theorem does not tell us how to find solutions anyway; it only tells us how many to expect.

Of particular interest are polynomial equations of the form

$$z^n = Z, \tag{1.29}$$

where $n > 1$ is an integer and Z is a given complex number. The n solutions are called the n nth roots of Z. For example, the two solutions of $z^2 = 1 + i$ are the square roots of $1 + i$, and the three solutions of $z^3 = -1 + 2i$ are the cube roots of $-1 + 2i$. DeMoivre's theorem can be used to advantage here. To illustrate we begin with the example

$$z^3 = 8i.$$

The three solutions are called the *cube roots* of $8i$. To find them we express z and $8i$ in polar form, $z = r(\cos\theta + \sin\theta\, i)$ and $8i = 8[\cos(\pi/2) + \sin(\pi/2)\, i]$, and substitute into $z^3 = 8i$:

$$[r(\cos\theta + \sin\theta\, i)]^3 = 8[\cos(\pi/2) + \sin(\pi/2)\, i].$$

Using DeMoivre's theorem on the left we obtain

$$r^3[\cos(3\theta) + \sin(3\theta)\, i] = 8[\cos(\pi/2) + \sin(\pi/2)\, i].$$

Conditions 1.21 for equality of complex numbers in polar form require

$$r^3 = 8 \quad \text{and} \quad 3\theta = \frac{\pi}{2} + 2k\pi, \qquad k = \text{integer}.$$

Hence,

$$r = 2 \quad \text{and} \quad \theta = \frac{\pi}{6} + \frac{2k\pi}{3}.$$

The values $k = 0$, 1, and 2 lead to distinct complex numbers:

$$z_0 = 2\left[\cos\left(\frac{\pi}{6}\right) + \sin\left(\frac{\pi}{6}\right) i\right] = 2\left(\frac{\sqrt{3}}{2} + \frac{i}{2}\right) = \sqrt{3} + i,$$

$$z_1 = 2\left[\cos\left(\frac{\pi}{6} + \frac{2\pi}{3}\right) + \sin\left(\frac{\pi}{6} + \frac{2\pi}{3}\right) i\right] = 2\left(\frac{-\sqrt{3}}{2} + \frac{i}{2}\right) = -\sqrt{3} + i,$$

$$z_2 = 2\left[\cos\left(\frac{\pi}{6} + \frac{4\pi}{3}\right) + \sin\left(\frac{\pi}{6} + \frac{4\pi}{3}\right) i\right] = -2i.$$

These are the cube roots of $8i$; the cube of each complex number is equal to $8i$. Other values of k simply repeat these cube roots. It is interesting to plot the cube roots in the complex plane (Figure 1.12). All three lie on a circle centered at the origin with radius 2 (the cube root of the modulus of $8i$). They are equally spaced around the circle, with angle $2\pi/3$ between each pair. The argument of z_0 is one-third the principal argument of $8i$.

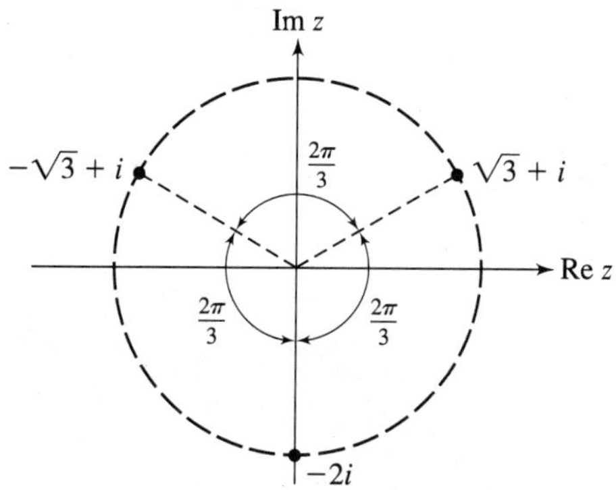

Figure 1.12

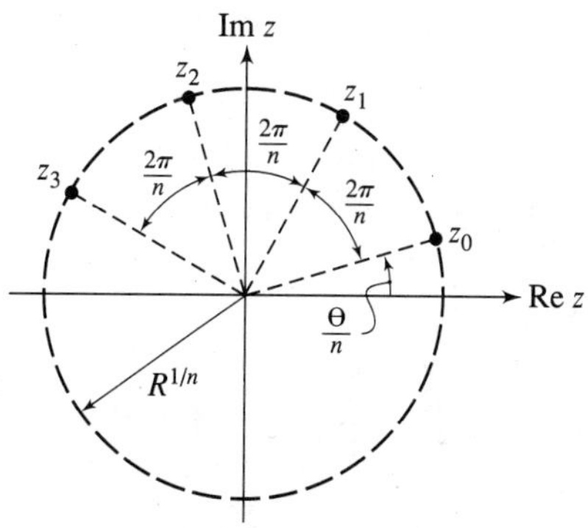

Figure 1.13

This process can be applied to general equation 1.29. If we set $z = r(\cos\theta + \sin\theta\, i)$ and $Z = R(\cos\Theta + \sin\Theta\, i)$, then

$$r^n[\cos(n\theta) + \sin(n\theta)\, i] = R(\cos\Theta + \sin\Theta\, i).$$

Conditions 1.21 require

$$r^n = R \quad \text{and} \quad n\theta = \Theta + 2k\pi, \qquad k \text{ an integer}.$$

Consequently,

$$r = R^{1/n} \quad \text{and} \quad \theta = \frac{\Theta + 2k\pi}{n}.$$

The n nth roots of $Z = R(\cos\Theta + \sin\Theta\, i)$ are, therefore,

$$z_k = R^{1/n}\left[\cos\left(\frac{\Theta + 2k\pi}{n}\right) + \sin\left(\frac{\Theta + 2k\pi}{n}\right) i\right], \tag{1.30}$$

where $k = 0, 1, \ldots, n-1$. Geometrically they are equally spaced around a circle of radius $R^{1/n}$ with an angle $2\pi/n$ between successive pairs (Figure 1.13). The first (z_0) has argument Θ/n.

In Section 1.2 we found square roots of complex numbers. The preceding discussion provides a more efficient method. Compare the following discussion for finding the square roots of $2+i$ with that in Exercise 35 in Section 1.2. To find the square roots of $2+i$, we set $z = r(\cos\theta + \sin\theta\, i)$ in $z^2 = 2+i$:

$$r^2(\cos 2\theta + \sin 2\theta\, i) = 2 + i = \sqrt{5}(\cos\Theta + \sin\Theta\, i)$$

where $\Theta = \mathrm{Tan}^{-1}(1/2)$. Conditions 1.21 require

$$r^2 = \sqrt{5} \quad \text{and} \quad 2\theta = \Theta + 2k\pi, \qquad k \text{ an integer.}$$

Hence $r = 5^{1/4}$ and $\theta = \Theta/2 + k\pi$, and we obtain

$$z_0 = 5^{1/4}[\cos(\Theta/2) + \sin(\Theta/2)\, i],$$
$$z_1 = 5^{1/4}[\cos(\Theta/2 + \pi) + \sin(\Theta/2 + \pi)\, i] = -z_0.$$

The square roots are $\pm 5^{1/4}[\cos(\Theta/2) + \sin(\Theta/2)\, i]$. We can use a calculator to approximate real and imaginary parts ($z \approx \pm(1.455 + 0.344i)$), or we can use trigonometry to find exact values. Because $\cos\Theta = 2/\sqrt{5}$ (Figure 1.14), it follows that

$$\sin(\Theta/2) = \sqrt{\frac{1 - \cos\Theta}{2}} = \sqrt{\frac{\sqrt{5} - 2}{2\sqrt{5}}}$$

and

$$\cos(\Theta/2) = \sqrt{\frac{1 + \cos\Theta}{2}} = \sqrt{\frac{\sqrt{5} + 2}{2\sqrt{5}}}.$$

Thus, the square roots are

$$\pm 5^{1/4}\left(\sqrt{\frac{\sqrt{5}+2}{2\sqrt{5}}} + \sqrt{\frac{\sqrt{5}-2}{2\sqrt{5}}}\, i\right) = \pm\left(\sqrt{\sqrt{5}/2 + 1} + \sqrt{\sqrt{5}/2 - 1}\, i\right).$$

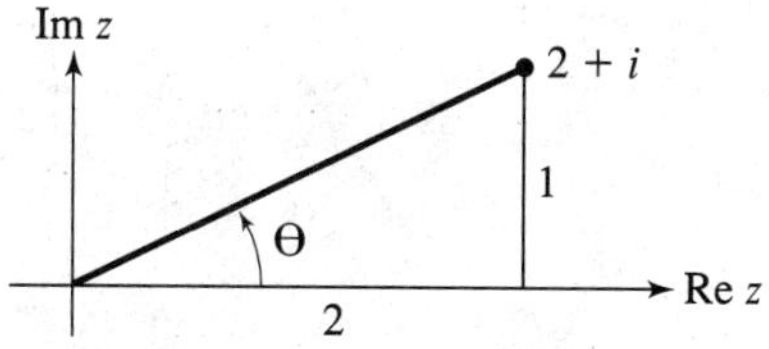

Figure 1.14

Triangle Inequality

It is a well-known fact from elementary geometry that the length of any side of a triangle is less than the sum of the lengths of the other two sides (called the *triangle inequality*). If the sides of the triangle are represented by the complex numbers z_1, z_2, and $z_1 + z_2$ (Figure 1.15), then the triangle inequality can be expressed in terms of the moduli $|z_1|$, $|z_2|$, and $|z_1 + z_2|$ as

$$|z_1 + z_2| < |z_1| + |z_2|. \tag{1.31}$$

When $z_2 = az_1$, where $a < 0$ is a real number (Figure 1.16), z_1, z_2, and $z_1 + z_2$ do not form a triangle, but inequality 1.31 is still valid. When $z_2 = az_1$, where $a > 0$ (Figure 1.17), there is once again no triangle, and in this case

$$|z_1 + z_2| = |z_1| + |z_2|.$$

Hence, we can say that for any two complex numbers z_1 and z_2,

$$|z_1 + z_2| \le |z_1| + |z_2|. \tag{1.32}$$

This continues to be called the triangle inequality in spite of the fact that when $z_2 = az_1$, there is no triangle. It is proved algebraically in Exercise 21.

An alternative geometric statement to the triangle inequality (but perhaps not quite so obvious) is that the length of any side of a triangle is greater than the difference between the lengths of the other two sides. When the sides of the triangle are represented by z_1, z_2, and $z_1 + z_2$, this leads to the inequality

$$|z_1 + z_2| \ge \big| |z_1| - |z_2| \big|. \tag{1.33}$$

This result can also be proved algebraically or derived from 1.32 (see Exercises 22 and 23).

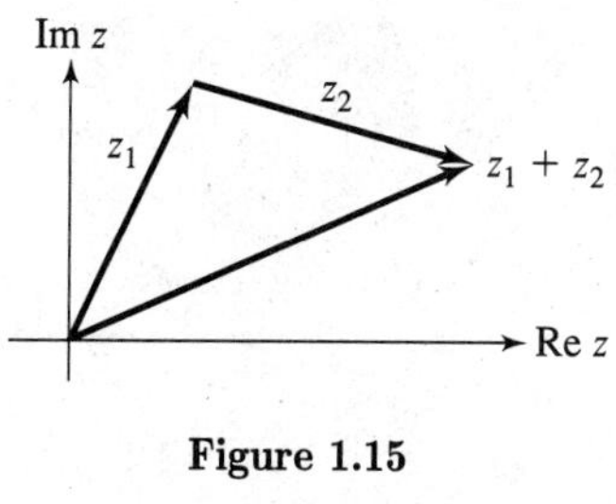

Figure 1.15

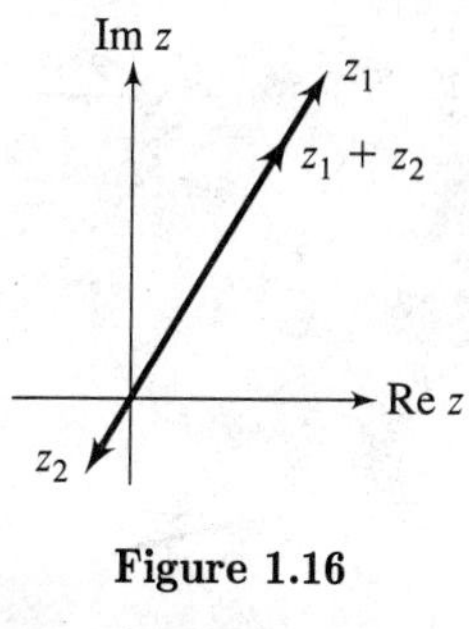

Figure 1.16

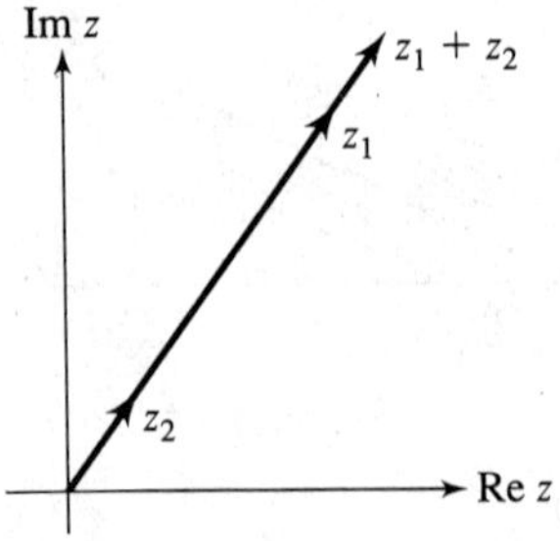

Figure 1.17

Ellipses and Circles

Moduli of complex numbers can be used to advantage in describing circles and ellipses in the complex plane (or the real plane). For example, consider the complex numbers that satisfy the inequality $|z - 2 + 3i| \leq 4$. Vector subtraction suggests that $z - 2 + 3i = z - (2 - 3i)$ may be regarded as the vector from the complex number $2 - 3i$ to the complex number z (Figure 1.18). Because $|z - 2 + 3i|$ is the length of this vector, the inequality $|z - 2 + 3i| \leq 4$ describes all complex numbers z within a distance 4 from $2 - 3i$; that is, all points inside and on a circle with center $2 - 3i$ and radius 4. We can also see this algebraically. If we set $z = x + yi$ in $|z - 2 + 3i| \leq 4$, we obtain

$$4 \geq |(x + yi) - 2 + 3i| = |(x - 2) + (y + 3)i| = \sqrt{(x - 2)^2 + (y + 3)^2}.$$

When both sides are squared, $(x - 2)^2 + (y + 3)^2 \leq 16$. This describes a circle with center $(2, -3)$ and radius 4 and its interior.

In general, for a fixed complex number z_0, the inequality $|z - z_0| \leq r$ describes all complex numbers inside and on a circle of radius r centered at z_0 (Figure 1.19).

An ellipse is the path traced out by a point that moves so that the sum of its distances from two fixed points called foci is a constant value. If the foci are the complex numbers z_0 and z_1 (Figure 1.20), then the ellipse traced out by a point that moves so that the sum of its distances from z_0 and z_1 is equal to a is

$$|z - z_0| + |z - z_1| = a.$$

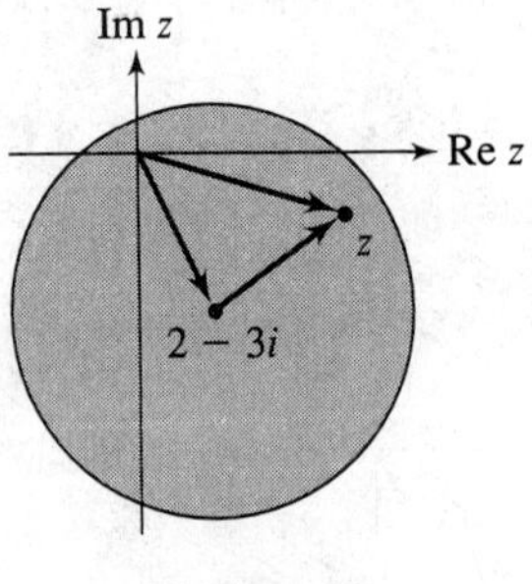

Figure 1.18

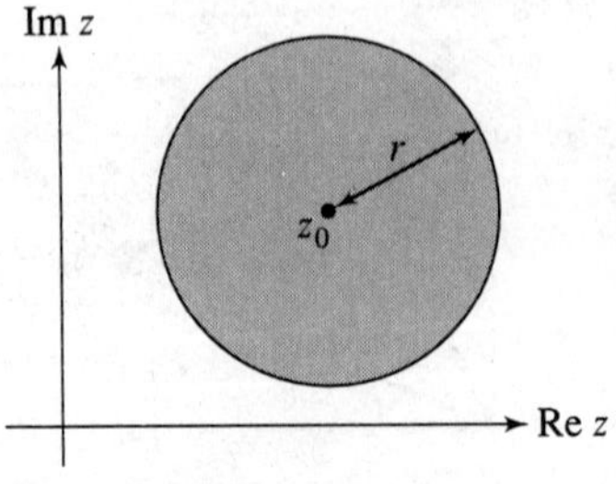

Figure 1.19

Naturally, a must be larger than the distance between z_0 and z_1. For example, the equality $|z - 3i| + |z - 4| = 7$ describes the ellipse in Figure 1.21. The inequality $|z - 3i| + |z - 4| > 7$ describes all points outside the ellipse (Figure 1.22). The fact that the ellipse itself is not described by the inequality is indicated by a dashed curve.

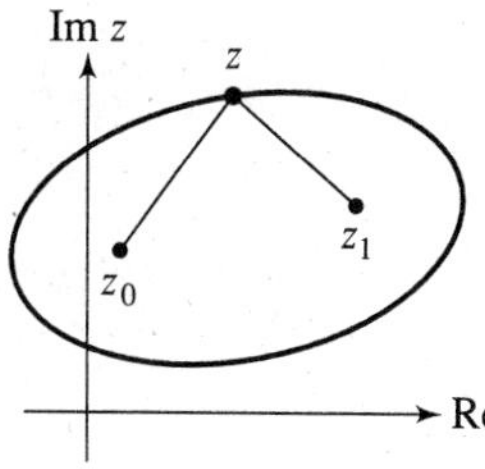

Figure 1.20

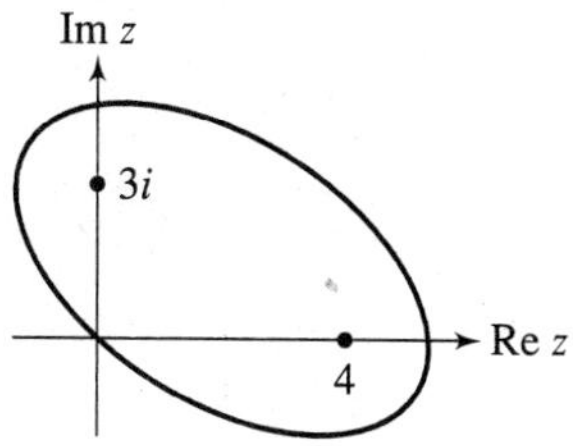

Figure 1.21

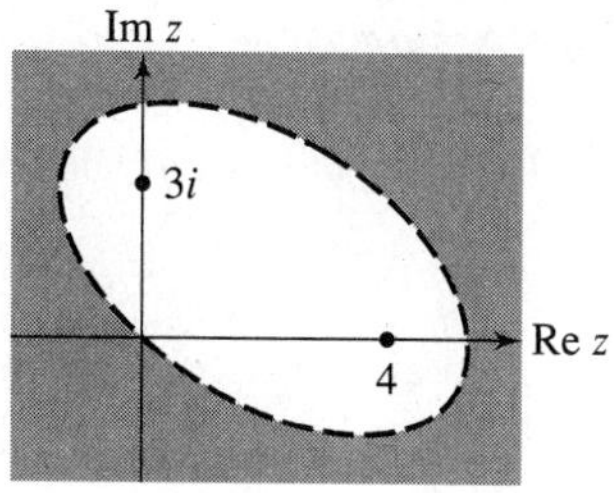

Figure 1.22

Exercises 1.4

In Exercises 1–3 use DeMoivre's theorem to find the square roots of the complex number. Compare the results and the difficulty in arriving at these results relative to Exercises 33, 34, and 36 in Section 1.2.

1. $-7 - 24i$
2. $21 - 20i$
3. $-1 + i$

In Exercises 4–11 find all solutions of the equation.

4. $z^3 = 1$
5. $z^4 = i$
6. $z^2 = 2 + 2\sqrt{3}i$
7. $z^5 = -32$
8. $z^3 = 4 - 3i$
9. $z^6 = -1$
10. $z^4 + 3z^2 + 5 = 0$
11. $z^6 - 2z^3 + 4 = 0$
12. The roots of the equation $z^n = 1$, where n is a positive integer, are called the roots of unity. Find them.
13. Prove that when $|z_3| > |z_4|$,

$$\left|\frac{z_1 + z_2}{z_3 + z_4}\right| \leq \frac{|z_1| + |z_2|}{|z_3| - |z_4|}.$$

Correct the inequality when $|z_3| < |z_4|$. What inequality encompasses both situations?

In Exercises 14–20 describe the set of points in the complex plane defined by the inequality.

14. $|z - 2 + i| \leq 3$
15. $|z + 3i| > 2$
16. $1 \leq |z + 3| \leq 4$
17. $2 < |z - 2 - i| \leq 5$

18. $|z+1|+|z-1| \leq 4$

19. $|z-2i|+|z+3i| \geq 8$

20. $4 \leq |z-1|+|z+1| \leq 6$

21. Give an algebraic proof of triangle inequality 1.32 by setting $z_1 = x_1 + y_1 i$ and $z_2 = x_2 + y_2 i$.

22. Give an algebraic proof of 1.33 by setting $z_1 = x_1 + y_1 i$ and $z_2 = x_2 + y_2 i$.

23. Use 1.32 to prove 1.33.

24. If Z is a given complex number and m and n are relatively prime, positive integers, find all solutions of the equation

$$z^n = Z^m.$$

25. Find all solutions of the equation $(z-2+3i)^4 = z^4$.

26. Find all solutions of the equation $(z+3)^5 = z^5$.

CHAPTER 2

Complex Functions and Their Derivatives

We are now ready to study the calculus of complex functions. In this chapter we introduce functions, limits, and derivatives. A complex function f is a rule that assigns a complex number $w = f(z)$ to each complex number z is some set. Because of the two-dimensional nature of both z and w, it is not possible to visualize a complex function geometrically in the same way that we visualize a real function as a curve in the xy-plane. Instead, we treat functions from a mapping perspective. A function f defines a mapping $w = f(z)$ from points in the z-plane to points in the w-plane. At first, this representation may seem far from adequate, but it will pay dividends in the application of conformal mapping to problems in electrostatics, heat flow, and fluid flow in Chapter 7. The definition of the limit of a function $f(z)$ as z approaches a complex number z_0 is identical to the limit of a real function $f(x)$ as x approaches x_0, but once again because of the two-dimensional nature of complex numbers, we find that the geometric interpretation of the limit is more reminiscent of the limit of a real function $f(x, y)$ as (x, y) approaches (x_0, y_0). Algebraically, the derivative $f'(z)$ is defined exactly the same way as $f'(x)$, but it is interpreted geometrically in the context of mappings (not as the slope of some tangent line). Existence of the derivative $f'(z)$ of a complex function places restrictions on the real and imaginary parts of $f(z)$. These conditions, called the Cauchy-Riemann equations, form the basis for all our studies in complex analysis.

2.1 Regions of the Complex Plane

An integral part of every real function $f(x)$ is its domain of definition. For example, the functions $f(x) = x^2 + 3$, $x \geq 0$, and $g(x) = x^2 + 3$, $x \leq 0$, perform the same mathematical operations, but because they operate on different sets of values of x, they are regarded as different functions. In spite of this, we seldom pay a great deal of attention to domains of definition of functions in real calculus. We usually say something to the effect that if the domain of definition of a function, specified algebraically as a formula $f(x)$, is not mentioned, it is assumed to be all values of x for which the formula makes sense. For complex functions we must be more careful. Properties of a complex function may depend on where the function is defined; different domains of definition may lead to different properties. It is therefore important that we not lose sight of the domain of definition of a function. In this section, we describe various possible domains of definition for functions, and fundamental to all of these sets is the *neighborhood*.

Definition 2.1
An **ϵ-neighborhood** of a point z_0 is the set of all complex numbers z satisfying

$$|z - z_0| < \epsilon. \tag{2.1}$$

It is the set of all complex numbers interior to a circle of radius ϵ and center z_0 (Figure 2.1). The fact that the circumference of the circle is not included in the ϵ-neighborhood is reflected in the dashed curve.

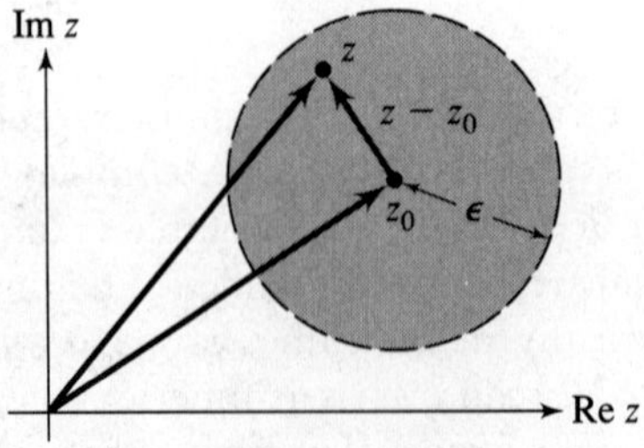

Figure 2.1

Definition 2.2
A point z of a set S of complex numbers is called an **interior point** of S if there exists an ϵ-neighborhood of z that contains only points of S; z is a **boundary point** of S if every ϵ-neighborhood of z contains at least one point of S and at least one point not in S. The set of all boundary points of a set S is called the **boundary** of S.

For example, consider the set of points described by the inequality $S_1 : |z+1|+|z-1| < 4$. Equality $|z+1|+|z-1| = 4$ describes all points on an ellipse with foci $z = \pm 1$. It follows therefore that S_1 consists of all points inside, but not on, the ellipse (Figure 2.2). Given any point z_0 in S_1, it is always possible to draw a circle centered at z_0 that contains only points of S_1 (Figure 2.3). Hence, every point in S_1 is an interior point of S_1. Since every neighborhood of every point z_1 on the ellipse itself contains points inside and outside S_1 (Figure 2.4), points on the ellipse $|z+1|+|z-1| = 4$ are boundary points of S_1; the ellipse is the boundary of S_1.

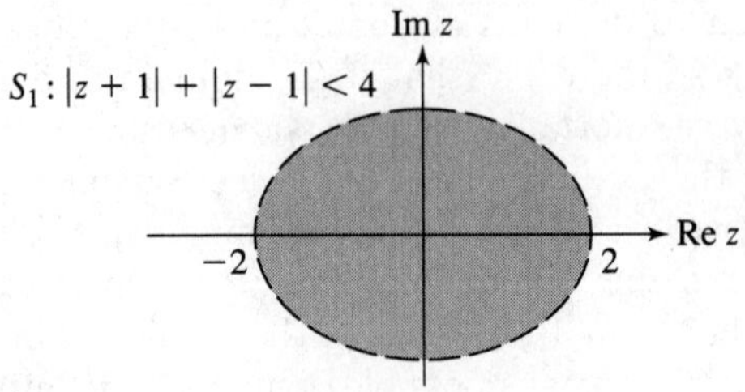

Figure 2.2

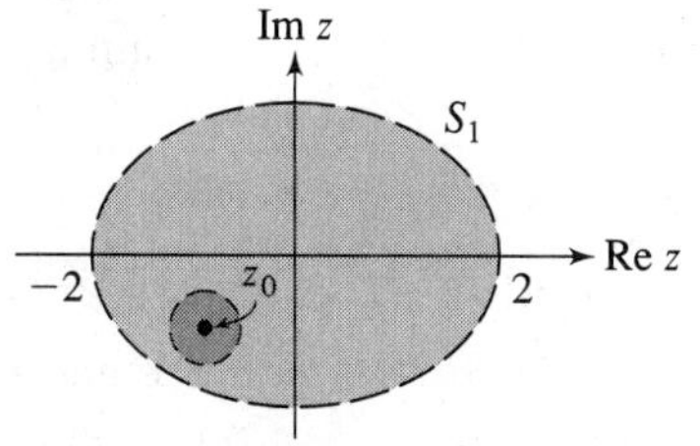

Figure 2.3

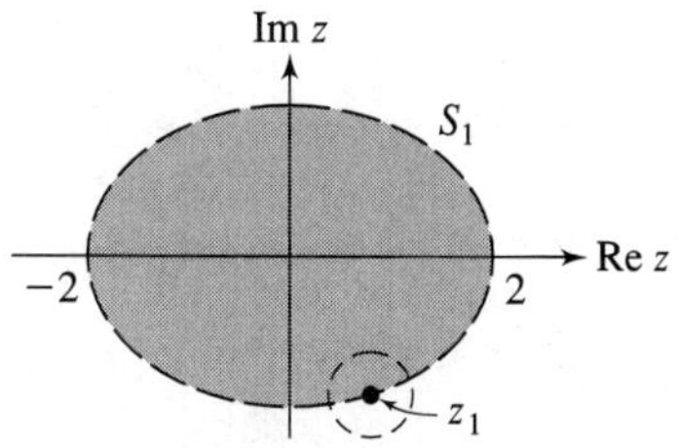

Figure 2.4

Consider now the set $S_2 : |z+1| + |z-1| \geq 4$ of all points on and outside the ellipse (Figure 2.5). Every point outside the ellipse is an interior point of S_2, and every point on the ellipse is a boundary point of S_2. The ellipse $|z+1|+|z-1| = 4$ is the boundary for S_2.

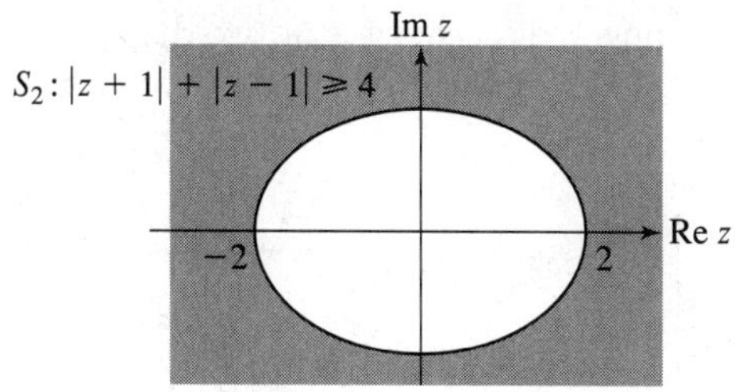

Figure 2.5

Definition 2.3
A set S of complex numbers is said to be **open** if all points in S are interior points.

Alternatively, a set is open if it contains none of its boundary points.

Definition 2.4
A set S of complex numbers is said to be **closed** if it contains all of its boundary points.

Set S_1 described earlier is open; it contains none of its boundary points (every point is an interior point). Set S_2 is closed, it contains all its boundary points. Suppose to S_1 we add the points on the upper half of the ellipse (Figure 2.6). This

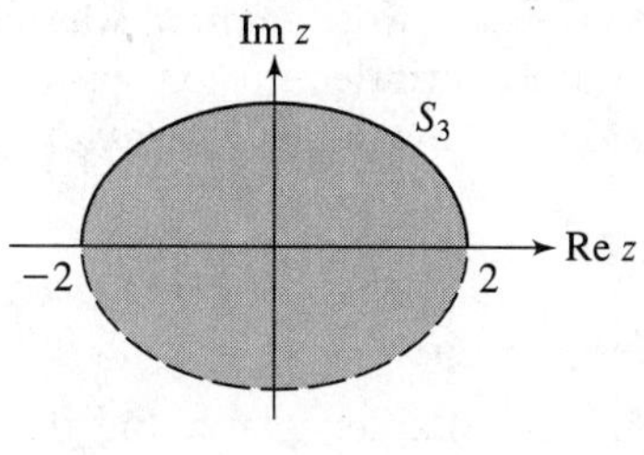

Figure 2.6

set, call it S_3, is neither open nor closed; it contains some of its boundary points but not all of them.

Definition 2.5
A set S of complex numbers is said to be **bounded** if there exists a positive (real) number R such that $|z| < R$ for all z in S.

Geometrically, a set is bounded if it can be located inside a circle with finite radius. Sets S_1 and S_3 are bounded; set S_2 is not.

Definition 2.6
A set S is said to be **connected** if every pair of points in S can be joined by a finite number of line segments joined end to end that lie entirely within S.

Sets S_1, S_2, and S_3 are all connected. Set S_4 in Figure 2.7 is also connected. Set S_5 in Figure 2.8 is not connected; it consists of two disjoint pieces. It is not possible to join point A to point B by line segments that never leave the set.

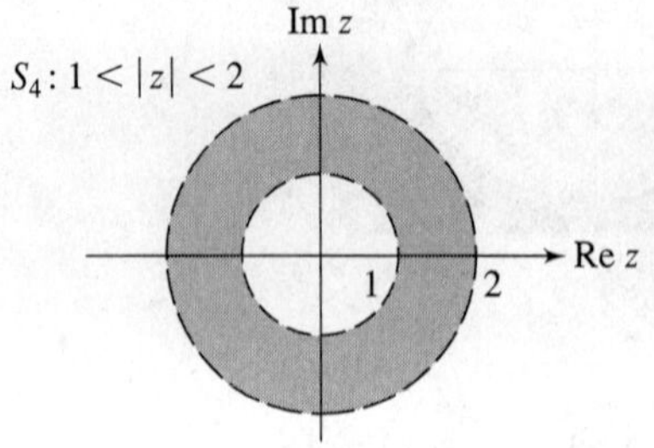

Figure 2.7

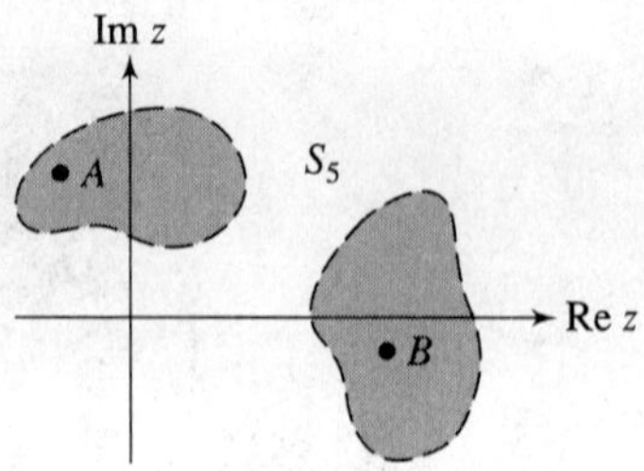

Figure 2.8

Definition 2.7
A **domain** is an open, connected set.

Sets S_1 and S_4 are domains; both are open and connected. Sets S_2, S_3, and S_5 are not domains; S_2 and S_3 fail to be open, and S_5 is not connected.

We might think that the counterpart of a real function being defined on an open interval is a complex function being defined on an open set. Such is not the case; the counterpart of an open interval is a domain. Like an interval, a domain is connected; an open set need not be connected.

It is perhaps unfortunate that the word *domain* was chosen to represent an open, connected set, because of its potential confusion with "domain" of definition of a function. We avoid confusion by always using *domain of definition* in the latter context.

EXAMPLE 2.1 ....

Sketch the following sets in the complex plane and discuss whether each is open, closed, bounded, connected, and a domain:

(a) $|z - i| < 2$ **(b)** $|z + 3i| \leq 2$ **(c)** $0 < |z + i| < 4$
(d) $1 < |z - 4 + i| \leq 3$ **(e)** $\text{Re}\, z < 3$ **(f)** $\text{Im}\, z \leq -2$

Solution **(a)** The set $|z - i| < 2$ is the interior of a circle of radius 2 centered at i (Figure 2.9). It is open, not closed, bounded, and connected. It is, therefore, a domain.

(b) The set $|z + 3i| \leq 2$ is the interior and edge of a circle with radius 2 and center $-3i$ (Figure 2.10). The circumference of the circle is solid to indicate that it is included in the set. The set is closed, not open, bounded, and connected. Because the set is not open, it cannot be a domain.

(c) The set $0 < |z + i| < 4$ is the interior of a circle centered at $-i$ with radius 4, excluding the center (Figure 2.11). The set is open but not closed; its boundary $|z + i| = 4$ and $z = -i$ are not contained inside the set. The set is bounded and connected. It is, therefore, a domain.

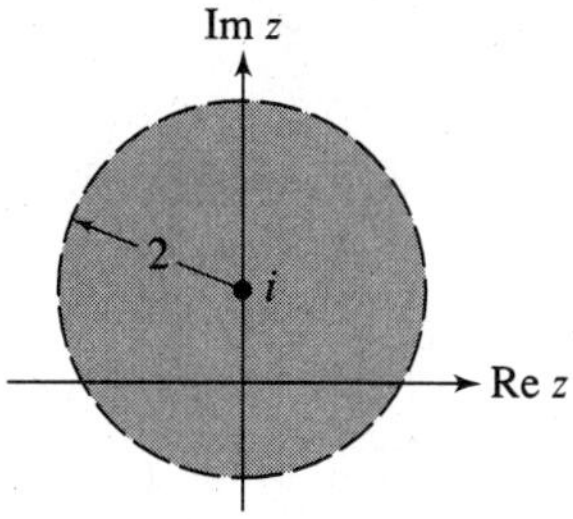

Figure 2.9

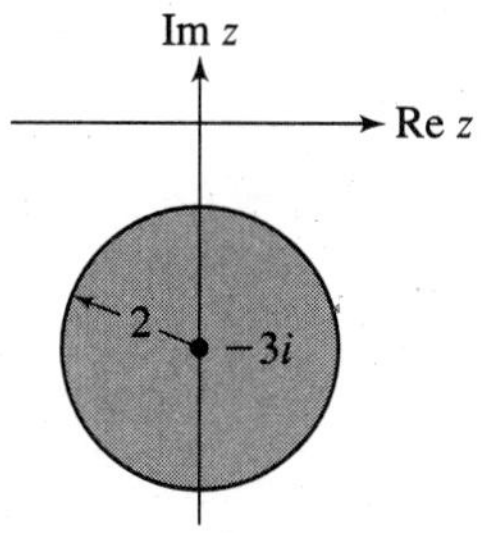

Figure 2.10

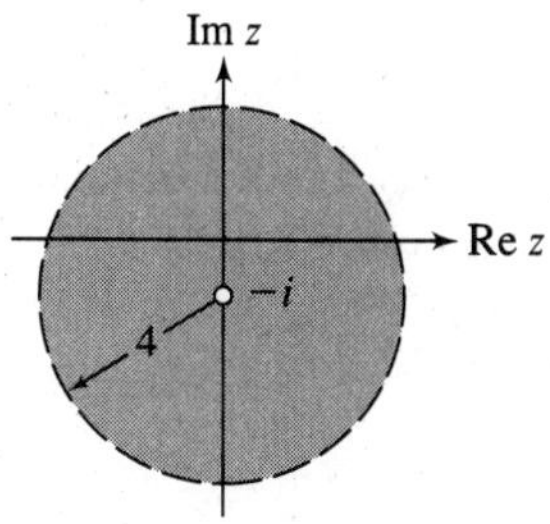

Figure 2.11

(d) The set $1 < |z - 4 + i| \leq 3$ is an annulus centered at $4 - i$ with inner and outer radii 1 and 3 (Figure 2.12). It includes the outer edge but not the inner. Because it includes some but not all of its boundary points, the set is neither open nor closed. It is bounded and connected but not a domain.

(e) The set $\text{Re}\, z < 3$ (Figure 2.13) is open and not closed. It is an unbounded domain.

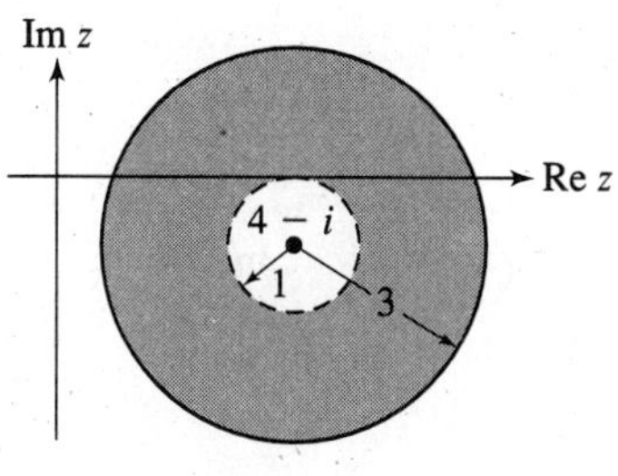

Figure 2.12

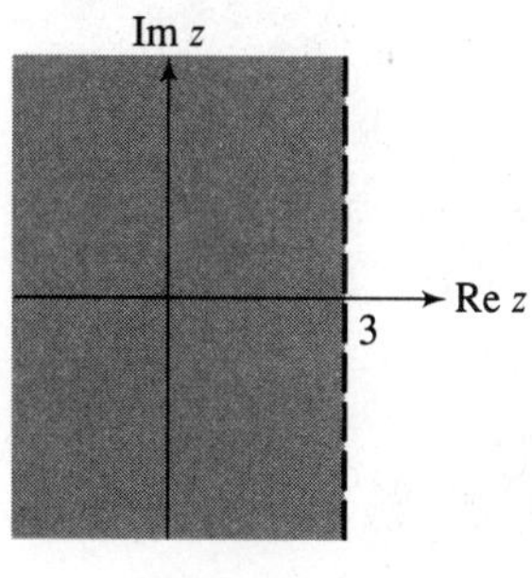

Figure 2.13

(f) The set $\operatorname{Im} z \leq -2$ (Figure 2.14) is closed, not open, unbounded, and connected. It is not a domain.

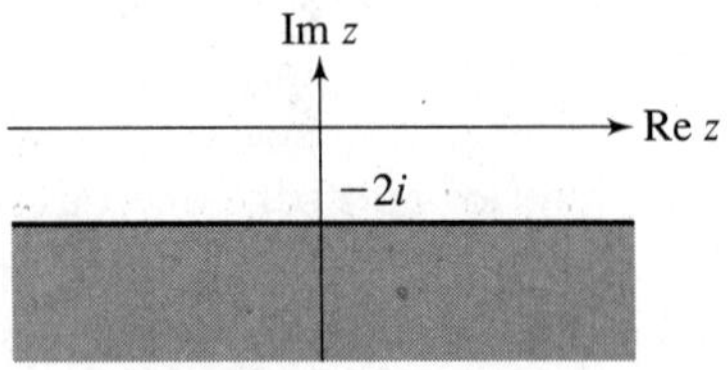

Figure 2.14

EXAMPLE 2.2 ....

Determine whether the set S of complex numbers satisfying $|z| < 1$ or $|z-2| < 1$ is a domain.

Solution The set consists of all points interior to either of the circles in Figure 2.15. It is an open set; it contains none of its boundary points on $|z| = 1$ and $|z-2| = 1$. Because the point $z = 1$ is not contained in S, the set is not connected and is, therefore, not a domain. If point $z = 1$ is added to S, the set becomes connected, but it is still not a domain. With $z = 1$ added to S, the set is no longer open.

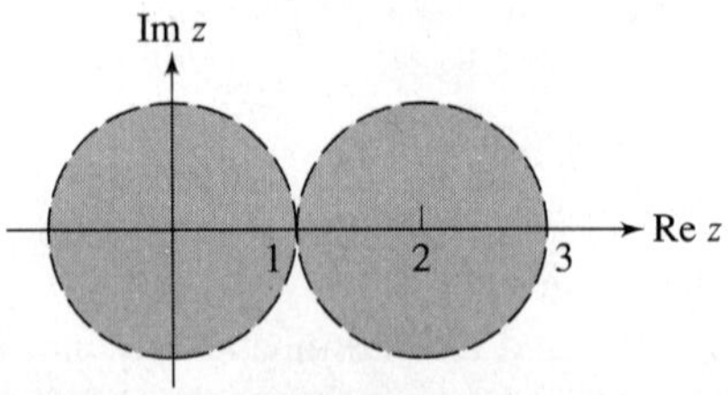

Figure 2.15

Exercises 2.1

In Exercises 1–10 sketch the set in the complex plane described by the inequalities. State whether the set is open, closed, bounded, connected, or a domain.

1. $|z - i + 3| \leq 4$
2. $|2z + 4 - 3i| > 5$
3. $|z + i| + |z - 4i| \geq 9$
4. $|z - 1 - i| + |z + 1 + i| < 6$
5. $2 < |z + 3 - 2i| < 4$
6. $\operatorname{Re}(z^2 + 2z) \leq 3$

7. $|z|^2 + |z| < 4\operatorname{Im} z$

8. $(2\operatorname{Re} z + \operatorname{Im} z)(\operatorname{Re} z - 3\operatorname{Im} z) \leq 0$

9. $(2\operatorname{Re} z + \operatorname{Im} z)(\operatorname{Re} z - 3\operatorname{Im} z) < 0$

10. $|z| < \operatorname{Arg} z + \pi$

11. Prove that the only nonempty set that is both open and closed is $\mathcal{C}$.

12. Is the set of points described by $|z| < 1 - \cos(\operatorname{Arg} z)$ a domain?

13. Suppose S is a subset of a domain D. The complement of S in D is the set of all points in D that are not in S. Prove that if the complement of an open set S in a domain D is also open, then $S = D$.

14. Is the set of points described by $|z|^2 < \cos(2\operatorname{Arg} z)$ a domain?

2.2 Complex Functions

A complex-valued function f of a complex variable z is a rule that assigns a complex number $f(z)$ to each complex number z in a set S. For example, the function $f(z) = z^2$ assigns the following complex numbers to $1 + i$ and $2 - 3i$:

$$f(1+i) = (1+i)^2 = 2i; \qquad f(2-3i) = (2-3i)^2 = -5 - 12i.$$

The complex function $g(z) = |z| + 3z$ assigns different numbers to $1 + i$ and $2 - 3i$:

$$g(1+i) = |1+i| + 3(1+i) = (\sqrt{2} + 3) + 3i;$$
$$g(2-3i) = |2-3i| + 3(2-3i) = (\sqrt{13} + 6) - 9i.$$

The set S of complex numbers on which a function operates is called the **domain of definition** of the function. It is often a domain in the sense of Definition 2.7, but it need not be so. When a function is specified algebraically by a formula such as $f(z) = z^3$ or $f(z) = (z+1)/(1-z)^2$ and the domain of definition of the function is not mentioned, it is assumed to be the set of all points in the complex plane for which the formula makes sense. For $f(z) = z^3$ this would be all $\mathcal{C}$, and for $f(z) = (z+1)/(1-z)^2$ this would be all $\mathcal{C}$ except $z = 1$. The set of complex numbers $f(z)$ that a function produces as z takes on all values in the domain of definition is called the range of the function. For example, when the domain of definition of the function $f(z) = 1/z$ is taken as the complex plane with $z = 0$ deleted, its range is also the complex plane with $0 + 0i$ deleted. If the domain of definition is only the set of points $0 < |z| < 1$, then its range is the set of complex numbers $f(z)$ for which $|f(z)| > 1$.

A real function $f(x)$ of a real variable x is illustrated graphically as the curve $y = f(x)$ in the Cartesian plane. Visualizations of complex functions $f(z)$ are not as revealing. If we set $w = f(z)$, it requires two dimensions to visualize the domain of definition of the function and another two dimensions to visualize its range. How do we represent four dimensions on a page (or even in space)? One way is to visualize the function as a mapping from the z-plane to the w-plane. Points in the z-plane are mapped by $f(z)$ onto points in the w-plane. By examining images in the w-plane of various regions in the z-plane, we may be able to appreciate the action of a function. We consider a number of examples. The function

$$w = z + b, \tag{2.2}$$

where b is a given complex number, is called a **translation**. The reason for this name becomes apparent when we draw the z-plane and the w-plane side by side (Figure 2.16).

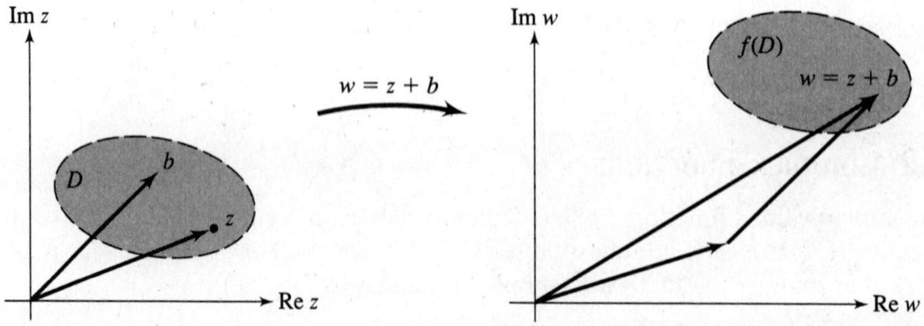

Figure 2.16

Given any point in the z-plane, its image $w = z+b$ in the w-plane is the complex number z translated by the vector corresponding to the complex number b. Any domain D in the z-plane is mapped by $w = z + b$ onto exactly the same domain in the w-plane translated by b.

Consider next the function,

$$w = az, \tag{2.3}$$

where a is a given complex number. If we write a and z in polar form, $a = \rho(\cos\phi + \sin\phi\, i)$ and $z = r(\cos\theta + \sin\theta\, i)$, then

$$w = \rho r[\cos(\theta + \phi) + \sin(\theta + \phi)\, i]. \tag{2.4}$$

This indicates that every point z is rotated through an angle ϕ and has its modulus multiplied by ρ (Figure 2.17).

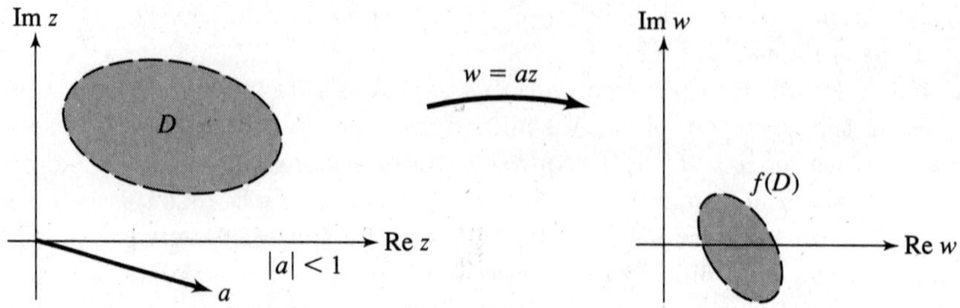

Figure 2.17

The mapping may be considered to be composed of two parts,

$$w^* = (\cos\phi + \sin\phi\, i)z, \tag{2.5a}$$

a **rotation** (see Example 1.8), and

$$w = \rho w^*, \tag{2.5b}$$

a **magnification.**

Consider now the function

$$w = \frac{1}{z}, \tag{2.6}$$

called a **reciprocation**. When we express z in polar form as $z = r(\cos\theta + \sin\theta\, i)$,

$$\begin{aligned} w = f(z) &= \frac{1}{r(\cos\theta + \sin\theta\, i)} = \frac{1}{r(\cos\theta + \sin\theta\, i)}\,\frac{\cos\theta - \sin\theta\, i}{\cos\theta - \sin\theta\, i} \\ &= \frac{1}{r}(\cos\theta - \sin\theta\, i) = \frac{1}{r}[\cos(-\theta) + \sin(-\theta)\, i]. \end{aligned}$$

In this form it is clear that $f(z)$ inverts the modulus of a complex number z and changes the sign of its argument. In Figure 2.18(a) we have drawn the z- and w-planes side by side and shown (via arrows) how three points z_1, z_2, and z_3 are mapped by $f(z)$ onto w_1, w_2, and w_3. The circle $|z| = 1$ is important for an appreciation of this function. Points on this circle are mapped by $f(z) = 1/z$ onto points on the circle $|w| = 1$. Points on the top half of $|z| = 1$ are mapped onto the bottom half of $|w| = 1$ and vice versa (since $f(z)$ changes signs of arguments). Points inside the circle $|z| = 1$ are mapped onto points outside $|w| = 1$. This is illustrated in Figure 2.18(b). Points outside $|z| = 1$ are mapped inside $|w| = 1$ (Figure 2.18(c)). Other regional mappings are shown in Figure 2.18(d)–(f).

A very important class of functions or mappings are **bilinear**, or **linear fractional mappings**. They are mappings of the form

$$w = \frac{az + b}{cz + d}, \tag{2.7}$$

where a, b, c, and d are complex numbers for which $ad - bc \neq 0$. This condition ensures that the mapping is not a constant. To see this, suppose that $ad - bc = 0$. If $a \neq 0$, we may solve for $d = bc/a$, and substitution into 2.7 gives

$$w = \frac{az + b}{cz + \dfrac{bc}{a}} = \frac{a}{c},$$

a constant. If $ad - bc = 0$ and $a = 0$, then either $b = 0$ or $c = 0$. When $a = b = 0$, it follows that $w = 0$, and when $a = c = 0$, we find $w = b/d$, a constant.

Every bilinear transformation is a succession of rotations, magnifications, reciprocations, and/or translations. If $c = 0$ in 2.7, then d may be absorbed into a and

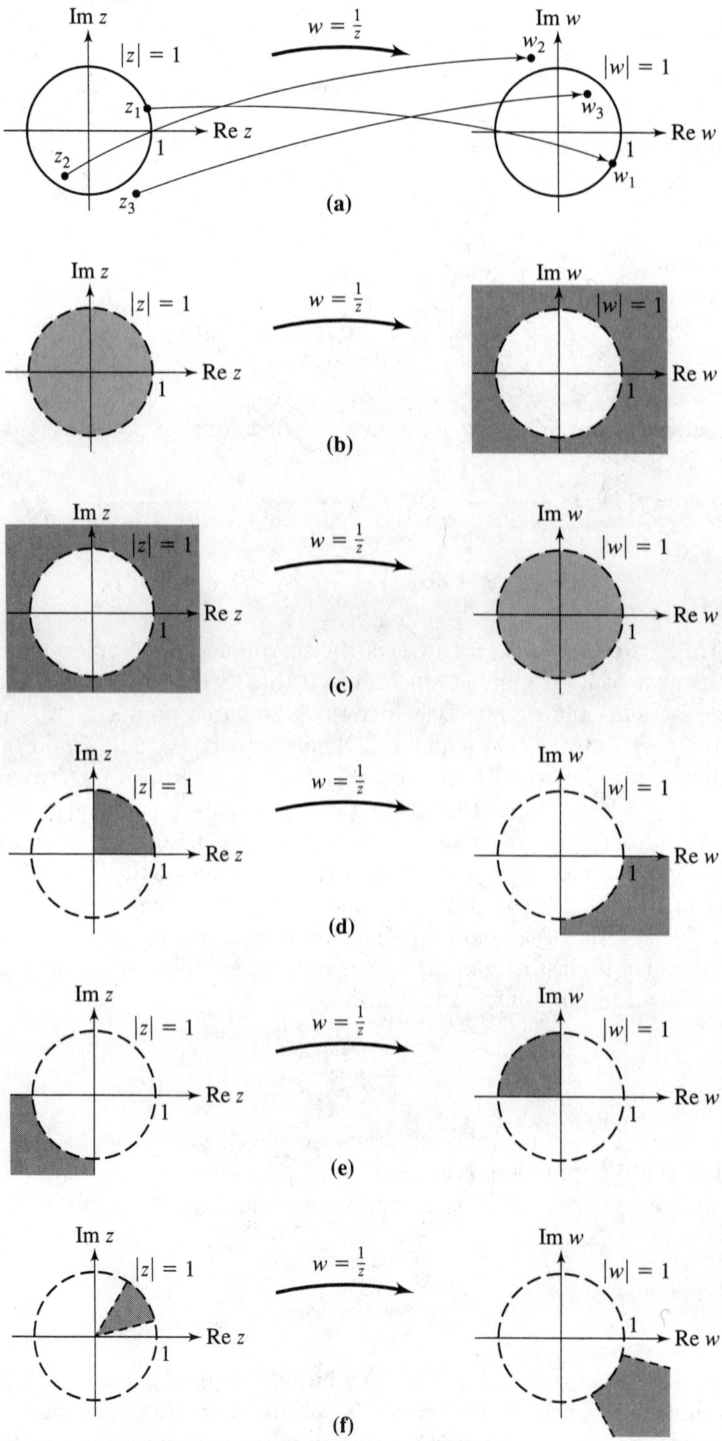

Figure 2.18

b, with the result that

$$w = az + b. \tag{2.8}$$

This is a rotation and magnification followed by a translation. If $c \neq 0$, then

$$w = \frac{az+b}{cz+d} = \frac{\dfrac{a}{c}(cz+d) + b - \dfrac{ad}{c}}{cz+d} = \frac{a}{c} + \frac{b - \dfrac{ad}{c}}{cz+d}. \tag{2.9}$$

This mapping may be regarded successively as a rotation and magnification, and a translation,

$$w_1 = cz + d,$$

a reciprocation,

$$w_2 = \frac{1}{w_1},$$

a rotation and magnification,

$$w_3 = \left(b - \frac{ad}{c}\right) w_2,$$

and, finally, a translation,

$$w = w_3 + \frac{a}{c}.$$

There are many properties of bilinear transformations that are important in physical problems. Some are illustrated in the exercises at the end of this section; they are discussed in detail in Chapter 7.

The complex number $w = f(z) = f(x + yi)$ has real and imaginary parts, which we denote by u and v. They are functions of x and y, the real and imaginary parts of z,

$$w = u + vi = u(x, y) + v(x, y)i. \tag{2.10}$$

For example, when $w = z^2 = (x + yi)^2 = (x^2 - y^2) + 2xyi$,

$$u(x, y) = x^2 - y^2, \qquad v(x, y) = 2xy,$$

and when $w = |z| + 3z = \sqrt{x^2 + y^2} + 3(x + yi)$,

$$u(x, y) = \sqrt{x^2 + y^2} + 3x, \qquad v(x, y) = 3y.$$

In other words, a complex function can be expressed as a pair of real-valued functions of two independent variables. We find this representation very useful. An illustration is contained in the following example.

EXAMPLE 2.3

Find the image of the strip $1 \le \operatorname{Re} z \le 2$ under the mapping $w = z^2$.

Solution The image of any point $z = x_0 + yi$ on the straight line $\operatorname{Re} z = x_0$ (Figure 2.19) is

$$w = (x_0 + yi)^2 = (x_0^2 - y^2) + 2x_0 yi.$$

As y varies along this line, the complex number w traces out a curve in the w-plane defined parametrically by

$$u = x_0^2 - y^2, \qquad v = 2x_0 y.$$

If we eliminate the parameter y, we have

$$u = x_0^2 - \left(\frac{v}{2x_0}\right)^2,$$

the parabola in Figure 2.19.

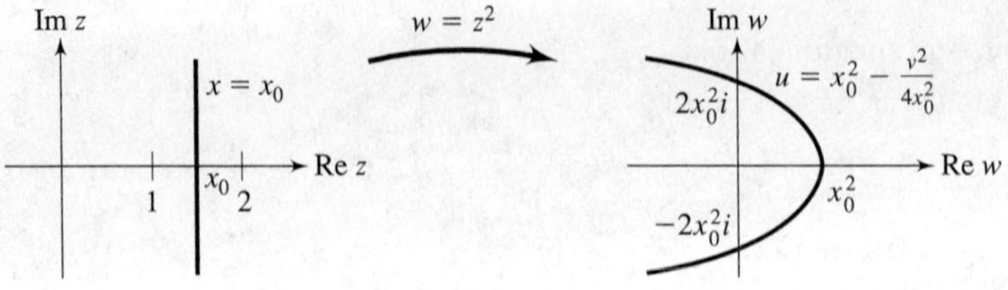

Figure 2.19

Vertical lines $\operatorname{Re} z = x_0$ in the z-plane are mapped by $w = z^2$ onto parabolas in the w-plane. The vertical line $\operatorname{Re} z = 1$ is mapped onto the parabola $u = 1 - v^2/4$, and $\operatorname{Re} z = 2$ is mapped to $u = 4 - v^2/16$ (Figure 2.20). Images of all vertical lines between $\operatorname{Re} z = 1$ and $\operatorname{Re} z = 2$ fill the parabolic strip between $u = 1 - v^2/4$ and $u = 4 - v^2/16$.

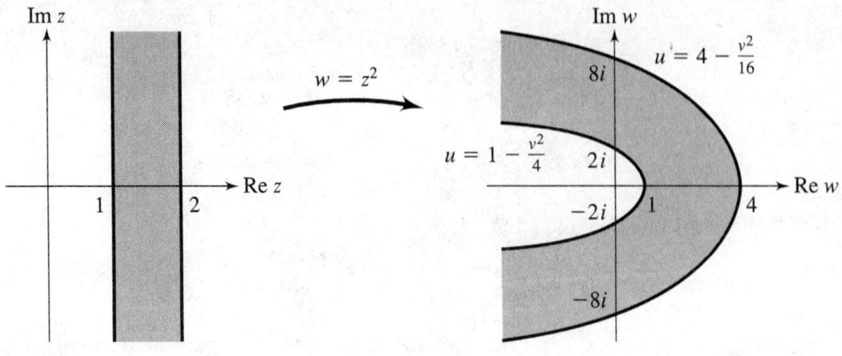

Figure 2.20

A complex function $w = f(z)$ determines a pair of real-valued functions $u(x, y)$ and $v(x, y)$, the real and imaginary parts of $f(z)$ (equation 2.10). It is important to realize that the converse is not in general true. Given a pair of real-valued functions $u(x, y)$ and $v(x, y)$, we can always define a complex function $u(x, y) + v(x, y)i$ of x and y. However, it is not usually possible to express $u(x, y) + v(x, y)i$ in the form $f(z)$, where $z = x + yi$. Let us illustrate. If $u(x, y) = x^2 - y^2$ and $v(x, y) = 2xy$, then the complex combination $u(x, y) + v(x, y)i = x^2 - y^2 + 2xyi$ can be expressed in terms of z as

$$x^2 - y^2 + 2xyi = (x + yi)^2 = z^2.$$

This cannot be done for $u(x, y) = x^2$ and $v(x, y) = y$. But how do we know this? And if $u(x, y) + v(x, y)i$ can be expressed in terms of z, how in general can we find the expression $f(z)$? One way to proceed is as follows. Given $u(x, y)$ and $v(x, y)$, we can always write $u(x, y) + v(x, y)i$. Now recall that if $z = x + yi$, then its complex conjugate is $\overline{z} = x - yi$. These equations can be solved for x and y in terms of z and $\overline{z}$:

$$x = \frac{z + \overline{z}}{2}, \qquad y = \frac{z - \overline{z}}{2i}. \tag{2.11}$$

When we substitute these values into $u(x, y) + v(x, y)i$, we obtain

$$u\left(\frac{z + \overline{z}}{2}, \frac{z - \overline{z}}{2i}\right) + v\left(\frac{z + \overline{z}}{2}, \frac{z - \overline{z}}{2i}\right)i. \tag{2.12}$$

If simplification of this expression leads to a result containing z and not $\overline{z}$, then we shall have proved that $u(x, y) + v(x, y)i$ can be expressed in terms of z and at the same time found $f(z)$. The following two examples demonstrate this procedure.

EXAMPLE 2.4 ....

Show that $(x^2 - y^2) + 2xyi$ can be expressed as a function of z.

Solution If we set $x = (z + \overline{z})/2$ and $y = (z - \overline{z})/(2i)$, then

$$\begin{aligned}(x^2 - y^2) + 2xyi &= \left[\left(\frac{z + \overline{z}}{2}\right)^2 - \left(\frac{z - \overline{z}}{2i}\right)^2\right] + 2\left(\frac{z + \overline{z}}{2}\right)\left(\frac{z - \overline{z}}{2i}\right)i \\ &= \left[\left(\frac{z^2 + 2z\overline{z} + \overline{z}^2}{4}\right) - \left(\frac{z^2 - 2z\overline{z} + \overline{z}^2}{-4}\right)\right] + \frac{1}{2}(z^2 - \overline{z}^2) \\ &= z^2.\end{aligned}$$

....

EXAMPLE 2.5 ....

Show that $x^2 + yi$ cannot be expressed as a function $f(z)$ of $z = x + yi$.

Solution If we substitute from 2.11, then

$$x^2 + yi = \left(\frac{z+\overline{z}}{2}\right)^2 + \left(\frac{z-\overline{z}}{2i}\right)i$$
$$= \frac{1}{4}(z^2 + 2z\overline{z} + \overline{z}^2) + \frac{1}{2}(z - \overline{z}).$$

Unlike Example 2.4, the $\overline{z}$ do not disappear, and therefore $x^2 + yi$ cannot be expressed solely in terms of z.

In order that the complex combination $u(x,y)+v(x,y)i$ of given functions $u(x,y)$ and $v(x,y)$ be expressible as a function $f(z)$ of $z = x + yi$, these functions must satisfy very stringent conditions. We shall see them in Section 2.4.

In Chapters 2–6 we concentrate on algebraic properties of functions; in Chapter 7 geometric aspects of mappings predominate.

Exercises 2.2

In Exercises 1–4 find real and imaginary parts $u(x,y)$ and $v(x,y)$ for the function.

1. $f(z) = z^3 + 2z - 1$
2. $f(z) = (3z+1)/(1-z)$
3. $f(z) = z^2 \operatorname{Re} z + 2z$
4. $f(z) = z|z|/(z+i)$

It is sometimes more convenient to express the real and imaginary parts of a complex function $f(z) = u + vi$ in terms of the modulus r and argument θ of z. Do this in Exercises 5 and 6.

5. $f(z) = z^7 + 3z^6$
6. $f(z) = z^5/(z-3)$
7. Show that the function $w = f(z) = z^2 + z$
   (a) Maps vertical lines $x = \operatorname{Re} z = x_0$ in the z-plane onto parabolas in the w-plane.
   (b) Has vertical lines $u = \operatorname{Re} w = u_0$ in the w-plane as images of hyperbolas in the z-plane.
8. Show that under reciprocation 2.6, real and imaginary parts of z and w are related by

$$x = \frac{u}{u^2+v^2}, \qquad y = \frac{-v}{u^2+v^2}.$$

In Exercises 9–14 use the expressions in Exercise 8 to verify the image of the curve under the reciprocation $w = 1/z$.

9. The image of $|z-1| = 1$ is $\operatorname{Re} w = 1/2$.
10. The image of $|z - 4 + 3i| = 5$ is $8 \operatorname{Re} w + 6 \operatorname{Im} w = 1$.
11. The image of $2 \operatorname{Re} z + 3 \operatorname{Im} z = 1$ is $|w - 1 + 3i/2| = \sqrt{13}/2$.
12. The image of $\operatorname{Re} z + 4 \operatorname{Im} z = 0$ is $\operatorname{Re} w - 4 \operatorname{Im} w = 0$.
13. The image of $|z+1| = 2$ is $|w - 1/3| = 2/3$.

14. The image of $|z - 2 + 3i| = 1$ is $|w - 1/6 - i/4| = 1/12$.
15. Find the image of $|z| = 1$ under the bilinear mapping $w = 1/(z+1)$.
16. Find the image of $|z| = 4$ under the bilinear mapping $w = 1/(z+2)$.
17. Find the image of $\operatorname{Re} z + 2\operatorname{Im} z = 4$ under the bilinear mapping $w = 3/(z-4)$.
18. Find images of circles $|z| = R$ under the mapping $w = f(z) = z + 1/z$.
19. Find the image of $|z| = 1$ under the mapping $w = z/(1 - z^2)$.
20. Find the image of $|z| = 1$ under the mapping $w = z/(1 + z^2)$.
21. Show that the reciprocation $w = 1/z$ maps
    (a) A straight line through $z = 0$ to a straight line.
    (b) A straight line not through $z = 0$ to a circle.
    (c) A circle through $z = 0$ to a straight line.
    (d) A circle not through $z = 0$ to a circle.

In Exercises 22–27 determine whether the expression can be written as a function $f(z)$ of $z = x + yi$.

22. $x^2 - 2xyi$

23. $(x^3 - 3xy^2) + (3x^2y - y^3)i$

24. $\dfrac{x}{x^2+y^2} - \dfrac{yi}{x^2+y^2}$

25. $\dfrac{x}{x^2+y^2} + \dfrac{yi}{x^2+y^2}$

26. $\dfrac{x^2+x+y^2}{(x+1)^2+y^2} + \dfrac{yi}{(x+1)^2+y^2}$

27. $(x^2+y^2)(x+yi)$

28. The Joukowski mapping, important in aeronautics, is

$$w = f(z) = z + \frac{a^2}{z}$$

where $a > 0$ is a constant. Show that all circles $|z| = R \neq a$ are mapped to ellipses with foci $(\pm 2a, 0)$. What is the image of $|z| = a$?

2.3 Limits and Continuity

Limits of complex functions are defined in exactly the same way as for real functions, but due to the two-dimensional nature of complex numbers, geometric interpretations are more reminiscent of limits of functions $f(x,y)$ of two real variables. Let S be an open set and z_0 be a point in S. Suppose that a complex function f is defined at every point in S except possibly at z_0. We say that f has limit L as z approaches z_0 if $f(z)$ can be made arbitrarily close to L by choosing z sufficiently close to z_0, but not equal to z_0. This is stated more precisely in the following definition.

Definition 2.8
A complex function f is said to have **limit** L as z approaches z_0, written

$$\lim_{z \to z_0} f(z) = L,$$

if given any $\epsilon > 0$, there exists a $\delta > 0$ such that $|f(z) - L| < \epsilon$ whenever $0 < |z - z_0| < \delta$.

What this definition demands geometrically is that given any circle of radius ϵ around the complex number L in the $w = f(z)$ plane (Figure 2.21), we must be able to find a circle with radius δ and center z_0 such that whenever z is inside the circle at z_0 but not equal to z_0, $f(z)$ is defined, and its value is inside the circle at L. Provided this can be done, no matter how small ϵ, then $f(z)$ must be getting closer to L as z gets closer to z_0, and $\lim_{z\to z_0} f(z) = L$.

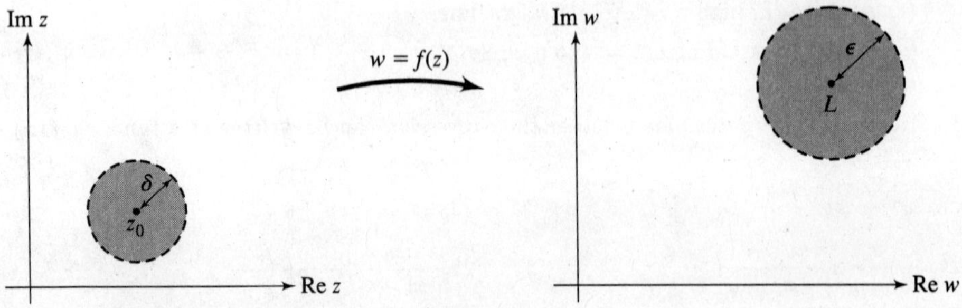

Figure 2.21

The limit of a real function $f(x)$ as x approaches x_0 must be independent of how x approaches x_0 along the x-axis. The same is true for the limit of $f(z)$ as z approaches z_0. Definition 2.8 implies that the limit of $f(z)$ is L as z approaches z_0 if, and only if, $f(z)$ approaches L for every possible mode of approach of z to z_0. But z could approach z_0 in many ways — along straight lines, parabolas, cubics, etc. (Figure 2.22). Thus, even though $f(z)$ is a function of only one variable, the fact that z requires the complex plane for its specification implies that the geometric interpretation of the limit of $f(z)$ as $z \to z_0$ is analogous to the limit of a real function of two independent variables. It follows, therefore, that if $f(z)$ can be made to approach two different complex numbers for different modes of approach of z to z_0, then $\lim_{z\to z_0} f(z)$ does not exist.

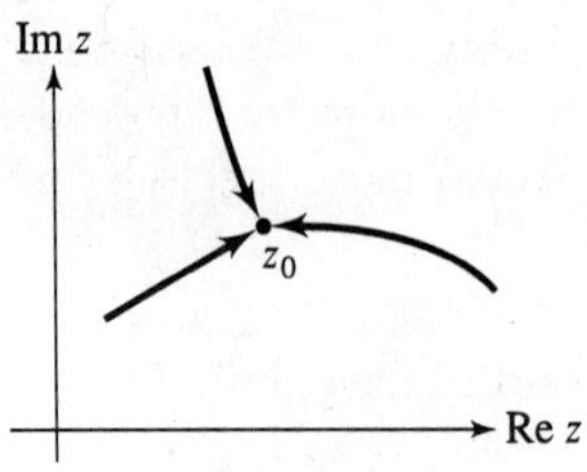

Figure 2.22

EXAMPLE 2.6 ....

Find and verify $\lim_{z \to 4+i} (2z + 1 + i)$.

Solution As z approaches $4 + i$, the function $2z + 1 + i$ approaches $9 + 3i$, and therefore

$$\lim_{z \to 4+i} (2z + 1 + i) = 9 + 3i.$$

To verify that this is indeed true, we suppose that $\epsilon > 0$ is any given real number and consider finding $\delta > 0$ so that

$$|(2z + 1 + i) - (9 + 3i)| < \epsilon$$

whenever $0 < |z - (4 + i)| < \delta$. Now,

$$|(2z + 1 + i) - (9 + 3i)| = |2z - 8 - 2i| = 2|z - (4 + i)|.$$

Thus, if we choose $\delta = \epsilon/2$, then whenever $0 < |z - (4 + i)| < \delta$,

$$|(2z + 1 + i) - (9 + 3i)| = 2|z - (4 + i)| < 2\left(\frac{\epsilon}{2}\right) = \epsilon.$$

This completes the proof that $\lim_{z \to 4+i} (2z + 1 + i) = 9 + 3i$.

....

EXAMPLE 2.7 ....

Find and verify $\lim_{z \to 1+i} (z^2 + 2)$.

Solution As z approaches $1+i$, it is anticipated that the function z^2+2 approaches $(1 + i)^2 + 2$, and therefore

$$\lim_{z \to 1+i} (z^2 + 2) = (1 + i)^2 + 2 = 2 + 2i.$$

To verify that this is correct, we suppose that $\epsilon > 0$ is any given real number and consider finding $\delta > 0$ such that

$$|(z^2 + 2) - (2 + 2i)| < \epsilon$$

whenever $0 < |z - (1 + i)| < \delta$. Now, $|(z^2 + 2) - (2 + 2i)| = |z^2 - 2i|$, and to find δ we must rework the z in $z^2 - 2i$ into the combination $z - (1 + i)$. To do this we note that $(z - 1 - i)^2 = z^2 - 2(1 + i)z + 2i$, so that

$$\begin{aligned} z^2 - 2i &= (z - 1 - i)^2 + 2(1 + i)z - 2i - 2i \\ &= (z - 1 - i)^2 + 2(1 + i)(z - 1 - i) + 2(1 + i)^2 - 4i \\ &= (z - 1 - i)^2 + 2(1 + i)(z - 1 - i). \end{aligned}$$

With triangle inequality 1.32, we may therefore write

$$\begin{aligned}|(z^2+2)-(2+2i)| = |z^2-2i| &= |(z-1-i)^2+2(1+i)(z-1-i)| \\ &\le |z-1-i|^2+2|1+i|\,|z-1-i| \\ &= |z-1-i|^2+2\sqrt{2}|z-1-i|.\end{aligned}$$

Suppose we set $p = |z-1-i|$ and consider finding p so that $p^2+2\sqrt{2}p < \epsilon$. Because $p^2+2\sqrt{2}p-\epsilon = 0$ when

$$p = \frac{-2\sqrt{2} \pm \sqrt{8+4\epsilon}}{2} = \pm\sqrt{2+\epsilon}-\sqrt{2},$$

it follows that $p^2+2\sqrt{2}p < \epsilon$ whenever $0 < p < \sqrt{2+\epsilon}-\sqrt{2}$. This result means that

$$|z-1-i|^2+2\sqrt{2}\,|z-1-i| < \epsilon \quad \text{whenever} \quad 0 < |z-1-i| < \sqrt{2+\epsilon}-\sqrt{2}.$$

Thus,

$$|(z^2+2)-(2+2i)| < \epsilon \quad \text{whenever} \quad 0 < |z-1-i| < \delta = \sqrt{2+\epsilon}-\sqrt{2},$$

and this completes the proof that $\lim_{z\to 1+i}(z^2+2) = 2+2i$.

Verification of limits of functions using Definition 2.8 can lead to very complicated calculations — witness Example 2.7, wherein z^2+2 is a very simple function, yet the calculations were far from trivial. Imagine using Definition 2.8 to verify the limit of $f(z) = (z^2+3z+2)/(z^3+3z-5)$ as z approaches $2-3i$. It would also appear that in determining a potential candidate for the limit of z^2+2 as z approaches $1+i$ in Example 2.7, we substituted $z = 1+i$ into z^2+2. We know (from real calculus) and Definition 2.8 that this should not be done; taking limits and evaluations are distinctly different operations. But recall from real calculus that when a function $f(x)$ is continuous at $x = a$, the limit of $f(x)$ as x approaches a is equal to $f(a)$. This suggests that we introduce continuity for complex functions. Before doing so, however, we consider an example wherein the function does not have a limit, and a basic theorem concerning limits of sums, differences, products, and quotients.

EXAMPLE 2.8

Show that $\lim_{z\to 0}\dfrac{\mathrm{Re}(z^2)}{|z|^2}$ does not exist.

Solution If the limit of $\mathrm{Re}(z^2)/|z|^2$ is to exist as $z \to 0$, then the function must approach the limit for all possible modes of approach of z to 0. On the other hand, if $\mathrm{Re}(z^2)/|z|^2$ approaches different values depending on how z approaches 0, then the function has no limit as $z \to 0$. We illustrate that the latter is the case by letting $z \to 0$ along the real and imaginary axes. To approach $z = 0$ along the real axis, we set $z = x$, in which case

$$\lim_{z\to 0} \frac{\text{Re}(z^2)}{|z|^2} = \lim_{x\to 0} \frac{\text{Re}(x^2)}{|x|^2} = \lim_{x\to 0} \frac{x^2}{x^2} = 1.$$

When we approach $z = 0$ along the imaginary axis, we set $z = yi$:

$$\lim_{z\to 0} \frac{\text{Re}(z^2)}{|z|^2} = \lim_{y\to 0} \frac{\text{Re}(-y^2)}{|yi|^2} = \lim_{y\to 0} \frac{-y^2}{y^2} = -1.$$

Because $\text{Re}(z^2)/|z|^2$ approaches different values as $z \to 0$, depending on the mode of approach, we conclude that $\lim_{z\to 0} \text{Re}(z^2)/|z|^2$ does not exist.

The following theorem has its counterpart in real function theory. It allows us to break complicated limits into component parts.

THEOREM 2.1 If $\lim\limits_{z\to z_0} f(z) = F$ and $\lim\limits_{z\to z_0} g(z) = G$, then

$$\lim_{z\to z_0} [f(z) \pm g(z)] = F \pm G; \qquad \textbf{(2.13a)}$$

$$\lim_{z\to z_0} [f(z)g(z)] = FG; \qquad \textbf{(2.13b)}$$

$$\lim_{z\to z_0} \frac{f(z)}{g(z)} = \frac{F}{G}, \qquad \text{provided } G \neq 0. \qquad \textbf{(2.13c)}$$

Proof We shall prove property 2.13(a) in the case of addition. The case of subtraction and the other two properties are verified in Exercises 21–23. Suppose $\epsilon > 0$ is any given real number. Because $\lim_{z\to z_0} f(z) = F$, there exists a $\delta_1 > 0$ such that whenever $0 < |z - z_0| < \delta_1$,

$$|f(z) - F| < \epsilon/2.$$

Similarly, there exists a $\delta_2 > 0$ such that whenever $0 < |z - z_0| < \delta_2$,

$$|g(z) - G| < \epsilon/2.$$

Now,

$$|[f(z) + g(z)] - (F + G)| = |[f(z) - F] + [g(z) - G]| \leq |f(z) - F| + |g(z) - G|.$$

If we choose δ as the smaller of δ_1 and δ_2, then we can say that whenever $0 < |z - z_0| < \delta$,

$$|[f(z) + g(z)] - (F + G)| < \frac{\epsilon}{2} + \frac{\epsilon}{2} = \epsilon.$$

This proves that $\lim_{z\to z_0} [f(z) + g(z)] = F + G$. ∎

EXAMPLE 2.9

Evaluate $\displaystyle\lim_{z\to 1-i}\frac{z-1}{(z+2i)\operatorname{Im} z}$.

Solution Because

$$\lim_{z\to 1-i}(z-1)=-i,\qquad \lim_{z\to 1-i}(z+2i)=1+i,\qquad \lim_{z\to 1-i}\operatorname{Im} z=-1,$$

it follows by Theorem 2.1 that

$$\lim_{z\to 1-i}\frac{z-1}{(z+2i)\operatorname{Im} z}=\frac{-i}{(1+i)(-1)}=\frac{i}{1+i}\frac{1-i}{1-i}=\frac{1}{2}+\frac{i}{2}.$$

The next theorem permits us to examine the limit of a complex function by considering limits of its real and imaginary parts. This has an advantage when the function is defined in the form $u(x,y)+v(x,y)i$ rather than in terms of z.

■ **THEOREM 2.2** If $f(z)=u(x,y)+v(x,y)i$, then $\lim_{z\to z_0} f(z)$ exists (where $z_0 = x_0+y_0 i$) if and only if

$$\lim_{(x,y)\to(x_0,y_0)} u(x,y) \quad\text{and}\quad \lim_{(x,y)\to(x_0,y_0)} v(x,y)$$

both exist, and when these limits exist,

$$\lim_{z\to z_0} f(z)=\left[\lim_{(x,y)\to(x_0,y_0)} u(x,y)\right]+\left[\lim_{(x,y)\to(x_0,y_0)} v(x,y)\right] i. \tag{2.14}$$

Proof Suppose first of all that

$$\lim_{(x,y)\to(x_0,y_0)} u(x,y)=u_0 \quad\text{and}\quad \lim_{(x,y)\to(x_0,y_0)} v(x,y)=v_0.$$

Then given any $\epsilon>0$, there exist constants δ_1 and δ_2 such that

$$|u(x,y)-u_0|<\epsilon/2 \quad\text{whenever}\quad 0<\sqrt{(x-x_0)^2+(y-y_0)^2}<\delta_1$$

and

$$|v(x,y)-v_0|<\epsilon/2 \quad\text{whenever}\quad 0<\sqrt{(x-x_0)^2+(y-y_0)^2}<\delta_2.$$

Now,

$$\begin{aligned}|f(z)-(u_0+v_0 i)| &= |u(x,y)+v(x,y)i-u_0-v_0 i|\\ &= |[u(x,y)-u_0]+[v(x,y)-v_0]i|\\ &\le |u(x,y)-u_0|+|v(x,y)-v_0|.\end{aligned}$$

If δ is the smaller of δ_1 and δ_2, then whenever

$$0 < |z - z_0| = \sqrt{(x - x_0)^2 + (y - y_0)^2} < \delta,$$

we can say that

$$|f(z) - (u_0 + v_0 i)| < \frac{\epsilon}{2} + \frac{\epsilon}{2} = \epsilon.$$

This shows that

$$\lim_{z \to z_0} f(z) = u_0 + v_0 i = \left[\lim_{(x,y)\to(x_0,y_0)} u(x,y)\right] + \left[\lim_{(x,y)\to(x_0,y_0)} v(x,y)\right] i.$$

Conversely, suppose that $\lim_{z \to z_0} f(z) = u_0 + v_0 i$. Given any $\epsilon > 0$, there exists a $\delta > 0$ such that whenever $0 < |z - z_0| < \delta$,

$$\begin{aligned} |f(z) - (u_0 + v_0 i)| &= |[u(x,y) + v(x,y)i] - (u_0 + v_0 i)| \\ &= |[u(x,y) - u_0] + [v(x,y) - v_0]i| < \epsilon. \end{aligned}$$

Thus, whenever $0 < |z - z_0| < \delta$,

$$\sqrt{[u(x,y) - u_0]^2 + [v(x,y) - v_0]^2} < \epsilon.$$

Because $|u(x,y) - u_0| < \sqrt{[u(x,y) - u_0]^2 + [v(x,y) - v_0]^2}$, it follows that whenever $0 < |z - z_0| = \sqrt{(x - x_0)^2 + (y - y_0)^2} < \delta$,

$$|u(x,y) - u_0| < \epsilon;$$

that is, $\lim_{(x,y)\to(x_0,y_0)} u(x,y) = u_0$. Similarly, $\lim_{(x,y)\to(x_0,y_0)} v(x,y) = v_0$. ■

EXAMPLE 2.10 ....

Given that $\sin x \cosh y + \cos x \sinh y\, i$ is a function of $z = x + yi$, evaluate $\lim_{z \to 2-3i} (\sin x \cosh y + \cos x \sinh y\, i)$, if it exists.

Solution Because

$$\lim_{(x,y)\to(2,-3)} \sin x \cosh y = \sin 2 \cosh(-3) = \sin 2 \cosh 3$$

and

$$\lim_{(x,y)\to(2,-3)} \cos x \sinh y = \cos 2 \sinh(-3) = -\cos 2 \sinh 3,$$

it follows by Theorem 2.2 that

$$\lim_{z \to 2-3i} (\sin x \cosh y + \cos x \sinh y\, i) = \sin 2 \cosh 3 - \cos 2 \sinh 3\, i \approx 9.15 + 4.16i.$$

....

Definition 2.9
A complex function f is said to be continuous at $z = z_0$ if

$$\lim_{z \to z_0} f(z) = f(z_0). \tag{2.15}$$

The function is said to be continuous on an open set if it is continuous at each point in the set.

As for real functions, Definition 2.9 imposes three conditions on the function f:

1. f must be defined at z_0.
2. The limit of f as $z \to z_0$ must exist.
3. The numbers in (1) and (2) must be identical.

According to the following theorem, continuous functions can be added, subtracted, multiplied, and divided; its proof is a straightforward application of Theorem 2.1.

■ **THEOREM 2.3** If $f(z)$ and $g(z)$ are continuous at z_0, then so also are the functions

$$f(z) \pm g(z), \qquad f(z)g(z), \qquad f(z)/g(z),$$

provided in the quotient case that $g(z_0) \neq 0$. ■

As a result of this theorem, we can state that polynomials, functions of the form $P_n(z) = a_n z^n + a_{n-1} z^{n-1} + \cdots + a_1 z + a_0$, where n is a nonnegative integer and the a_j $(j = 0, \ldots, n)$ are complex constants, are continuous functions in the whole complex plane. Rational functions, functions of the form $P(z)/Q(z)$, where $P(z)$ and $Q(z)$ are polynomials, are continuous except where $Q(z) = 0$.

EXAMPLE 2.11

Find $\displaystyle\lim_{z \to 2-i} \frac{z^2 + 3z + i}{2z - 4i}$.

Solution Because the rational function $(z^2 + 3z + i)/(2z - 4i)$ is continuous at $z = 2 - i$, the limit is equal to the value of the function at $z = 2 - i$:

$$\lim_{z \to 2-i} \frac{z^2 + 3z + i}{2z - 4i} = \frac{(2-i)^2 + 3(2-i) + i}{2(2-i) - 4i} = \frac{9 - 6i}{4 - 6i} = \frac{9 - 6i}{4 - 6i}\frac{4 + 6i}{4 + 6i} = \frac{72 + 30i}{52}$$
$$= \frac{36 + 15i}{26}.$$

EXAMPLE 2.12 ....

The function $f(z) = (z^2+1)/(z-i)$ is discontinuous at $z = i$. Can it be assigned a value at $z = i$ to make it continuous there?

Solution Because $\lim_{z\to i} \dfrac{z^2+1}{z-i} = \lim_{z\to i} \dfrac{(z+i)(z-i)}{z-i} = \lim_{z\to i}(z+i) = 2i$, if we define $f(i) = 2i$, the function becomes continuous for all z including $z = i$. We say that $f(z)$ has a *removable discontinuity* at $z = i$. ....

EXAMPLE 2.13 ....

Determine whether discontinuities of $f(z) = \operatorname{Im} z/\operatorname{Re} z$ are removable.

Solution The functions $\operatorname{Re} z$ and $\operatorname{Im} z$, which take real and imaginary parts of complex numbers, are continuous for all z. It follows (by Theorem 2.3) that $f(z)$ is continuous for all z with $\operatorname{Re} z \neq 0$. The function is discontinuous at all points on the imaginary axis (where $\operatorname{Re} z = 0$), because $f(z)$ is undefined there. Each such discontinuity is removable if $f(z)$ has a limit as z approaches the discontinuity. Consider then whether the limit of $f(z)$ exists as z approaches a point $z_0 = y_0 i$ on the imaginary axis (Figure 2.23). For any point $z = x + yi$ close to z_0,

$$f(z) = \frac{\operatorname{Im} z}{\operatorname{Re} z} = \frac{y}{x}.$$

When $y_0 \neq 0$, the limit of this function as $x \to 0$ does not exist. In other words, discontinuities $z = y_0 i$ where $y_0 \neq 0$ are not removable. Consider now the discontinuity $z = 0$. If we approach $z = 0$ along the line $y = mx$, then

$$\lim_{z\to 0} f(z) = \lim_{x\to 0} \frac{mx}{x} = m.$$

Because $f(z)$ approaches different numbers as $z \to 0$, depending on the mode of approach of $z \to 0$, we conclude that $\lim_{z\to 0} f(z)$ does not exist; that is, $z = 0$ is not a removable discontinuity. Thus, no discontinuity of $f(z)$ is removable.

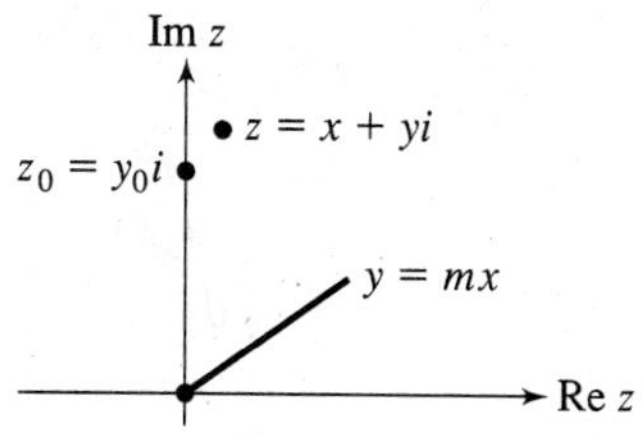

Figure 2.23

....

The function $\operatorname{Arg} z$ takes the principal value of the argument of a complex number; its discontinuities have serious ramifications for many of the functions that we encounter in Chapter 3. The value of $\operatorname{Arg} z$ at every point on the negative real axis is π, but its limit as z approaches a point on the negative real axis does not exist. As z approaches z_0 along the curve C in Figure 2.24, $\operatorname{Arg} z$ approaches π, whereas it approaches $-\pi$ along the curve C'. Consequently $\operatorname{Arg} z$ is discontinuous at every point on the negative real axis, and these discontinuities are not removable. The function is also discontinuous at $z = 0$, because $\operatorname{Arg}(0)$ is undefined. This discontinuity is not removable either, because $\lim_{z\to 0} \operatorname{Arg} z$ does not exist. To see this, suppose we let z approach 0 along the positive real axis. Then $z = x$, and

$$\lim_{z\to 0} \operatorname{Arg} z = \lim_{x\to 0+} \operatorname{Arg}(x) = \lim_{x\to 0+} (0) = 0.$$

But if z approaches 0 along the positive imaginary axis, then $z = yi$ $(y > 0)$, and

$$\lim_{z\to 0} \operatorname{Arg} z = \lim_{y\to 0+} \operatorname{Arg}(yi) = \lim_{y\to 0+} (\pi/2) = \pi/2.$$

Other modes of approach lead to other values. Thus, $\operatorname{Arg} z$ is discontinuous at $z = 0$ and every point on the negative real axis. As we stated earlier, these discontinuities play a fundamental role in properties of many functions in Chapter 3.

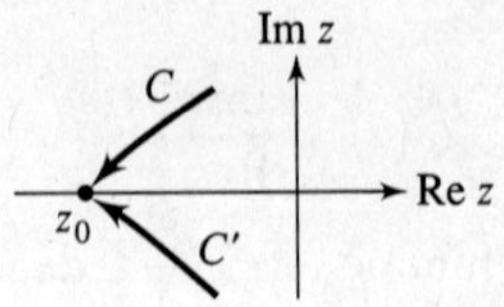

Figure 2.24

EXAMPLE 2.14 ....

Where is $f(z) = (\operatorname{Arg} z)^2$ discontinuous? Are discontinuities removable?

Solution Because the only discontinuities of $\operatorname{Arg} z$ are $z = 0$ and points on the negative real axis, these are the only potential discontinuities for $f(z)$. Certainly $z = 0$ is a discontinuity, because $f(0)$ is undefined. This discontinuity is not removable, because $\lim_{z\to 0} (\operatorname{Arg} z)^2$ does not exist. For any point z_0 on the negative real axis, however,

$$f(z_0) = \pi^2 = \lim_{z\to z_0} f(z),$$

and $f(z)$ is continuous at z_0. ....

EXAMPLE 2.15 ....

Does the function $f(z) = (z+1) \operatorname{Arg} z$ have any discontinuities? Are they removable?

Solution Because $z + 1$ is continuous for all z and $\operatorname{Arg} z$ is continuous for all z except $z = 0$ and points on the negative real axis, it follows (by Theorem 2.3) that

$f(z) = (z+1)\operatorname{Arg} z$ is continuous for all z except possibly $z = 0$ and points on the negative real axis. Because $f(0)$ is undefined, $f(z)$ is discontinuous at $z = 0$. This discontinuity is not removable, because $\lim_{z\to 0}(z+1)\operatorname{Arg} z$ does not exist. (If $\lim_{z\to 0}(z+1)\operatorname{Arg} z$ were to exist, then property 2.13(c) of Theorem 2.1 could be used with $\lim_{z\to 0}(z+1) = 1$ to prove that $\lim_{z\to 0}\operatorname{Arg} z$ exists, a fact that is not true.) For any point z_0 on the negative real axis, except $z = -1$, $\lim_{z\to z_0}(z+1) = z_0 + 1$, but $\lim_{z\to z_0}\operatorname{Arg} z$ does not exist. It follows that $\lim_{z\to z_0} f(z)$ does not exist, and $f(z)$ has a discontinuity at z_0 that is not removable. This leaves only the point $z = -1$. Because $f(-1) = 0$, continuity of $f(z)$ at $z = -1$ comes down to the existence of $\lim_{z\to -1} f(z)$. Although $\lim_{z\to -1}\operatorname{Arg} z$ does not exist, the fact that $\lim_{z\to -1}(z+1) = 0$ implies that $\lim_{z\to -1} f(z) = 0$. To prove this, we use Definition 2.8. Suppose $\epsilon > 0$ is given. Because the maximum value of $\operatorname{Arg} z$ is π, consider all points z satisfying $0 < |z+1| < \delta = \epsilon/\pi$ (Figure 2.25). For all such z,

$$|f(z) - 0| = |(z+1)\operatorname{Arg} z| \le |z+1|\pi < \frac{\epsilon}{\pi}\pi = \epsilon.$$

Thus, $\lim_{z\to -1} f(z) = 0$, and $f(z)$ is continuous at $z = -1$. The function $f(z) = (z+1)\operatorname{Arg} z$ is therefore discontinuous at $z = 0$ and every point on the negative real axis except $z = -1$.

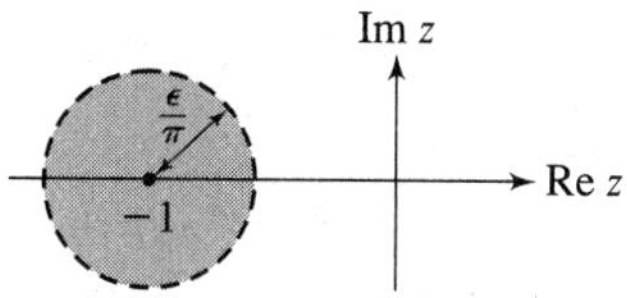

Figure 2.25

Theorem 2.4 implies that continuity of a complex function can be discussed in terms of continuity of its real and imaginary parts.

■ **THEOREM 2.4** A function $f(z) = u(x,y) + v(x,y)i$ is continuous at $z_0 = x_0 + y_0 i$ if and only if $u(x,y)$ and $v(x,y)$ are continuous at (x_0, y_0). ■

For example, the complex function $f(z) = \cos x \cosh y - \sin x \sinh y\, i$ is continuous at every point in the complex plane, because its real and imaginary parts are continuous everywhere in the xy-plane.

Exercises 2.3

In Exercises 1–8 evaluate the limit, if it exists.

1. $\lim_{z\to 2-i} z^2 + 3$

2. $\lim_{z\to i} \dfrac{z^2-1}{z+i}$

3. $\lim_{z \to -3i} \frac{z^7 + 4}{1 - z}$

4. $\lim_{z \to 0} \frac{\operatorname{Im} z}{z}$

5. $\lim_{z \to 1} (z-1) \operatorname{Re}(z-1)$

6. $\lim_{z \to 1} \operatorname{Arg} z$

7. $\lim_{z \to -1} \operatorname{Arg} z$

8. $\lim_{z \to 0} z \operatorname{Arg} z$

In Exercises 9–18, find domains (open, connected sets) in which the function is continuous.

9. $f(z) = \frac{1}{z+1}$

10. $f(z) = \frac{z^2 + 2}{z^2 + 3z + 8}$

11. $f(z) = \frac{z^3 + 2z}{z^4 + 1}$

12. $f(z) = e^x(\cos y + \sin y\, i)$

13. $f(z) = \sin x \cosh y + \cos x \sinh y\, i$

14. $f(z) = \cos y \cosh x + \sin y \sinh x\, i$

15. $f(z) = \operatorname{Re} z$

16. $f(z) = |z|$

17. $f(z) = \frac{z}{|z|}$

18. $f(z) = \operatorname{Arg} z$

19. Show that the function $f(z) = \ln|z| + \operatorname{Arg} z\, i$ is discontinuous at $z = 0$ and all points on the negative real axis.

20. Show that the function $f(z) = \sqrt{r}[\cos(\theta/2) + \sin(\theta/2)\, i]$, where $z = r(\cos\theta + \sin\theta\, i)$, $-\pi < \theta \le \pi$, is discontinuous at all points on the negative real axis. Is it continuous at $z = 0$?

In Exercises 21–23 we use Definition 2.8 to prove Theorem 2.1. In each exercise assume that $\lim_{z\to z_0} f(z) = F$ and $\lim_{z\to z_0} g(z) = G$.

21. Show that given any $\epsilon > 0$, there exist numbers $\delta_1 > 0$ and $\delta_2 > 0$ such that

$$|f(z) - F| < \epsilon/2 \quad \text{whenever} \quad 0 < |z - z_0| < \delta_1, \quad \text{and}$$
$$|g(z) - G| < \epsilon/2 \quad \text{whenever} \quad 0 < |z - z_0| < \delta_2.$$

Use these results along with inequality 1.32 to prove property 2.13(a) of Theorem 2.1 in the subtraction case.

22. **(a)** Verify that

$$|f(z)g(z) - FG| \le |f(z)|\,|g(z) - G| + |G|\,|f(z) - F|.$$

(b) Show that given any $\epsilon > 0$, there exist numbers $\delta_1 > 0$, $\delta_2 > 0$, and $\delta_3 > 0$ such that

$$|f(z)| < |F| + 1 \quad \text{whenever} \quad 0 < |z - z_0| < \delta_1,$$
$$|g(z) - G| < \frac{\epsilon}{2(|F| + 1)} \quad \text{whenever} \quad 0 < |z - z_0| < \delta_2,$$
$$|f(z) - F| < \frac{\epsilon}{2|G| + 1} \quad \text{whenever} \quad 0 < |z - z_0| < \delta_3.$$

(c) Use these results to prove property 2.13(b) of Theorem 2.1.

23. **(a)** Verify that when $G \neq 0$,

$$\left|\frac{f(z)}{g(z)} - \frac{F}{G}\right| \leq \frac{|f(z) - F|}{|g(z)|} + \frac{|F|\,|G - g(z)|}{|G|\,|g(z)|}.$$

(b) Show that given any $\epsilon > 0$, there exist numbers $\delta_1 > 0$, $\delta_2 > 0$, and $\delta_3 > 0$ such that

$$|g(z)| > \frac{|G|}{2} \qquad \text{whenever} \quad 0 < |z - z_0| < \delta_1,$$

$$|f(z) - F| < \frac{\epsilon|G|}{4} \qquad \text{whenever} \quad 0 < |z - z_0| < \delta_2,$$

$$|g(z) - G| < \frac{\epsilon|G|^2}{4|F| + 1} \qquad \text{whenever} \quad 0 < |z - z_0| < \delta_3.$$

(c) Now prove property 2.13(c) of Theorem 2.1.

2.4 Derivatives of Complex Functions

We are now ready to take up the main topic of study, differentiation of complex functions. Although the derivative of a complex function has the same algebraic definition as the derivative of a real function, we shall find that existence of the derivative of a complex function has far-reaching consequences that have no counterpart in real analysis. Much of our work in succeeding chapters is an investigation of these consequences.

Definition 2.10

The **derivative** of a complex function f with respect to z at z_0 is defined as

$$f'(z_0) = \lim_{\Delta z \to 0} \frac{f(z_0 + \Delta z) - f(z_0)}{\Delta z}, \qquad \textbf{(2.16)}$$

provided the limit exists. If the limit exists, f is said to be **differentiable** at the point z_0.

When a function f is differentiable at a point z_0, then it is continuous at z_0. This is most easily proved by multiplying both sides of 2.16 by $\lim_{\Delta z \to 0} \Delta z$:

$$f'(z_0)\left[\lim_{\Delta z \to 0} \Delta z\right] = \left[\lim_{\Delta z \to 0} \frac{f(z_0 + \Delta z) - f(z_0)}{\Delta z}\right]\left[\lim_{\Delta z \to 0} \Delta z\right].$$

Because $\lim_{\Delta z \to 0} \Delta z = 0$, the left side vanishes, and property 2.13(b) for limits allows us to bring the product of the limits on the right to the limit of a product,

$$0 = \lim_{\Delta z \to 0} \left[\frac{f(z_0 + \Delta z) - f(z_0)}{\Delta z} \Delta z\right] = \lim_{\Delta z \to 0} [f(z_0 + \Delta z) - f(z_0)].$$

This can be rewritten as

$$\lim_{\Delta z \to 0} f(z_0 + \Delta z) = f(z_0),$$

which is equivalent to condition 2.15 for continuity of $f(z)$ at z_0.

The converse is not necessarily true. Continuous functions need not be differentiable. For example, the function $\operatorname{Re} z$ is continuous everywhere, but we show (in Example 2.17) that it is nowhere differentiable.

For limit 2.16 to exist as $\Delta z \to 0$, the quotient $[f(z_0 + \Delta z) - f(z_0)]/\Delta z$ must be defined in an open set around $\Delta z = 0$ but not necessarily at $\Delta z = 0$ (see the paragraph before Definition 2.8). This implies that f must be defined in an open set around z_0. Furthermore, the fact that 2.16 contains $f(z_0)$ means that f must be defined at z_0. Thus, for a function f to have a derivative at z_0, the function must be defined in an open set containing z_0. What this means is that if a function f is defined at all points inside and on the curve C in Figure 2.26, there is no derivative at points on C. We can consider f' only at points inside C. These requirements are captured in the following definition.

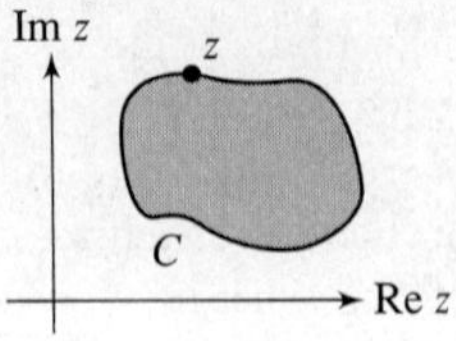

Figure 2.26

Definition 2.11
A complex function f is said to be **analytic** (regular, or holomorphic) in an open set S if it has a derivative at every point of S.

The theory of complex functions is a study of the properties of analytic functions. In other words, differentiability on a point-by-point basis is replaced by differentiability in open sets.

It is also convenient to define analyticity at a point.

Definition 2.12
A complex function f is said to be **analytic at a point** z if it is analytic in some neighborhood of z.

Differentiability is a point concept; analyticity is differentiability at all points in an open set. Even in Definition 2.12, analyticity at a point means differentiability in a circle around the point.

Definition 2.13
A function that is analytic in the whole complex plane is said to be an **entire** function.

The fact that differentiability implies continuity gives us our first property of analytic functions.

■ **THEOREM 2.5** If a complex function f is analytic in an open set S, then f is continuous in S. ■

When we drop the subscript in 2.16, we obtain

$$f'(z) = \lim_{\Delta z \to 0} \frac{f(z + \Delta z) - f(z)}{\Delta z}, \tag{2.17}$$

the derivative of f at the point z. We call the function f' defined by 2.17 the derivative function. As might be expected from equation 2.17, derivatives of complex functions can be calculated in much the same way as derivatives of real functions. In particular, we have the usual rules:

$$[f(z) + g(z)]' = f'(z) + g'(z) \qquad \text{(sum rule);} \tag{2.18a}$$

$$[f(z)g(z)]' = f'(z)g(z) + f(z)g'(z) \qquad \text{(product rule);} \tag{2.18b}$$

$$\left[\frac{f(z)}{g(z)}\right]' = \frac{g(z)f'(z) - f(z)g'(z)}{[g(z)]^2} \qquad \text{(quotient rule);} \tag{2.18c}$$

$$\{[f(z)]^n\}' = n[f(z)]^{n-1}f'(z) \qquad \text{(power rule);} \tag{2.18d}$$

if $w = f(\zeta)$ and $\zeta = g(z)$,

$$\frac{dw}{dz} = \frac{dw}{d\zeta}\frac{d\zeta}{dz} \qquad \text{(chain rule).} \tag{2.18e}$$

The sum and power rules guarantee that polynomials are entire functions. Because of the quotient rule, rational functions are analytic in open sets that do not contain points at which denominators vanish. Other than these functions, we have encountered $\operatorname{Re} z$, $\operatorname{Im} z$, $|z|$, and $\operatorname{Arg} z$. We shall discuss analyticity for these functions shortly. This represents the totality of complex functions with which we are familiar, a very limited source of examples to illustrate the theory of this chapter. In Chapter 3, we rectify this situation when we introduce exponential and logarithm functions, trigonometric and inverse trigonometric, hyperbolic and inverse hyperbolic, and root functions.

EXAMPLE 2.16 ....

Find derivatives for the following functions:

$$\textbf{(a)}\ f(z) = \left(z^2 + \frac{1}{z}\right)^3 \qquad \textbf{(b)}\ f(z) = \frac{z^2+2}{i-3z}$$

In what open sets are the functions analytic?

Solution **(a)** With power rule 2.18(d),

$$f'(z) = 3\left(z^2 + \frac{1}{z}\right)^2 \left(2z - \frac{1}{z^2}\right).$$

Because this is defined for all $z \neq 0$, the function is analytic in any open set that does not contain the point $z = 0$.

(b) With quotient rule 2.18(c),

$$f'(z) = \frac{(i-3z)(2z) - (z^2+2)(-3)}{(i-3z)^2} = \frac{-3z^2 + 2iz + 6}{(i-3z)^2}.$$

Because f' exists at all points except $z = i/3$, the function f is analytic in any open set that does not contain the point $z = i/3$. ....

Everything seems to be unfolding very smoothly. With the sum, product, quotient, power, and chain rules, it seems that differentiation of complex functions is an easy process; it's just like real calculus. However, this may not always be the case. A complex function f may be specified in terms of its real and imaginary parts rather than in terms of z; that is, it may be given in the form $f(z) = u(x,y) + v(x,y)i$. For example, $f(z)$ might be

$$f(z) = \left(x + \frac{x}{x^2+y^2}\right) + \left(y - \frac{y}{x^2+y^2}\right)i.$$

When f is defined in this way, how shall we calculate f' or even know that f' exists? One method is to find $f(z)$ in terms of z and then utilize differentiation rules 2.18; this process could be very complicated. In addition, let us not forget a major question that surfaced in Section 2.2. Under what conditions does a complex combination $u(x,y) + v(x,y)i$ of real functions $u(x,y)$ and $v(x,y)$ define a function $f(z)$ of $z = x + yi$? Both these questions are answered in Theorem 2.6, the most important theorem in this chapter. Because the theorem is quite complicated, intricately using equation 2.17 for the derivative of a function, a few remarks concerning this definition are worthwhile.

What is Δz in equation 2.17? Analogous to $x + \Delta x$ in the definition for the real derivative $f'(x) = \lim_{\Delta x \to 0} [f(x+\Delta x) - f(x)]/\Delta x$, we think of $z + \Delta z$ as a complex number close to z (Figure 2.27). Quantity Δz is a complex number, the difference between $z + \Delta z$ and z. It has real and imaginary parts that we denote by $\Delta z = \Delta x + \Delta y\, i$, so that $z + \Delta z = (x + yi) + (\Delta x + \Delta y\, i) = (x + \Delta x) + (y + \Delta y)i$.

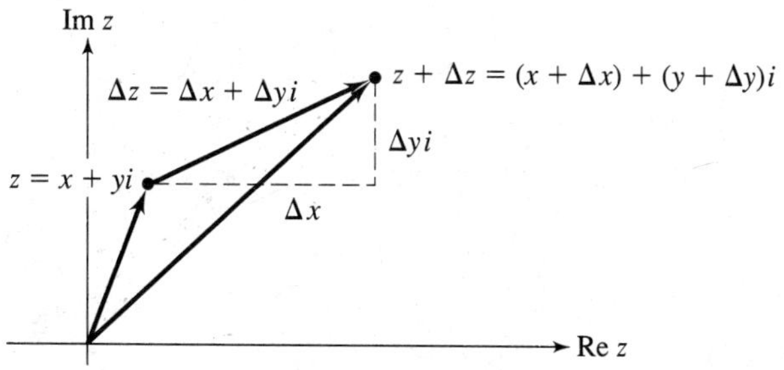

Figure 2.27

For Δz to approach 0 in 2.17, we must have $\Delta x \to 0$ and $\Delta y \to 0$. How Δx and Δy approach 0 dictates how $z + \Delta z$ approaches z. For example, if we set $\Delta x = 0$, then $\Delta z = \Delta y\, i$, and $z + \Delta z$ approaches z along the vertical line through z (Figure 2.28(a)). If we set $\Delta y = 0$, then $\Delta z = \Delta x$, and $z + \Delta z$ approaches z along the horizontal line through z (Figure 2.28(b)). With $\Delta x = \Delta y$, we have $\Delta z = \Delta x + \Delta x\, i = (1 + i)\Delta x$, and the mode of approach of $z + \Delta z$ to z is along a line through z making an angle of $\pi/4$ radians with the horizontal (Figure 2.28(c)).

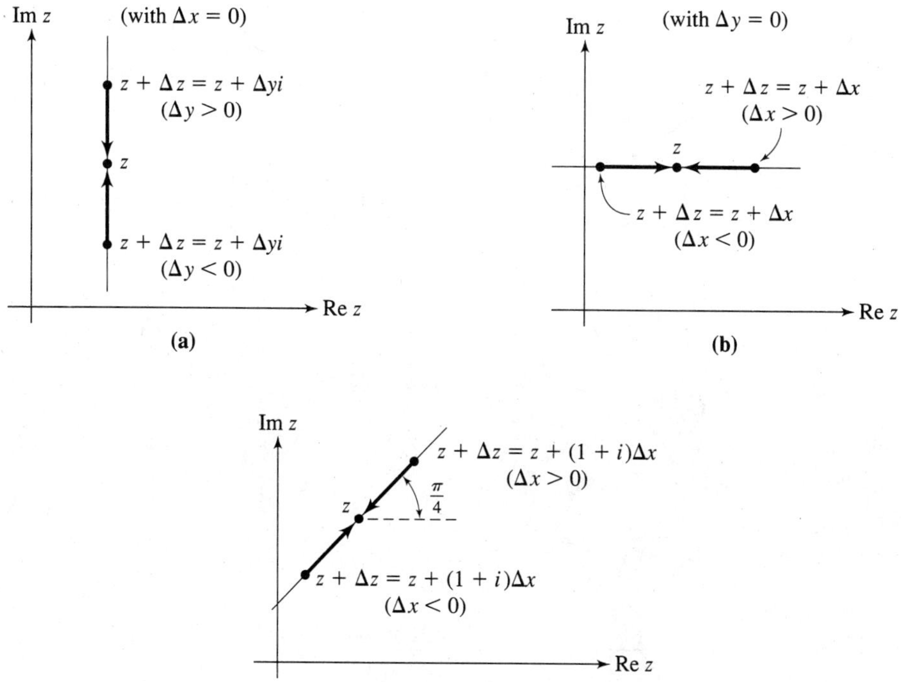

Figure 2.28

Derivative f' exists if and only if the limit in 2.17 exists and is the same for all modes of approach of $z + \Delta z$ to z, be it vertical, horizontal, or any other mode, straight line, or otherwise.

We now state and prove Theorem 2.6.

■ **THEOREM 2.6** A function $f(z) = u(x, y) + v(x, y)i$ is analytic in an open set S if and only if the first partial derivatives of $u(x, y)$ and $v(x, y)$ are continuous in S and satisfy the Cauchy-Riemann equations therein:

$$\frac{\partial u}{\partial x} = \frac{\partial v}{\partial y}, \qquad \frac{\partial v}{\partial x} = -\frac{\partial u}{\partial y}. \tag{2.19}$$

Proof Suppose $f(z)$ is analytic in S and z is any point in S. The limit in equation 2.17 must be independent of the mode of approach of $\Delta z = \Delta x + \Delta y\, i$ to zero. If we set $\Delta y = 0$ and let $\Delta z \to 0$ by letting $\Delta x \to 0$ (Figure 2.28(b)), then

$$\begin{aligned} f'(z) &= \lim_{\Delta x \to 0} \frac{f(x + yi + \Delta x) - f(x + yi)}{\Delta x} \\ &= \lim_{\Delta x \to 0} \frac{u(x + \Delta x, y) + v(x + \Delta x, y)i - u(x, y) - v(x, y)i}{\Delta x}. \end{aligned}$$

Because $f'(z)$ exists, it follows that limits of real and imaginary parts of the right side both exist (Theorem 2.2); that is,

$$\lim_{\Delta x \to 0} \frac{u(x + \Delta x, y) - u(x, y)}{\Delta x} = \frac{\partial u}{\partial x} \quad \text{and} \quad \lim_{\Delta x \to 0} \frac{v(x + \Delta x, y) - v(x, y)}{\Delta x} = \frac{\partial v}{\partial x}$$

both exist, and

$$f'(z) = \frac{\partial u}{\partial x} + \frac{\partial v}{\partial x} i.$$

We now let $\Delta z \to 0$ by setting $\Delta x = 0$ in $\Delta z = \Delta x + \Delta y\, i$, and letting $\Delta y \to 0$ (Figure 2.28(a)),

$$\begin{aligned} f'(z) &= \lim_{\Delta y \to 0} \frac{f(x + yi + \Delta y\, i) - f(x + yi)}{\Delta y\, i} \\ &= \lim_{\Delta y \to 0} \frac{u(x, y + \Delta y) + v(x, y + \Delta y)i - u(x, y) - v(x, y)i}{\Delta y\, i}. \end{aligned}$$

Once again existence of $f'(z)$ implies existence of real and imaginary parts,

$$\lim_{\Delta y \to 0} \frac{v(x, y + \Delta y) - v(x, y)}{\Delta y} = \frac{\partial v}{\partial y}$$

and
$$\lim_{\Delta y \to 0} \frac{u(x, y + \Delta y) - u(x, y)}{\Delta y} = \frac{\partial u}{\partial y};$$

and, therefore,

$$f'(z) = \frac{\partial v}{\partial y} - \frac{\partial u}{\partial y} i.$$

If we equate these two expressions for $f'(z)$, we obtain

$$\frac{\partial u}{\partial x} + \frac{\partial v}{\partial x} i = \frac{\partial v}{\partial y} - \frac{\partial u}{\partial y} i.$$

Real and imaginary parts of this equation give the Cauchy-Riemann equations 2.19. Continuity of the partial derivatives will be verified later.

Conversely, suppose that the first partial derivatives of u and v are continuous in S and satisfy the Cauchy-Riemann equations therein. Then for $\Delta z = \Delta x + \Delta y\, i$,

$$\frac{f(z + \Delta z) - f(z)}{\Delta z}$$
$$= \frac{u(x + \Delta x, y + \Delta y) + v(x + \Delta x, y + \Delta y)i - u(x, y) - v(x, y)i}{\Delta x + \Delta y\, i}.$$

Because $u(x, y)$ and $v(x, y)$ have first partial derivatives in S, Taylor's remainder formula implies that $u(x + \Delta x, y + \Delta y)$ and $v(x + \Delta x, y + \Delta y)$ can be expressed in the forms

$$u(x + \Delta x, y + \Delta y) = u(x, y) + \frac{\partial u}{\partial x}_{|(x', y')} \Delta x + \frac{\partial u}{\partial y}_{|(x', y')} \Delta y,$$

$$v(x + \Delta x, y + \Delta y) = v(x, y) + \frac{\partial v}{\partial x}_{|(x'', y'')} \Delta x + \frac{\partial v}{\partial y}_{|(x'', y'')} \Delta y,$$

where (x', y') and (x'', y'') are points on the line segment joining (x, y) and $(x + \Delta x, y + \Delta y)$. Because the first partial derivatives are continuous at (x, y), it follows that

$$\frac{\partial u}{\partial x}_{|(x', y')} = \frac{\partial u}{\partial x}_{|(x, y)} + \epsilon_1, \qquad \frac{\partial u}{\partial y}_{|(x', y')} = \frac{\partial u}{\partial y}_{|(x, y)} + \epsilon_2,$$

$$\frac{\partial v}{\partial x}_{|(x'', y'')} = \frac{\partial v}{\partial x}_{|(x, y)} + \epsilon_3, \qquad \frac{\partial v}{\partial y}_{|(x'', y'')} = \frac{\partial v}{\partial y}_{|(x, y)} + \epsilon_4,$$

where $\epsilon_1, \epsilon_2 \to 0$ as $(x', y') \to (x, y)$ and $\epsilon_3, \epsilon_4 \to 0$ as $(x'', y'') \to (x, y)$. Thus, the quotient $[f(z+\Delta z)-f(z)]/\Delta z$ can be written as

$$\begin{aligned}\frac{f(z+\Delta z)-f(z)}{\Delta z} &= \frac{1}{\Delta x+\Delta y\, i}\left[\left(\frac{\partial u}{\partial x}+\epsilon_1\right)\Delta x+\left(\frac{\partial u}{\partial y}+\epsilon_2\right)\Delta y\right.\\ &\qquad \left.+\left(\frac{\partial v}{\partial x}+\epsilon_3\right)\Delta x\, i+\left(\frac{\partial v}{\partial y}+\epsilon_4\right)\Delta y\, i\right]\\ &= \frac{1}{\Delta x+\Delta y\, i}\left[\left(\frac{\partial u}{\partial x}+\frac{\partial v}{\partial x}i\right)\Delta x+\left(\frac{\partial u}{\partial y}+\frac{\partial v}{\partial y}i\right)\Delta y\right.\\ &\qquad \left.+(\epsilon_1+\epsilon_3 i)\Delta x+(\epsilon_2+\epsilon_4 i)\Delta y\right],\end{aligned}$$

where all partial derivatives are evaluated at (x, y). If we substitute from the Cauchy-Riemann equations,

$$\begin{aligned}\frac{f(z+\Delta z)-f(z)}{\Delta z} &= \frac{1}{\Delta x+\Delta y\, i}\left[\left(\frac{\partial u}{\partial x}+\frac{\partial v}{\partial x}i\right)\Delta x+\left(-\frac{\partial v}{\partial x}+\frac{\partial u}{\partial x}i\right)\Delta y\right.\\ &\qquad \left.+(\epsilon_1+\epsilon_3 i)\Delta x+(\epsilon_2+\epsilon_4 i)\Delta y\right]\\ &= \frac{1}{\Delta x+\Delta y\, i}\left[\left(\frac{\partial u}{\partial x}+\frac{\partial v}{\partial x}i\right)(\Delta x+\Delta y\, i)\right.\\ &\qquad \left.+(\epsilon_1+\epsilon_3 i)\Delta x+(\epsilon_2+\epsilon_4 i)\Delta y\right]\\ &= \frac{\partial u}{\partial x}+\frac{\partial v}{\partial x}i+\frac{(\epsilon_1+\epsilon_3 i)\Delta x+(\epsilon_2+\epsilon_4 i)\Delta y}{\Delta x+\Delta y\, i}.\end{aligned}$$

Now,

$$\begin{aligned}\left|\frac{(\epsilon_1+\epsilon_3 i)\Delta x+(\epsilon_2+\epsilon_4 i)\Delta y}{\Delta x+\Delta y\, i}\right| &\le \frac{|(\epsilon_1+\epsilon_3 i)\Delta x|}{|\Delta x+\Delta y\, i|}+\frac{|(\epsilon_2+\epsilon_4 i)\Delta y|}{|\Delta x+\Delta y\, i|}\\ &\le \frac{|(\epsilon_1+\epsilon_3 i)|\,|\Delta x|}{|\Delta x|}+\frac{|(\epsilon_2+\epsilon_4 i)|\,|\Delta y|}{|\Delta y|}\\ &= |\epsilon_1+\epsilon_3 i|+|\epsilon_2+\epsilon_4 i|,\end{aligned}$$

and the right side approaches zero as Δx, $\Delta y \to 0$. It follows therefore that

$$\lim_{\Delta z\to 0}\frac{f(z+\Delta z)-f(z)}{\Delta z}=\frac{\partial u}{\partial x}+\frac{\partial v}{\partial x}i,$$

and $f'(z)$ exists. ■

Theorem 2.6 is most important; the Cauchy-Riemann equations are fundamental to all complex analysis. Careful attention to their role in differentiation is essential.

From the proof of the theorem, we also obtain two formulas for calculating f' when f is specified in terms of its real and imaginary parts.

■ **COROLLARY** When $f(z) = u(x, y) + v(x, y)i$, and $f'(z)$ exists,

$$f'(z) = \frac{\partial u}{\partial x} + \frac{\partial v}{\partial x}i = \frac{\partial v}{\partial y} - \frac{\partial u}{\partial y}i. \qquad (2.20)$$ ■

Thus, when a complex function is specified in terms of z, we calculate its derivative using differentiation rules 2.18; when the function is specified in terms of its real and imaginary parts, we may calculate its derivative with formulas 2.20.

Theorem 2.6 also emphasizes a fact that we pointed out in Section 2.2, namely, that real and imaginary parts of a complex function are not independent; they satisfy the Cauchy-Riemann equations. Furthermore, we now have an answer to the question, Under what conditions does a complex combination $u(x, y) + v(x, y)i$ of real functions $u(x, y)$ and $v(x, y)$ constitute a complex function $f(z)$ of $z = x + yi$? They constitute an analytic function when partial derivatives of $u(x, y)$ and $v(x, y)$ satisfy the Cauchy-Riemann equations. For example, real and imaginary parts of $u(x, y) + v(x, y)i = (x^2 - y^2) + 2xyi$ satisfy the Cauchy-Riemann equations in the whole complex plane. This function is, therefore, an entire function of $z = x + yi$. It is, in fact, the function $f(z) = z^2$. On the other hand, real and imaginary parts of $u + vi = x - yi$ do not satisfy the Cauchy-Riemann equations anywhere. Thus, $u + vi = x - yi$ is a complex-valued function of x and y, but it is not an analytic function of $z = x + yi$.

EXAMPLE 2.17 ....

Determine open sets in which the following functions are analytic:

(a) $f(z) = \left(x + \frac{x}{x^2 + y^2}\right) + \left(y - \frac{y}{x^2 + y^2}\right)i$ **(b)** $f(z) = |z|$ **(c)** $f(z) = \operatorname{Re} z$

Solution **(a)** For this function, $u(x, y) = x + x/(x^2 + y^2)$ and $v(x, y) = y - y/(x^2 + y^2)$, and therefore

$$\frac{\partial u}{\partial x} = 1 + \frac{1}{x^2 + y^2} + \frac{-2x^2}{(x^2 + y^2)^2} = 1 + \frac{y^2 - x^2}{(x^2 + y^2)^2}; \qquad \frac{\partial u}{\partial y} = \frac{-2xy}{(x^2 + y^2)^2}$$

$$\frac{\partial v}{\partial x} = \frac{2xy}{(x^2 + y^2)^2}; \qquad \frac{\partial v}{\partial y} = 1 - \frac{1}{x^2 + y^2} - \frac{-2y^2}{(x^2 + y^2)^2} = 1 + \frac{y^2 - x^2}{(x^2 + y^2)^2}$$

Because these partial derivatives are continuous and satisfy the Cauchy-Riemann equations at every point except $(0, 0)$, $f(z)$ is analytic in any open set not containing $z = 0$. Its derivative is

$$f'(z) = \frac{\partial u}{\partial x} + \frac{\partial v}{\partial x}i = \left[1 + \frac{y^2 - x^2}{(x^2 + y^2)^2}\right] + \frac{2xyi}{(x^2 + y^2)^2}.$$

(b) Even though $f(z) = |z|$ is specified in terms of z, differentiation rules 2.18 are not applicable. We therefore resort to the Cauchy-Riemann equations. With $f(z) = |z| = \sqrt{x^2 + y^2}$, we have $u(x, y) = \sqrt{x^2 + y^2}$ and $v(x, y) = 0$. Thus,

$$\frac{\partial u}{\partial x} = \frac{x}{\sqrt{x^2 + y^2}}; \qquad \frac{\partial u}{\partial y} = \frac{y}{\sqrt{x^2 + y^2}}; \qquad \frac{\partial v}{\partial x} = \frac{\partial v}{\partial y} = 0.$$

Because the Cauchy-Riemann equations are never satisfied, there is no open set in which $f(z) = |z|$ is analytic.

(c) For $f(z) = \operatorname{Re} z = x$, we obtain $\partial u/\partial x = 1$ and $\partial v/\partial y = 0$, and therefore $f(z)$ is nowhere analytic.

If a real function has a derivative equal to zero on an open interval, the function must be constant on that interval. The complex analogue is also valid.

■ **THEOREM 2.7** If $f'(z) = 0$ at every point of a domain D, then $f(z)$ must be constant in D.

Proof If $f'(z) = 0$ at every point in D, then

$$0 = \frac{\partial u}{\partial x} + \frac{\partial v}{\partial x} i = \frac{\partial v}{\partial y} - \frac{\partial u}{\partial y} i,$$

and these equations imply that $\frac{\partial u}{\partial x} = \frac{\partial u}{\partial y} = \frac{\partial v}{\partial x} = \frac{\partial v}{\partial y} = 0$ at every point in D. It follows that u and v must be constant in D, and hence $f(z)$ is constant in D. ■

We demanded that D be a domain, not just an open set, in this theorem. Were D only an open set, it would not necessarily be connected; it could perhaps consist of two disjoint pieces D_1 and D_2, as in Figure 2.29. In this case the proof of Theorem 2.7 indicates that $f(z) = C_1$ in D_1 and $f(z) = C_2$ in D_2. But it would not be necessary for $C_1 = C_2$. Thus, Theorem 2.7 can be generalized to open sets, but it is necessary to distinguish among disjoint parts.

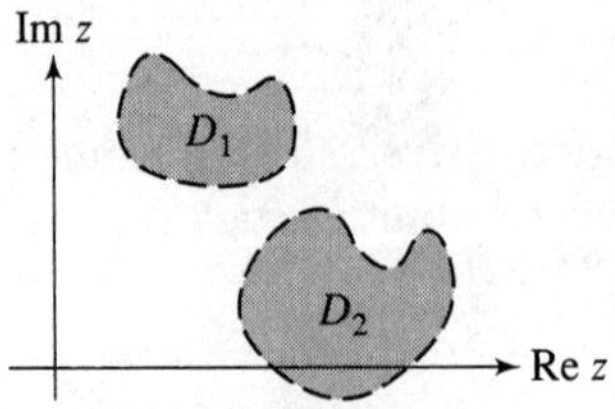

Figure 2.29

An immediate consequence of Theorem 2.7 is the following corollary (see Exercise 19 for a proof).

■ **COROLLARY** If $f(z) = u + vi$ is analytic in a domain D, and if either $u(x, y)$ or $v(x, y)$ is constant in D, then $f(z)$ is constant in D. ■

For some functions, it is easier to express real and imaginary parts in terms of the modulus r and argument θ of z, $f(z) = u(r, \theta) + v(r, \theta)i$. For example, if $g(z) = z^{10}$, then DeMoivre's theorem quickly gives $g(z) = [r(\cos\theta + \sin\theta\, i)]^{10} = (r^{10}\cos 10\theta) + (r^{10}\sin 10\theta)\, i$. Expressing this in terms of x and y is particularly uninviting. For functions expressed in this way, the Cauchy-Riemann equations take the form

$$r\frac{\partial u}{\partial r} = \frac{\partial v}{\partial \theta}, \qquad r\frac{\partial v}{\partial r} = -\frac{\partial u}{\partial \theta}, \qquad r \neq 0. \tag{2.21}$$

They are verified in Exercises 24 and 25. For example, had we not known that $g(z) = (r^{10}\cos 10\theta) + (r^{10}\sin 10\theta)\, i$ represents z^{10}, a short calculation shows that it satisfies 2.21 for all z. Hence the function is entire. The function $f(z) = r^{1/3}[\cos(\theta/3) + \sin(\theta/3)i]$ satisfies equations 2.21 at every point except $z = 0$ (where $r = 0$). There is a hidden difficulty associated with θ, however. Angle θ cannot be just any argument of z, because $f(z)$ would not then be a function. For example, arguments $\pi/4$ and $9\pi/4$ of $z = 1 + i$ lead to different values for $f(1 + i)$. A natural way out of this dilemma is to choose θ as the principal value of the argument of z, so that in effect $f(z) = |z|^{1/3}[\cos(\operatorname{Arg} z/3) + \sin(\operatorname{Arg} z/3)\, i]$. This result makes $f(z)$ a function, but because $\operatorname{Arg} z$ is discontinuous at points on the negative real axis, so also are the real and imaginary parts of $f(z)$. Theorem 2.4 implies that $f(z)$ is discontinuous at these points, and hence $f(z)$ cannot have a derivative there. Thus, the function is analytic in the domain $r > 0$, $-\pi < \theta < \pi$, the complex plane with the origin and negative real axis deleted. Remember our warning that discontinuities of $\operatorname{Arg} z$ would have serious consequences for many functions in Chapter 3. We have just seen an example of this. The function $f(z) = r^{1/3}[\cos(\operatorname{Arg} z/3) + \sin(\operatorname{Arg} z/3)\, i]$ is defined for all $z \neq 0$. It is discontinuous and does not have a derivative at $z = 0$ and points on the negative real axis. This is a direct result of the discontinuities of $\operatorname{Arg} z$.

In Exercise 20 we ask you to verify the following analogue of the corollary to Theorem 2.7.

■ **THEOREM 2.8** If f is analytic in a domain D, and if $|f(z)|$ is constant in D, then f is constant in D. ■

Notably absent from our discussions has been a geometric interpretation of the derivative f' of a complex function f. We mention it here, but because justification is delayed until Section 7.4, the geometric interpretation of f' does not play the same fundamental role as for the derivative of a real function in elementary calculus. Naturally the geometric interpretation of f' must be in terms of the geometric interpretation of f as a mapping $w = f(z)$ from the z-plane to the w-plane. Suppose a curve C in the z-plane is mapped by $w = f(z)$ onto a curve $f(C)$ in the w-plane (Figure 2.30). A point z_0 on C has image $f(z_0)$ on $f(C)$, and a small length ds along C at z_0 has image $dw = f(ds)$ along $f(C)$. When $f'(z_0) \neq 0$, the modulus $|f'(z_0)|$ is the ratio of the length of dw to that of ds; that is, $|dw| = |f'(z_0)|\,|ds|$. In other words, the modulus of $f'(z_0)$ is a length-magnification factor of the mapping $w = f(z)$. Naturally, then, $|f'(z_0)|^2$ is an area-magnification factor; a small area A at z_0 is magnified to $|f'(z_0)|^2 A$ at $f(z_0)$. Furthermore, angles of inclination θ and ϕ of the tangent vectors to C and $f(C)$ with the positive real directions are related by $\phi = \theta + \arg[f'(z_0)]$. Thus, the argument of $f'(z_0)$ is a measure of the rotational effect of the mapping $w = f(z)$. As we said, these geometric interpretations are discussed more fully in Chapter 7.

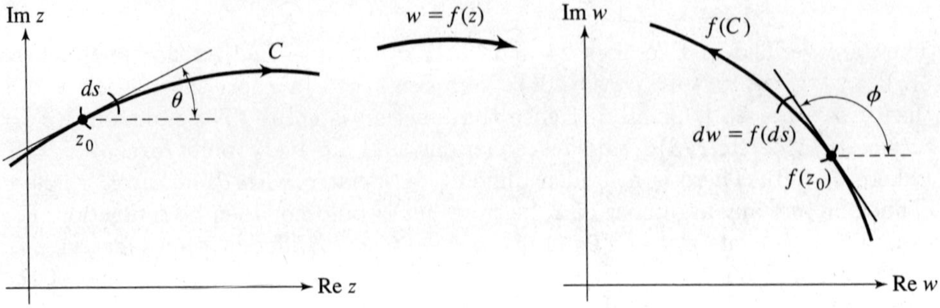

Figure 2.30

Exercises 2.4

In Exercises 1–8 find $f'(z)$. Where is $f(z)$ analytic?

1. $f(z) = z^3 + 1/z^2 - 2z$
2. $f(z) = z^2(1 + z^3)^{10}$
3. $f(z) = (z^3 + 1)/(z^2 - 2)$
4. $f(z) = z^3(1 - z)^3/(3z + 2i)$
5. $f(z) = \operatorname{Im} z$
6. $f(z) = z|z|^2$
7. $f(z) = \operatorname{Arg} z$
8. $f(z) = \overline{z}$

In Exercises 9–14 determine whether the complex function is a differentiable function $f(z)$ of $z = x + yi$. Find $f'(z)$, if it exists, and indicate where $f(z)$ is analytic.

9. $(2x^2 - 2y^2 + 3x) + (4xy + 3y + 4)i$
10. $(x^3 - 3xy^2 - 2x + 3) + (3x^2y - y^3 - 2y)i$
11. $(x + 1 - yi)/[(x + 1)^2 + y^2]$
12. $1/x - i/y$
13. $\cos x \cosh y - \sin x \sinh y\, i$
14. $\cos y \sinh x + \sin y \cosh x\, i$

In Exercise 24 we ask you to verify that when the real and imaginary parts of a function $f(z)$ are expressed in terms of r and θ, where $z = r(\cos\theta + \sin\theta\, i)$, the Cauchy-Riemann equations take the form 2.21. We also verify that formulas that express $f'(z)$ in terms of derivatives of u and v with respect to r and θ are

$$f'(z) = (\cos\theta - \sin\theta\, i)\left(\frac{\partial u}{\partial r} + \frac{\partial v}{\partial r} i\right)$$

$$= \frac{\cos\theta - \sin\theta\, i}{r}\left(\frac{\partial v}{\partial \theta} - \frac{\partial u}{\partial \theta} i\right).$$

Use these results to determine whether the function in Exercises 15–18 is analytic in its domain of definition. What is $f'(z)$?

15. $f(z) = (r^8 \cos 8\theta + 1) + (r^8 \sin 8\theta)i$
16. $f(z) = r + \theta\, i$
17. $f(z) = \sqrt{r}[\cos(\theta/2) + \sin(\theta/2)\, i]$, $r > 0$, $-\pi < \theta < \pi$
18. $f(z) = \ln r + \theta\, i$, $r > 0$, $-\pi < \theta < \pi$
19. Prove the corollary to Theorem 2.7.
20. Prove Theorem 2.8.
21. Is $f(z) = |z|$ analytic at $z = 0$? Does $f'(0)$ exist?
22. Is $f(z) = |z|^2$ analytic at $z = 0$? Does $f'(0)$ exist?
23. Show that the complex function $(x^2 - y^3) + (x^3 + y^2)i$ satisfies the Cauchy-Riemann equations at every point on the line $y = x$. Is it analytic at these points?
24. (a) Verify equations 2.21 by transforming 2.19 from Cartesian to polar coordinates.
    (b) Verify the expressions for $f'(z)$ in the instructions for Exercises 15–18.
25. Verify equations 2.21 by using Definition 2.17 and taking limits as $z + \Delta z \to z$ radially and along a circular arc.
26. Prove that if $f(z) = u + vi$ and $\overline{f(z)} = u - vi$ are both analytic in a domain D, then $f(z)$ is constant in D.

2.5 Harmonic Functions

The real and imaginary parts of a complex function f constitute a unique pair of real-valued functions $u(x, y)$ and $v(x, y)$ of two variables x and y. On the other hand, every pair of real-valued functions does not necessarily determine a complex analytic function $f(z) = u(x, y) + v(x, y)i$. They do so if and only if they have continuous first partial derivatives that satisfy the Cauchy-Riemann equations. What do these equations say about u and v individually? If the second partial derivatives of u and v are continuous, then the Cauchy-Riemann equations can be used to write

$$\frac{\partial^2 u}{\partial x^2} = \frac{\partial}{\partial x}\left(\frac{\partial u}{\partial x}\right) = \frac{\partial}{\partial x}\left(\frac{\partial v}{\partial y}\right) = \frac{\partial}{\partial y}\left(\frac{\partial v}{\partial x}\right) = \frac{\partial}{\partial y}\left(-\frac{\partial u}{\partial y}\right);$$

that is,

$$\frac{\partial^2 u}{\partial x^2} + \frac{\partial^2 u}{\partial y^2} = 0.$$

Similarly,

$$\frac{\partial^2 v}{\partial x^2} + \frac{\partial^2 v}{\partial y^2} = 0.$$

In other words, the real and imaginary parts of an analytic function satisfy Laplace's equation. Laplace's equation is found in many areas of applied mathematics. We shall see it in problems of heat flow, fluid flow, and electrostatics in Sections 2.6, 2.7, and 2.8 of this chapter and again in Chapter 7. There are others. What we have shown here is that real and imaginary parts of complex analytic functions are a limitless source of solutions to Laplace's equation. According to the following definition, solutions of Laplace's equation are called harmonic functions.

Definition 2.14
A real-valued function $\phi(x, y)$ is said to be harmonic in a domain D if its second partial derivatives are continuous in D and if at each point of D, ϕ satisfies Laplace's equation

$$\frac{\partial^2 \phi}{\partial x^2} + \frac{\partial^2 \phi}{\partial y^2} = 0. \tag{2.22}$$

We have essentially established the following property of analytic functions.

■ **THEOREM 2.9** The real and imaginary parts of a function analytic in a domain D are harmonic in D. ■

The proof of the theorem is not quite complete, because the preceding discussion assumed continuity of second partial derivatives of u and v. In Section 4.6 we show that the real and imaginary parts of analytic functions have continuous derivatives not just of second order, but of all orders.

When $u(x, y)$ is harmonic in a domain D, a function $v(x, y)$ such that $u + vi$ is analytic in D is called a **harmonic conjugate** of $u(x, y)$. This definition suggests that given a harmonic function $u(x, y)$, we can always find harmonic conjugates $v(x, y)$.* Alternatively, given harmonic $v(x, y)$, we can find $u(x, y)$ for which $v(x, y)$ is a harmonic conjugate. We illustrate with an example.

EXAMPLE 2.18 ....

Construct an analytic function whose real part is $u(x, y) = x^3 - 3xy^2 + y$.

....

*This statement is not quite true. Only if $u(x, y)$ is harmonic in certain kinds of domains can we guarantee the existence of a harmonic conjugate $v(x, y)$. This topic is discussed in Section 4.8.

Solution Because $\partial^2 u/\partial x^2 + \partial^2 u/\partial y^2 = 6x - 6x = 0$, the function $u(x, y)$ is harmonic in the whole complex plane. To find a harmonic conjugate $v(x, y)$, we have, by the Cauchy-Riemann equations,

$$\frac{\partial v}{\partial y} = \frac{\partial u}{\partial x} = 3x^2 - 3y^2, \qquad \frac{\partial v}{\partial x} = -\frac{\partial u}{\partial y} = 6xy - 1.$$

When the first of these is integrated with respect to y,

$$v(x, y) = 3x^2 y - y^3 + h(x),$$

where $h(x)$ is an arbitrary differentiable function. If we substitute this into the second equation,

$$6xy + h'(x) = 6xy - 1,$$

which implies that $h(x) = -x + k$, where k is a real constant. Thus,

$$v(x, y) = 3x^2 y - y^3 - x + k,$$

and the required analytic function is

$$f(z) = (x^3 - 3xy^2 + y) + (3x^2 y - y^3 - x + k)i.$$

Should we want the right side in terms of z, we could write

$$\begin{aligned} f(z) &= (x^3 + 3x^2 yi - 3xy^2 - y^3 i) - i(x + yi) + ki \\ &= (x + yi)^3 - (x + yi)i + ki \\ &= z^3 - (z - k)i. \end{aligned}$$

Alternatively, by setting $x = (z + \overline{z})/2$ and $y = (z - \overline{z})/(2i)$ (see equations 2.11),

$$\begin{aligned} f(z) = &\left[\left(\frac{z+\overline{z}}{2}\right)^3 - 3\left(\frac{z+\overline{z}}{2}\right)\left(\frac{z-\overline{z}}{2i}\right)^2 + \left(\frac{z-\overline{z}}{2i}\right)\right] \\ &+ \left[3\left(\frac{z+\overline{z}}{2}\right)^2\left(\frac{z-\overline{z}}{2i}\right) - \left(\frac{z-\overline{z}}{2i}\right)^3 - \left(\frac{z+\overline{z}}{2}\right) + k\right] i, \end{aligned}$$

which simplifies to $z^3 - (z - k)i$. ....

An important property of harmonic functions and their conjugates, as far as applications are concerned, relates to orthogonal curves. To derive this property, suppose that $u(x, y)$ is harmonic in a domain D of the xy-plane, as in Figure 2.31, and $v(x, y)$ is a harmonic conjugate of $u(x, y)$ in D. The equation $u(x, y) = C_1$ where C_1 is a constant defines a one-parameter family of curves, one of which passes through each point of D. Similarly, $v(x, y) = C_2$ defines a family of curves that also covers D. Thus, through each point (x, y) of D passes exactly one curve from each family.

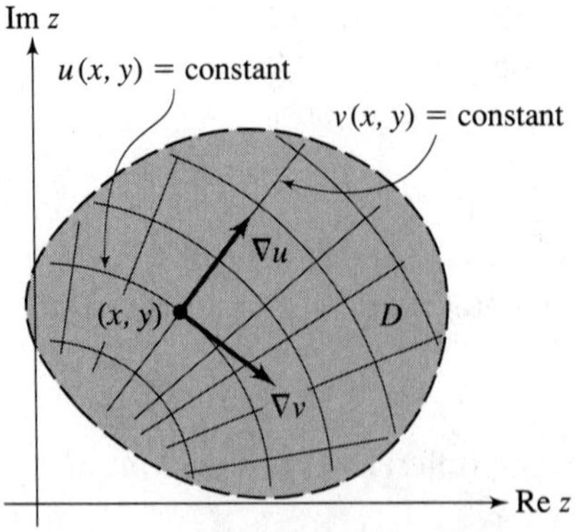

Figure 2.31

A normal vector to $u(x,y) = C_1$ at any point is the gradient of $u(x,y)$, $\nabla u = (\partial u/\partial x)\hat{\imath} + (\partial u/\partial y)\hat{\jmath}$, and a vector perpendicular to $v(x,y) = C_2$ at the same point is $\nabla v = (\partial v/\partial x)\hat{\imath} + (\partial v/\partial y)\hat{\jmath}$. Because u and v satisfy the Cauchy-Riemann equations, it follows that the scalar product of these vectors is

$$\nabla u \cdot \nabla v = \frac{\partial u}{\partial x}\frac{\partial v}{\partial x} + \frac{\partial u}{\partial y}\frac{\partial v}{\partial y} = \frac{\partial u}{\partial x}\left(-\frac{\partial u}{\partial y}\right) + \frac{\partial u}{\partial y}\left(\frac{\partial u}{\partial x}\right) = 0.$$

Consequently, at every point (x,y) in D, the curves $u = C_1$ and $v = C_2$ through (x,y) are perpendicular, or orthogonal. We have shown then that harmonic conjugates define orthogonal families of curves. We shall see the importance of this property in the applications of harmonic functions. Figure 2.32(a)–(c) illustrates this orthogonality for the analytic function $f(z) = z^2 - z$. Its real and imaginary parts are $u(x,y) = x^2 - y^2 - x$ and $v(x,y) = 2xy - y$. The one-parameter family of curves $u(x,y) = C_1$ is shown in Figure 2.32(a), and the curves $v(x,y) = C_2$ are shown in Figure 2.32(b). Superposing these curves in Figure 2.32(c) clearly demonstrates that each curve from one family intersects each curve from the other family at right angles.

Exercises 2.5

In Exercises 1–6 find domains in which $u(x,y)$ is harmonic. Find all harmonic conjugates.

1. $u(x,y) = 4 + x - x^2 + y^2$
2. $u(x,y) = x^3 - 3xy^2 + 2x + 5$
3. $u(x,y) = (x+1)/[(x+1)^2 + y^2]$
4. $u(x,y) = \sin x \cosh y$
5. $u(x,y) = \cos y \cosh x$
6. $u(x,y) = (x^2 + x + y^2)/(x^2 + y^2)$
7. Find all functions $u(x,y)$ to which $v(x,y) = 2xy + 3y$ is conjugate.
8. **(a)** Find all domains in which $u(x,y) = (2x^2 + 2y^2 - 6y)/(9 + x^2 + y^2 - 6y)$ is harmonic. Find all harmonic conjugates.

   (b) If $f(z) = u + vi$, find $f(z)$ in terms of z.

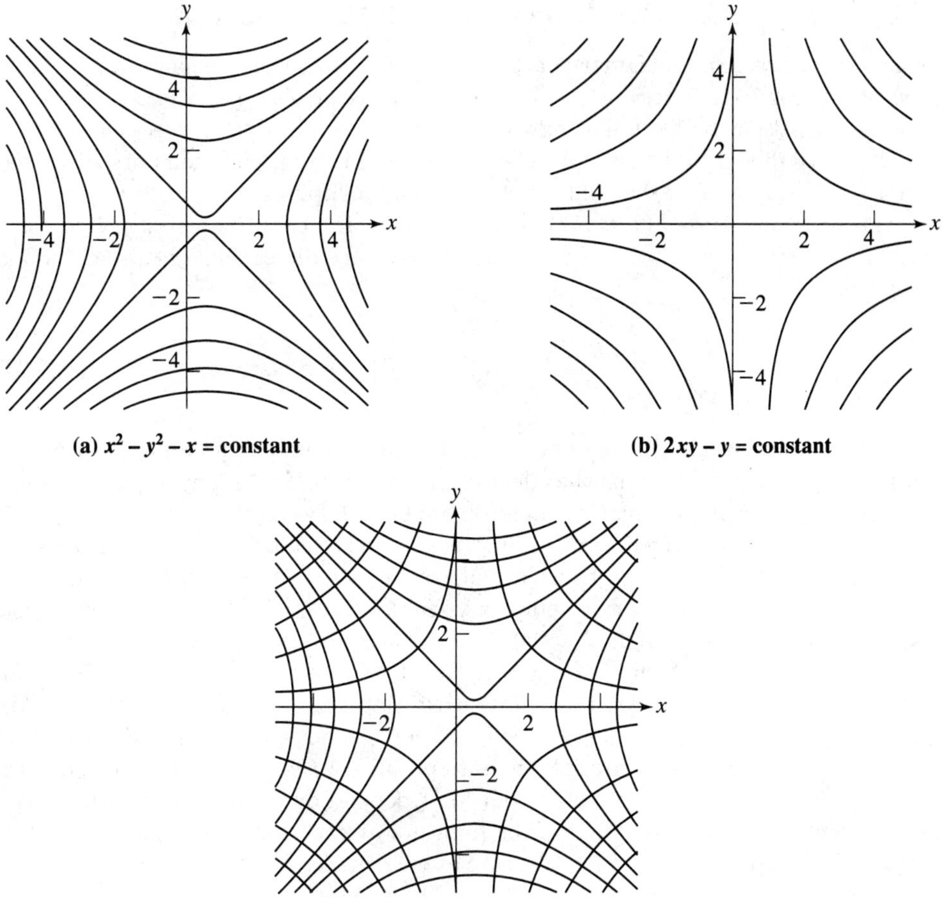

(a) $x^2 - y^2 - x$ = constant

(b) $2xy - y$ = constant

(c)

Figure 2.32

9. Verify that when $u(x, y)$ and $v(x, y)$ are harmonic conjugates in a domain D, then their product $u(x, y)v(x, y)$ is harmonic in D. Is it true that if $u(x, y)$ and $v(x, y)$ are harmonic in D, then their product is harmonic in D?

10. Find all harmonic conjugates of $u(x, y) = x^2 - y^2$. Draw the orthogonal families of curves defined by $u(x, y)$ and its harmonic conjugates.

11. Harmonic conjugates $u(x, y)$ and $v(x, y)$ define orthogonal families of curves $u(x, y) = C$ and $v(x, y) = C$. Do orthogonal families of curves $u(x, y) = C$ and $v(x, y) = C$ always define harmonic conjugates?

12. Prove that if $v(x, y)$ is the harmonic conjugate of $u(x, y)$ in some domain D, and $u(x, y)$ is the harmonic conjugate of $v(x, y)$ in D, then $u(x, y)$ and $v(x, y)$ are constant functions. Must they be the same constant function?

2.6 Steady-State Two-Dimensional Heat Conduction

The fact that real and imaginary parts of an analytic function are harmonic suggests the use of complex functions in physical problems involving the two-dimensional Laplace equation. In the following three sections we introduce three such applications — heat flow, fluid flow, and electrostatics. These applications are discussed further in Chapter 7 in the context of conformal mappings.

When temperature $T(x, y, z, t)$ is a function of the coordinates x, y, z in some region V of space and time t, it must satisfy the three-dimensional heat-conduction equation

$$\frac{\partial T}{\partial t} = k\left(\frac{\partial^2 T}{\partial x^2} + \frac{\partial^2 T}{\partial y^2} + \frac{\partial^2 T}{\partial z^2}\right) + g(x, y, z, t), \tag{2.23}$$

where k is the thermal diffusivity of the material in the region (assumed constant). Function $g(x, y, z, t)$ contains the effect on temperature due to any sources or sinks of heat in V. Accompanying this equation are boundary conditions that $T(x, y, z, t)$ must satisfy on the surfaces surrounding V and an initial condition specifying temperature in V at some initial time (usually taken as $t = 0$). Such problems must be specialized considerably before complex variables can be used to advantage. First, we consider partial differential equation 2.23 only in regions that contain no heat sources or sinks, in which case $g(x, y, z, t) = 0$. This is not to say that there can be no heat sources or sinks; they must exist or there would be no heat flow. We are simply considering regions V in which they are not to be found. Secondly, we eliminate time by considering steady-state problems; that is, we consider only situations in which temperature in V does not depend on time. Temperature varies from point to point in V, but the temperature at each point remains constant in time. The temperature function $T(x, y, z)$ then satisfies Laplace's equation,

$$\frac{\partial^2 T}{\partial x^2} + \frac{\partial^2 T}{\partial y^2} + \frac{\partial^2 T}{\partial z^2} = 0. \tag{2.24}$$

Finally, we assume that temperature is independent of the z-coordinate; it depends only on x and y. Then $T(x, y)$ satisfies the two-dimensional Laplace equation

$$\frac{\partial^2 T}{\partial x^2} + \frac{\partial^2 T}{\partial y^2} = 0. \tag{2.25}$$

Although temperature is a function of only two variables, x and y, region V is still three-dimensional. We visualize this in one of two ways. Think of V as a thin flat plate lying on the xy-plane (Figure 2.33); its edge is defined by the curve C. If perfect insulation is applied to top and bottom faces of the plate, then no heat can enter or leave the plate through these faces. With appropriate boundary conditions on the edges of the plate, heat flows only in the x- and y-directions; temperature does not vary with z, and $T(x, y)$ satisfies 2.25. Because temperature is the same in all cross sections parallel to the xy-plane, we usually draw the cross section D in the xy-plane (Figure 2.34) and visualize heat flowing in D. However, do not lose

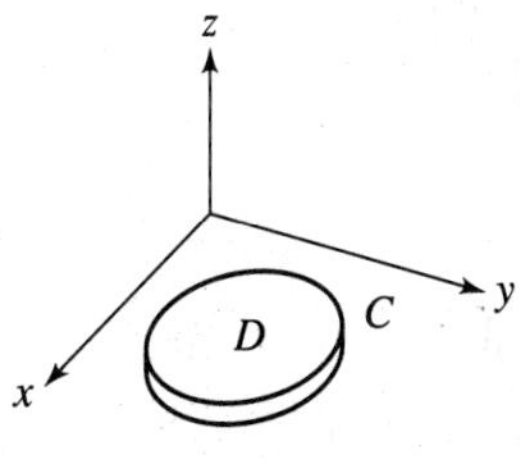

Figure 2.33

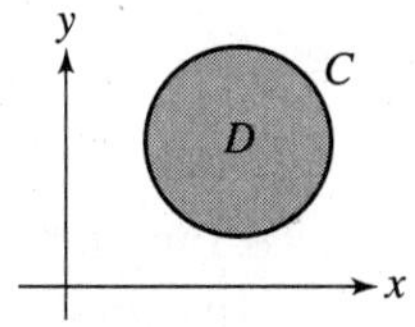

Figure 2.34

sight of the fact that D represents one of many cross sections, all with the same temperature.

Alternatively, we can think of V as an infinite cylinder in the z-direction with cross-sectional area D (Figure 2.35). Provided boundary conditions on the surface of the cylinder are independent of z, heat does not flow in the z-direction; temperature is identical in all cross-sections parallel to the xy-plane, and we therefore solve 2.25 in D.

If $\tilde{T}(x,y)$ is a harmonic conjugate of $T(x,y)$ in D, then $f(z) = T(x,y) + \tilde{T}(x,y)i$ is an analytic function of the complex variable z in D. (Do not confuse the complex variable $z = x + yi$ with the suppressed z-coordinate in space.) Curves $T(x,y) =$ constant are called **isothermal curves**. Each curve joins all points in D with the same temperature, and exactly one such curve passes through each point of D. The curves $\tilde{T}(x,y) =$ constant are orthogonal trajectories of isothermal curves; at each point of D, the curves from the two families intersect at right angles (Figure 2.36).

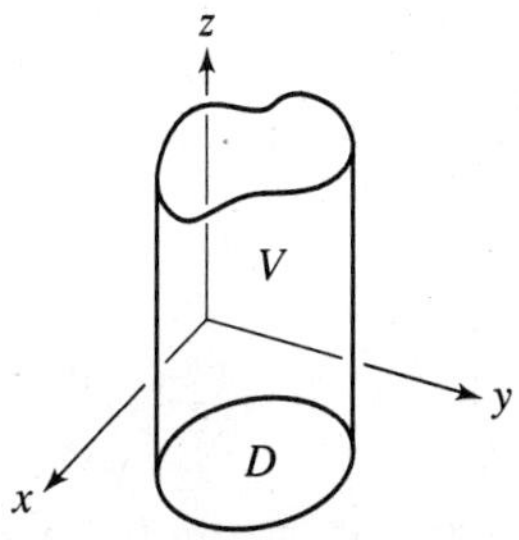

Figure 2.35

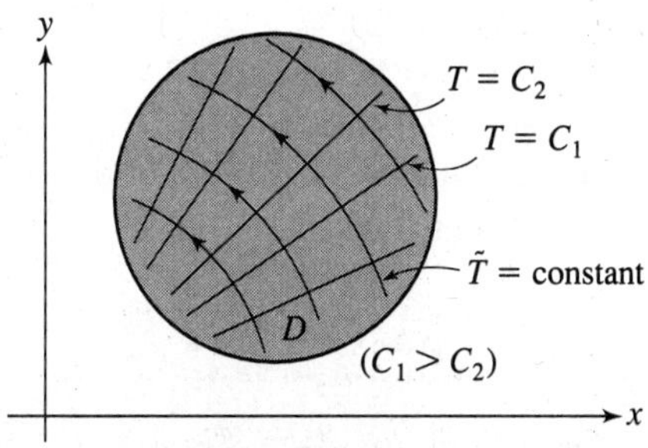

Figure 2.36

According to Fourier's law of heat conduction, heat flows in the direction opposite to the gradient of $T(x,y)$. Because the gradient of $T(x,y)$ is perpendicular to isothermal curves, it follows that at each point in D, heat flows in the direction of the tangent to $\tilde{T}(x,y) =$ constant. In other words, heat flows along the curves $\tilde{T}(x,y) =$ constant.

Laplace's equation 2.25 will always be accompanied by boundary conditions that specify either $T(x,y)$, its normal derivative, or a combination of these on the boundary of D. A **Dirichlet boundary condition** specifies temperature on the boundary of D:

$$T(x,y) = F(x,y), \qquad (x,y) \text{ on boundary of } D, \tag{2.26a}$$

where $F(x,y)$ is a given function. A **Neumann boundary condition** specifies the derivative of $T(x,y)$ in the direction perpendicular to the boundary of D. This condition is often written in the form

$$\nabla T \cdot \hat{\mathbf{n}} = F(x,y), \qquad (x,y) \text{ on boundary of } D, \tag{2.26b}$$

where $F(x,y)$ is some given function and $\hat{\mathbf{n}}$ is the unit vector perpendicular to the boundary of D (Figure 2.37). Because heat flows in direction $-\nabla T$, a Neumann boundary condition essentially specifies the amount of heat flowing across the boundary of D. In particular, if the boundary is perfectly insulated (so that no heat can enter or leave D), 2.26(b) takes the form

$$\nabla T \cdot \hat{\mathbf{n}} = 0, \qquad (x,y) \text{ on boundary of } D. \tag{2.26c}$$

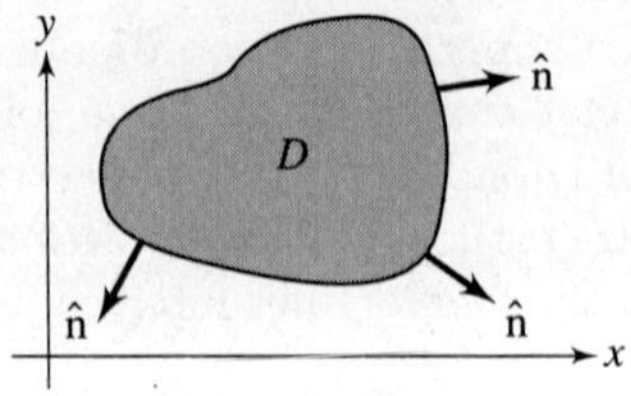

Figure 2.37

In the examples to follow, we illustrate isothermal curves and heat-flow lines for some simple problems. More difficult problems are discussed in Chapter 7 in the context of conformal mapping.

EXAMPLE 2.19 ....

Find steady-state temperature $T(x,y)$ in the rectangle $0 < x < a$, $0 < y < b$, given that it must satisfy the boundary conditions in Figure 2.38, where T_0 and T_a are constants. The fact that $\partial T/\partial y$ vanishes on $y = 0$ and $y = b$ means that these edges are perfectly insulated. What are isothermal curves and lines of heat flow?

Solution Because no heat can pass through $y = 0$ or $y = b$ and edges $x = 0$ and $x = a$ are held at constant temperature, heat flows only in the x-direction. As a result, $T(x,y)$ should be a function only of x, in which case 2.25 reduces to the ordinary differential equation

$$\frac{d^2T}{dx^2} = 0.$$

The general solution of this equation is $T = Ax + B$, and boundary conditions on $x = 0$ and $x = a$ require

$$T_0 = B, \qquad T_a = Aa + B.$$

Hence, $T(x) = (T_a - T_0)x/a + T_0$. Isothermal curves $T =$ constant are vertical lines, and heat-flow lines are horizontal (Figure 2.39).

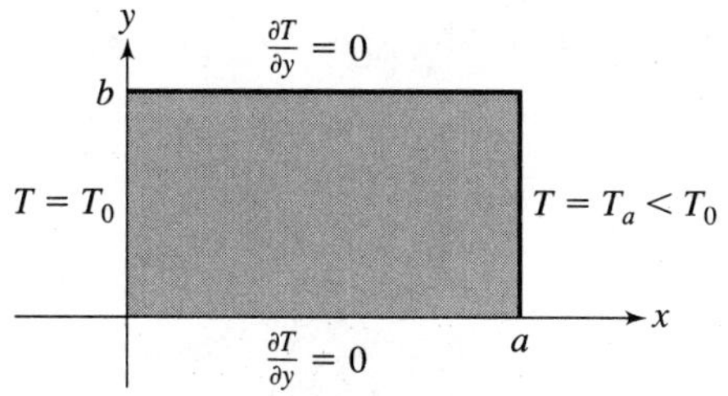

Figure 2.38

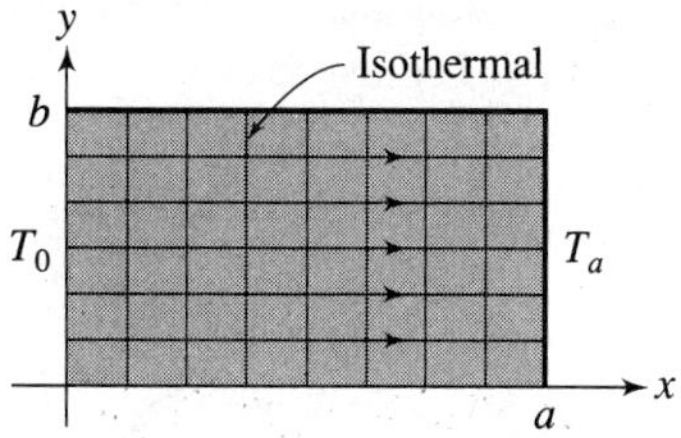

Figure 2.39

EXAMPLE 2.20

Find steady-state temperature $T(x, y)$ between two concentric circles $a^2 < x^2+y^2 < b^2$ given that temperatures on inner and outer boundaries are constant values T_a and T_b (Figure 2.40). What are the isothermal curves and lines of heat flow?

Solution In a steady-state situation with constant inner and outer boundary temperatures, temperature does not vary with polar coordinate angle θ; it depends only on radial distance r from the center of the concentric circles. Laplace's equation in polar coordinates is

$$\frac{\partial^2 T}{\partial r^2} + \frac{1}{r}\frac{\partial T}{\partial r} + \frac{1}{r^2}\frac{\partial^2 T}{\partial \theta^2} = 0, \tag{2.27}$$

and when T is independent of θ, it reduces to

$$\frac{d^2T}{dr^2} + \frac{1}{r}\frac{dT}{dr} = 0.$$

When multiplied by r, this equation can be expressed in the form

$$\frac{d}{dr}\left(r\frac{dT}{dr}\right) = 0,$$

and integration gives

$$r\frac{dT}{dr} = C.$$

Hence, $T(r) = C \ln r + D$. The boundary conditions require

$$T_a = C \ln a + D, \qquad T_b = C \ln b + D,$$

from which

$$T(r) = T_a + (T_b - T_a)\frac{\ln (r/a)}{\ln (b/a)}.$$

Isothermal curves are concentric circles centered at the origin, and heat flows along radial lines (Figure 2.41).

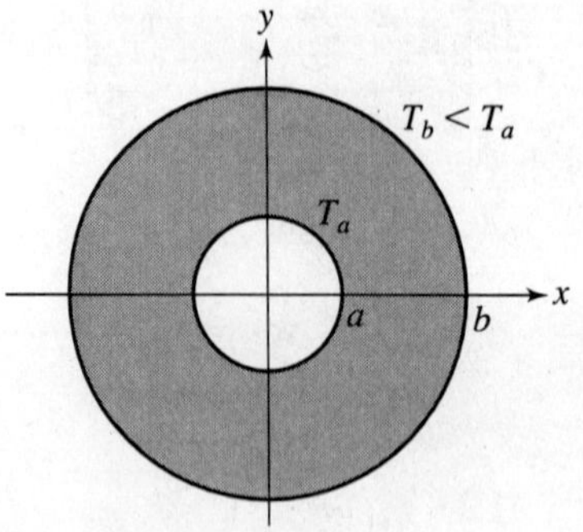

Figure 2.40

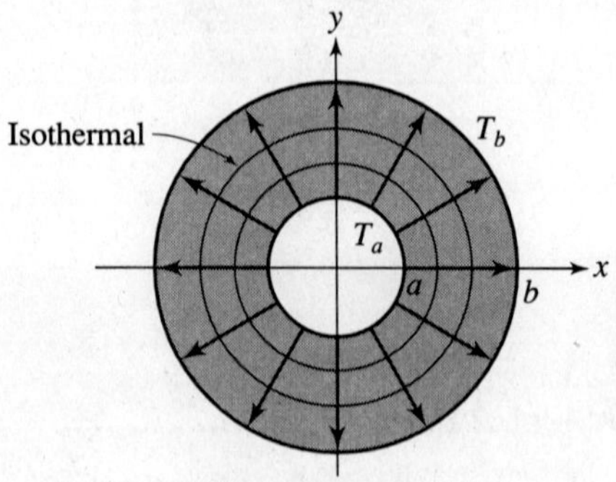

Figure 2.41

Exercises 2.6

1. Find the steady-state temperature in the half-plane $y > 0$ given that temperature on the negative x-axis is a constant value T_1 and temperature along the positive x-axis is a constant value T_2. Describe isothermal curves and heat-flow lines. *Hint:* Think about radial lines as isothermal curves.

2. Find the steady-state temperature in the quarter-plane $x > 0$, $y > 0$ given that temperature on the positive y-axis is a constant value T_1 and temperature on the positive x-axis is a constant value T_2. Describe isothermal curves and heat-flow lines.

3. When temperature on the x-axis between $x = \pm 1$ is a constant value T_0 and is zero outside this interval, the steady-state temperature in the half-plane $y > 0$ is

$$T(x, y) = \frac{T_0}{\pi}\left[\operatorname{Tan}^{-1}\left(\frac{1+x}{y}\right) + \operatorname{Tan}^{-1}\left(\frac{1-x}{y}\right)\right].$$

   (a) Show that isothermal curves are circles passing through $(\pm 1, 0)$ with centers on the y-axis.

(b) Show that the harmonic conjugate of $T(x, y)$ is

$$\tilde{T}(x,y) = \frac{T_0}{2\pi} \ln\left[\frac{y^2 + (1+x)^2}{y^2 + (1-x)^2}\right].$$

(c) Find lines of heat flow.

4. When temperature on the x-axis between $x = \pm 1$ is zero and is a constant value T_0 outside this interval, the steady-state temperature in the half-plane $y > 0$ is

$$T(x,y) = T_0 - \frac{T_0}{\pi}\left[\text{Tan}^{-1}\left(\frac{1+x}{y}\right) + \text{Tan}^{-1}\left(\frac{1-x}{y}\right)\right].$$

(a) Show that isothermal curves are circles passing through $(\pm 1, 0)$ with centers on the y-axis.

(b) Show that the harmonic conjugate of $T(x, y)$ is

$$\tilde{T}(x,y) = \frac{T_0}{2\pi} \ln\left[\frac{y^2 + (1-x)^2}{y^2 + (1+x)^2}\right].$$

(c) Find lines of heat flow.

5. When temperature on the semicircle $y = -\sqrt{R^2 - x^2}$ is a constant value T_1 and temperature on $y = \sqrt{R^2 - x^2}$ is a constant T_2, the steady-state temperature in the circle $x^2 + y^2 < R^2$ is

$$T(x,y) = \frac{T_1 + T_2}{2} + \frac{1}{\pi}(T_2 - T_1)\,\text{Tan}^{-1}\left(\frac{2Ry}{R^2 - x^2 - y^2}\right).$$

(a) Show that isothermal curves are circular arcs through the points $(\pm R, 0)$ with centers on the y-axis.

(b) Show that heat-flow lines are circular arcs with centers on the x-axis.

6. When temperature on the semicircle $x^2 + y^2 = R^2$, $y > 0$ is a constant value T_0 and temperature on the straight edge $y = 0$ is 0, the steady-state temperature in the semicircle $x^2 + y^2 < R^2$, $y > 0$ is

$$T(x,y) = T_0 - \frac{2T_0}{\pi}\,\text{Tan}^{-1}\left(\frac{R^2 - x^2 - y^2}{2Ry}\right).$$

(a) Show that isothermal curves are circular arcs through the points $(\pm R, 0)$ with centers on the y-axis.

(b) Show that heat-flow lines are circular arcs with centers on the x-axis.

2.7 Electrostatics

When an electrostatic charge distribution (continuous, discrete, or combination thereof) in a vacuum is symmetric in one direction (which we take as the z-direction), the electrostatic potential $V(x, y)$ due to the charges must satisfy the two-dimensional Laplace equation

$$\frac{\partial^2 V}{\partial x^2} + \frac{\partial^2 V}{\partial y^2} = 0 \tag{2.28}$$

at all points exterior to the charge. Because $V(x, y)$ has continuous derivatives of all orders, it must be harmonic in every domain D of the xy-plane that is free of charge.

If $W(x, y)$ is a harmonic conjugate of $V(x, y)$ in D, then

$$f(z) = V(x, y) + W(x, y)i \tag{2.29}$$

is an analytic function of $z = x + yi$ called the **complex electrostatic potential**. Thus, with every two-dimensional electrostatic field in a vacuum, we associate an analytic complex electrostatic potential $f(z)$.

Now the electric field intensity due to the charge distribution is given by

$$\mathbf{E} = -\nabla V = -\frac{\partial V}{\partial x}\hat{\mathbf{i}} - \frac{\partial V}{\partial y}\hat{\mathbf{j}}. \tag{2.30}$$

But

$$f'(z) = \frac{\partial V}{\partial x} + \frac{\partial W}{\partial x}i = \frac{\partial V}{\partial x} - \frac{\partial V}{\partial y}i, \tag{2.31}$$

and hence

$$-\overline{f'(z)} = -\frac{\partial V}{\partial x} - \frac{\partial V}{\partial y}i. \tag{2.32}$$

In other words, the real and imaginary parts of $-\overline{f'(z)}$ are the components of the electric field intensity $\mathbf{E}$.

Curves in the one-parameter family $V(x, y) = C$ are called **equipotential curves**. (In actual fact, they represent equipotential cylindrical surfaces.) At every point, the electric field intensity is perpendicular to the equipotential curve through that point (Figure 2.42). Because $W(x, y) = C$ defines an orthogonal family of curves, it follows that at every point, $\mathbf{E}$ is tangent to the curve $W(x, y) = C$ through that point; indeed, by the Cauchy-Riemann equations,

$$\mathbf{E} = -\frac{\partial V}{\partial x}\hat{\mathbf{i}} - \frac{\partial V}{\partial y}\hat{\mathbf{j}} = -\frac{\partial W}{\partial y}\hat{\mathbf{i}} + \frac{\partial W}{\partial x}\hat{\mathbf{j}},$$

which is tangent to $W(x, y) = C$. Because $\mathbf{E}$ is proportional to the force that would be exerted on a test charge placed at a point, the force field is always tangent to $W(x, y) = C$, and we therefore call these curves **lines of force**.

Laplace's equation 2.28 will always be accompanied by boundary conditions that specify either $V(x, y)$ or its normal derivative on the boundary of D. A Dirichlet boundary condition specifies $V(x, y)$ on the boundary of D,

$$V(x, y) = F(x, y), \qquad (x, y) \text{ on boundary of } D, \tag{2.33a}$$

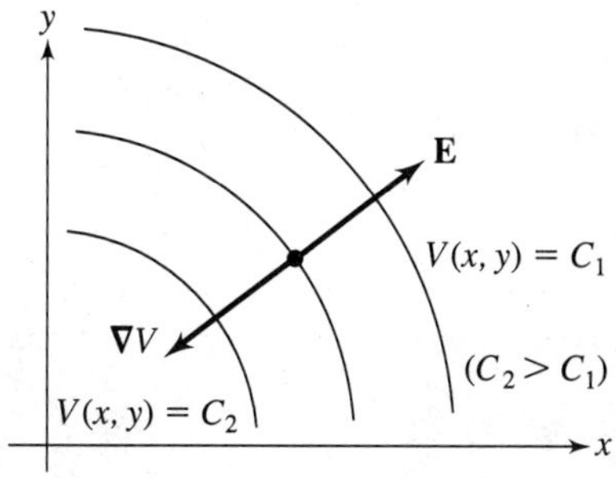

Figure 2.42

where $F(x, y)$ is a given function. When the boundary of D is a conductor, $F(x, y) = 0$. A Neumann boundary condition specifies the derivative of $V(x, y)$ in the direction perpendicular to the boundary of D. This is expressed in the form

$$\nabla V \cdot \hat{\mathbf{n}} = F(x, y), \qquad (x, y) \text{ on boundary of } D, \tag{2.33b}$$

where $F(x, y)$ is a given function and $\hat{\mathbf{n}}$ is the unit vector perpendicular to the boundary of D. Neumann boundary conditions arise when the boundary of D is a dielectric.

In the following examples we illustrate equipotentials and lines of force for some simple problems. More difficult problems will be discussed in Chapter 7 in the context of conformal mapping.

EXAMPLE 2.21 ....

Find the electrostatic potential in the half-plane $y > 0$ given that potential on the negative x-axis is a constant value V_1 and potential on the positive x-axis is a constant value V_2 (Figure 2.43). Describe equipotentials and lines of force.

Solution The negative and positive x-axis are equipotentials, and we might suspect that the y-axis is also an equipotential (perhaps $(V_1+V_2)/2$). Because these lines are coordinate curves in polar coordinates, we consider solving 2.28 in polar coordinates:

$$\frac{\partial^2 V}{\partial r^2} + \frac{1}{r}\frac{\partial V}{\partial r} + \frac{1}{r^2}\frac{\partial^2 V}{\partial \theta^2} = 0$$

(see equation 2.27). The fact that $\theta = 0$, $\theta = \pi/2$, and $\theta = \pi$ are equipotentials might suggest further that V is a function only of θ, in which case Laplace's equation reduces to

$$\frac{d^2 V}{d\theta^2} = 0,$$

the general solution of which is $V(\theta) = A\theta + B$. The boundary conditions on $\theta = 0$ and $\theta = \pi$ require

$$V_2 = B, \qquad V_1 = A\pi + B.$$

Thus, $V(\theta) = \left(\dfrac{V_1 - V_2}{\pi}\right)\theta + V_2$. Equipotentials are radial lines, and lines of force are semicircles centered at the origin (Figure 2.44).

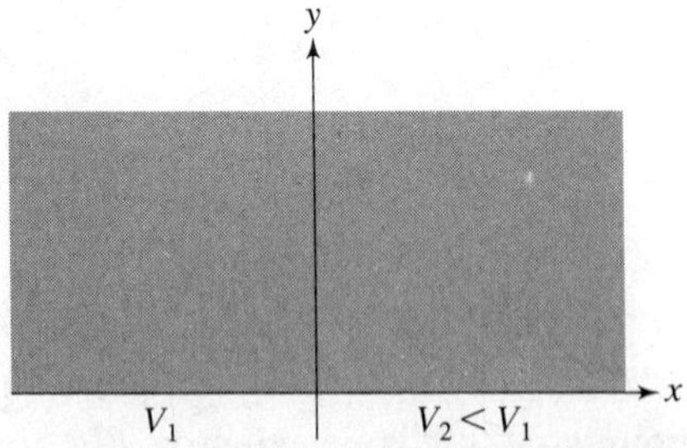

Figure 2.43

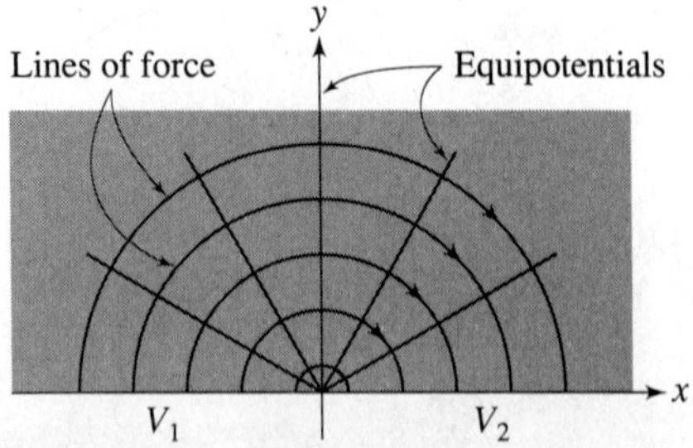

Figure 2.44

EXAMPLE 2.22

When potential on the x-axis between $x = \pm 1$ is a constant value V_0 and is zero outside this interval, the electrostatic potential satisfying Laplace's equation in the half-plane $y > 0$ is

$$V(x,y) = \frac{V_0}{\pi}\left[\operatorname{Tan}^{-1}\left(\frac{1+x}{y}\right) + \operatorname{Tan}^{-1}\left(\frac{1-x}{y}\right)\right].$$

Find equipotentials and lines of force.

Solution Equipotentials are given by

$$C = \frac{V_0}{\pi}\left[\operatorname{Tan}^{-1}\left(\frac{1+x}{y}\right) + \operatorname{Tan}^{-1}\left(\frac{1-x}{y}\right)\right],$$

or

$$\operatorname{Tan}^{-1}\left(\frac{1+x}{y}\right) + \operatorname{Tan}^{-1}\left(\frac{1-x}{y}\right) = \frac{\pi C}{V_0}.$$

If we take tangents of both sides,

$$\frac{\dfrac{1+x}{y} + \dfrac{1-x}{y}}{1 - \left(\dfrac{1+x}{y}\right)\left(\dfrac{1-x}{y}\right)} = \tan\left(\frac{\pi C}{V_0}\right).$$

This equation simplifies to

$$x^2 + \left[y - \cot\left(\frac{\pi C}{V_0}\right)\right]^2 = \csc^2\left(\frac{\pi C}{V_0}\right).$$

Described are circles that pass through the points $(\pm 1, 0)$ and have centers on the y-axis. Equipotentials for $C = V_0/8$, $3V_0/16$, and $V_0/4$ are shown in Figure 2.45(a).

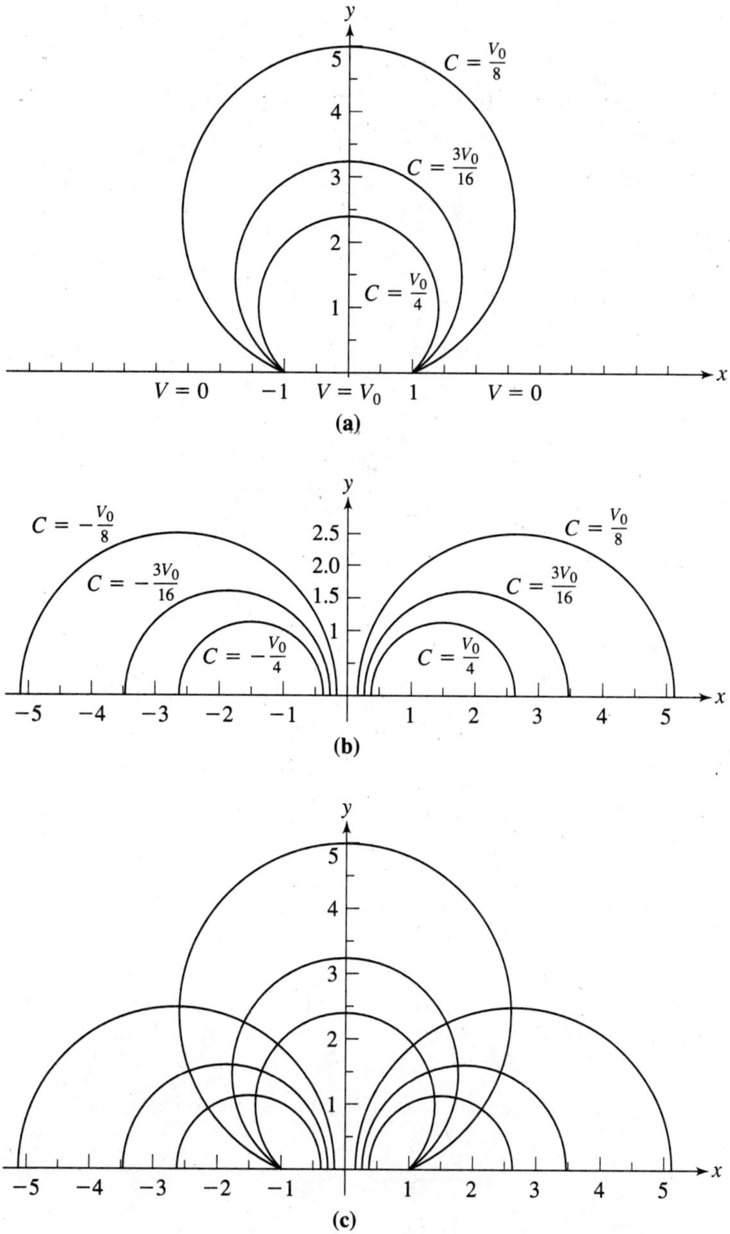
$C = \frac{V_0}{8}$
$C = \frac{3V_0}{16}$
$C = \frac{V_0}{4}$
$V = 0$
$V = V_0$
$V = 0$
(a)
$C = -\frac{V_0}{8}$
$C = -\frac{3V_0}{16}$
$C = -\frac{V_0}{4}$
$C = \frac{V_0}{8}$
$C = \frac{3V_0}{16}$
$C = \frac{V_0}{4}$
(b)
(c)

Figure 2.45

To determine lines of force, we find the harmonic conjugate $W(x,y)$ of $V(x,y)$. The Cauchy-Riemann equations require

$$\frac{\partial W}{\partial y} = \frac{\partial V}{\partial x} = \frac{V_0}{\pi}\left[\frac{y}{y^2+(1+x)^2} - \frac{y}{y^2+(1-x)^2}\right],$$

$$\frac{\partial W}{\partial x} = -\frac{\partial V}{\partial y} = \frac{V_0}{\pi}\left[\frac{1+x}{y^2+(1+x)^2} + \frac{1-x}{y^2+(1-x)^2}\right].$$

These give

$$W(x,y) = \frac{V_0}{2\pi}\ln\left[\frac{y^2+(1+x)^2}{y^2+(1-x)^2}\right].$$

Lines of force are therefore defined by

$$C = \frac{V_0}{2\pi}\ln\left[\frac{y^2+(1+x)^2}{y^2+(1-x)^2}\right],$$

and this equation can be rewritten in the form

$$\left[x - \coth\left(\frac{\pi C}{V_0}\right)\right]^2 + y^2 = \operatorname{csch}^2\left(\frac{\pi C}{V_0}\right).$$

Described are circles with centers on the x-axis. The circles for $C = \pm V_0/8$, $\pm 3V_0/16$, and $\pm V_0/4$ are shown in Figure 2.45(b). Figure 2.45(a) and (b) are superposed in Figure 2.45(c), illustrating orthogonality of equipotentials and lines of force.

Exercises 2.7

1. Express the potential function in Example 2.21 in terms of x and y.
2. In this exercise we determine electrostatic potential due to a line of charge (q coulombs per meter) perpendicular to the xy-plane at the origin, by working from the electric field intensity rather than Laplace's equation.

   (a) Use integration in conjunction with Coulomb's law to show that

   $$\mathbf{E} = \frac{q}{2\pi\epsilon_0(x^2+y^2)}(x\hat{\imath} + y\hat{\jmath}).$$

   (b) Use $\mathbf{E} = -\nabla V$ to find V.

   (c) What is the harmonic conjugate of $V(x,y)$?
3. Is there a complex electrostatic potential associated with the real electrostatic potential due to a point charge at the origin?

The exercises in Section 2.6 can also be treated as potential problems. Simply replace the word temperature with potential. Isothermal curves become equipotentials and lines of heat flow become lines of force.

2.8 Two-Dimensional Fluid Flow

Consider fluid flowing over a domain D of the xy-plane in such a way that motion is identical in all planes parallel to the xy-plane. In this situation, it is sufficient to consider motion of the sheet of fluid "in" the xy-plane itself. Velocity of the fluid in this sheet is parallel to the xy-plane, and it is normally a function of position and time. We consider only the case in which the velocity is independent of time, called **steady-state flow**. In this case, velocity of the fluid at point (x, y) can be expressed in the form

$$\mathbf{q} = u(x, y)\hat{\mathbf{\imath}} + v(x, y)\hat{\mathbf{\jmath}} \tag{2.34}$$

where, as indicated, u and v depend only on x and y.

Harmonic functions do not fit into the most general of two-dimensional, steady-state flows; further specializations must be made. In the absence of *sources* and *sinks* of fluid, points at which fluid appears or disappears, the equation of continuity for fluid flow is

$$\nabla \cdot (\rho \mathbf{q}) = -\frac{\partial \rho}{\partial t}, \tag{2.35}$$

where ρ is the density of the fluid. The time rate of change of density, $\partial \rho/\partial t$, measures how much more mass of fluid enters unit volume in the fluid than leaves it. Because $\rho \mathbf{q}$ is the mass flow rate of the fluid, its divergence $\nabla \cdot (\rho \mathbf{q})$ measures how much more mass leaves unit volume in the fluid than enters. Clearly then, $\partial \rho/\partial t$ and $\nabla \cdot (\rho \mathbf{q})$ measure the same quantity (but in opposite directions) and are, therefore, related by the equation of continuity.

Fluid flow is said to be **incompressible** if density of the fluid is the same at every point of the fluid and is also independent of time. With constant ρ, the equation of continuity becomes $0 = \nabla \cdot (\rho \mathbf{q}) = \rho \nabla \cdot \mathbf{q}$, or, in terms of u and v,

$$\frac{\partial u}{\partial x} + \frac{\partial v}{\partial y} = 0. \tag{2.36}$$

We also restrict consideration to regions of flow in which

$$\nabla \times \mathbf{q} = \mathbf{0}, \tag{2.37}$$

called **irrotational flows**. The x- and y-components of $\nabla \times \mathbf{q}$ vanish identically, and the component perpendicular to the xy-plane yields

$$\frac{\partial v}{\partial x} - \frac{\partial u}{\partial y} = 0. \tag{2.38}$$

It is in the study of incompressible, irrotational, steady-state, two-dimensional flows that complex functions become useful. In suitable domains D, irrotational condition 2.37 or 2.38 implies that there exists a function $\phi(x, y)$, called the **velocity potential**, such that

$$u = \frac{\partial \phi}{\partial x}, \qquad v = \frac{\partial \phi}{\partial y}. \tag{2.39}$$

But then equation 2.36 implies that

$$\frac{\partial^2 \phi}{\partial x^2} = \frac{\partial u}{\partial x} = -\frac{\partial v}{\partial y} = -\frac{\partial^2 \phi}{\partial y^2}; \tag{2.40}$$

that is, $\phi(x,y)$ satisfies Laplace's equation. When u and v have continuous first partial derivatives, ϕ has continuous second derivatives, and we conclude that the velocity potential $\phi(x,y)$ is harmonic in D. Therefore, a harmonic conjugate $\psi(x,y)$ exists in D such that $\phi + \psi i$ is an analytic function $f(z)$ of the complex variable $z = x + yi$,

$$f(z) = \phi(x,y) + \psi(x,y)i. \tag{2.41}$$

This conjugate function $\psi(x,y)$ is called the **stream function** for the flow, and $f(z)$ is called the **complex potential**.

Analyticity of $f(z)$ in D implies that ϕ and ψ satisfy the Cauchy-Riemann equations

$$\frac{\partial \phi}{\partial x} = \frac{\partial \psi}{\partial y}, \qquad \frac{\partial \phi}{\partial y} = -\frac{\partial \psi}{\partial x}. \tag{2.42}$$

With these we may write

$$f'(z) = \frac{\partial \phi}{\partial x} + \frac{\partial \psi}{\partial x}i = \frac{\partial \phi}{\partial x} - \frac{\partial \phi}{\partial y}i = u - vi, \tag{2.43}$$

or,

$$u + vi = \overline{f'(z)}. \tag{2.44}$$

Thus, with incompressible, irrotational, steady-state, two-dimensional flow, we associate an analytic complex potential $f(z)$ such that the components $u(x,y)$ and $v(x,y)$ of the velocity $\mathbf{q}$ are the real and imaginary parts of $\overline{f'(z)}$. Conversely, given an analytic function $f(z)$, the real and imaginary parts of $\overline{f'(z)}$ define an incompressible, irrotational, steady-state, two-dimensional flow.

Associated with velocity potential $\phi(x,y)$ is a one-parameter family of curves $\phi(x,y) = C$ called **equipotential curves**. At any point, the velocity of the flow, $\mathbf{q} = \nabla\phi$, is perpendicular to the equipotential curve through that point (Figure 2.46(a)).

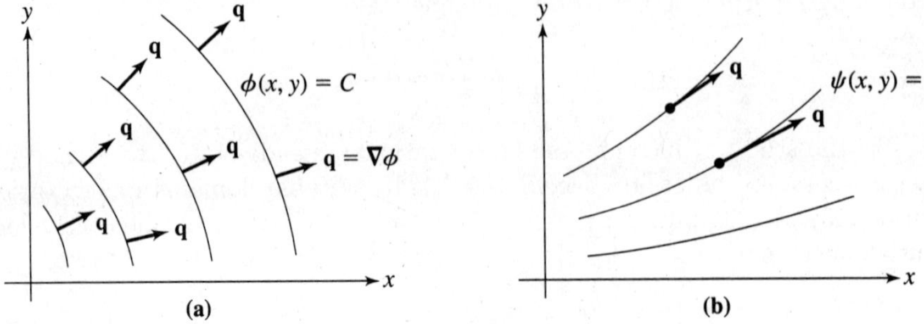

Figure 2.46

But $\psi(x,y) = C$ defines a one-parameter family of curves that is orthogonal to the family $\phi(x,y) = C$. Thus, at every point in the fluid, $\mathbf{q}$ is tangent to the curve $\psi(x,y) = C$ through that point (Figure 2.46(b)). In other words, particles of fluid move along the curve $\psi(x,y) = C$, which are therefore called **stream curves.**

In the preceding discussion we assumed that there were no sources or sinks. We may do so because at a source or sink, equation of continuity 2.36 is violated, as may be irrotationality condition 2.37 or 2.38. On the other hand, no difficulty arises in introducing sources and sinks as long as discussion is restricted to domains in which conditions 2.36 and 2.38 are satisfied.

EXAMPLE 2.23 ....

Find a complex potential for flow with constant velocity field $\mathbf{q} = \alpha\hat{\mathbf{i}} + \beta\hat{\mathbf{j}}$.

Solution If the complex potential for this parallel straight-line flow is $f(z) = \phi + \psi i$, then velocity potential $\phi(x,y)$ is defined by

$$\alpha = \frac{\partial\phi}{\partial x}, \qquad \beta = \frac{\partial\phi}{\partial y}.$$

Consequently, $\phi(x,y) = \alpha x + \beta y + C$, where C is a real constant. To find $\psi(x,y)$, the harmonic conjugate of $\phi(x,y)$, we use the Cauchy-Riemann equations,

$$\frac{\partial\psi}{\partial y} = \frac{\partial\phi}{\partial x} = \alpha, \qquad \frac{\partial\psi}{\partial x} = -\frac{\partial\phi}{\partial y} = -\beta.$$

Thus, $\psi(x,y) = -\beta x + \alpha y + D$, where D is a real constant. The complex potential associated with this flow is, therefore,

$$\begin{aligned} f(z) = \phi + \psi i &= (\alpha x + \beta y + C) + (-\beta x + \alpha y + D)i \\ &= \alpha(x + yi) - \beta i(x + yi) + (C + Di) \\ &= (\alpha - \beta i)(x + yi) + (C + Di) \\ &= (\alpha - \beta i)z + (C + Di). \end{aligned}$$

If δ is the angle of inclination of the flow with the positive x-axis (Figure 2.47), then $\sin\delta = \beta/\sqrt{\alpha^2+\beta^2}$ and $\cos\delta = \alpha/\sqrt{\alpha^2+\beta^2}$. Consequently, we may write

$$\alpha - \beta i = \sqrt{\alpha^2+\beta^2}\cos\delta - \sqrt{\alpha^2+\beta^2}\sin\delta\, i = \sqrt{\alpha^2+\beta^2}(\cos\delta - \sin\delta\, i).$$

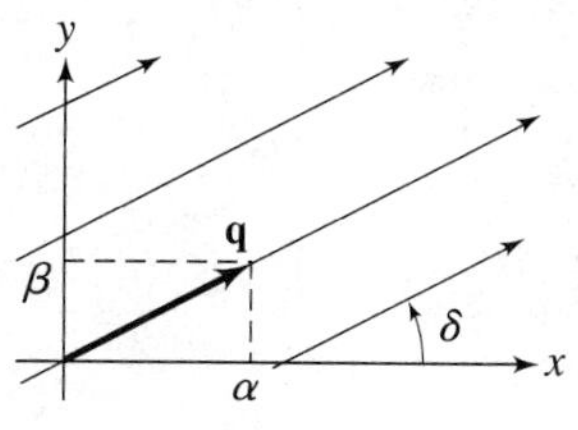

Figure 2.47

In other words, we may also write $f(z)$ in the form

$$f(z) = |\mathbf{q}|(\cos\delta - \sin\delta\, i)z + C + Di.$$

EXAMPLE 2.24

Describe the fluid flow defined by the complex potential

$$f(z) = V_0\left(z + \frac{a^2}{z}\right), \qquad |z| \geq a,$$

where V_0 and a are positive constants.

Solution If we set $z = r(\cos\theta + \sin\theta\, i)$, then

$$f(z) = V_0\left[r(\cos\theta + \sin\theta\, i) + \frac{a^2}{r}(\cos\theta - \sin\theta\, i)\right].$$

Stream curves are therefore defined by

$$C = \operatorname{Im}[f(z)] = V_0\left(r\sin\theta - \frac{a^2}{r}\sin\theta\right), \qquad C \text{ a constant.}$$

The $C = 0$ stream curve consists of the circle $r = a$ and those parts of the rays $\theta = 0$ and $\theta = \pi$ for which $r \geq a$ (Figure 2.48). Various other stream curves are also shown. The direction arrows are determined by calculating

$$\overline{f'(z)} = V_0\left(1 - \frac{a^2}{\overline{z}^2}\right) = V_0\left[1 - \frac{a^2}{r^2}(\cos 2\theta + \sin 2\theta\, i)\right].$$

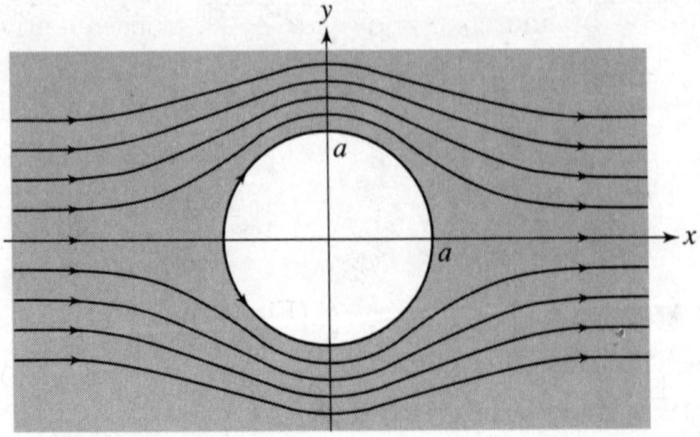

Figure 2.48

It then follows that

$$\mathbf{q} = \text{Re}\left[\overline{f'(z)}\right]\hat{\mathbf{\imath}} + \text{Im}\left[\overline{f'(z)}\right]\hat{\mathbf{\jmath}} = V_0\left(1 - \frac{a^2}{r^2}\cos 2\theta\right)\hat{\mathbf{\imath}} + V_0\left(-\frac{a^2}{r^2}\sin 2\theta\right)\hat{\mathbf{\jmath}}.$$

Notice that for large r, $\mathbf{q}$ is essentially equal to $V_0\hat{\mathbf{\imath}}$, thus corroborating in Figure 2.48 that far from the origin, velocity is essentially parallel to the x-axis. Because $r = a$ is a stream curve and no stream curves are to be considered inside the circle, we interpret the flow as that due to a circular obstacle (actually a cylinder) placed in a uniform flow.

Points in the flow where the velocity of the fluid vanishes are called **stagnation points**. In Example 2.24, $(\pm a, 0)$ are stagnation points.

It is important to realize that because fluid flows along stream curves, physical boundaries of the flow must correspond to stream curves $\psi = C$. Such is the case in Example 2.24, where the circular boundary $|z| = a$ to the flow is defined by the stream curve $\psi = 0$. Notice also that by treating a stream curve $\psi =$ constant in a given flow field as a boundary, other flow fields can be obtained. For instance, the stream curve $\psi = 0$ in Example 2.24 corresponds not just to $|z| = a$, but also to the real axis outside the circle. If we treat this stream curve as a boundary to a flow field and consider only that part of the flow above the real axis (Figure 2.49), then $f(z) = V_0(z + a^2/z)$ is the complex potential for uniform flow in the half-plane $y > 0$ over a semicircular object of radius a at the origin.

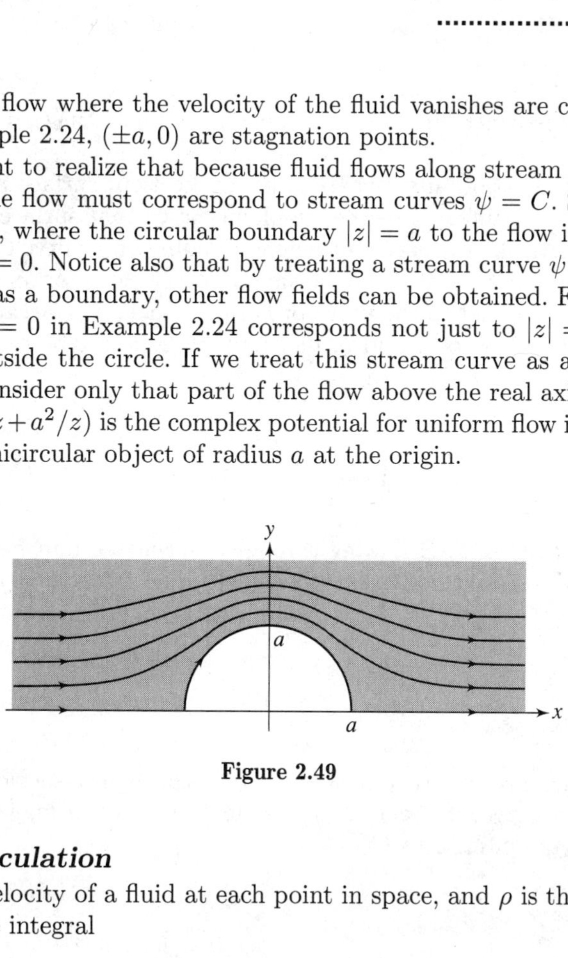

Figure 2.49

Flux and Circulation

When $\mathbf{q}$ is the velocity of a fluid at each point in space, and ρ is the density of the fluid, the surface integral

$$\iint_S \rho\mathbf{q}\cdot\hat{\mathbf{n}}\,dS, \tag{2.45}$$

where $\hat{\mathbf{n}}$ is the unit normal to the surface S (Figure 2.50), is called **flux**. It measures the mass of fluid flowing through the surface in direction $\hat{\mathbf{n}}$ per unit time. If S is a closed surface, and $\hat{\mathbf{n}}$ is the unit outer normal to S, flux is a measure of the net outward flow of fluid through S. It is likely that fluid is flowing into the volume bounded by S at some points on S and flowing out at other points. When the flux

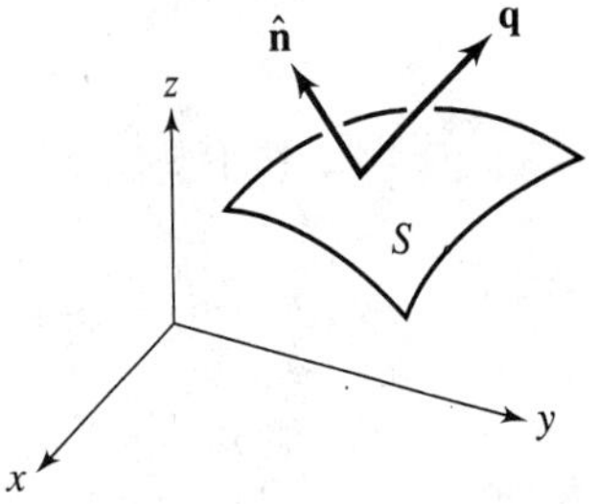

Figure 2.50

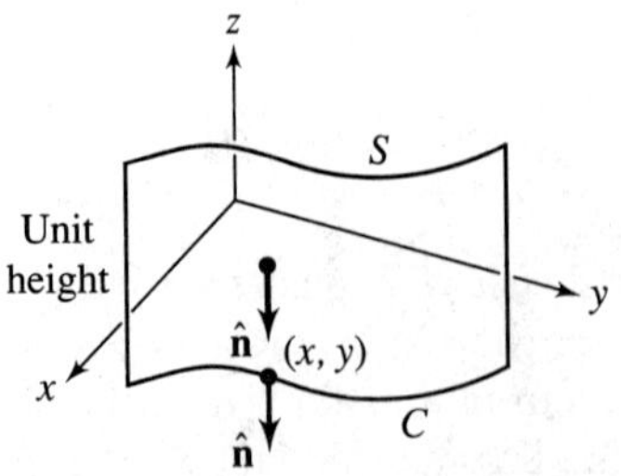

Figure 2.51

is positive, there is a net outward flow, and when flux is negative, there is a net inward flow.

When the flow is two-dimensional, as in this section, surface integral 2.45 can be reduced to a line integral in the xy-plane for certain surfaces. In particular, suppose S is a vertical surface of unit height standing on a curve C in the xy-plane (Figure 2.51). In this case the unit normal at every point on S directly above a point (x, y) on C is equal to the unit normal to C at (x, y). Because ρ is constant and $\mathbf{q}$ is only a function of x and y, flux integral 2.45 can be expressed as a line integral:

$$\text{flux} = \rho \int_C \mathbf{q} \cdot \hat{\mathbf{n}}\, ds. \tag{2.46}$$

It measures the mass of fluid flowing through a surface of unit height standing on C. When C is closed and positively oriented (Figure 2.52), $\hat{\mathbf{n}} = (dy/ds, -dx/ds)$, and

$$\text{flux} = \rho \oint_C u\, dy - v\, dx. \tag{2.47}$$

This integral measures the net mass flow rate through a surface of unit height standing on C. It is zero if C contains no sources or sinks; it may be nonzero when sources or sinks are interior to C.

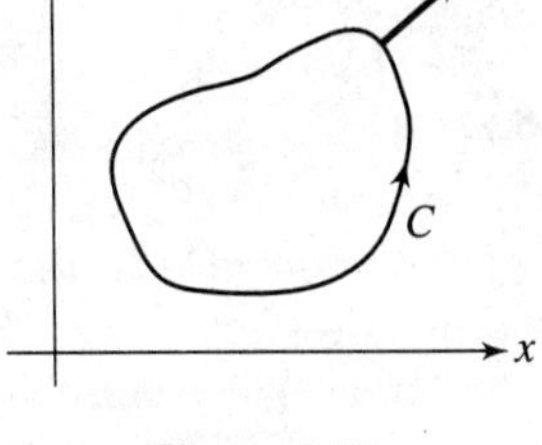

Figure 2.52

EXAMPLE 2.25 ....

Verify that flux through the curve $C : x^2 + y^2 = 4a^2$ for the flow of Example 2.24 vanishes.

Solution In Example 2.24, we calculated the velocity

$$\mathbf{q} = V_0\left(1 - \frac{a^2}{r^2}\cos 2\theta\right)\hat{\mathbf{i}} - V_0\left(\frac{a^2}{r^2}\sin 2\theta\right)\hat{\mathbf{j}}.$$

Using the polar form $r = 2a$ for C, we calculate the flux as

$$\begin{aligned}
\text{flux} &= \rho\oint_C V_0\left(1 - \frac{a^2}{r^2}\cos 2\theta\right)dy + V_0\left(\frac{a^2}{r^2}\sin 2\theta\right)dx \\
&= \rho V_0\int_0^{2\pi}\left(1 - \frac{a^2}{r^2}\cos 2\theta\right)(r\cos\theta\,d\theta) + \frac{a^2}{r^2}\sin 2\theta\,(-r\sin\theta\,d\theta) \\
&= 2\rho a V_0\int_0^{2\pi}\left[\left(1 - \frac{a^2}{4a^2}\cos 2\theta\right)\cos\theta - \frac{a^2}{4a^2}\sin 2\theta\,\sin\theta\right]d\theta \\
&= 2a\rho V_0\int_0^{2\pi}\frac{3}{4}\cos\theta\,d\theta \\
&= 0.
\end{aligned}$$

....

A second useful quantity in fluid flow is called *circulation*. For a closed curve C in a fluid, **circulation** Γ is defined by the line integral

$$\Gamma = \oint_C \mathbf{q}\cdot d\mathbf{r} = \oint_C u\,dx + v\,dy. \tag{2.48}$$

It is a measure of the tendency of the fluid to circulate around the curve C. If C contains only points at which irrotationality condition 2.38 is satisfied, then circulation vanishes; otherwise, it may not vanish.

EXAMPLE 2.26 ....

Calculate circulation for the curve and flow in Example 2.25.

Solution Using definition 2.48,

$$\begin{aligned}
\Gamma &= \oint_C V_0\left(1 - \frac{a^2}{r^2}\cos 2\theta\right)dx - \frac{V_0 a^2}{r^2}\sin 2\theta\,dy \\
&= V_0\int_0^{2\pi}\left(1 - \frac{a^2}{r^2}\cos 2\theta\right)(-r\sin\theta\,d\theta) - \frac{a^2}{r^2}\sin 2\theta\,(r\cos\theta\,d\theta)
\end{aligned}$$

$$= 2aV_0 \int_0^{2\pi} \left[-\left(1 - \frac{a^2}{4a^2}\cos 2\theta\right)\sin\theta - \frac{a^2}{4a^2}\sin 2\theta \cos\theta \right] d\theta$$

$$= 2aV_0 \int_0^{2\pi} -\frac{5}{4}\sin\theta \, d\theta$$

$$= 0.$$

In Chapter 4 we shall see that both flux and circulation integrals can be combined into a single integral of the complex potential function $f(z)$ associated with the flow.

Exercises 2.8

In Exercises 1–4 describe the fluid flow defined by the complex potential. Identify any stagnation points in the flow.

1. $f(z) = kz^2, \quad \operatorname{Im} z \geq 0, \quad \text{where } k > 0$
2. $f(z) = kz^2, \quad \operatorname{Im} z \geq 0, \quad \operatorname{Re} z \geq 0, \quad \text{where } k > 0$
3. $f(z) = z^2 + 2z, \quad \operatorname{Re} z \geq -1$
4. $f(z) = z^4, \quad 0 \leq \operatorname{Arg} z \leq \pi/4$
5. **(a)** What stream curves result by setting $\psi(x, y) = 0$ for the complex potential $f(z) = z^2 + 1/z^2$ defined only for $x^2 + y^2 \geq 1$, $x \geq 0$, $y \geq 0$?

   (b) Show that other stream curves have equations expressible in the form

$$\theta = \begin{cases} \dfrac{1}{2}\operatorname{Sin}^{-1}\left(\dfrac{Cr^2}{r^4 - 1}\right) \\ \dfrac{\pi}{2} - \dfrac{1}{2}\operatorname{Sin}^{-1}\left(\dfrac{Cr^2}{r^4 - 1}\right). \end{cases}$$

   Sketch these curves for $C = 1$ and $C = 2$ to obtain an idea of the flow pattern.

6. Find the complex potential for flow with constant speed U in the positive x-direction over the half-plane $y \geq 0$. Choose $y = 0$ as the stream function $\psi = 0$ and $x = 0$ as the velocity potential $\phi = 0$.
7. Show that flux and circulation for the flow of Exercise 2 vanish around every closed curve.
8. Show that flux and circulation for the flow of Exercise 4 vanish around every closed curve.
9. **(a)** Show that stream curves for the flow defined by the complex potential $f(z) = 1/z$ are circles with centers on the y-axis that "pass" through the origin.

   (b) Show that velocity potentials are circles with centers on the x-axis that "pass" through the origin.

(c) Prove that flux and circulation vanish for every curve not containing $z = 0$.

(d) Find flux and circulation for $x^2 + y^2 = a^2$ directed counterclockwise.

10. Show that when A and B are any two points in a flow and C is a piecewise-smooth curve in the flow joining A and B, the flux through C is $\rho[\psi(B) - \psi(A)]$, where $\psi(x, y) = \psi(A)$ is the stream curve through A and $\psi(x, y) = \psi(B)$ is the stream curve through B.

CHAPTER 3

Elementary Complex Functions

In this chapter we extend definitions of exponential, logarithm, trigonometric, inverse trigonometric, hyperbolic, and inverse hyperbolic functions to the complex plane. If $f(x)$ is any one of these functions, we shall want to extend $f(x)$ to a complex function $f(z)$ that agrees with $f(x)$ in the case that z is real and, we hope, has the same properties in the complex plane as does $f(x)$ on the real line. Of particular importance when it comes to applications are *zeros* and *singularities* of functions, and we therefore pay careful attention to zeros and singularities of each newly defined function.

3.1 Zeros and Singularities

Definition 3.1

A point z_0 is called a **zero** of order m of a complex function f if f is analytic at z_0, and

$$0 = f(z_0) = f'(z_0) = \cdots = f^{(m-1)}(z_0), \qquad f^{(m)}(z_0) \neq 0. \tag{3.1}$$

Recall that f is analytic "at z_0" if it is analytic in a neighborhood of z_0. For a zero of order m, the function and its first $m-1$ derivatives vanish at z_0, but the mth derivative is not zero. If the mth derivative does vanish, then the order of the zero is greater than m.

For example, the polynomial $P(z) = (z-1+2i)^3$ has a zero of order 3 at $z = 1-2i$, because $P(1-2i) = P'(1-2i) = P''(1-2i) = 0$, but $P'''(1-2i) = 6 \neq 0$. The rational function $R(z) = (z+1)^2(z-4i)/(z^2+1)$ has a zero of order 2 at $z = -1$ and a zero of order 1 at $z = 4i$. Zeros of order 1 are often called **simple zeros**. We denote zeros of a function with small circles in the complex plane (Figure 3.1 for $R(z)$).

Points at which a function fails to be analytic are called *singularities*. More precisely, we make the following definitions.

Definition 3.2

A point z_0 is called a **singularity** of a complex function f if f is not analytic at z_0, but every neighborhood of z_0 contains at least one point at which f is analytic.

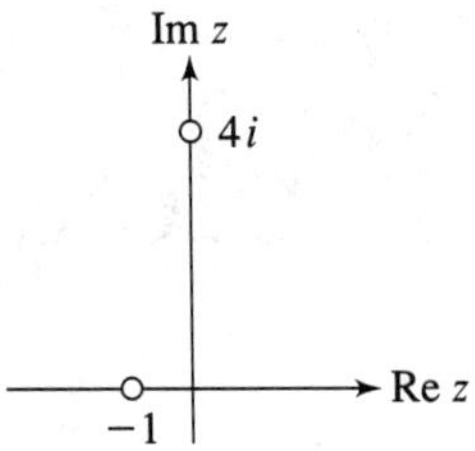

Figure 3.1

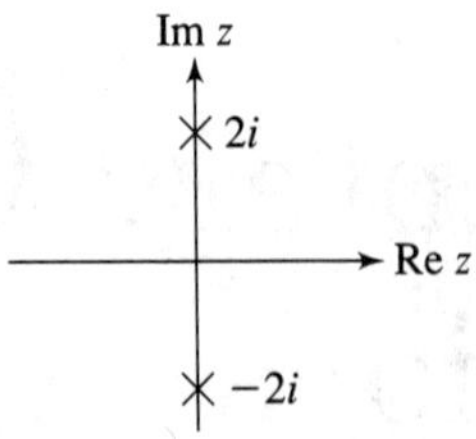

Figure 3.2

Definition 3.3
A singularity z_0 of a complex function f is said to be **isolated** if there exists a neighborhood of z_0 in which z_0 is the only singularity of f.

Entire functions have no singularities. Rational functions $P(z)/Q(z)$, where $P(z)$ and $Q(z)$ are polynomials, have isolated singularities at points where $Q(z) = 0$. For example, $f(z) = z^2/(z^2+4)$ has isolated singularities at $z = \pm 2i$. We denote singularities of a function $f(z)$ by an $\times$. We therefore have $\times$'s at $z = \pm 2i$ for the function $f(z) = z^2/(z^2+4)$ (Figure 3.2).

EXAMPLE 3.1 ....

Find and illustrate the zeros and singularities of the function $f(z) = \dfrac{z^4 + 2z^2 + 1}{z^3 - 8}$.

Solution Because $z^4 + 2z^2 + 1 = (z^2+1)^2 = (z+i)^2(z-i)^2$, $f(z)$ has zeros of order 2 at $z = \pm i$. With $z^3 - 8 = (z-2)(z^2+2z+4) = (z-2)(z+1-\sqrt{3}i)(z+1+\sqrt{3}i)$, we conclude that $f(z)$ has isolated singularities at $z = 2, -1 \pm \sqrt{3}i$. These zeros and singularities are shown in Figure 3.3.

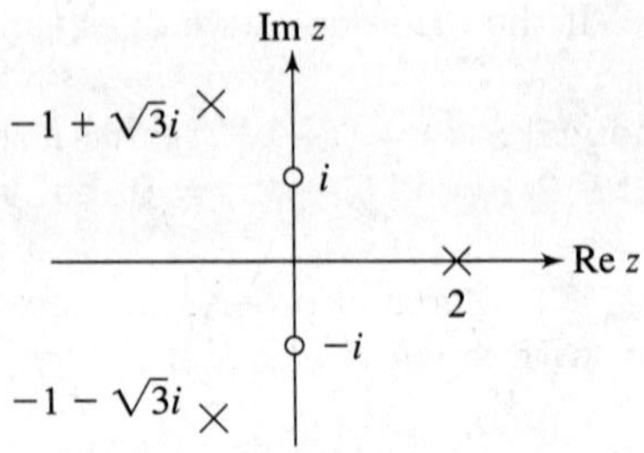

Figure 3.3

....

EXAMPLE 3.2 ....

What are the singularities of $f(z) = \operatorname{Re} z$?

Solution According to Example 2.17, this function is nowhere analytic. Does this mean that every point in the complex plane is a singularity? No, the function has no singularities, because every neighborhood of a singularity must contain a point at which the function is analytic. We do not deny that the function fails to have a derivative at every point in the complex plane, but we cannot call them singularities according to Definition 3.2.

Exercises 3.1

In Exercises 1–22 identify all zeros and their orders for the function. Locate all singularities and determine which are isolated.

1. $f(z) = 3z - 2$
2. $f(z) = \dfrac{1}{3z-2}$
3. $f(z) = z^2 + 3z - 2$
4. $f(z) = z^2 + 4$
5. $f(z) = z^3 + 8$
6. $f(z) = z^2 + 2z + 1$
7. $f(z) = \dfrac{1}{z^2+2z+4}$
8. $f(z) = \dfrac{z^2-3}{z^2+5z+1}$
9. $f(z) = \dfrac{z^3+2z}{z^2-2z+1}$
10. $f(z) = \dfrac{z^3}{3z+4}$
11. $f(z) = \dfrac{z}{(3z+1)^3}$
12. $f(z) = \operatorname{Arg} z$
13. $f(z) = |z|$
14. $f(z) = \left(\dfrac{z-1}{z+2}\right)^2$
15. $f(z) = \dfrac{z^3-3z^2+3z-1}{z+5}$
16. $f(z) = \dfrac{z^3+1}{z^3-1}$
17. $f(z) = \dfrac{z^2+i}{2z-3}$
18. $f(z) = z^2 + 4iz - 2$
19. $f(z) = \dfrac{z^2-2iz+i}{z+4}$
20. $f(z) = \dfrac{z^3-2z^2}{3z^2+4z-i}$
21. $f(z) = z^2 + \dfrac{4i}{z}$
22. $f(z) = z^3 + \dfrac{2+z^2-5z^3}{z+5}$

3.2 Complex Exponential Function

In this section we define the complex exponential function e^z. Unlike real analysis, where the trigonometric, inverse trigonometric, and functions with fractional powers have no relation whatsoever to e^x, we find that in complex analysis, these functions, as well as logarithm, hyperbolic, and inverse hyperbolic functions are all defined in terms of e^z. In other words, complex exponentials are fundamental to our studies; master e^z and its properties before proceeding to further sections in this chapter.

Many approaches lead to a definition of e^z. Although our approach may not be the quickest, it has the advantage of being intuitive and complete. It utilizes differentiation, and it emphasizes once again the importance of the Cauchy-Riemann equations in complex analysis. We shall want to define e^z in such a way that whenever $z = x$ is real, e^z will be e^x as we understand it now. We ask, therefore, what algebraic and/or differentiation properties separate the real exponential function e^x from all other real functions. Such properties might then be used to define the complex exponential function e^z. Certainly we cannot forget our first calculus instructor emphasizing the fact that e^x is a function that differentiates to give itself; that is, e^x satisfies the property that $f'(x) = f(x)$. But so also does any constant times e^x. What distinguishes e^x from other solutions of $f'(x) = f(x)$ is the additional requirement that $f(0) = 1$. In other words, e^x is the one and only solution of

$$f'(x) = f(x), \qquad f(0) = 1.$$

It is this characterization that we use to define e^z. We define e^z as the solution of the complex differential equation

$$f'(z) = f(z), \qquad f(0) = 1. \tag{3.2}$$

(In this way we ensure that $f(z)$ reduces to e^x when z is real.) What do we know about complex differential equations? Nothing, but this can be remedied by separating 3.2 into real and imaginary parts. To do this, we set $f(z) = u(x,y)+v(x,y)i$, in which case 3.2 becomes

$$\frac{\partial u}{\partial x} + \frac{\partial v}{\partial x} i = u + vi, \qquad u(0,0) + v(0,0)i = 1.$$

Taking real and imaginary parts gives

$$\frac{\partial u}{\partial x} = u, \qquad u(0,0) = 1; \qquad \frac{\partial v}{\partial x} = v, \qquad v(0,0) = 0,$$

and we have two separate systems, one in $u(x,y)$ and one in $v(x,y)$. They imply that

$$u(x,y) = p(y)e^x \quad \text{where } p(0) = 1; \qquad v(x,y) = q(y)e^x \quad \text{where } q(0) = 0.$$

We have not quite exhausted the information in 3.2. The fact that $f'(z)$ must exist (that is, $f(z)$ is analytic) requires u and v to satisfy the Cauchy-Riemann equations. This requirement means that

$$p(y)e^x = q'(y)e^x, \qquad q(y)e^x = -p'(y)e^x,$$

and from these $p = q'$ and $q = -p'$. It therefore follows that

$$p'' = -q' = -p \quad \text{and} \quad q'' = p' = -q.$$

Thus, $p(y)$ and $q(y)$ each satisfy the real differential equation

$$\frac{d^2\phi}{dy^2} + \phi = 0.$$

But all solutions of this differential equation are of the form $A\cos y + B\sin y$ for some constants A and B. Because $p(y)$ satisfies $p(0) = 1$ and $p'(0) = -q(0) = 0$, it follows that $p(y) = \cos y$. Similarly, for $q(0) = 0$ and $q'(0) = p(0) = 1$, we have $q(y) = \sin y$. Hence, the solution of differential equation 3.2, which we have agreed to call e^z, is defined as follows

$$e^z = e^x \cos y + e^x \sin y\, i = e^x(\cos y + \sin y\, i). \tag{3.3}$$

Because this complex exponential function satisfies the Cauchy-Riemann equations (by construction; it does so for all z) and first partial derivatives of u and v are clearly continuous, it follows that e^z is an entire function. Based on 3.2 we can state that

$$\frac{d}{dz}e^z = e^z. \tag{3.4}$$

EXAMPLE 3.3 ....

Express e^{3-i} and e^{2i-1} in Cartesian form.

Solution By Definition 3.3,

$$\begin{aligned} e^{3-i} &= e^3[\cos(-1) + \sin(-1)\, i] = e^3(\cos 1 - \sin 1\, i) \\ &= (e^3 \cos 1) - (e^3 \sin 1)i \approx 10.9 - 16.9i, \end{aligned}$$

and

$$\begin{aligned} e^{2i-1} &= e^{-1}(\cos 2 + \sin 2\, i) = (e^{-1}\cos 2) + (e^{-1}\sin 2)i \\ &\approx -0.153 + 0.335i. \end{aligned}$$

....

EXAMPLE 3.4 ....

Find $f''(-2+4i)$ if $f(z) = e^{3z}$.

Solution The chain rule gives $f''(z) = 9e^{3z}$, and therefore

$$f''(-2+4i) = 9e^{3(-2+4i)} = 9e^{-6+12i} = 9e^{-6}(\cos 12 + \sin 12\, i) \approx 0.0188 - 0.0120i.$$

....

Algebraic properties of the complex exponential function e^z must be based on Definition 3.3. It is straightforward to show that

$$e^{z_1}e^{z_2} = e^{z_1+z_2}, \tag{3.5a}$$

$$e^{z_1}/e^{z_2} = e^{z_1-z_2}. \tag{3.5b}$$

For instance, if $z_1 = x_1 + y_1 i$ and $z_2 = x_2 + y_2 i$, then

$$
\begin{aligned}
e^{z_1}e^{z_2} &= [e^{x_1}(\cos y_1 + \sin y_1\, i)][e^{x_2}(\cos y_2 + \sin y_2\, i)] && \text{(by 3.3)}\\
&= e^{x_1+x_2}[(\cos y_1\, \cos y_2 - \sin y_1\, \sin y_2)\\
&\quad + (\cos y_1\, \sin y_2 + \sin y_1\, \cos y_2)i] && \text{(by ordinary algebra)}\\
&= e^{x_1+x_2}[\cos(y_1+y_2) + \sin(y_1+y_2)\, i] && \text{(by real trigonometric identities)}\\
&= e^{x_1+x_2+(y_1+y_2)i} && \text{(by 3.3)}\\
&= e^{z_1+z_2}.
\end{aligned}
$$

We might have expected a third identity to accompany 3.5(a) and (b), namely, $(e^{z_1})^{z_2} = e^{z_1 z_2}$. But the left side of this equation is a complex number — e^{z_1} raised to a complex number z_2, an operation as yet undefined. When it is defined in Section 3.6, we will discuss the identity.

When we set $x = 0$ in 3.3, e^z becomes e^{yi}, and therefore

$$e^{yi} = \cos y + \sin y\, i. \tag{3.6}$$

This result is called **Euler's identity**. It immediately leads to the exponential form for complex numbers, a more compact and far more suggestive form for complex numbers than the polar representation. When r and θ are modulus and argument of a complex number z, then $z = r(\cos\theta + \sin\theta\, i)$. But because of Euler's identity, we may write

$$z = r(\cos\theta + \sin\theta\, i) = re^{\theta i}. \tag{3.7}$$

Rules 1.23 and 1.25 for multiplication and division of complex numbers are simpler in exponential form and follow naturally from properties 3.5:

$$z_1 z_2 = (r_1 e^{\theta_1 i})(r_2 e^{\theta_2 i}) = r_1 r_2 e^{(\theta_1+\theta_2)i}, \tag{3.8a}$$

$$\frac{z_1}{z_2} = \frac{re^{\theta_1 i}}{r_2 e^{\theta_2 i}} = \frac{r_1}{r_2} e^{(\theta_1-\theta_2)i}. \tag{3.8b}$$

In exponential notation, De Moivre's theorem (equation 1.27) seems most natural:

$$z^n = \left(re^{\theta i}\right)^n = r^n e^{n\theta i}. \tag{3.8c}$$

EXAMPLE 3.5 ....

Express the complex numbers -1 and $3 - 2i$ in exponential notation.

Solution Because $|-1| = 1$ and $\mathrm{Arg}(-1) = \pi$, it follows that

$$-1 = (1)e^{\pi i} = e^{\pi i}.$$

With $|3-2i| = \sqrt{13}$ and $\theta = \text{Arg}(3-2i) = \text{Tan}^{-1}(-2/3) \approx -0.588$,

$$3 - 2i = \sqrt{13}e^{\theta i} = \sqrt{13}e^{i\,\text{Tan}^{-1}(-2/3)} \approx \sqrt{13}e^{-0.588i}.$$

Neither of these representations is unique; it is not necessary to use principal values of arguments. We could write $-1 = e^{(\pi+2n\pi)i}$ and $3 - 2i \approx \sqrt{13}e^{(2n\pi-0.588)i}$ for any integer n.

EXAMPLE 3.6

Find $(1-2i)^9 e^{3i}$ in Cartesian form.

Solution Because $|1-2i| = \sqrt{5}$ and $\theta = \text{Arg}(1-2i) = \text{Tan}^{-1}(-2)$, it follows that

$$(1-2i)^9 e^{3i} = (\sqrt{5}e^{\theta i})^9 e^{3i} = 5^{9/2}e^{9\theta i}e^{3i} = 5^{9/2}e^{(9\theta+3)i}$$
$$= 5^{9/2}[\cos(9\theta+3) + \sin(9\theta+3)\,i] \approx 1085.7 - 880.0i\,.$$

In the introduction to this chapter we indicated that we would investigate zeros and singularities of each newly defined function. Being entire, e^z has no singularities. It also has no zeros, as can be seen from the fact that

$$|e^z| = |e^x(\cos y + \sin y\, i)| = |e^x|\,|\cos y + \sin y\, i| = e^x,$$

and $e^x \neq 0$. What have profound significance about e^z are the complex numbers z for which e^z is equal to 1. If $1 = e^z = e^x(\cos y + \sin y\, i)$, then real and imaginary parts imply that

$$e^x \cos y = 1, \qquad e^x \sin y = 0.$$

The second equation requires $y = n\pi$, where n is an integer, which substituted into the first gives $1 = e^x(-1)^n$. Consequently, n must be an even integer, and in this case $x = 0$. Thus, for any integer n, we have

$$e^{2n\pi i} = 1. \tag{3.9}$$

Equation 3.9 indicates that the exponential function $w = e^z$ maps each of the complex numbers $z = 2n\pi i$ onto the real number $w = 1$. Let us examine the mapping further by considering the infinite strip $-\pi < y \le \pi$ in the complex z-plane (Figure 3.4). Points on the imaginary axis ($x = 0$) are mapped by $w = e^z$ onto points $w = e^z = e^{yi} = \cos y + \sin y\, i$—i.e., to points on the unit circle $|w| = 1$. For points in the strip with negative real parts ($x < 0$), we find that $|e^z| = e^x < 1$. Hence, points in the left half of the strip are mapped inside the unit circle. Similarly, points with positive real parts ($x > 0$) are mapped onto points outside the unit circle.

We now show that every point $w = u + vi$ in the w-plane (except $w = 0$) is the image of exactly one point z in the strip $-\pi < y \le \pi$. To find $z = x + yi$ such that

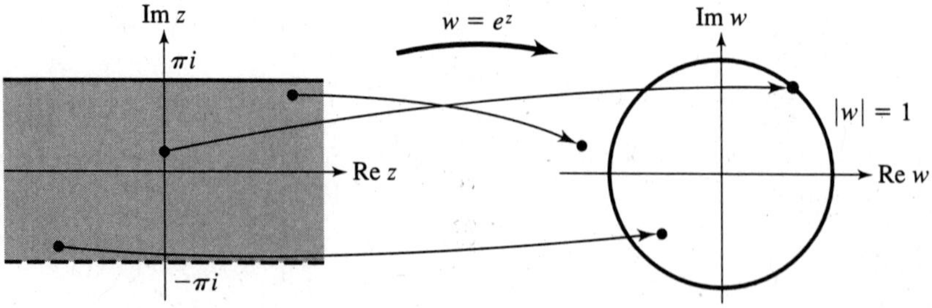

Figure 3.4

$w = u + vi = e^z$ for given w, we set

$$u + vi = e^z = e^x(\cos y + \sin y\, i).$$

Real and imaginary parts of this equation require

$$e^x \cos y = u \quad \text{and} \quad e^x \sin y = v.$$

These imply that $e^{2x} = u^2 + v^2$, or $x = (1/2)\ln(u^2 + v^2)$; and y is then defined by

$$\cos y = \frac{u}{\sqrt{u^2+v^2}}, \qquad \sin y = \frac{v}{\sqrt{u^2+v^2}}.$$

Because the value of x is unique and there is exactly one value of y in the interval $-\pi < y \leq \pi$ satisfying these equations, it follows that the image of the strip of Figure 3.4 is the entire w-plane (except for the point $w = 0$).

Because y is defined only to a multiple of 2π, it also follows that every strip $y_0 < y \leq y_0 + 2\pi$ is mapped in a similar fashion onto the w-plane. Another way of saying the same thing is that because

$$e^{z+2\pi i} = e^z, \tag{3.10}$$

e^z is a periodic complex function with complex period $2\pi i$. The real exponential function e^x is certainly not periodic, but its extension $e^z = e^x(\cos y + \sin y\, i)$ to the complex plane is. Let us examine this periodicity in a slightly different way.

Points $z = x$ on the real z-axis are mapped by $w = e^z$ onto $w = e^x$, points on the positive real w-axis. Points on a horizontal line $y = y_0$, where y_0 is just greater than zero, are mapped onto the points $w = e^x(\cos y_0 + \sin y_0\, i)$; they all lie on the half-line with argument y_0 in the w-plane (Figure 3.5).

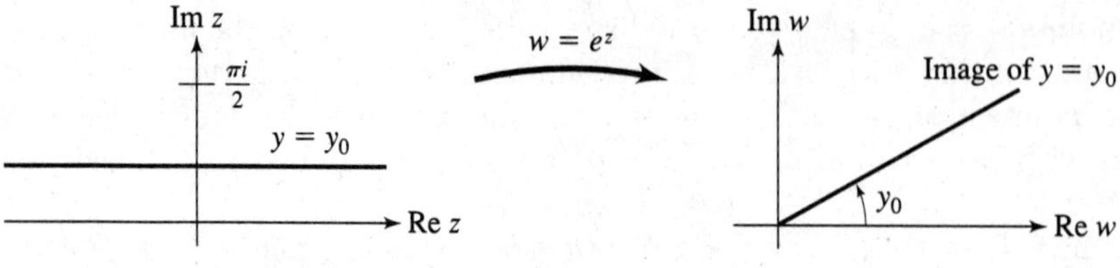

Figure 3.5

As we raise the line $y = y_0$, its image in the w-plane rotates. When $y_0 = \pi/2$, its image is the upper half of the imaginary axis in the w-plane. When $y_0 = \pi$, the image is the negative real axis. Finally, when $y_0 = 2\pi$, the image is once again the positive real w-axis. As we see, then, e^z retains the nonperiodic nature of e^x along the real z-axis but introduces an imaginary period $2\pi i$ with the extension of e^x into the complex plane.

EXAMPLE 3.7 ....

Find the image of the strip $0 \le \operatorname{Im} z \le \pi$ under the mapping $w = e^z$.

Solution The real z-axis C_1 is mapped to the positive real w-axis (Figure 3.6). Points $z = x + \pi i$ on $C_2 : \operatorname{Im} z = \pi$ are mapped to $w = e^{x+\pi i} = e^x(\cos \pi + \sin \pi\, i) = -e^x$, points on the negative real axis in the w-plane. When $z = x + yi$, where $0 < y < \pi$, we find that $w = e^{x+yi} = e^x(\cos y + \sin y\, i)$, and the imaginary part $e^x \sin y > 0$. In other words, $w = e^z$ maps the strip $0 \le \operatorname{Im} z \le \pi$ into the half-plane $\operatorname{Im} w \ge 0$ (less, of course, the point $w = 0$).

It is straightforward to show that every point in $\operatorname{Im} w > 0$ is the image of exactly one point in $0 < \operatorname{Im} z < \pi$. Hence, the strip $0 \le \operatorname{Im} z \le \pi$ is mapped onto the half-plane $\operatorname{Im} w \ge 0$ (less $w = 0$).

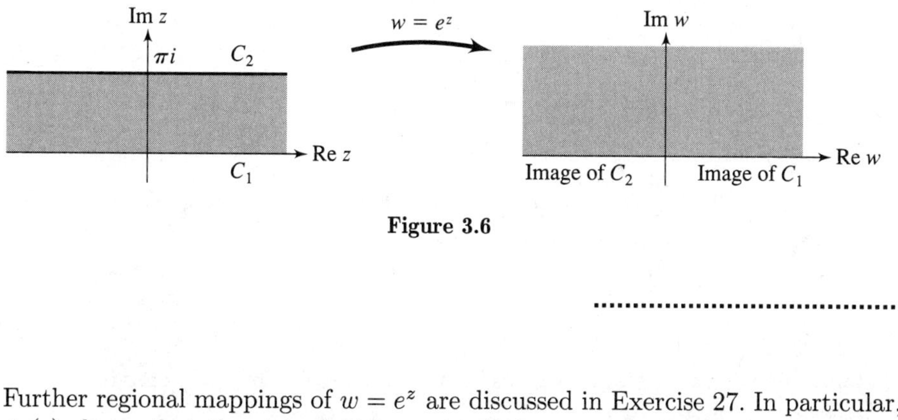

Figure 3.6

....

Further regional mappings of $w = e^z$ are discussed in Exercise 27. In particular, part (c) shows that the strip $0 \le \operatorname{Im} z \le \pi/2$ is mapped onto the quarter plane $\operatorname{Re} w \ge 0$, $\operatorname{Im} w \ge 0$ (Figure 3.7), again less $w = 0$.

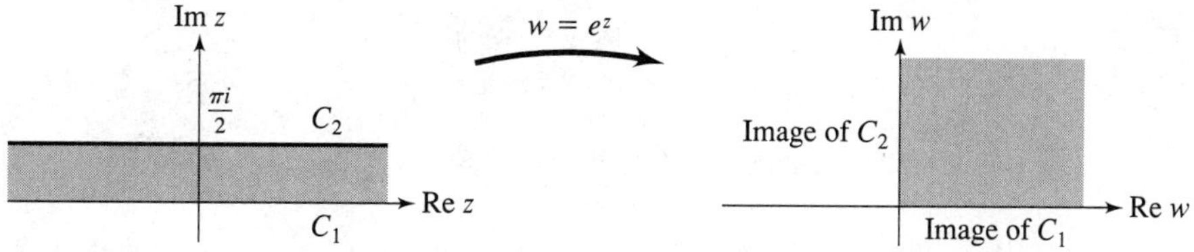

Figure 3.7

Exercises 3.2

In Exercises 1–10 express the complex number in exponential form.

1. $1 - i$
2. $-1 + \sqrt{3}i$
3. $(2 + 2i)^2$
4. $3 + 4i$
5. $-2 + 6i$
6. $(1 - 2i)^2$
7. $4[\cos(\pi/3) + \sin(\pi/3)\, i]$
8. $-4[\cos(-21\pi/10) + \sin(-21\pi/10)\, i]$
9. $2[\cos(\pi/12) - \sin(\pi/12)\, i]$
10. $3[\cos(\pi/4) + \sin(\pi/4)\, i]^3$

In Exercises 11–14 express the complex number in Cartesian form.

11. e^{2+3i}
12. e^{-1-4i}
13. $e^{(2-i)/(3+i)}$
14. e^{i^3+2i}

In Exercises 15–22 find all solutions of the equation.

15. $e^z = 4$
16. $e^z = 4i$
17. $e^z = 4 + 4i$
18. $e^z = 2 - 3i$
19. $e^{2-z} = -1 + i$
20. $e^{2z} + 2e^z + 1 = 0$
21. $e^{2z} + e^z + 1 = 0$
22. $e^{2z} + 3e^z + 5 = 0$

In Exercises 23–26, find the indicated derivative.

23. $f'(2 + i)$ if $f(z) = e^{3z+2}$
24. $f''(-i)$ if $f(z) = z^2e^{-z}$
25. $f'(3 + 2i)$ if $f(z) = ze^{-2z}$
26. $f'''(3i + 1)$ if $f(z) = e^{iz}$
27. When $w = e^z$ is regarded as a mapping from the z-plane to the w-plane show that
    (a) Vertical lines in the z-plane are mapped to circles in the w-plane.
    (b) A rectangle $-\pi < y_0 \le y \le y_1 < \pi$, $x_0 \le x \le x_1$ is mapped to a partial circular annulus.
    (c) The strip $0 < y < \pi/2$ is mapped to the quarter-plane $u > 0$, $v > 0$.
28. Find a function $w = f(z)$ that maps the infinite strip $0 < \operatorname{Im} z < a$ onto the half-plane $\operatorname{Im} w > 0$.
29. Find a function $w = f(z)$ that maps the infinite strip $0 < \operatorname{Im} z < a$ onto the quarter-plane $\operatorname{Re} w > 0$, $\operatorname{Im} w > 0$.
30. Describe the fluid flow defined by the complex potential

$$f(z) = V_0 \left(\frac{z}{e^{\alpha i}} + \frac{a^2 e^{\alpha i}}{z} \right),$$

where a, V_0, and α are positive constants and $|z| \ge a$. (See Example 2.24 in Section 2.8.)

31. Based on Definition 3.2 we know that the derivative of e^z is e^z (see equation 3.4). Show this using 3.3 and the Cauchy-Riemann equations.

3.3 Complex Trigonometric Functions

Most introductions to real trigonometric functions begin with trigonometric ratios of sides of right-angled triangles. This leads immediately to a wide variety of applications, and it also provides a straightforward extension to trigonometric ratios of arbitrary angles, the birth of the trigonometric functions. The appeal of this approach is its constant reference to triangles and associated visualizations. (A picture is worth a thousand words.) We soon realize, however, that trigonometric functions, especially the sine and cosine, enter many applications with no associated triangles. We accept that what is important about a function is that it has values and that we can find these values, and we gradually push into the background ratios of sides of triangles unless they fit naturally into a particular application. But how do we find values for trigonometric functions, such as sin(1.356)? As a student, I used tables of trigonometric functions, not particularly accurate and not particularly efficient. You undoubtedly use an electronic calculator, more efficient and definitely more accurate. There is an alternative. Maclaurin series for the sine and cosine functions can provide values of these functions with whatever accuracy is required. These series can also be used to verify trigonometric identities, and from many perspectives, the sine and cosine functions should be defined in this way.

What we are suggesting here is that as you acquired more and more mathematical knowledge, we asked you to regard trigonometric functions in a less and less geometric way. In this section, we take the final step. We apply trigonometric functions to complex numbers. There is no geometry. We don't ask you to suppress it; there just isn't any.

The real trigonometric functions are unrelated to the real exponential function e^x; the trigonometric functions are periodic, but e^x is not. Just the opposite is true for complex trigonometric functions. We actually define complex trigonometric functions in terms of the complex exponential function e^z. To see how, recall that Euler's identity implies that $e^{\theta i} = \cos\theta + \sin\theta\, i$ and $e^{-\theta i} = \cos\theta - \sin\theta\, i$ for any real number θ. By adding and subtracting these equations, we obtain

$$\cos\theta = \frac{e^{\theta i} + e^{-\theta i}}{2} \quad \text{and} \quad \sin\theta = \frac{e^{\theta i} - e^{-\theta i}}{2i}. \tag{3.11}$$

These equations express the real functions $\sin\theta$ and $\cos\theta$ in terms of the complex exponentials $e^{\theta i}$ and $e^{-\theta i}$. Now, the extension of a real trigonometric function to the complex plane should be done in such a way that when the argument of the complex trigonometric function is real, the function should return to its real form. With this and 3.11 in mind, it is natural to extend the sine and cosine functions to the complex plane according to

$$\cos z = \frac{e^{zi} + e^{-zi}}{2} \quad \text{and} \quad \sin z = \frac{e^{zi} - e^{-zi}}{2i}. \tag{3.12}$$

Because e^z is $2\pi i$-periodic and definitions of $\sin z$ and $\cos z$ involve e^{iz} and e^{-iz}, it follows that $\sin z$ and $\cos z$, like their real counterparts, are 2π-periodic. They also satisfy the usual trigonometric identities. For example,

$$\sin^2 z + \cos^2 z = \left(\frac{e^{zi} - e^{-zi}}{2i}\right)^2 + \left(\frac{e^{zi} + e^{-zi}}{2}\right)^2$$
$$= -\frac{1}{4}(e^{2zi} - 2 + e^{-2zi}) + \frac{1}{4}(e^{2zi} + 2 + e^{-2zi}) = 1$$

and

$$\sin z \cos w + \cos z \sin w = \frac{e^{zi} - e^{-zi}}{2i}\frac{e^{wi} + e^{-wi}}{2} + \frac{e^{zi} + e^{-zi}}{2}\frac{e^{wi} - e^{-wi}}{2i}$$
$$= \frac{1}{4i}[e^{(z+w)i} - e^{(w-z)i} + e^{(z-w)i} - e^{-(z+w)i}$$
$$+ e^{(z+w)i} - e^{(z-w)i} + e^{(w-z)i} - e^{(z+w)i}]$$
$$= \frac{e^{(z+w)i} - e^{-(z+w)i}}{2i} = \sin(z + w).$$

Because $\cos z$ and $\sin z$ are sums of entire functions, they are also entire functions; they have no singularities. It is straightforward to find their derivatives,

$$\frac{d}{dz}(\sin z) = \frac{d}{dz}\left(\frac{e^{zi} - e^{-zi}}{2i}\right) = \frac{ie^{zi} + ie^{-zi}}{2i} = \frac{e^{zi} + e^{-zi}}{2} = \cos z,$$
$$\frac{d}{dz}(\cos z) = \frac{d}{dz}\left(\frac{e^{zi} + e^{-zi}}{2}\right) = \frac{ie^{zi} - ie^{-zi}}{2} = \frac{-e^{zi} + e^{-zi}}{2i} = -\sin z.$$

EXAMPLE 3.8 ....

Express $\sin(3 + 2i)$ and $\cos(-1 + 4i)$ in Cartesian form.

Solution By Definition 3.12,

$$\sin(3 + 2i) = \frac{e^{(3+2i)i} - e^{-(3+2i)i}}{2i} = -\frac{i}{2}[e^{-2+3i} - e^{2-3i}]$$
$$= -\frac{i}{2}[e^{-2}(\cos 3 + \sin 3\, i) - e^{2}(\cos 3 - \sin 3\, i)]$$
$$= \frac{1}{2}\sin 3\,(e^2 + e^{-2}) + \frac{1}{2}\cos 3\,(e^2 - e^{-2})i$$
$$\approx 0.531 - 3.59i$$

and

$$\cos(-1 + 4i) = \frac{e^{(-1+4i)i} + e^{-(-1+4i)i}}{2} = \frac{1}{2}[e^{-4-i} + e^{4+i}]$$
$$= \frac{1}{2}[e^{-4}(\cos 1 - \sin 1\, i) + e^{4}(\cos 1 + \sin 1\, i)]$$

$$= \frac{1}{2}\cos 1\,(e^4 + e^{-4}) + \frac{1}{2}\sin 1\,(e^4 - e^{-4})i$$
$$\approx 14.8 + 23.0i.$$

You may have noticed hyperbolic functions rearing their heads in Example 3.8. Trigonometric and hyperbolic functions are intimately related in complex variable work, and we constantly work simultaneously with both sets of functions. Exercise 17 asks you to show that real and imaginary parts of $\sin z$ and $\cos z$ can be expressed as follows:

$$\sin z = \sin(x + yi) = \sin x\,\cosh y + \cos x\,\sinh y\,i, \qquad \textbf{(3.13a)}$$

$$\cos z = \cos(x + yi) = \cos x\,\cosh y - \sin x\,\sinh y\,i. \qquad \textbf{(3.13b)}$$

With these we can be more expedient in Example 3.8:

$$\sin(3 + 2i) = \sin 3\,\cosh 2 + \cos 3\,\sinh 2\,i \approx 0.531 - 3.59i,$$
$$\cos(-1 + 4i) = \cos(-1)\,\cosh 4 - \sin(-1)\,\sinh 4\,i \approx 14.8 + 23.0i.$$

Zeros of $\sin z$ are defined by

$$0 = \sin z = \sin x\,\cosh y + \cos x\,\sinh y\,i.$$

When we equate real and imaginary parts to zero,

$$0 = \sin x\,\cosh y, \qquad 0 = \cos x\,\sinh y.$$

The first of these implies that $x = n\pi$, and substitution of this into the second yields $y = 0$. Consequently, the zeros of $\sin z$ are the same as those of $\sin x$, namely, $z = n\pi$. Because the derivative of $\sin z$ at $z = n\pi$ is not zero, these zeros are simple zeros. Similarly, $\cos z$ has simple zeros $z = (2n+1)\pi/2$. These zeros are illustrated in Figures 3.8 and 3.9. Thus, although we have extended the sine and cosine functions from the real axis to the entire complex plane, no new zeros are introduced.

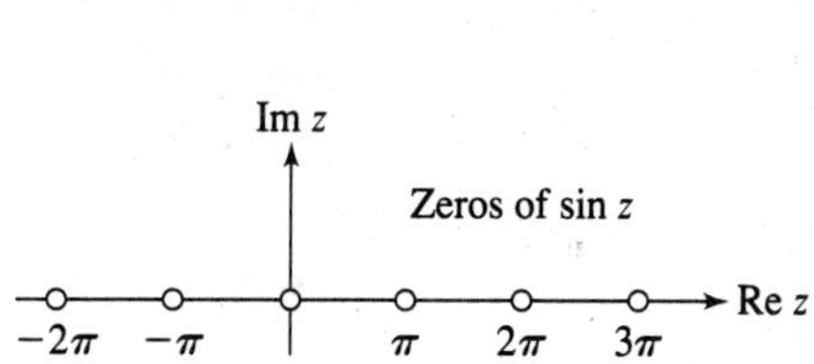

Figure 3.8

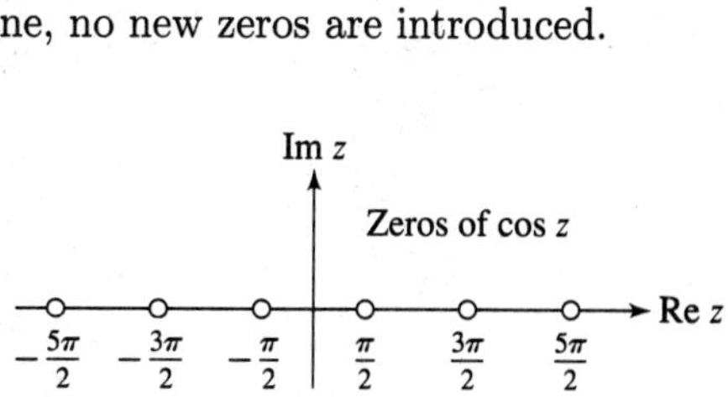

Figure 3.9

According to 3.13(a), the modulus of $\sin z$ is

$$|\sin z| = \sqrt{\sin^2 x\,\cosh^2 y + \cos^2 x\,\sinh^2 y}.$$

This expression becomes arbitrarily large for large positive and negative values of y. In other words, unlike $\sin x$, which satisfies $|\sin x| \leq 1$, the modulus of $\sin z$ is unbounded as $y \to \pm\infty$ for any x whatsoever. The modulus of $\cos z$ is also unbounded as $y \to \pm\infty$.

The remaining complex trigonometric functions are defined as usual:

$$\tan z = \frac{\sin z}{\cos z}; \qquad \cot z = \frac{\cos z}{\sin z}; \qquad \sec z = \frac{1}{\cos z}; \qquad \csc z = \frac{1}{\sin z}. \tag{3.14}$$

They are analytic functions in the complex plane except where their denominators vanish; in other words, isolated singularities of $\tan z$ and $\sec z$ occur at the zeros $z = (2n+1)\pi/2$ of $\cos z$, and $\cot z$ and $\csc z$ have isolated singularities at the zeros $z = n\pi$ of $\sin z$. Derivatives of these functions are as expected,

$$\frac{d}{dz}(\tan z) = \sec^2 z; \qquad \frac{d}{dz}(\cot z) = -\csc^2 z; \tag{3.15a}$$

$$\frac{d}{dz}(\sec z) = \sec z \tan z; \qquad \frac{d}{dz}(\csc z) = -\csc z \cot z. \tag{3.15b}$$

EXAMPLE 3.9 ....

Express $\tan i$ in Cartesian form.

Solution Using 3.14 and 3.13,

$$\tan i = \frac{\sin i}{\cos i} = \frac{\sinh 1\, i}{\cosh 1} \approx 0.762i.$$

....

EXAMPLE 3.10 ....

Find all solutions of the equation $\sec z = 4i$.

Solution If $\sec z = 4i$, then $\cos z = -i/4$. If we set $z = x + yi$, then

$$-\frac{i}{4} = \cos x \cosh y - \sin x \sinh y\, i.$$

When we equate real and imaginary parts, x and y satisfy

$$0 = \cos x \cosh y, \qquad -\frac{1}{4} = -\sin x \sinh y.$$

The first requires $x = (2n+1)\pi/2$, where n is an integer, which substituted into the second gives

$$\frac{1}{4} = \sin\left[\frac{(2n+1)\pi}{2}\right] \sinh y = (-1)^n \sinh y.$$

Hence, $y = \text{Sinh}^{-1}[(-1)^n/4] = (-1)^n \text{Sinh}^{-1}(1/4) \approx 0.247(-1)^n$, and the required points are

$$z \approx \frac{(2n+1)\pi}{2} + 0.247(-1)^n i.$$

EXAMPLE 3.11

Show that $w = \sin z$ maps the semi-infinite strip $-\pi/2 \leq \text{Re}\, z \leq \pi/2$, $\text{Im}\, z \geq 0$ onto the half-plane $\text{Im}\, w \geq 0$.

Solution Points on $C_2 : z = x$, $-\pi/2 \leq x \leq \pi/2$, are mapped to $w = \sin x$ — that is, points on the real axis in the w-plane with $-1 \leq \text{Re}\, w \leq 1$ (Figure 3.10). Points on $C_1 : z = -\pi/2 + yi$, $y \geq 0$ are mapped to

$$w = \sin\left(-\frac{\pi}{2} + yi\right) = \sin\left(-\frac{\pi}{2}\right)\cosh y + \cos\left(-\frac{\pi}{2}\right)\sinh y\, i = -\cosh y.$$

Because $1 \leq \cosh y < \infty$ for $y \geq 0$, it follows that the image of C_1 is the real w-axis to the left of $w = -1$. Similarly, the image of C_3 is the real w-axis to the right of $w = 1$. For $z = x + yi$, where $-\pi/2 < x < \pi/2$ and $y > 0$, the imaginary part of $w = \sin z = \sin x \cosh y + \cos x \sinh y\, i$ is positive. We have shown therefore that the semi-infinite strip is mapped into the half-plane $\text{Im}\, w \geq 0$. To show that the strip is mapped onto $\text{Im}\, w \geq 0$, we must show that every point $w = u + vi$ in $\text{Im}\, w > 0$ is the image of at least one point in the semi-infinite strip. Exercise 19 asks you to show that every point in $\text{Im}\, w > 0$ is the image of exactly one point in the strip.

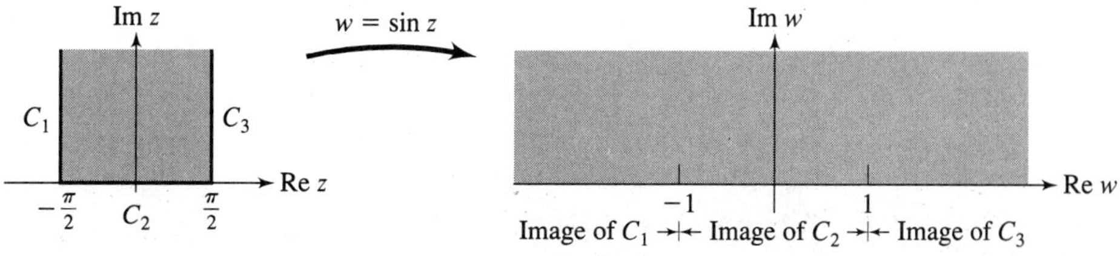

Figure 3.10

Exercises 3.3

In Exercises 1–12 express the complex number in Cartesian form.

1. $\sin i$
2. $\cos(1 - i)$
3. $\cos(1.3 + 2i)$
4. $\sin(5.6 + 4.1i)$
5. $e^{2+i} \cos i$
6. $\cos i \sin i$

7. $\cos^2(1+i)$

8. $(2-3i)\sin(2i^2+1)$

9. $\tan 2i$

10. $\cot(-i)$

11. $\csc(2/i)$

12. $\sec(2-3i)$

In Exercises 13–16 find all solutions of the equation.

13. $\sin z = 2i$

14. $\cos z = 4$

15. $\sin(z+i) = 1$

16. $\tan z = i$

17. Verify 3.13.

18. Show that

$$|\sin z|^2 = \cosh^2 y - \cos^2 x, \qquad |\cos z|^2 = \cosh^2 y - \sin^2 x.$$

19. Show that every point $w = u + vi$ in the half-plane $\operatorname{Im} w > 0$ is the image of exactly one point in the semi-infinite strip $-\pi/2 < \operatorname{Re} z < \pi/2$, $\operatorname{Im} z > 0$, under the mapping $w = \sin z$.

In Exercises 20–24 find the indicated derivative. Express the answer in Cartesian form.

20. $f'(2+i)$ if $f(z) = \sin iz$
21. $f''(i)$ if $f(z) = z\tan z$
22. $f'(-i)$ if $f(z) = \csc(2z+i)$
23. $f'(3)$ if $f(z) = \sin(z^2+i)$
24. $f''(-2i)$ if $f(z) = i\cos^2(2iz)$
25. Verify that $\sin^2 z + \cos^2 z = 1$ by showing that the derivative of the function $f(z) = \sin^2 z + \cos^2 z$ is equal to $f'(z) = 0$ and $f(0) = 1$.
26. Prove that under the mapping $w = \sin z$
    - **(a)** Horizontal lines in the z-plane are mapped to ellipses in the w-plane.
    - **(b)** Vertical lines in the z-plane are mapped to hyperbolas in the w-plane.
    - **(c)** A rectangle $0 < x_0 \le x \le x_1 < \pi/2$, $0 < y_0 \le y \le y_1$ is mapped to a strip bounded by two ellipses and two hyperbolas.
    - **(d)** The strip $0 \le x \le \pi/2$, $y \ge 0$ is mapped to the quarter-plane $u \ge 0$, $v \ge 0$.
    - **(e)** A rectangle $-\pi/2 \le x \le \pi/2$, $0 \le y \le y_1$ is mapped to a semielliptic region.
27. Prove that under the mapping $w = \cos z$
    - **(a)** Horizontal lines in the z-plane are mapped to ellipses in the w-plane.
    - **(b)** Vertical lines in the z-plane are mapped to hyperbolas in the w-plane.
    - **(c)** A rectangle $0 < x_0 \le x \le x_1 < \pi/2$, $0 < y_0 \le y \le y_1$ is mapped to a strip bounded by two ellipses and two hyperbolas.
    - **(d)** The strip $0 \le x \le \pi$, $y \ge 0$ is mapped to the half-plane $v \le 0$.
    - **(e)** The strip $0 \le x \le \pi/2$, $y \ge 0$ is mapped to the quarter-plane $u \ge 0$, $v \le 0$.
    - **(f)** A rectangle $0 \le x \le \pi$, $0 \le y \le y_1$ is mapped to a semielliptic region.
28. What is the image of the semi-infinite strip $-a \le \operatorname{Re} z \le a$, $\operatorname{Im} z \ge 0$ ($a > 0$ a constant) under the mapping $w = \sin[\pi z/(2a)]$?

29. For what values of z is $\cos z$ (a) real (b) purely imaginary?

30. For what values of z is $\sin z$ (a) real (b) purely imaginary?

31. Prove that all solutions of the equation $z \tan z = k$, where $k > 0$ is real, must be real.

32. Consider $f(z) = \sin z$ as the complex potential for a fluid flow.

(a) What is the velocity of the flow at each point (x, y)?

(b) Draw the stream curve $\psi(x, y) = 0$.

(c) Noting that the lines $x = \pm\pi/2$ for $y > 0$ and $y = 0$ for $-\pi/2 \leq x \leq \pi/2$ are part of the stream curve in (b), we can consider flow in the domain $y > 0$, $-\pi/2 < x < \pi/2$ with boundary points of the domain forming the physical boundary for the flow. Describe the flow by drawing stream curves ψ = constant in the flow domain. What are the stagnation points?

33. Repeat Exercise 32 with $f(z) = \cos z$. In part (c) use the domain $y > 0$, $0 < x < \pi$.

34. A function $w = f(z)$, defined implicitly by the equation

$$z = a \cos w, \qquad a > 0 \text{ constant},$$

is regarded as the complex potential for fluid flow.

(a) Show that stream curves are ellipses with foci at $(\pm a, 0)$.

(b) By choosing a specific ellipse in (a), say, $\psi = \psi_1$, as a physical boundary, the flow may be regarded as that about an elliptic object. Find the velocity and speed of the flow at any point z in terms of values of the velocity potential ϕ and stream function ψ at that point. Is the flow clockwise or counterclockwise?

3.4 Complex Hyperbolic Functions

Because the real hyperbolic functions $\sinh x$ and $\cosh x$ are defined in terms of e^x, it is natural to carry these definitions directly into the complex plane,

$$\sinh z = \frac{e^z - e^{-z}}{2}, \qquad \cosh z = \frac{e^z + e^{-z}}{2}. \tag{3.16}$$

They are entire functions with derivatives

$$\frac{d}{dz}\sinh z = \cosh z, \qquad \frac{d}{dz}\cosh z = \sinh z. \tag{3.17}$$

Because these functions have the same definitions as their real counterparts, they satisfy the familiar hyperbolic identities. For example,

$$\cosh^2 z - \sinh^2 z = \left(\frac{e^z + e^{-z}}{2}\right)^2 - \left(\frac{e^z - e^{-z}}{2}\right)^2 = 1.$$

Unlike the real hyperbolic functions, the complex hyperbolic sine and cosine are periodic, the period being $2\pi i$ (the period for e^z). Furthermore, unlike their real counterparts, the complex hyperbolic sine and cosine functions are related to the trigonometric sine and cosine functions according to

$$\sin(iz) = i \sinh z, \qquad \cos(iz) = \cosh z, \tag{3.18a}$$

$$\sinh (iz) = i \sin z, \qquad \cosh(iz) = \cos z. \tag{3.18b}$$

These, or Definition 3.16, can be used to find real and imaginary parts of $\sinh z$ and $\cosh z$:

$$\sinh z = \cos y \, \sinh x + \sin y \, \cosh x \, i, \tag{3.19a}$$

$$\cosh z = \cos y \, \cosh x + \sin y \, \sinh x \, i. \tag{3.19b}$$

EXAMPLE 3.12 ....

Express $3\cosh(2+3i) - 2\sinh(i/2)$ in Cartesian form.

Solution Using 3.19,

$$\begin{aligned} 3\cosh(2+3i) - 2\sinh(i/2) &= 3(\cos 3 \, \cosh 2 + \sin 3 \sinh 2 \, i) - 2i \sin(1/2) \\ &= 3\cos 3 \, \cosh 2 + [3 \sin 3 \, \sinh 2 - 2\sin(1/2)]i \\ &\approx -11.2 + 0.577i. \end{aligned}$$

....

Because $\sinh z = -i\sin(iz)$ and $\sin(iz) = 0$ when $iz = n\pi$, it follows that $\sinh z$ vanishes when $z = -n\pi i$; that is, the zeros of $\sinh z$ are $z = n\pi i$ (Figure 3.11). Similarly, the zeros of $\cosh z$ are $z = (2n+1)\pi i/2$ (Figure 3.12). All such zeros are simple ones.

The remaining four hyperbolic functions are defined by

$$\tanh z = \frac{\sinh z}{\cosh z}, \qquad \operatorname{csch} z = \frac{1}{\sinh z}, \tag{3.20a}$$

$$\coth z = \frac{\cosh z}{\sinh z}, \qquad \operatorname{sech} z = \frac{1}{\cosh z}. \tag{3.20b}$$

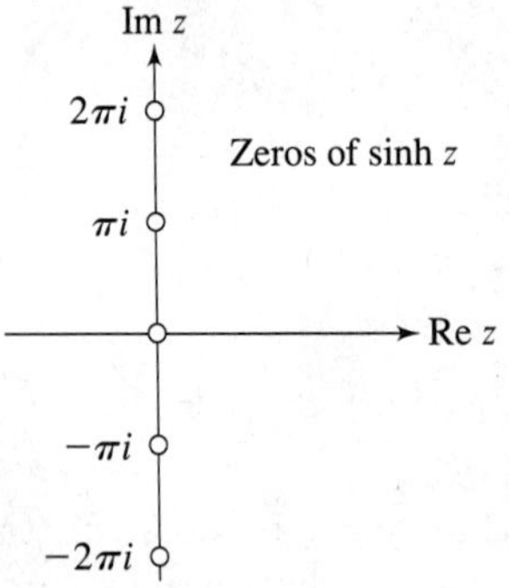

Figure 3.11

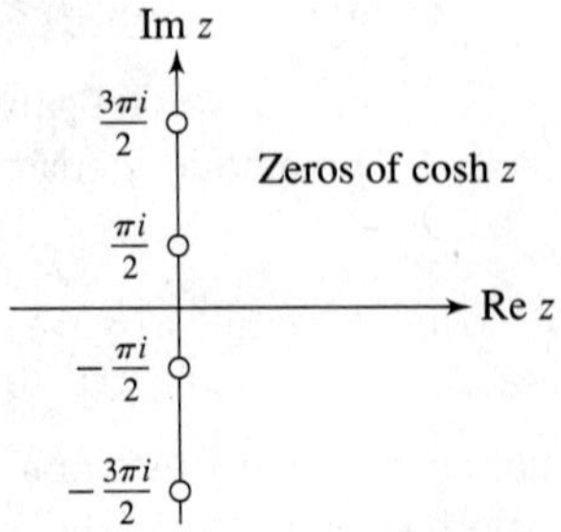

Figure 3.12

The $\tanh z$ and $\operatorname{sech} z$ functions have isolated singularities at $z = (2n+1)\pi i/2$, the zeros of $\cosh z$ (Figure 3.12), and $\coth z$ and $\operatorname{csch} z$ have isolated singularities at the zeros $z = n\pi i$ of $\sinh z$ (Figure 3.11). Their derivatives are as expected:

$$\frac{d}{dz}(\tanh z) = \operatorname{sech}^2 z; \qquad \frac{d}{dz}(\coth z) = -\operatorname{csch}^2 z; \tag{3.21a}$$

$$\frac{d}{dz}(\operatorname{sech} z) = -\operatorname{sech} z \tanh z; \qquad \frac{d}{dz}(\operatorname{csch} z) = -\operatorname{csch} z \coth z. \tag{3.21b}$$

EXAMPLE 3.13 ....

Show that $w = \sinh z$ maps the semi-infinite strip $-\pi/2 \le \operatorname{Im} z \le \pi/2$, $\operatorname{Re} z \ge 0$ to the half-plane $\operatorname{Re} w \ge 0$.

Solution We could verify this result with an analysis parallel to that in Example 3.11. Alternatively, we can use the result of Example 3.11 and the fact that $w = \sinh z = -i\,\sin(iz)$. The semi-infinite strip $-\pi/2 \le \operatorname{Im} z \le \pi/2$, $\operatorname{Re} z \ge 0$ is mapped by $z^* = iz$ onto the semi-infinite strip $-\pi/2 \le \operatorname{Re} z^* \le \pi/2$, $\operatorname{Im} z^* \ge 0$ (Figure 3.13), because multiplication by i rotates complex numbers $\pi/2$ radians counterclockwise. According to Example 3.11, $w^* = \sin z^*$ maps the semi-infinite strip $-\pi/2 \le \operatorname{Re} z^* \le \pi/2$, $\operatorname{Im} z^* \ge 0$ onto the half-plane $\operatorname{Im} w^* \ge 0$. Finally, $w = -iw^*$ rotates the half-plane $\operatorname{Im} w^* \ge 0$ clockwise through $\pi/2$ radians. Thus,

$$w = -iw^* = -i\,\sin z^* = -i\,\sin(iz) = \sinh z$$

maps the semi-infinite strip $-\pi/2 \le \operatorname{Im} z \le \pi/2$, $\operatorname{Re} z \ge 0$ to the half-plane $\operatorname{Re} w \ge 0$.

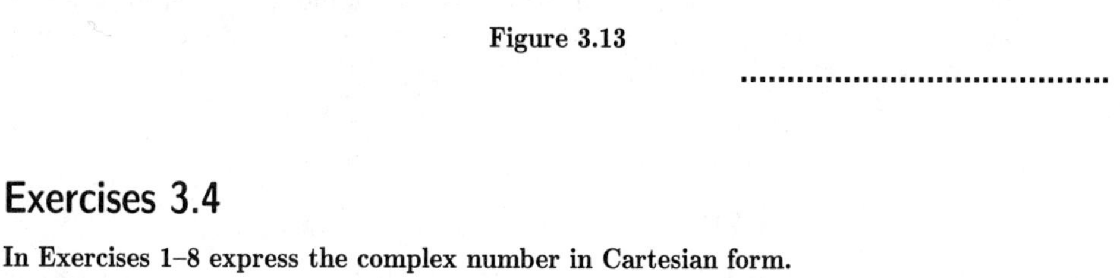

Figure 3.13

....

Exercises 3.4

In Exercises 1–8 express the complex number in Cartesian form.

1. $\sinh 3i$
2. $\cosh(1-2i)$
3. $4\sinh(\pi i/3)$
4. $\coth(3\pi i/4)$
5. $\tanh(1+i)$
6. $\operatorname{csch}(2i+1)$
7. $\sinh^2(-6i)$
8. $\operatorname{sech}(4+5i)$

In Exercises 9–12 find all solutions of the equation.

9. $\cosh z = 1$

10. $\sinh z = -3i$

11. $\tanh z = i$

12. $\cosh^2 z + \sinh^2 z = 1$

13. Verify 3.19.

14. Show that

$$|\sinh z|^2 = \sinh^2 x + \sin^2 y, \qquad |\cosh z|^2 = \sinh^2 x + \cos^2 y,$$

and hence deduce that $|\sinh z|$ and $|\cosh z|$ are unbounded as $x \to \pm\infty$ for fixed y.

In Exercises 15–18 find the indicated derivative. Express the answer in Cartesian form.

15. $f'(2+i)$ if $f(z) = \sinh(3z+1+i)$

16. $f'(i)$ if $f(z) = \tanh 3z$

17. $f'(1)$ if $f(z) = z \cosh iz$

18. $f''(-2i)$ if $f(z) = \operatorname{sech}(3z+i)$

19. Use Exercise 27 in Section 3.3 to describe the geometry of the mapping $w = \cosh z$.

20. Use Exercise 26 in Section 3.3 to describe the geometry of the mapping $w = \sinh z$.

21. Are $\sinh z$ and $\cosh z$ periodic?

22. Find a function that maps the semi-infinite strip $-a < \operatorname{Im} z < a$, $\operatorname{Re} z > 0$ onto the half-plane $\operatorname{Im} w > 0$.

23. A function $w = f(z)$, defined implicitly by the equation

$$z = a \cosh w, \qquad a > 0 \text{ constant},$$

is regarded as the complex potential for fluid flow.

(a) Show that stream curves are hyperbolas with foci $(\pm a, 0)$.

(b) By choosing a specific hyperbola in (a), such as $\psi = \psi_1$, as a physical boundary, the flow may be regarded as that through a hyperbolic aperature. Show that the speed of the flow at any point is given by

$$|\mathbf{q}|^2 = \frac{2}{a^2(\cosh 2\phi - \cos 2\psi)}, \qquad \text{where } w = \phi + \psi i.$$

(c) Show that by choosing the physical boundary as the limiting hyperbolas $\psi \to 0$ and $\psi \to \pi$, the flow becomes that through a slit $y = 0$, $-a < x < a$ in the x-axis. What is the speed of this flow at $(\pm a, 0)$?

3.5 Complex Logarithm Function

The real logarithm function $\ln x$ is defined as the inverse of the exponential function; $y = \ln x$ is the unique solution of the equation $x = e^y$. This works because e^x is a one-to-one function; if $x_1 \neq x_2$, then $e^{x_1} \neq e^{x_2}$. This is not the case for e^z; we have seen that e^z is $2\pi i$-periodic, so that all complex numbers of the form $z + 2n\pi i$ are mapped by $w = e^z$ onto the same complex number as z. To define the logarithm function, $\log z$, as the inverse of e^z is clearly going to lead to difficulties, and these difficulties are much like those encountered when finding the inverse function of $\sin x$ in real-variable calculus. Let us proceed. We call w the logarithm of z and write $w = \log z$, if $z = e^w$. To find w we let $w = u + vi$ be the Cartesian form for w and $z = re^{\theta i}$ be the exponential form for z. When we substitute these into $z = e^w$,

$$re^{\theta i} = e^{u+vi} = e^u e^{vi}.$$

According to equation 1.21, equality of these complex numbers implies that

$$e^u = r \quad \text{or} \quad u = \ln r,$$

and

$$v = \theta = \arg z.$$

Thus, $w = \ln r + \theta i$, and the logarithm of a complex number z is

$$\log z = \ln|z| + (\arg z)i. \tag{3.22}$$

We use ln only for logarithms of real numbers; log is used for logarithms of complex numbers.

Because 3.22 yields logarithms of every nonzero complex number, we have defined the complex logarithm function. It is defined for all $z \neq 0$, and because $\arg z$ is determined only to a multiple of 2π, each nonzero complex number has an infinite number of logarithms. For example,

$$\log(1+i) = \ln\sqrt{2} + (\pi/4 + 2k\pi)i = (\tfrac{1}{2})\ln 2 + (8k+1)\pi i/4.$$

Thus, to the complex number $1 + i$, the logarithm function assigns an infinite number of values, $\log(1+i) = (\frac{1}{2})\ln 2 + (8k+1)\pi i/4$. They all have the same real part, but their imaginary parts differ by multiples of 2π. In real-variable work we have not permitted this; a function must assign exactly one value to every number in its domain of definition. In complex analysis, it is sometimes, but not always, convenient to allow a function to take on more than one value for each complex number in its domain of definition. Such a function is said to be **multiple-valued**. The complex logarithm function is multiple-valued; to each complex number in its domain, it assigns infinitely many values.

EXAMPLE 3.14 ....

Express $\log(2-3i)$ in Cartesian form.

Solution Because $\arg(2-3i) = 2k\pi - \text{Tan}^{-1}(\frac{3}{2})$,

$$\log(2-3i) = \ln\sqrt{13} + [2k\pi - \text{Tan}^{-1}(\tfrac{3}{2})]i = \tfrac{1}{2}\ln 13 + [2k\pi - \text{Tan}^{-1}(\tfrac{3}{2})]i.$$

....

Some of the properties of the real logarithm function have their counterpart in the complex logarithm. For example, if $z = x + yi = re^{\theta i}$, then

$$e^{\log z} = e^{\ln r + \theta i} = e^{\ln r} e^{\theta i} = re^{\theta i} = z, \tag{3.23a}$$

(as should be expected from the definition of $\log z$), and

$$\begin{aligned} \log(e^z) = \log(e^{x+yi}) &= \ln(e^x) + (y + 2k\pi)i \qquad (k \text{ an integer}) \\ &= x + yi + 2k\pi i \\ &= z + 2k\pi i. \end{aligned} \tag{3.23b}$$

In real analysis the counterpart of this equation is $\log e^x = x$. The $z + 2k\pi$ on the right side of 3.23(b) is a reflection of the facts that $\log z$ is multiple-valued and the logarithm is the last operation on the left side of the equation.

If $z_1 = re^{\theta i}$ and $z_2 = Re^{\phi i}$, then

$$\begin{aligned} \log(z_1 z_2) &= \log[rRe^{(\theta+\phi)i}] = \ln(rR) + (\theta + \phi + 2p\pi)i \qquad (p \text{ an integer}) \\ &= (\ln r + \theta i) + (\ln R + \phi i) + 2p\pi i. \end{aligned}$$

But $\log z_1 = \ln r + (\theta + 2n\pi)i$ and $\log z_2 = \ln R + (\phi + 2m\pi)i$. Hence,

$$\begin{aligned} \log(z_1 z_2) &= (\log z_1 - 2n\pi i) + (\log z_2 - 2m\pi i) + 2p\pi i \\ &= \log z_1 + \log z_2 + 2(p - n - m)\pi i \\ &= \log z_1 + \log z_2 + 2k\pi i. \end{aligned} \tag{3.24a}$$

Similarly,

$$\log\left(\frac{z_1}{z_2}\right) = \log z_1 - \log z_2 + 2k\pi i. \tag{3.24b}$$

The last two results must be approached with care. Because the logarithm function is multiple-valued, each equation must be interpreted as saying that given values for the logarithm terms, there is a value of k for which the equation holds. It is also possible to write these equations in the forms

$$\log(z_1 z_2) = \log z_1 + \log z_2, \tag{3.25a}$$

$$\log\left(\frac{z_1}{z_2}\right) = \log z_1 - \log z_2. \tag{3.25b}$$

We interpret them as saying that given values for two of the logarithm terms, there is a value of the third logarithm for which the equation is valid.

Multiple-valued functions cannot be analytic. To see why, consider the derivative of $\log z$,

$$\frac{d}{dz} \log z = \lim_{\Delta z \to 0} \frac{\log(z + \Delta z) - \log z}{\Delta z}.$$

This limit must exist, be unique, and be independent of the mode of approach of Δz to 0. But these conditions are impossible if there is an infinity of possible choices for $\log z$ and for $\log(z+\Delta z)$ for each value of Δz. Thus, only single-valued functions can have derivatives. We therefore ask if it is possible to restrict the range of the logarithm function to obtain an analytic function; that is, can we make $\log z$ single-valued in such a way that it will have a derivative? The answer is yes, and there are many ways to do it. The most natural way to make $\log z$ single-valued is to restrict $\arg z$ in Definition 3.22 to its principal value $\operatorname{Arg} z$. When this is done, we denote the resulting single-valued function by

$$\operatorname{Log} z = \ln|z| + (\operatorname{Arg} z)i, \qquad z \neq 0. \tag{3.26}$$

Is $\operatorname{Log} z$ an analytic function? The answer is yes, in a suitably restricted domain. To see this, we note that at any point on the negative real axis, the imaginary part of $\operatorname{Log} z$ is a discontinuous function. Theorem 2.4 implies therefore that $\operatorname{Log} z$ is discontinuous at points on the negative real axis, and $\operatorname{Log} z$ cannot be differentiable thereon. Suppose we consider the domain $|z| > 0$, $-\pi < \operatorname{Arg} z < \pi$. If we express $\operatorname{Log} z$ in the form

$$\operatorname{Log} z = \ln r + \theta i, \qquad -\pi < \theta < \pi,$$

partial derivatives of its real and imaginary parts are

$$\frac{\partial u}{\partial r} = \frac{1}{r}, \qquad \frac{\partial v}{\partial \theta} = 1, \qquad \frac{\partial u}{\partial \theta} = 0, \qquad \frac{\partial v}{\partial r} = 0.$$

Because these derivatives satisfy Cauchy-Riemann equations 2.21 and are continuous in the domain $|z| > 0$, $-\pi < \operatorname{Arg} z < \pi$, it follows that $\operatorname{Log} z$ is an analytic function in this domain. According to the formula in Exercise 14 of Section 2.4, the derivative of $\operatorname{Log} z$ is

$$(\cos\theta - \sin\theta\, i)\left(\frac{1}{r}\right) = \frac{1}{r}(\cos\theta - \sin\theta\, i)\left(\frac{\cos\theta + \sin\theta\, i}{\cos\theta + \sin\theta\, i}\right) = \frac{1}{r(\cos\theta + \sin\theta\, i)};$$

that is,

$$\frac{d}{dz}\operatorname{Log} z = \frac{1}{z}. \tag{3.27}$$

Thus, $\operatorname{Log} z$ is analytic in the domain $|z| > 0$, $-\pi < \operatorname{Arg} z < \pi$. It is defined for all $z \neq 0$ but analytic only in the aforementioned domain. Points on the negative real axis and $z = 0$ are singularities of $\operatorname{Log} z$, but they are not isolated singularities (Figure 3.14).

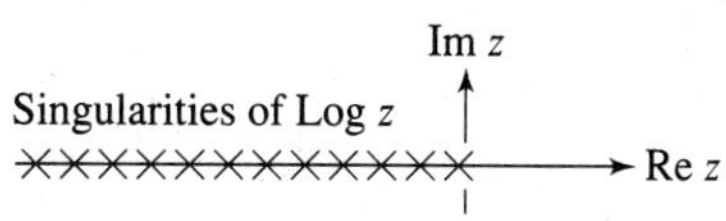

Figure 3.14

The restriction of $\arg z$ to $\operatorname{Arg} z$ in the definition of $\operatorname{Log} z$ was arbitrary; other choices are certainly possible. For example, we could restrict the range of $\log z$ by

$$\log z = \ln|z| + (\arg z)i, \qquad 0 < \arg z \leq 2\pi,$$

and this new function would be analytic in the domain $|z| > 0$, $0 < \arg z < 2\pi$.

Given a multiple-valued function f defined in a domain D, specific values of f can always be chosen to create a single-valued function F. When these values are selected in an attempt to define an analytic function F, it may be necessary to restrict differentiation from a line (or curve) of points in D. Such a line is called a **branch cut** for F; it is a line along which F is not differentiable (but may be defined). The function F so obtained is called a **branch** of f. An end of a branch cut is called a **branch point**. We do not consider a branch point as part of the branch cut, but it would be an equally acceptable point of view.

In particular, $\log z$ has branch point $z = 0$. Branches of the function can be created by choosing (half) straight lines eminating from $z = 0$ as branch cuts. $\operatorname{Log} z$ is one branch, with the negative real axis as branch cut (Figure 3.15). This branch of $\log z$ is called the **principal branch** (because principal values $\operatorname{Arg} z$ for $\arg z$ were chosen in its definition). The function

$$f(z) = \ln|z| + (\arg z)i, \qquad |z| > 0, \quad 0 < \arg z \leq 2\pi,$$

is a different branch of the multiple-valued $\log z$ function with the positive real axis as branch cut (Figure 3.16). The function

$$f(z) = \ln|z| + (\arg z)i, \quad |z| > 0, \quad 2\pi < \arg z \leq 4\pi,$$

is yet another branch, but it also has the positive real axis as branch cut.

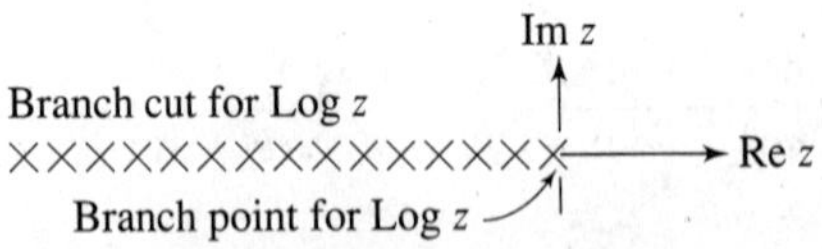

Figure 3.15

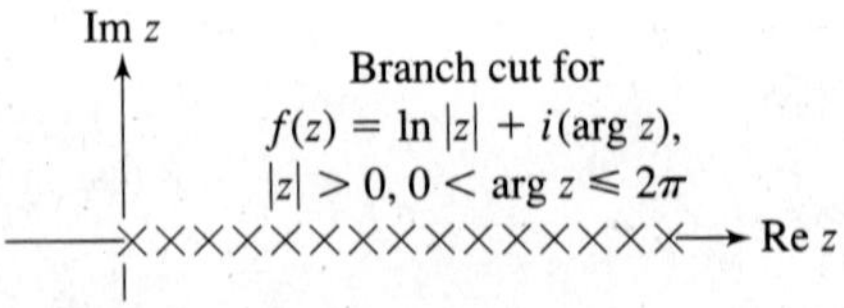

Figure 3.16

We use the following notation to identify some of the more commonly used branches of $\log z$:

$$\log_\phi z = \ln|z| + (\arg z)i, \tag{3.28a}$$

where

$$|z| > 0, \qquad \phi < \arg z \leq \phi + 2\pi. \tag{3.28b}$$

The branch cut is a half-line making angle ϕ with the positive real axis (Figure 3.17). Each point on the branch cut is a singularity, but not an isolated singularity, of $\log_\phi z$.

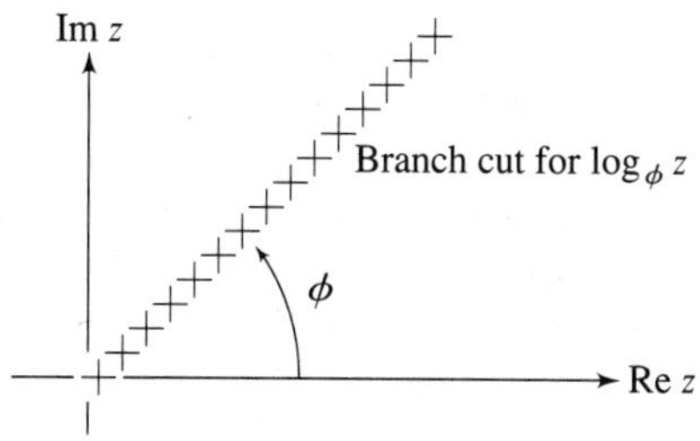

Figure 3.17

With this notation,

$$\log_0 z = \ln|z| + (\arg z)i, \qquad \text{where } |z| > 0, \quad 0 < \arg z \le 2\pi; \tag{3.29a}$$

$$\log_{-\pi} z = \operatorname{Log} z = \ln|z| + (\operatorname{Arg} z)i, \qquad \text{where } |z| > 0, \quad -\pi < \operatorname{Arg} z \le \pi; \tag{3.29b}$$

$$\log_{2\pi} z = \ln|z| + (\arg z)i, \qquad \text{where } |z| > 0, \quad 2\pi < \arg z \le 4\pi. \tag{3.29c}$$

In real-variable work, the notation $\log_a x$ indicates that a is the base of the logarithm function. This is not the case in complex function theory. Only e is used as the base for logarithms, and ϕ in $\log_\phi z$ indicates a particular branch of the $\log z$ function.

EXAMPLE 3.15 ....

Express $\operatorname{Log}(-1+\sqrt{3}i)$, $\log_{\pi/4}(-1+\sqrt{3}i)$, and $\log_{-3\pi/2}(-1+\sqrt{3}i)$ in Cartesian form.

Solution The complex number is shown in Figure 3.18. We see that

$$\operatorname{Log}(-1+\sqrt{3}i) = \ln 2 + (2\pi/3)i,$$

$$\log_{\pi/4}(-1+\sqrt{3}i) = \ln 2 + (2\pi/3)i,$$

$$\log_{-3\pi/2}(-1+\sqrt{3}i) = \ln 2 + (-4\pi/3)i.$$

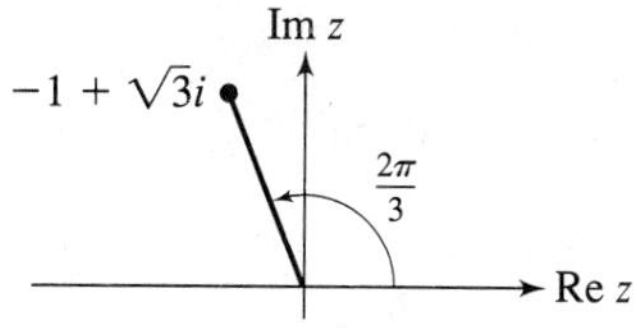

Figure 3.18

....

There are other branches of $\log z$ besides those in 3.28. For instance,

$$\log z = \ln|z| + (\arg z)i, \qquad |z| > 0, \quad 0 \le \arg z < 2\pi$$

has the same branch cut as $\log_0 z$. The only difference is the values that it assigns to its imaginary part at points on the branch cut. Branch cuts need not be straight lines. We could stipulate that the branch cut of a branch of $\log z$ is the parabolic curve in Figure 3.19. We simply agree that at each point on this curve, arguments of z will be specified in a certain way, perhaps as $0 < \arg z < \pi/2$, that arguments will increase to the left of the curve, and that they will jump by 2π across the curve.

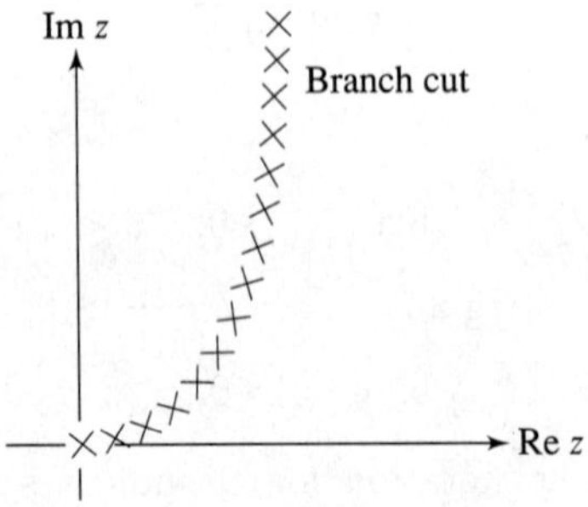

Figure 3.19

To understand branches of a multiple-valued function f, it is sometimes helpful to visualize ranges of the branches in the $w = f(z)$ plane. The three branches of $\log z$ in equations 3.29 are illustrated in Figures 3.20–22. Each branch maps the z-plane (less $z = 0$) onto a horizontal strip of width 2π. The choice of ϕ in $\log_\phi z$ indicates the lower edge of the strip. Notice once again that each branch is defined on its branch cut (but not at the branch point $z = 0$).

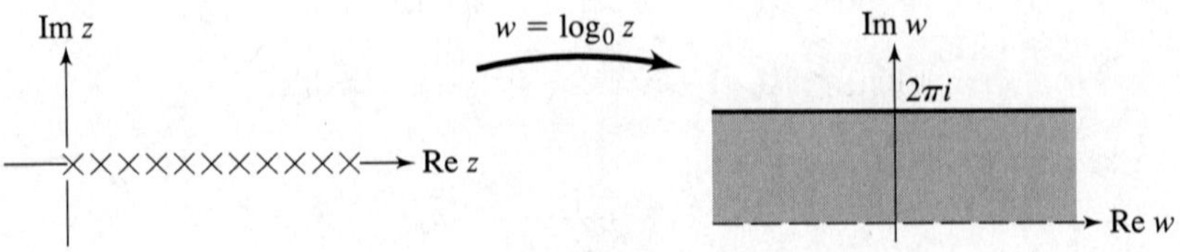

Figure 3.20

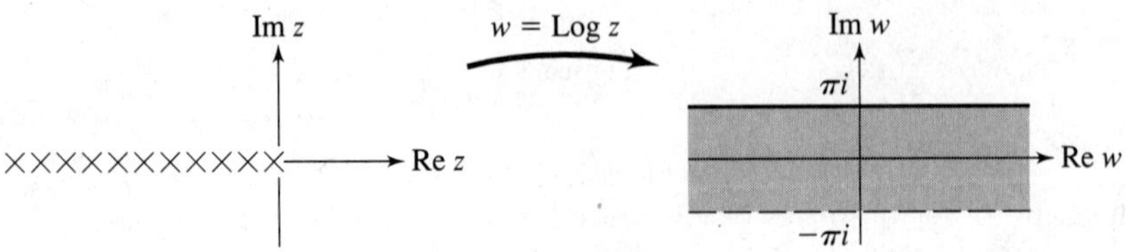

Figure 3.21

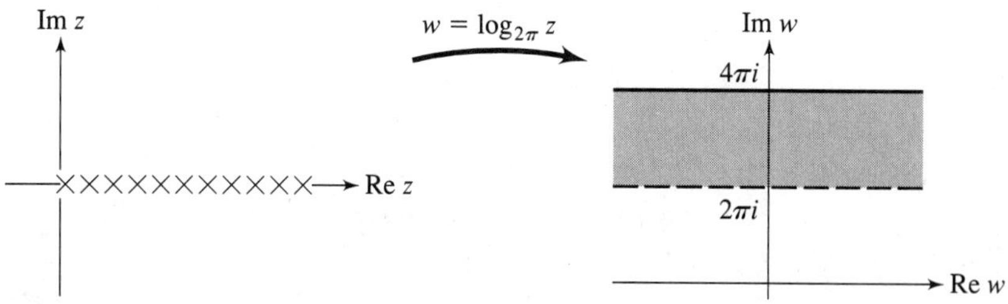

Figure 3.22

Defining branches of $\log z$ is reminiscent of the derivation of the inverse sine function in real-variable work. The situations are very similar. Because $\sin x$ is 2π-periodic, defining $y = \sin^{-1} x$ when $x = \sin y$ results in many values for $\sin^{-1} x$. Principal values $-\pi/2 \le \mathrm{Sin}^{-1} x \le \pi/2$ are chosen in order to create a single-valued function. The difference in the two situations is that we always use principal values of $\mathrm{Sin}^{-1} x$; other choices are possible, but they are never used. In order that $\log z$ be single-valued, we restrict $\arg z$ to an interval of length 2π. This leads to branches of $\log z$ — the principal branch $\mathrm{Log}\, z$ as well as other branches. Because we have taken the effort to discuss other branches of $\log z$ and have even given some of them special notations ($\log_\phi z$), the implication is that they are important. This is indeed true. Different problems require different branches of $\log z$.

Exercises 3.5

In Exercises 1–4 express the complex number(s) in Cartesian form.

1. $\log i$
2. $\mathrm{Log}(2 - 6i)$
3. $\log_{\pi/2}(1 + i)$
4. $\log_{-3}(-2 + 3i)$

In Exercises 5–8 find all solutions of the equation.

5. $\mathrm{Log}\, z = \pi i/2$
6. $\log_1 z = 2 + 3i$
7. $z^4 = i$
8. $e^{z+2} = 4$

Use the complex logarithm function to find all solutions of the equations in Exercises 9–11.

9. Exercises 15–22 in Section 3.2
10. Exercises 13–15 in Section 3.3
11. Exercises 9–11 in Section 3.4
12. You may have noticed that we have not stated the complex analogue of $\ln(x^a) = a \ln x$. The reason for this is that, in general,

$$\log(z^a) \ne a \log z.$$

Illustrate this with $z = 1 + i$ and $a = 4$.

13. Is $\log_\phi z_1 z_2 = \log_\phi z_1 + \log_\phi z_2$?
14. Find all zeros of (a) $\operatorname{Log} z$ (b) $\log_{2\pi} z$.
15. Find $f'(3+i)$ if $f(z) = \operatorname{Log}(2z + 3 - i)$.
16. Find $f''(2i)$ if $f(z) = \log_{3\pi/8}(-z + i)$.
17. Show that, in appropriate domains, the complex electrostatic potential $f(z)$ in Exercise 2 of Section 2.7 can be expressed in the form

$$f(z) = \frac{-q}{2\pi\epsilon_0} \log_\phi z + (C + Di).$$

18. **(a)** Use Exercise 17 to find the electrostatic potential due to a line of charge q coulombs per meter perpendicular to the xy-plane at (x_0, y_0), and a line of charge $-q$ coulombs per meter perpendicular to the xy-plane at (x_1, y_1).
    (b) Find equipotentials and lines of force when $y_0 = y_1 = 0$ and $x_0 = -x_1 = a > 0$.
19. When $f(z) = k \log_\phi z$ is a complex potential for fluid flow (Section 2.8), the flow is said to have a line source at the origin of strength k if $k > 0$ and a line sink if $k < 0$.
    (a) Show that flow is radially outward for $k > 0$ and radially inward for $k < 0$. Find the speed of the flow at any point.
    (b) Calculate flux and circulation when C is a closed curve not containing the origin.
    (c) Calculate flux and circulation when C is a circle centered at the origin.
20. When $f(z) = -ki \log_\phi z$ is a complex potential for fluid flow (Section 2.8), the flow is said to have a line vortex at the origin.
    (a) Show that stream curves are circles centered at the origin. Find the speed of the flow at any point.
    (b) Calculate flux and circulation when C is a closed curve not containing the origin.
    (c) Calculate flux and circulation when C is a circle centered at the origin.
21. Show that stream curves for fluid flow described by the complex potential

$$f(z) = k[\log_\phi(z + a) - \log_\phi(z - a)], \qquad a > 0, \quad k > 0,$$

are circles centered on the y-axis that pass through $z = \pm a$. What is the velocity and speed of the flow?

22. Is $\operatorname{Log} e^z = z$?
23. Prove that under the mapping $w = \operatorname{Log} z$
    (a) Circles $|z| =$ constant are mapped onto vertical line segments.
    (b) Rays $\operatorname{Arg} z =$ constant are mapped onto horizontal half-lines.
    (c) The domain $R_0 < |z| < R_1$, $\Theta_0 < \operatorname{Arg} z < \Theta_1$ is mapped onto a rectangle.
    (d) A sector $\Theta_0 \le \operatorname{Arg} z \le \Theta_1$ is mapped onto a strip.

The function $\operatorname{Log}[f(z)]$ has branch points where $f(z) = 0$ and branch cuts where $f(z)$ is real and negative. In Exercises 24–27 identify branch points and branch cuts for the function.

24. $\operatorname{Log}(z + i)$
25. $\operatorname{Log}(z - 1)$
26. $\operatorname{Log}(3 - 2z)$
27. $\operatorname{Log}(3z - 2 + 4i)$

28. **(a)** What are the branch points for the function $f(z) = \text{Log}(z^2 - 4)$?

(b) Show that the imaginary axis and that part of the real axis between the branch points are branch cuts.

29. Find branch points and branch cuts for $f(z) = \text{Log}(z^2 + 1)$.

The function $\log_\phi[f(z)]$ has branch points where $f(z) = 0$ and branch cuts where one of the arguments of $f(z)$ is equal to ϕ. In Exercises 30–33 identify branch points and branch cuts for the function.

30. $\log_0(z - 2i)$

31. $\log_{\pi/2}(3 - z)$

32. $\log_{-\pi/2}(4 - 2z)$

33. $\log_2(2z + 1)$

34. **(a)** What are the branch points for the function $f(z) = \log_0(z^2 + 1)$?

(b) Show that the real axis and the line segment joining the branch points are branch cuts.

35. Find branch points and branch cuts for $f(z) = \log_{3\pi/2}(1 - z^2)$.

36. In this exercise we show that logarithm functions cannot always be combined in what might be thought to be natural ways.

(a) Show that in general

$$\log_\phi(z^2 - 1) \neq \log_\phi(z + 1) + \log_\phi(z - 1).$$

Illustrate with $\phi = -\pi$ and $z = -1 + i$.

(b) Show that in general

$$\log_\phi\left(\frac{z+1}{z-1}\right) \neq \log_\phi(z + 1) - \log_\phi(z - 1).$$

Illustrate with $\phi = 0$ and $z = -1 + i$.

37. Compare branch cuts for $f(z) = \text{Log}(z^2 - 1)$ and $g(z) = \text{Log}(z + 1) + \text{Log}(z - 1)$.

38. Compare branch cuts for $f(z) = \text{Log}(4 - z^2)$ and $g(z) = \text{Log}(2 + z) + \text{Log}(2 - z)$.

39. Compare branch cuts for $f(z) = \log_{-\pi/2}(z^2 - 1)$ and $g(z) = \log_{-\pi/2}(z + 1) + \log_{-\pi/2}(z - 1)$.

40. Compare branch cuts for $f(z) = \log_{-\pi/2}(4 - z^2)$ and $g(z) = \log_{-\pi/2}(2 + z) + \log_{-\pi/2}(2 - z)$.

41. **(a)** Prove that for $b > a > 1$, the function $\text{Log}\left(\frac{z-a}{z-b}\right)$ is analytic except for points on the $\text{Re}\, z$ axis for which $a \leq \text{Re}\, z \leq b$.

(b) Verify that for $|z| > b$,

$$\text{Log}\left(\frac{z-a}{z-b}\right) = \text{Log}\left(1 - \frac{a}{z}\right) - \text{Log}\left(1 - \frac{b}{z}\right).$$

3.6 Complex Powers

The meaning of z^n when n is an integer is clear. When n is a rational number, say $n = \frac{2}{3}$, we also have no conceptual difficulty. We write $z^{2/3} = (z^{1/3})^2$, and $z^{1/3}$ is interpreted as the cube root of z. There are three cube roots, which can be obtained by methods discussed in Section 1.4, and these should be squared to give values for $z^{2/3}$. Note that we just said values, not value; that is, we are saying that $z^{2/3}$ is not single-valued. It probably has three values for given z. In this section, we clarify the situation; we define z^α for complex α. When α is a real integer, z^α is single-valued (as is to be expected); when α is not an integer, z^α has many values — sometimes infinitely many, sometimes finitely many.

DeMoivre's theorem 3.8(c) implies that for any positive integer n,

$$z^n = r^n e^{n\theta i} = e^{n \ln r} e^{n\theta i} = e^{n(\ln r + \theta i)} = e^{n \log z}. \tag{3.30}$$

We use this identity to define the function z^α for complex α and $z \neq 0$:

$$z^\alpha = e^{\alpha \log z}. \tag{3.31}$$

When it makes sense to do so, we define $0^\alpha = 0$. For example, this would certainly be done when α is a (real) positive integer, but it would be pointless to do so when α is a negative integer.

Because $\log z$ is a multiple-valued function, it follows that so also, in general, is z^α. We shall show, however, that for certain values of α, z^α is single-valued, double-valued, etc.

EXAMPLE 3.16 ....

Find all possible values for **(a)** $(2i)^i$ and **(b)** $(1-i)^{1/2+i}$.

Solution **(a)** According to 3.31,

$$(2i)^i = e^{i \log(2i)} = e^{i[\ln 2 + (\pi/2 + 2k\pi)i]} = e^{-(\pi/2+2k\pi)} e^{i \ln 2}$$

where k is an integer. These complex numbers, which vary in modulus $(e^{-(\pi/2+2k\pi)})$ but have the same argument $(\ln 2)$, are shown in Figure 3.23. What we are saying is that the function z^i has infinitely many values for each value of z.

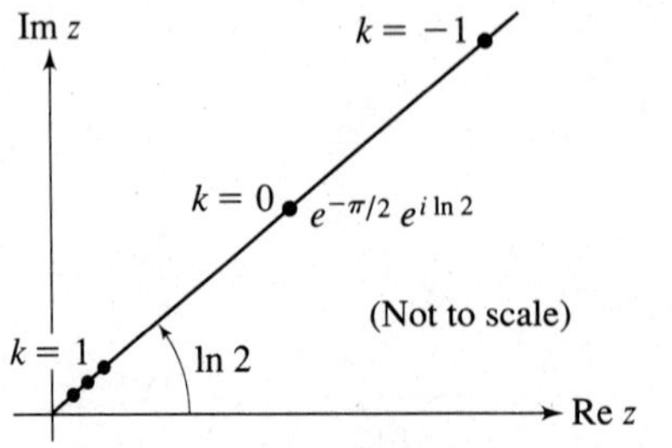

Figure 3.23

(b) Because

$$(1-i)^{1/2+i} = e^{(1/2+i)\log(1-i)} = e^{(1/2+i)[\ln\sqrt{2}+(-\pi/4+2k\pi)i]}$$
$$= e^{[(1/2)\ln\sqrt{2}+\pi/4-2k\pi]+(\ln\sqrt{2}-\pi/8+k\pi)i},$$

the function $z^{1/2+i}$ also has infinitely many values for each value of z.

EXAMPLE 3.17

When n is an integer, we certainly hope that z^n is single-valued. Verify this.

Solution According to equation 3.31,

$$\begin{aligned} z^n = e^{n\log z} &= e^{n(\ln|z|+i\arg z)} \\ &= e^{n\ln|z|}e^{in(\operatorname{Arg} z+2k\pi)} \qquad (k \text{ an integer}) \\ &= |z|^n e^{in\operatorname{Arg} z}e^{2kn\pi i} \\ &= |z|^n e^{in\operatorname{Arg} z} \qquad (e^{2kn\pi i}=1), \end{aligned}$$

and z^n is single-valued.

Definition 3.31 for raising complex numbers to complex powers is inconsistent with the exponential function. The exponential function e^z has a unique value for any given z, namely, $e^z = e^x(\cos y + \sin y\, i)$. On the other hand, Definition 3.31 leads, in general, to multiple values of e^z. To avoid this confusion, many authors use the notation $\exp(z)$ for the exponential function. Because we never use e^z in the sense of 3.31, no confusion can arise, and e^z always denotes the single-valued exponential function.

The Square Root Function

According to equation 3.31, the definition of the function $z^{1/2}$ is

$$z^{1/2} = e^{(1/2)\log z} = e^{(1/2)(\ln|z|+i\arg z)} = e^{(1/2)\ln|z|}e^{(i/2)\arg z} = \sqrt{|z|}\,e^{(i/2)\arg z}. \tag{3.32}$$

Because $\arg z = \operatorname{Arg} z + 2k\pi$, it follows that

$$z^{1/2} = \sqrt{|z|}\,e^{(i/2)(\operatorname{Arg} z+2k\pi)} = \pm\sqrt{|z|}\,e^{(i/2)\operatorname{Arg} z}, \tag{3.33}$$

because $e^{k\pi i} = \pm 1$. Thus, every nonzero complex number has two square roots, and $z^{1/2}$ is a double-valued function. When $z=0$, we define $0^{1/2}=0$, in which case $z^{1/2}$ has only one value, at $z=0$.

EXAMPLE 3.18 ....

Find, in Cartesian form, the square roots of $3 - 4i$.

Solution According to 3.33,

$$\begin{aligned}(3-4i)^{1/2} &= \pm\sqrt{|3-4i|}\,e^{(i/2)\,\mathrm{Arg}(3-4i)} = \pm\sqrt{5}\,e^{(i/2)\,\mathrm{Tan}^{-1}(-4/3)} \\ &= \pm\sqrt{5}\,\{\cos\,[(\tfrac{1}{2})\,\mathrm{Tan}^{-1}\,(-\tfrac{4}{3})] + \sin[(\tfrac{1}{2})\,\mathrm{Tan}^{-1}(-\tfrac{4}{3})]\,i\} \\ &= \pm(2-i).\end{aligned}$$

....

To create an analytic function from $z^{1/2}$, we first make it single-valued. This can be done by restricting $\arg z$ in 3.32 to any interval of length 2π — i.e., by choosing a branch of $\log z$. When principal values of $\arg z$ are chosen, the resulting single-valued function, denoted by

$$\sqrt{z} = \sqrt{|z|}\,e^{(i/2)\,\mathrm{Arg}\,z}, \tag{3.34}$$

is called the **principal square root function**. This function maps the complex z-plane onto the right half-plane shown in Figure 3.24; that is, it finds that square root of a complex number that has a nonnegative real part. In Section 1.2 we made the agreement that the square root sign over a complex number ($\sqrt{1+i}$, for example) would always denote a number with positive real part. We now see the reason for this; it creates a continuous single-valued function $\sqrt{z}$. This function is not only continuous, it is analytic in the domain consisting of all points in the complex plane with $z = 0$ and the negative real axis deleted. In other words, $z = 0$ and points on the negative real axis are singularities of $\sqrt{z}$.

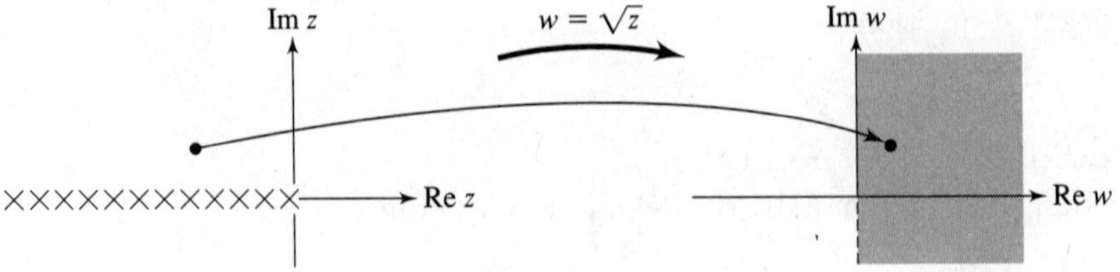

Figure 3.24

The derivative of the principal square root function is

$$\frac{d}{dz}\sqrt{z} = \frac{d}{dz}e^{(1/2)\,\mathrm{Log}\,z} = e^{(1/2)\,\mathrm{Log}\,z}\,\frac{1}{2z} = \frac{\sqrt{z}}{2z} = \frac{1}{2\sqrt{z}}. \tag{3.35}$$

Other branches of $z^{1/2}$ are obtained by restricting $\arg z$ to other intervals of length

2π, such as

$$z^{1/2} = \sqrt{|z|}e^{(i/2)\arg z}, \qquad \phi < \arg z \leq \phi + 2\pi.$$

For this branch, the branch cut is the line $\arg z = \phi$ (Figure 3.25).

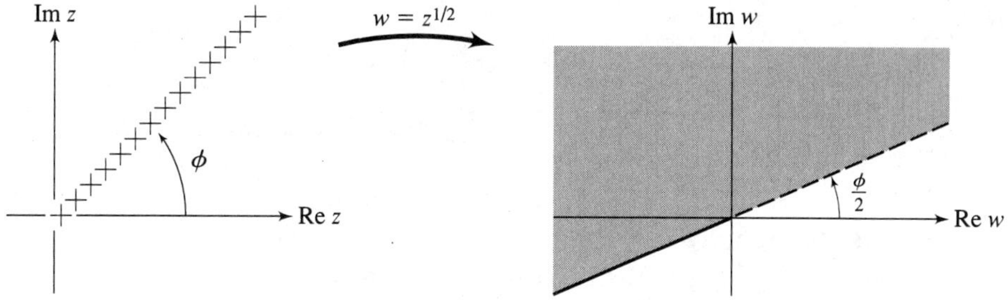

Figure 3.25

The derivative of this square root function is

$$\frac{d}{dz}z^{1/2} = \frac{1}{2z^{1/2}}, \tag{3.36}$$

where $z^{1/2}$ is the same branch on each side of the equation.

When x_1 and x_2 are nonnegative real numbers, $\sqrt{x_1 x_2} = \sqrt{x_1}\sqrt{x_2}$, and this result is valid whether we interpret the $\sqrt{\ }$ as the real square root function or the principal branch of the complex root function. Unfortunately, the principal branch of the complex root function does not satisfy this property in general; that is, in general

$$\sqrt{z_1 z_2} \neq \sqrt{z_1}\sqrt{z_2}.$$

For example, when $z_1 = z_2 = -1$,

$$\sqrt{(-1)(-1)} = \sqrt{1} = 1 \quad \text{whereas} \quad \sqrt{-1}\sqrt{-1} = i \cdot i = -1.$$

In other words, we must be careful when performing algebraic manipulations with complex root functions not to assume properties of their real counterparts. For another easy trap, see Exercise 17(a).

EXAMPLE 3.19 ....

Does $\sqrt{z}$ have any zeros?

Solution The only value of z for which $\sqrt{z} = 0$ is $z = 0$. But $z = 0$ is not a zero, because $\sqrt{z}$ is not analytic at $z = 0$. It is a singularity (but not an isolated singularity). ....

EXAMPLE 3.20 ....

Evaluate $f'(-i)$ if $f(z) = -\sqrt{z}$.

Solution According to 3.35, $f'(z) = -1/(2\sqrt{z})$, so that $f'(-i) = -1/(2\sqrt{-i})$. Because

$$\sqrt{-i} = \sqrt{|-i|}\, e^{(i/2)\,\mathrm{Arg}(-i)} = e^{(i/2)(-\pi/2)} = e^{-\pi i/4},$$

it follows that

$$f'(-i) = -\frac{1}{2e^{-\pi i/4}} = -\frac{1}{2}e^{\pi i/4} = -\frac{1}{2}\left(\frac{1}{\sqrt{2}} + \frac{i}{\sqrt{2}}\right) = -\frac{\sqrt{2}}{4}(1+i).$$

The nth Root Function

When n is a positive integer, the function $z^{1/n}$ is defined by

$$z^{1/n} = e^{(1/n)\log z} = e^{(1/n)[\ln|z| + i\arg z]} = \sqrt[n]{|z|}\, e^{(i/n)\arg z}, \tag{3.37}$$

provided $z \neq 0$. We define $0^{1/n} = 0$. If we set $\arg z = \mathrm{Arg}\, z + 2k\pi$, where k is an integer, then

$$z^{1/n} = \sqrt[n]{|z|}\, e^{(i/n)(\mathrm{Arg}\, z + 2k\pi)} = \sqrt[n]{|z|}\, e^{(i/n)\,\mathrm{Arg}\, z} e^{2k\pi i/n}. \tag{3.38}$$

For given z, the quantity $\sqrt[n]{|z|}\, e^{(i/n)\,\mathrm{Arg}\, z}$ is unique, but as k takes on the values $0, 1, \ldots, n-1$, $e^{2k\pi i/n}$ has n distinct values. In other words, $z^{1/n}$ is an n-valued function; there are n nth roots of a nonzero complex number. For example, when $n = 4$ and $z = -16$,

$$(-16)^{1/4} = \sqrt[4]{16}\, e^{(i/4)(\pi)} e^{2k\pi i/4} = 2e^{(2k+1)\pi i/4}.$$

For $k = 0, 1, 2, 3$, we obtain the four complex numbers

$$2e^{\pi i/4} = \sqrt{2}\,(1+i), \qquad 2e^{3\pi i/4} = \sqrt{2}\,(-1+i),$$
$$2e^{5\pi i/4} = \sqrt{2}\,(-1-i), \qquad 2e^{7\pi i/4} = \sqrt{2}\,(1-i).$$

Thus, the four fourth roots of -16 are $\sqrt{2}\,(1 \pm i)$ and $\sqrt{2}\,(-1 \pm i)$. We have shown these roots in Figure 3.26. They lie on a circle of radius 2 (the fourth root of the modulus of -16). The first $\sqrt{2}\,(1+i)$ (from $k = 0$) has an argument equal to one-quarter that of -16, and all others are equally spaced around the circle (separated by angles of $\pi/2$).

In general, $z^{1/n}$, for given n and z, defines n complex numbers, which lie on a circle of radius $|z|^{1/n}$. One has argument equal to $(\mathrm{Arg}\, z)/n$, and the $n-1$ remaining roots are equally spaced around the circle, each pair separated by an angle of $2\pi/n$ radians (Figure 3.27).

In Section 1.4 we used polar representations of complex numbers to find square roots, cube roots, etc. Functions $z^{1/2}$, $z^{1/3}$, etc., perform the same task. The only difference is that we now have a functional representation for the roots, as opposed to simply a procedure by which to find them. In addition, the exponential representation in 3.37 and 3.38 is vastly superior to the polar representation in Section 1.4.

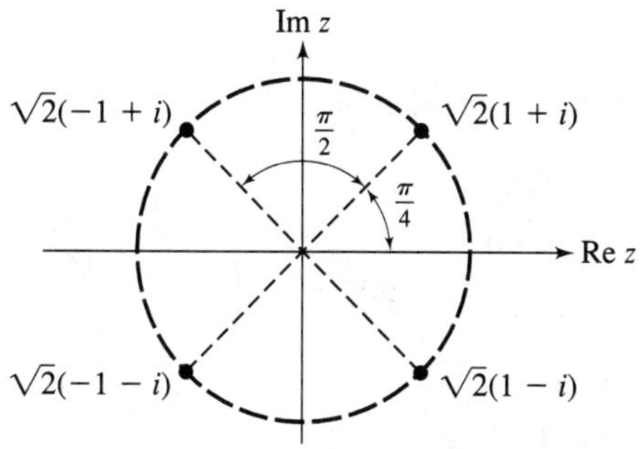

Figure 3.26

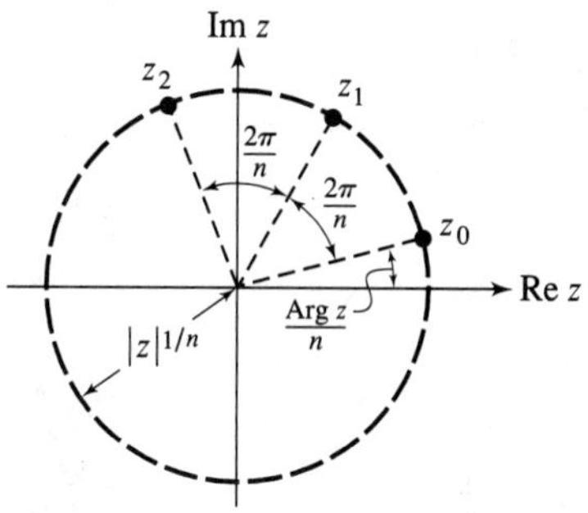

Figure 3.27

By choosing a branch of the logarithm function, $z^{1/n}$ becomes a single-valued function. In particular, when the principal branch $\operatorname{Log} z$ is chosen, we denote the resulting branch of $z^{1/n}$ by

$$\sqrt[n]{z} = \sqrt[n]{|z|}e^{(i/n)\operatorname{Arg} z}, \tag{3.39}$$

and call it the **principal branch** of $z^{1/n}$. It is defined for all z, and it is analytic in the domain $|z| > 0$, $-\pi < \operatorname{Arg} z < \pi$, with derivative

$$\frac{d}{dz}\sqrt[n]{z} = \frac{d}{dz}e^{(1/n)\operatorname{Log} z} = e^{(1/n)\operatorname{Log} z}\left(\frac{1}{nz}\right) = \frac{\sqrt[n]{z}}{n\left(\sqrt[n]{z}\right)^n} = \frac{1}{n\left(\sqrt[n]{z}\right)^{n-1}}. \tag{3.40}$$

In fact, no matter what branch of $z^{1/n}$ is chosen, we may write that

$$\frac{d}{dz}z^{1/n} = \frac{1}{n}z^{1/n-1}. \tag{3.41}$$

The only proviso is that branches of $z^{1/n}$ and $z^{1/n-1}$ must correspond; that is, both must use the same branch of $\log z$.

EXAMPLE 3.21 ....

Find the derivative of $\sqrt[3]{z}$ at $z = -1 - i$.

Solution According to 3.40, $\dfrac{d}{dz}\sqrt[3]{z} = \dfrac{1}{3\left(\sqrt[3]{z}\right)^2}$. We use 3.39 to evaluate $\left(\sqrt[3]{z}\right)^2$ at $-1 - i$,

$$\left(\sqrt[3]{-1-i}\right)^2 = \left[(\sqrt{2})^{1/3}e^{(i/3)(-3\pi/4)}\right]^2 = 2^{1/3}e^{-\pi i/2} = -2^{1/3}i.$$

Thus, $\dfrac{d}{dz}\left(\sqrt[3]{z}\right)_{|z=-1-i} = \dfrac{1}{3(-2^{1/3}i)} = \dfrac{i}{3\cdot 2^{1/3}}$. ....

Exercises 3.6

In Exercises 1–10 express the complex number(s) in Cartesian form.

1. 2^i
2. $(2i)^{3i}$
3. $\sqrt{1+i}$
4. $(1+i)^{1/2}$
5. $\sqrt[3]{8i}$
6. $(8i)^{1/3}$
7. $(\sqrt[4]{i})^3$
8. $i^{3/4}$
9. $(1-i)^{1+i}$
10. $(1+i)^{1-i}$
11. Find the derivative of $\sqrt{z}$ at $z = i$.
12. Find the derivative of $\sqrt[3]{z}$ at $z = 1 - i$.
13. Is there a difference between $(\sqrt{z})^3$ and $\sqrt{z^3}$?
14. Is there a difference between $(z^{1/2})^3$ and $(z^3)^{1/2}$?
15. Find the second derivative of $(\sqrt{z})^3$ at $z = -2 - 3i$.
16. Evaluate the derivative of $\sqrt{z-i}$ at $z = 2$.
17. **(a)** When is $\sqrt{z^2} = z$?
    (b) Is $(z^2)^{1/2} = z$?
18. Is $(e^{z_1})^{z_2} = e^{z_1 z_2}$?
19. Is $z = 0$ a zero of $\sqrt[n]{z}$?
20. What is $f(1)$ if $f(z) = z^{\sqrt{2}}$?
21. Explain why z^n, which is defined as $e^{n \log z}$, does not have a branch cut when n is a positive integer. Is it an entire function?
22. **(a)** It is a property of reals that $(a^b)^c = a^{bc}$. This property does not in general hold for complex numbers. Illustrate this by calculating $(1^2)^i$ and 1^{2i} using 3.31.
    (b) Show that when n is an integer, $(z^\alpha)^n = z^{n\alpha}$.

The function $\sqrt{f(z)}$ has branch points where $f(z) = 0$ and branch cuts where $f(z)$ is real and negative. In Exercises 23–31 identify branch points and branch cuts for the function.

23. $\sqrt{z-1}$
24. $\sqrt{2-3z}$
25. $\sqrt{2z+i}$
26. $\sqrt{3z-2+3i}$
27. $\sqrt{z^2-4}$
28. $\sqrt{4-z^2}$
29. $\sqrt{z^2+4}$
30. $\sqrt{z^2+4z+5}$
31. $\sqrt{z^4-16}$

The function $[f(z)]^{1/2} = e^{(1/2)\log_\phi[f(z)]}$ has branch points where $f(z) = 0$ and branch cuts where one of the arguments of $f(z)$ is equal to ϕ. In Exercises 32–42 identify branch points and branch cuts for the function.

32. $(z+3)^{1/2} = e^{(1/2)\log_0(z+3)}$
33. $(2i-z)^{1/2} = e^{(1/2)\log_0(2i-z)}$
34. $(3z+2)^{1/2} = e^{(1/2)\log_{-\pi/2}(3z+2)}$
35. $(z-3i)^{1/2} = e^{(1/2)\log_{\pi/2}(z-3i)}$

36. $(2-i-z)^{1/2} = e^{(1/2)\log_{2\pi}(2-i-z)}$

37. $(z^2-1)^{1/2} = e^{(1/2)\log_0(z^2-1)}$

38. $(z^2+4)^{1/2} = e^{(1/2)\log_0(z^2+4)}$

39. $(1-z^2)^{1/2} = e^{(1/2)\log_{\pi/2}(1-z^2)}$

40. $(z^2+9)^{1/2} = e^{(1/2)\log_{-\pi/2}(z^2+9)}$

41. $(z^2+2z-3)^{1/2} = e^{(1/2)\log_{3\pi/2}(z^2+2z-3)}$

42. $(z^2+4z+5)^{1/2} = e^{(1/2)\log_0(z^2+4z+5)}$

43. In this exercise we illustrate that care must be taken in combining square root functions.

(a) Show that, in general,

$$\sqrt{z^2-a^2} \neq \sqrt{z+a}\sqrt{z-a}, \qquad a>0 \text{ a constant.}$$

Illustrate with $z=-2a$.

(b) Show that branch cuts of $\sqrt{z^2-a^2}$ are the imaginary axis and the real axis between $z=\pm a$.

(c) Show that the branch cut of $\sqrt{z+a}\sqrt{z-a}$ is the real axis between $z=\pm a$.

44. What are branch cuts for the following functions, where $a>0$ is a constant?

(a) $\sqrt{z^2+a^2}$

(b) $\sqrt{z+ai}\sqrt{z-ai}$

(c) $(z^2+a^2)^{1/2} = e^{(1/2)\log_{-\pi/2}(z^2+a^2)}$

(d) $(z+ai)^{1/2}(z-ai)^{1/2} = e^{(1/2)\log_{-\pi/2}(z+ai)}e^{(1/2)\log_{-\pi/2}(z-ai)}$

45. Show that branches of the functions $(z-z_0)^{1/2}$ and $(z-z_1)^{1/2}$ can be chosen so that $(z-z_0)^{1/2}(z-z_1)^{1/2}$ is discontinuous only along the line segment joining z_0 and z_1.

46. Consider $f(z)=z^{\pi/\alpha}$ (principal branch), where $\pi/2<\alpha<\pi$, as the complex potential for fluid flow in the domain $0<\operatorname{Arg} z<\alpha$.

(a) Draw the stream curve $\psi(x,y)=0$.

(b) What is the fluid velocity at every point in the flow? Are there any stagnation points?

(c) Are the components of the velocity positive or negative for $0<\operatorname{Arg} z<\alpha/2$?

(d) Are the components of the velocity positive or negative for $\alpha/2<\operatorname{Arg} z<\alpha$?

(e) What is the velocity when $\operatorname{Arg} z=\alpha/2$? Is it perpendicular to the line $\operatorname{Arg} z = \alpha/2$?

47. Repeat Exercise 46 if $0<\alpha<\pi/2$.

48. Consider $f(z)=e^{-i\pi\alpha/(\pi-\alpha)}z^{\pi/(\pi-\alpha)}$ (principal branch), where $0<\alpha<\pi/2$, as the complex potential for fluid flow in the domain $\alpha<|\operatorname{Arg} z|\leq\pi$.

(a) Draw the stream curve $\psi(x,y)=0$. Is it consistent with flow past a wedge of angle 2α symmetrically positioned around the positive real z-axis at $z=0$?

(b) What is the velocity of the flow? Show that the speed of the flow is the same for all points equidistant from the vertex of the wedge. Is there a stagnation point?

49. Prove that z^α has finitely many values only when α is a rational number.

3.7 Inverse Trigonometric and Hyperbolic Functions

Because the exponential function e^z is $2\pi i$-periodic, defining the logarithm function as its inverse led to a multiple-valued function $\log z$. Branches of $\log z$ evolved as analytic parts of this function. Because the trigonometric and hyperbolic functions are periodic, it should come as no surprise that similar situations will arise when we define their inverses. Let us proceed.

We call w the inverse sine of z if $z = \sin w$. We find w by solving

$$z = \sin w = \frac{e^{wi} - e^{-wi}}{2i}$$

for w in terms of z. Multiplication by e^{wi} gives a quadratic equation in e^{wi}:

$$\left(e^{wi}\right)^2 - 2iz(e^{wi}) - 1 = 0.$$

Application of quadratic formula 1.13(b) yields

$$e^{wi} = \frac{2iz + (-4z^2 + 4)^{1/2}}{2} = iz + (1 - z^2)^{1/2},$$

where $(1 - z^2)^{1/2}$ has two values. When we take logarithms of both sides, the result is

$$w = -i \log[iz + (1 - z^2)^{1/2}].$$

This function is called the **inverse sine function** and is denoted by $\sin^{-1} z$:

$$\sin^{-1} z = -i \log[iz + (1 - z^2)^{1/2}]. \qquad \textbf{(3.42)}$$

Because the complex logarithm function is multiple-valued and $(1-z^2)^{1/2}$ is double-valued, $\sin^{-1} z$ is multiple-valued; that is, for given z, there are infinitely many values for $\sin^{-1} z$. This fact is illustrated in the following example.

EXAMPLE 3.22 ....

Find the values of $\sin^{-1} i$.

Solution By Definition 3.42,

$$\begin{aligned}\sin^{-1} i = -i \log[i(i) + (1 - i^2)^{1/2}] &= -i \log[\pm\sqrt{2} - 1] \\ &= -i \begin{cases} \ln(\sqrt{2} - 1) + 2n\pi i \\ \ln(\sqrt{2} + 1) + (2n + 1)\pi i \end{cases} \\ &= \begin{cases} 2n\pi + i \ln(\sqrt{2} + 1) \\ (2n + 1)\pi - i \ln(\sqrt{2} + 1). \end{cases}\end{aligned}$$

....

The inverse sine function can be made single-valued by choosing a particular branch of the logarithm function, and we use this branch both explicitly in 3.42 and implicitly in $(1-z^2)^{1/2}$. When the principal branch $\operatorname{Log} z$ is chosen, the resulting single-valued function is called the **principal branch** of $\sin^{-1} z$, denoted by

$$\operatorname{Sin}^{-1} z = -i \operatorname{Log}[iz + \sqrt{1-z^2}]. \tag{3.43}$$

In Exercise 20 you are asked to show that

$$\sin^{-1} z = \begin{cases} \operatorname{Sin}^{-1} z + 2n\pi \\ (2n+1)\pi - \operatorname{Sin}^{-1} z. \end{cases} \tag{3.44}$$

In other words, all complex numbers $\sin^{-1} z$ whose sines are equal to given z can be obtained from the principal value $\operatorname{Sin}^{-1} z$. (This is precisely the situation for the inverse sine function in real-variable work.)

EXAMPLE 3.23

Use 3.44 to solve Example 3.22.

Solution Because

$$\operatorname{Sin}^{-1} i = -i \operatorname{Log}[i(i) + \sqrt{1-i^2}] = -i \operatorname{Log}[\sqrt{2}-1] = -i \ln(\sqrt{2}-1) = i \ln(\sqrt{2}+1),$$

it follows from 3.44 that

$$\sin^{-1} i = \begin{cases} 2n\pi + i \ln(\sqrt{2}+1) \\ (2n+1)\pi - i \ln(\sqrt{2}+1). \end{cases}$$

EXAMPLE 3.24

Does the value of $\operatorname{Sin}^{-1} z$ coincide with $\operatorname{Sin}^{-1} x$ when z is a real number between -1 and 1?

Solution When z is a real number, say x, and $-1 \le x \le 1$,

$$\begin{aligned} \operatorname{Sin}^{-1} z &= -i \operatorname{Log}[iz + \sqrt{1-z^2}] \\ &= -i \operatorname{Log}[ix + \sqrt{1-x^2}] \\ &= -i[\ln \sqrt{1-x^2+x^2} + i \operatorname{Sin}^{-1} x] \qquad \text{(see Figure 3.28)} \\ &= \operatorname{Sin}^{-1} x. \end{aligned}$$

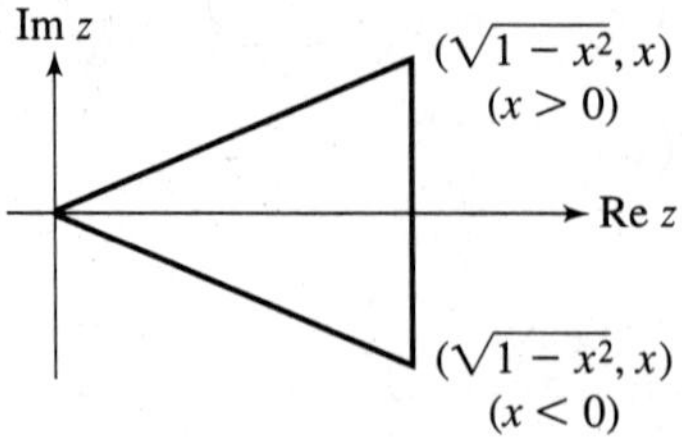

Figure 3.28

This example illustrates that our definition of the inverse sine function adheres to the principle stated in the first paragraph of this chapter — when a function $f(x)$ is extended to a complex function $f(z)$, the values of $f(z)$ for real z should coincide with those of $f(x)$.

Similar derivations lead to the inverse cosine and tangent functions:

$$\cos^{-1} z = -i \log[z + i(1-z^2)^{1/2}]. \tag{3.45}$$

$$\tan^{-1} z = \frac{i}{2} \log\left(\frac{i+z}{i-z}\right). \tag{3.46}$$

Derivatives of the inverse trigonometric functions can be obtained by direct differentiation of 3.42, 3.45, and 3.46 or by implicit differentiation:

$$\frac{d}{dz} \sin^{-1} z = \frac{1}{(1-z^2)^{1/2}}, \tag{3.47}$$

$$\frac{d}{dz} \cos^{-1} z = \frac{-1}{(1-z^2)^{1/2}}, \tag{3.48}$$

$$\frac{d}{dz} \tan^{-1} z = \frac{1}{1+z^2}. \tag{3.49}$$

These equations are not valid in the strictest sense, because it is not possible to differentiate multiple-valued functions. What we mean is that given a branch for the inverse trigonometric function on the left, its derivative is the corresponding branch of the function on the right; that is, the same branch of $\log z$ must be chosen on the right and left sides of these equations.

The inverse hyperbolic functions are obtained in an analogous way to the inverse trigonometric functions. The inverse hyperbolic sine, cosine, and tangent are

$$\sinh^{-1} z = \log[z + (z^2+1)^{1/2}], \tag{3.50}$$

$$\cosh^{-1} z = \log[z + (z^2-1)^{1/2}], \tag{3.51}$$

$$\tanh^{-1} z = \frac{1}{2} \log\left(\frac{1+z}{1-z}\right) \tag{3.52}$$

Their derivatives are

$$\frac{d}{dz} \sinh^{-1} z = \frac{1}{(z^2+1)^{1/2}}, \tag{3.53}$$

$$\frac{d}{dz}\cosh^{-1} z = \frac{1}{(z^2-1)^{1/2}}, \tag{3.54}$$

$$\frac{d}{dz}\tanh^{-1} z = \frac{1}{1-z^2}. \tag{3.55}$$

EXAMPLE 3.25 ....

Evaluate $\cosh^{-1}(1/2)$.

Solution From 3.51,

$$\cosh^{-1}(\tfrac{1}{2}) = \log[\tfrac{1}{2} + (\tfrac{1}{4}-1)^{1/2}] = \log[\tfrac{1}{2} \pm \sqrt{3}i/2] = \ln(1) + i(2n\pi \pm \pi/3)$$
$$= (6n \pm 1)\pi i/3.$$

....

EXAMPLE 3.26 ....

What is the derivative of $\operatorname{Sinh}^{-1} z$ at $z = 1+i$?

Solution The capital letter S indicates the principal branch of the inverse hyperbolic sine function, in which case

$$\frac{d}{dz}\operatorname{Sinh}^{-1} z_{|1+i} = \frac{1}{\sqrt{z^2+1}}_{|1+i} = \frac{1}{\sqrt{(1+i)^2+1}} = \frac{1}{\sqrt{1+2i}}.$$

Because $|1+2i| = \sqrt{5}$ and $\operatorname{Arg}(1+2i) = \operatorname{Tan}^{-1} 2$,

$$\frac{d}{dz}\operatorname{Sinh}^{-1} z_{|1+i} = \frac{1}{\sqrt[4]{5}} e^{-(i/2)\operatorname{Tan}^{-1} 2} \approx 0.569 - 0.352i.$$

....

Exercises 3.7

In Exercises 1–10 express the complex number(s) in Cartesian form.

1. $\sin^{-1}(\sqrt{2})$
2. $\tan^{-1}(1-i)$
3. $\cos^{-1}(4i)$
4. $\sinh^{-1} 2$
5. $\tanh^{-1}(3+6i)$
6. $\cosh^{-1}(\sqrt{2}i)$
7. $(2+i)\sin^{-1} i$
8. $[\tanh^{-1} i]^2$
9. $\tan^{-1}(i^{1/2})$
10. $\sin^{-1} 1 + \cos^{-1} 1$

In Exercise 11–16 find the derivative of the function.

11. $f(z) = \operatorname{Cosh}^{-1}[(2+i)z]$
12. $f(z) = \operatorname{Log}[\operatorname{Sin}^{-1}(3z)]$
13. $f(z) = \operatorname{Tan}^{-1}\sqrt{z}$
14. $f(z) = e^{\operatorname{Sinh}^{-1}(3iz+6)}$
15. $f(z) = \tan[\operatorname{Cosh}^{-1}(z^2)]$
16. $f(z) = \{\operatorname{Cos}^{-1}[\operatorname{Log}(3z)]\}^2$

17. Solve Exercises 13–16 in Section 3.3 using inverse trigonometric functions.

18. Solve Exercises 9–11 in Section 3.4 using inverse hyperbolic functions.

19. Verify that when z_1 and z_2 satisfy $\sin z_1 = \sin z_2$, then either $z_2 = z_1 + 2n\pi$, or, $z_2 = (\pi - z_1) + 2n\pi$, where n is an integer.

20. Verify 3.44.

21. **(a)** Verify that $\cos^{-1} z = 2n\pi \pm \mathrm{Cos}^{-1} z$.

(b) What is the value of $\mathrm{Cos}^{-1} z$ when z is real?

(c) Verify that when z is a real number x between -1 and 1, the values of $\mathrm{Cos}^{-1} z$ coincide with those of $\mathrm{Cos}^{-1} x$.

22. **(a)** Verify that $\tan^{-1} z = n\pi + \mathrm{Tan}^{-1} z$.

(b) What is the value of $\mathrm{Tan}^{-1} z$ when z is real? Does it coincide with the real function $\mathrm{Tan}^{-1} x$?

23. Prove that for any branch of the logarithm function, the derivative of $\sin^{-1} z$ is given by 3.47.

24. Show that the function $w = \mathrm{Sin}^{-1} z$ maps the half-plane $y > 0$ onto the semi-infinite strip $-\pi/2 < u < \pi/2$, $v > 0$.

25. **(a)** If $w = u + vi = \mathrm{Sin}^{-1} z = \mathrm{Sin}^{-1}(x + yi)$, verify that

$$x = \sin u \cosh v, \qquad y = \cos u \sinh v.$$

(b) Show that for any fixed value of u, the equations in (a) are parametric equations for the hyperbola

$$\frac{x^2}{\sin^2 u} - \frac{y^2}{\cos^2 u} = 1.$$

Verify that $u = \mathrm{Sin}^{-1}\left[\dfrac{\sqrt{(x+1)^2+y^2} - \sqrt{(x-1)^2+y^2}}{2}\right]$.

(c) Show that for any fixed value of v, the equations in (a) are parametric equations for the ellipse

$$\frac{x^2}{\cosh^2 v} + \frac{y^2}{\sinh^2 v} = 1.$$

Verify that

$$v = (\mathrm{sgn}\, y)\,\mathrm{Cosh}^{-1}\left[\frac{\sqrt{(x+1)^2+y^2} + \sqrt{(x-1)^2+y^2}}{2}\right],$$

where $\mathrm{sgn}\, y = \begin{cases} 1, & y \geq 0 \\ -1, & y < 0. \end{cases}$

26. For what values of z is $\mathrm{Sin}^{-1} z$ real? *Hint:* Use Exercise 25.

27. Find inverse functions for $\csc z$, $\sec z$, and $\cot z$.

28. Find inverse functions for $\mathrm{csch}\, z$, $\mathrm{sech}\, z$, and $\coth z$.

In Exercises 29–35 we investigate branch cuts for principal branches of the inverse trigonometric and inverse hyperbolic functions.

29. Show that branch points of $\sqrt{1-z^2}$ are $z = \pm 1$ and the branch cut consists of the real axis $y = 0$ to the right of $x = 1$ and to the left of $x = -1$; that is, $|x| > 1$.

30. Verify that $iz + \sqrt{1-z^2}$ is never zero, and therefore $\operatorname{Sin}^{-1} z = -i \operatorname{Log}(iz + \sqrt{1-z^2})$ has no further branch points or cuts other than those introduced by $\sqrt{1-z^2}$.

31. Verify that $z + i\sqrt{1-z^2}$ never vanishes, and therefore $\operatorname{Cos}^{-1} z = -i \operatorname{Log}(z + i\sqrt{1-z^2})$ has no further branch points or cuts other than those introduced by $\sqrt{1-z^2}$.

32. Find branch point(s) and cut(s) for $\operatorname{Tan}^{-1} z$.

33. Show that $z = \pm i$ are branch points for $\operatorname{Sinh}^{-1} z$ and that branch cuts are the imaginary axis above $z = i$ and below $z = -i$.

34. Show that $z = \pm 1$ are branch points for $\operatorname{Cosh}^{-1} z$ and that branch cuts are the imaginary axis and that part of the real axis between $z = \pm 1$.

35. Verify that $z = \pm 1$ are branch points for $\operatorname{Tanh}^{-1} z$ and that branch cuts are the real axis to the right of $z = 1$ and to the left of $z = -1$.

CHAPTER 4

Complex Integrals

We now turn our attention to integration of complex functions. As for real functions we define two kinds of integrals. The antiderivative of a complex function f is completely analogous to that for a real function; it is a complex function whose derivative is f. What corresponds to the definite integral of a real function is called a *contour* integral of a complex function. Because complex numbers are two-dimensional, having real and imaginary parts, the definition of a contour integral more closely resembles that of a line integral in the xy-plane than a definite integral along the x-axis. In spite of this, we find that strategies for evaluating real definite integrals are as prevalent in evaluating contour integrals as are strategies for calculating real line integrals.

Contour integrals lead to a host of properties for analytic functions. These in turn lead to properties of harmonic functions because real and imaginary parts of analytic functions are harmonic.

4.1 Indefinite Integrals, or Antiderivatives

Definition 4.1

A complex function F is said to be an **antiderivative**, or an **indefinite integral**, of a complex function f in a domain D if

$$F'(z) = f(z) \tag{4.1}$$

at every point in D.

For example, because $d(z^4)/dz = 4z^3$, we say that z^4 is an antiderivative or an indefinite integral of $4z^3$. But so also is $z^4 + 10i$ an antiderivative. In fact, for any complex number C whatsoever, $z^4 + C$ is an antiderivative of $4z^3$.

We have already shown (Theorem 2.7) that if a function F has a derivative equal to zero at every point in some domain D, then F must be a constant function in D. This implies that once we have found one antiderivative of a function, we have essentially found them all.

■ **THEOREM 4.1** If F is an indefinite integral of f in some domain D, then every indefinite integral of f is of the form $F + C$, where C is a complex constant. ■

When F is an indefinite integral of f, we call $F + C$ *the* indefinite integral of f and write

$$\int f(z)\,dz = F(z) + C. \tag{4.2}$$

It is a simple exercise in differentiation and use of Definition 4.1 to verify the following properties of indefinite integrals.

■ **THEOREM 4.2** If $\int f(z)\,dz = F(z) + C$ and $\int g(z)\,dz = G(z) + C$, then

$$\int [f(z) + g(z)]\,dz = F(z) + G(z) + C, \tag{4.3}$$

$$\int k\,f(z)\,dz = kF(z) + C, \qquad k \text{ a complex constant.} \tag{4.4}$$ ■

EXAMPLE 4.1 ....

Evaluate $\displaystyle\int \frac{1}{z-3-2i}\,dz$.

Solution Because $\dfrac{d}{dz}\log_\phi(z-3-2i) = \dfrac{1}{z-3-2i}$, we can say that

$$\int \frac{1}{z-3-2i}\,dz = \log_\phi(z-3-2i) + C,$$

where ϕ is any real number. This antiderivative is valid at every point except $2+3i$ and points on the branch cut in Figure 4.1. These are not the only antiderivatives of $(z-3-2i)^{-1}$, however. Any branch of $\log(z-3-2i)$ is an antiderivative in its domain of analyticity. We might even write

$$\int \frac{1}{z-3-2i}\,dz = \log(z-3-2i) + C,$$

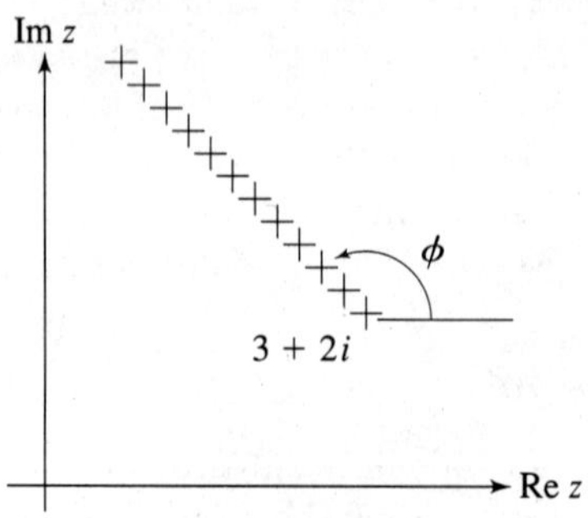

Figure 4.1

but it would be clearly understood that a branch of the logarithm function is required on the right, any branch. ....

Exercises 4.1

In Exercises 1–20 evaluate the indefinite integral. Specify the domain of validity of the result.

1. $\displaystyle\int (z^3 + 3z + 2)\, dz$

2. $\displaystyle\int \left(\frac{1}{z^2} - 2z\right) dz$

3. $\displaystyle\int z(z^2 + 3)^4\, dz$

4. $\displaystyle\int z^2(4z^3 + 2)^6\, dz$

5. $\displaystyle\int \frac{z}{(z^2 + 1)^2}\, dz$

6. $\displaystyle\int \frac{4z - 6}{(z^2 - 3z + 5)^3}\, dz$

7. $\displaystyle\int \cos 3z\, dz$

8. $\displaystyle\int z \sin(4z^2)\, dz$

9. $\displaystyle\int \cos^2 z\, dz$

10. $\displaystyle\int \sec^2 z\, dz$

11. $\displaystyle\int \frac{1}{z + 5}\, dz$

12. $\displaystyle\int \frac{1}{3z + 4i}\, dz$

13. $\displaystyle\int z \sin z\, dz$

14. $\displaystyle\int \sin^3 z\, dz$

15. $\displaystyle\int \frac{z}{(z^2 - 4)^2}\, dz$

16. $\displaystyle\int \frac{z^2}{z^3 - 3z + 2}\, dz$

17. $\displaystyle\int z^2 e^{2z}\, dz$

18. $\displaystyle\int \frac{z}{\sqrt{1 - z}}\, dz$

19. $\displaystyle\int \operatorname{Log} z\, dz$

20. $\displaystyle\int \frac{1}{z^2} \sin\left(\frac{1}{z}\right) dz$

21. Real antidifferentiation would suggest that

$$\int \frac{z}{z^2 - 4}\, dz = \frac{1}{2} \log_\phi (z^2 - 4) + C.$$

Show that partial fractions lead to the antiderivative

$$\int \frac{z}{z^2 - 4}\, dz = \frac{1}{2} \log_\phi (z + 2) + \frac{1}{2} \log_\psi (z - 2) + C.$$

Are these antiderivatives identical?

4.2 Curves and Line Integrals

In Section 4.3 we define how to integrate a complex function from one complex number z_0 to another complex number z_1 called a contour integral. Because there are many paths from z_0 to z_1 in the complex plane, it will be necessary to specify the particular curve to be followed. This is reminiscent of line integrals in the real xy-plane. Properties of real line integrals are also prominent in the evaluation of contour integrals in the complex plane. In preparation, then, for contour integrals we give a quick review of curves, their parametrizations, and real line integrals.

A curve is defined parametrically in the xy-plane if it is specified in the form

$$x = x(t), \quad y = y(t), \qquad \alpha \leq t \leq \beta. \tag{4.5}$$

We regard $\big(x(\alpha), y(\alpha)\big)$ as the initial point on the curve and $\big(x(\beta), y(\beta)\big)$ as the final point, and the direction of the curve is from initial point to final point. For example, the equations

$$x = 2 - 3t, \qquad y = 4 + t, \qquad 0 \leq t \leq 1$$

describe the line segment in Figure 4.2 directed (as shown by the arrowhead) from initial point $(2, 4)$ to final point $(-1, 5)$. The equations

$$x = 3\cos t, \qquad y = 3\sin t, \qquad 0 \leq t \leq \pi$$

describe the semicircle from $(3, 0)$ to $(-3, 0)$ in Figure 4.3.

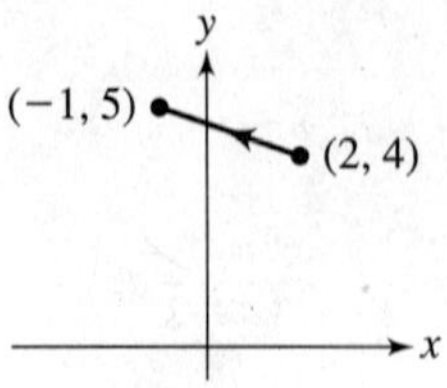

Figure 4.2

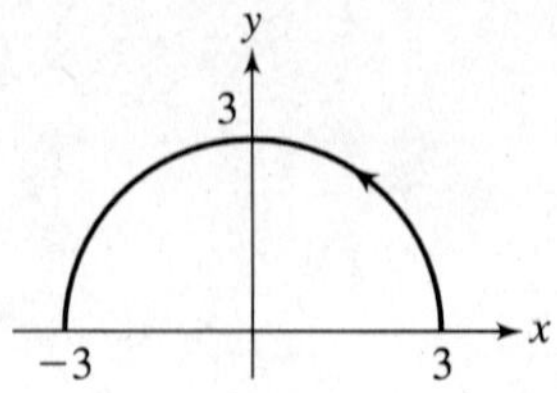

Figure 4.3

Curves can be parametrized in countless ways. Two additional parametrizations for the semicircle in Figure 4.3 are

$$\begin{matrix} x = 3\cos 2t \\ y = 3\sin 2t \end{matrix} \quad 0 \leq t \leq \pi/2; \qquad \begin{matrix} x = -t \\ y = \sqrt{9 - t^2} \end{matrix} \quad -3 \leq t \leq 3.$$

A curve is a set of points from some initial point to a final point. Its properties depend only on the set of points, not on the parametric equations that describe the points. All properties discussed in this section are independent of parametric representation of the curve.

Curve 4.5 is **continuous** when both functions $x(t)$ and $y(t)$ are continuous for $\alpha \leq t \leq \beta$. What this means is that at no point is there a separation in the curve. The curves in Figures 4.2 and 4.3 are continuous; the curve in Figure 4.4 is not. All other curves in this chapter are continuous.

A curve is said to be **closed** if its initial and final points are the same point. Parametric equations $x = 3\cos t$, $y = -2\sin t$, $0 \leq t \leq 2\pi$ define the ellipse in Figure 4.5. It is a closed curve; initial and final points are both equal to $(3, 0)$. The

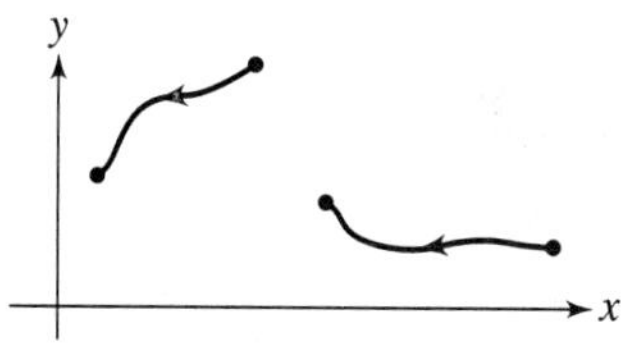

Figure 4.4

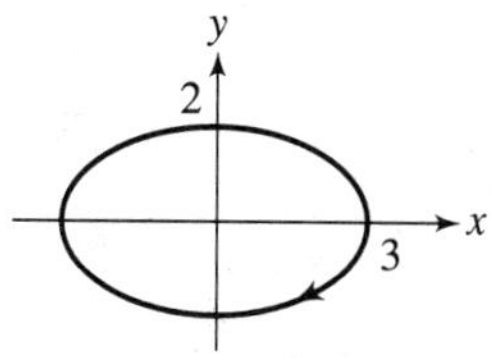

Figure 4.5

curve in Figure 4.6 has a loop but it is not a closed curve; initial and final points are not the same.

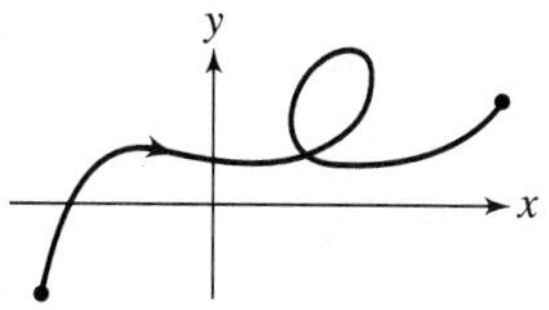

Figure 4.6

The position vector of points on a continuous curve C with parametric equations 4.5 is

$$\mathbf{r} = \mathbf{r}(t) = x(t)\hat{\mathbf{i}} + y(t)\hat{\mathbf{j}} \tag{4.6}$$

where $\hat{\mathbf{i}}$ and $\hat{\mathbf{j}}$ are unit vectors in the positive x- and y-directions. It is customary to place the tail of this vector at the origin, in which case its tip is at the point on the curve with coordinates $\big(x(t), y(t)\big)$ (Figure 4.7). A tangent vector to C at each point on C is

$$\frac{d\mathbf{r}}{dt} = \frac{dx}{dt}\hat{\mathbf{i}} + \frac{dy}{dt}\hat{\mathbf{j}}, \tag{4.7}$$

provided dx/dt and dy/dt both exist and do not simultaneously vanish. This tangent vector always points in the direction in which the parameter increases along the curve. If the parameter increases from initial point to final point (as in 4.5), then $d\mathbf{r}/dt$ points in the direction of motion along C (Figure 4.8).

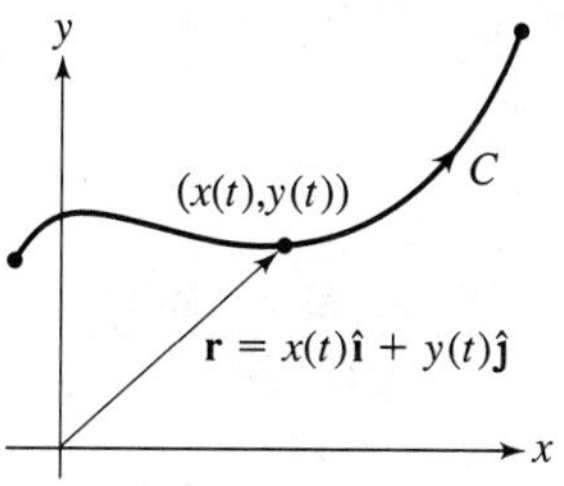

Figure 4.7

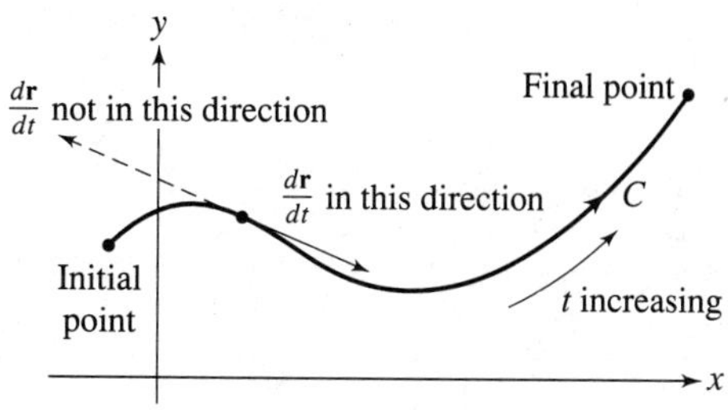

Figure 4.8

EXAMPLE 4.2

Find a tangent vector at each point on the curve

$$C : x = t^3,\ y = t^5, \qquad -1 \le t \le 1.$$

Solution According to 4.7,

$$\frac{d\mathbf{r}}{dt} = 3t^2\hat{\mathbf{\imath}} + 5t^4\hat{\mathbf{\jmath}}.$$

This is a tangent vector at each point on C except $(0, 0)$, where it vanishes. The curve is shown in Figure 4.9, and it is clear that a tangent vector at $(0, 0)$ is $\mathbf{T} = \hat{\mathbf{\imath}}$. This tangent vector can be obtained from $d\mathbf{r}/dt$ by writing

$$\frac{d\mathbf{r}}{dt} = t^2(3\hat{\mathbf{\imath}} + 5t^2\hat{\mathbf{\jmath}}).$$

It follows that $\mathbf{T} = 3\hat{\mathbf{\imath}} + 5t^2\hat{\mathbf{\jmath}}$ is a tangent vector at each point on C (we have simply removed the scale factor t^2 from $d\mathbf{r}/dt$). At $t = 0$, we obtain $\mathbf{T} = 3\hat{\mathbf{\imath}}$, which is indeed tangent to C at $(0, 0)$.

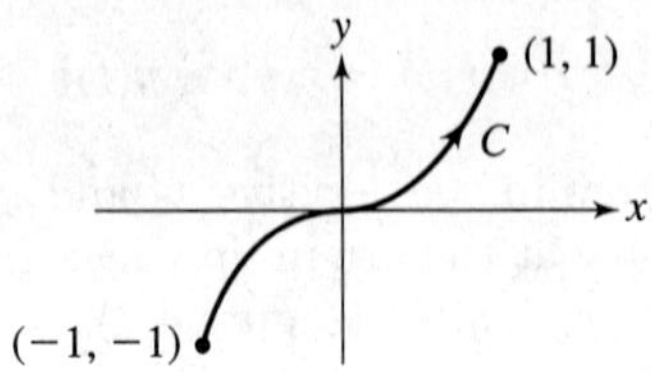

Figure 4.9

A curve is said to be **smooth** if it is continuous and there can be assigned a tangent vector at each point on the curve that varies continuously along the curve. By saying that the tangent vector varies "continuously" along the curve, we mean that small changes in t produce small changes in the tangent vector. When the curve is defined by 4.5, this is the case if $x'(t)$ and $y'(t)$ are both continuous for $\alpha < t < \beta$, and $x'(t)$ and $y'(t)$ do not vanish simultaneously for $\alpha < t < \beta$.

EXAMPLE 4.3

Sketch the curve $C : x = 2\cosh t,\ y = 3\sinh t,\ -\infty < t < \infty$. Is it smooth?

Solution Because

$$9x^2 - 4y^2 = 9(4\cosh^2 t) - 4(9\sinh^2 t) = 36$$

and x is always positive, these equations define the right half of the hyperbola in Figure 4.10 directed from fourth quadrant to first quadrant. A tangent vector is

$$\frac{d\mathbf{r}}{dt} = 2\sinh t\,\hat{\mathbf{\imath}} + 3\cosh t\,\hat{\mathbf{\jmath}}.$$

Because components of this vector are continuous and the y-component never vanishes, the curve is smooth. Figure 4.10 makes this equally clear. The tangent direction to the hyperbola changes continuously from point to point.

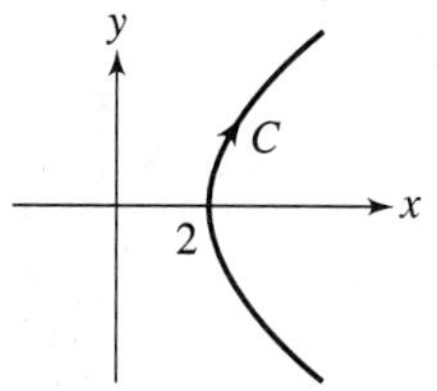

Figure 4.10

EXAMPLE 4.4

Is the curve in Example 4.2 smooth?

Solution Figure 4.9 indicates that the curve is indeed smooth. Notice, however, that the components of $d\mathbf{r}/dt = 3t^2\hat{\mathbf{\imath}} + 5t^4\hat{\mathbf{\jmath}}$ both vanish at $t = 0$. What this means is that the requirement of simultaneous nonvanishing of $x'(t)$ and $y'(t)$ is a sufficient condition for smoothness of a curve, but it is not necessary. We have already noted that $\mathbf{T} = 3\hat{\mathbf{\imath}} + 5t^2\hat{\mathbf{\jmath}}$ is also a tangent vector at each point; this vector varies continuously along the curve and does not vanish at $t = 0$.

EXAMPLE 4.5

Sketch the curve $C : x = t^3,\ y = t^2,\ -1 \le t \le 1$. Is it smooth?

Solution The curve is shown in Figure 4.11. Due to the *cusp* at $(0, 0)$, it is not smooth. Tangent vectors reverse direction at $(0, 0)$. Notice that $d\mathbf{r}/dt = 3t^2\hat{\mathbf{\imath}} + 2t\hat{\mathbf{\jmath}}$ vanishes at $(0, 0)$.

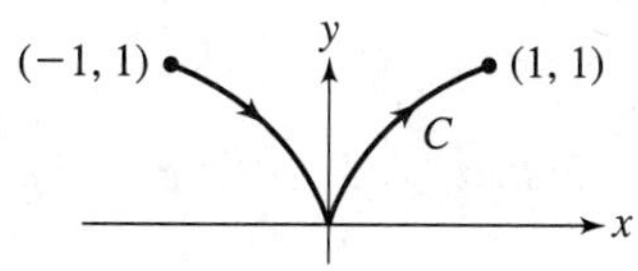

Figure 4.11

Most curves with which we deal in this chapter fall into the following category. A curve is said to be **piecewise smooth** if it is continuous and can be divided into a finite number of smooth subcurves. A rectangle is piecewise smooth (Figure 4.12); it can be divided into four smooth subcurves (the sides of the rectangle). The curve in Figure 4.11 is piecewise smooth; parts on either side of the y-axis are smooth. The curve in Figure 4.13 that joins the points $(1/(2n), 1)$ and $(1/(2n-1), -1)$ where $n \geq 1$ is an integer, is not piecewise smooth. It can be divided into smooth line segments, but not a finite number of them.

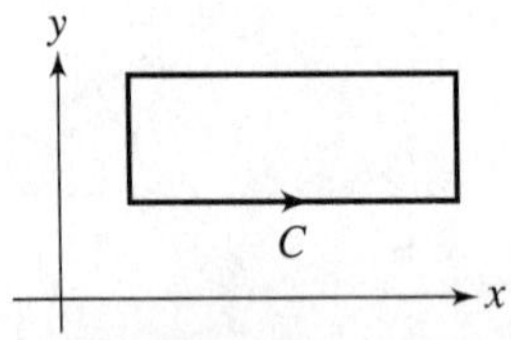

Figure 4.12

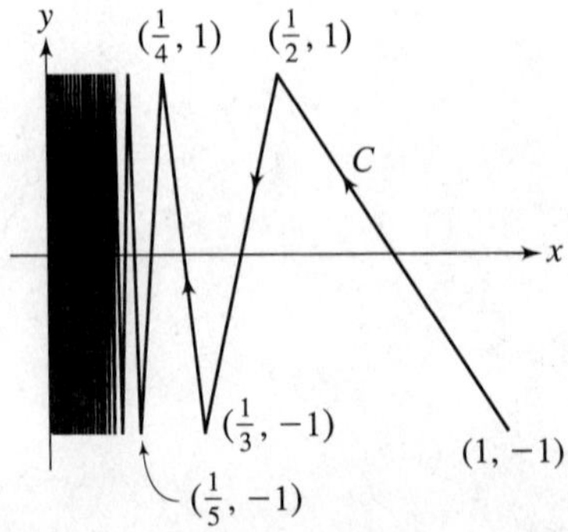

Figure 4.13

Continuous curve 4.5 is said to be **simple** if $\mathbf{r}(t_1) = \mathbf{r}(t_2)$ implies that $t_1 = t_2$ for all $\alpha < t_1 < t_2 < \beta$. What this says is that a simple curve cannot cross itself; two different values of t cannot give the same point on the curve, except the initial and final point when the curve is closed. The figure eight in Figure 4.14 is not simple; it crosses itself at the origin. The curve in Figure 4.6 is not simple.

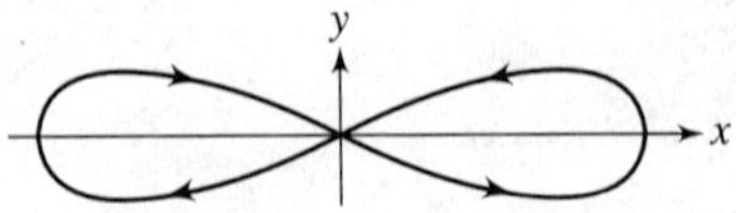

Figure 4.14

Line Integrals in the Plane

Line integrals in the xy-plane that correspond to integrals in the complex plane take the form

$$\int_C P(x, y)\, dx + Q(x, y)\, dy. \qquad \textbf{(4.8)}$$

When $P(x, y)$ and $Q(x, y)$ are continuous and C is smooth, we can calculate the value of the line integral by expressing P, Q, dx, and dy in terms of any parametric representation 4.5 of C and evaluating the resulting definite integral,

$$\int_C P(x,y)\,dx + Q(x,y)\,dy = \int_\alpha^\beta \left\{ P[x(t),y(t)]\frac{dx}{dt} + Q[x(t),y(t)]\frac{dy}{dt} \right\} dt. \tag{4.9}$$

All parametrizations of the curve lead to the same value. When C is piecewise smooth, we evaluate the line integral along each smooth subcurve and add the results.

EXAMPLE 4.6 ....

Evaluate $\int_C xy\,dx + x^2y\,dy$, where C is the curve $y = x^3$ from $(-1,-1)$ to $(1,1)$.

Solution Using the parametric representation $x = t$, $y = t^3$, $-1 \le t \le 1$,

$$\int_C xy\,dx + x^2y\,dy = \int_{-1}^1 t(t^3)(dt) + t^2(t^3)(3t^2\,dt) = \int_{-1}^1 (t^4 + 3t^7)\,dt$$

$$= \left\{ \frac{t^5}{5} + \frac{3t^8}{8} \right\}_{-1}^1 = \frac{2}{5}.$$

....

Normally the value of line integral 4.8 from one point to another depends on the curve taken between the points. For instance, if we follow the straight line $C' : x = -1 + 2t$, $y = -1 + 2t$, $0 \le t \le 1$, from $(-1,-1)$ to $(1,1)$ in Example 4.6 (Figure 4.15), the value of the line integral is

$$\int_{C'} xy\,dx + x^2y\,dy = \int_0^1 (-1+2t)(-1+2t)(2\,dt) + (-1+2t)^2(-1+2t)(2\,dt)$$

$$= 2\int_0^1 [(-1+2t)^2 + (-1+2t)^3]\,dt$$

$$= 2\left\{ \frac{(-1+2t)^3}{6} + \frac{(-1+2t)^4}{8} \right\}_0^1 = \frac{2}{3}.$$

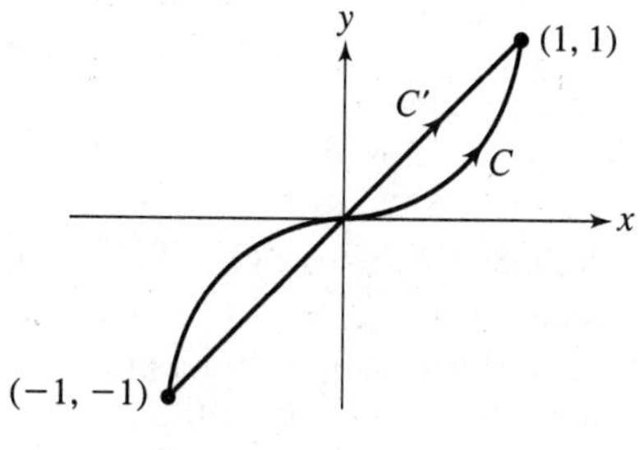

Figure 4.15

The integration is simpler if C' is parametrized with $x = t$, $y = t$, $-1 \le t \le 1$,

$$\int_{C'} xy\,dx + x^2y\,dy = \int_{-1}^{1} t(t)(dt) + t^2(t)(dt) = \int_{-1}^{1} (t^2 + t^3)\,dt$$
$$= \left\{\frac{t^3}{3} + \frac{t^4}{4}\right\}_{-1}^{1} = \frac{2}{3}.$$

For some line integrals, it does not matter what path is taken between two points; the value is always the same. Such line integrals are said to be *independent of path.* Before making this more formal, we digress to introduce a special kind of domain. Open, closed, and connected sets and domains were introduced as special regions in the complex plane. But these concepts are equally valid for regions in the xy-plane. A domain in the xy-plane is therefore an open connected set.

Definition 4.2
A domain is said to be **simply connected** if every closed curve in the domain contains in its interior only points of the domain.

For example, the domain D_1 in Figure 4.16(a) is simply connected. Every closed curve in D_1 contains only points of the set. Examples are curves C_1, C_2, and C_3 in Figure 4.16(b).

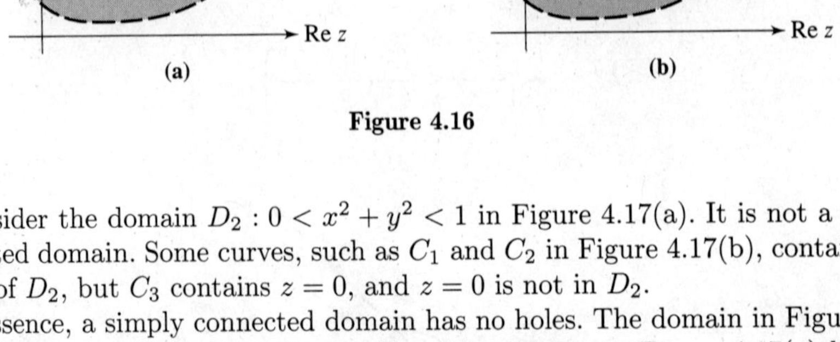

Figure 4.16

Consider the domain $D_2 : 0 < x^2 + y^2 < 1$ in Figure 4.17(a). It is not a simply connected domain. Some curves, such as C_1 and C_2 in Figure 4.17(b), contain only points of D_2, but C_3 contains $z = 0$, and $z = 0$ is not in D_2.

In essence, a simply connected domain has no holes. The domain in Figure 4.18 is not simply connected; it has two holes. The domain in Figure 4.17(a) has one hole.

We now return to the main discussion — independence of path for line integrals of form 4.8. We say that line integral 4.8 is **independent of path** in a domain D if for any two points A and B in D, the value of the line integral is the same for all

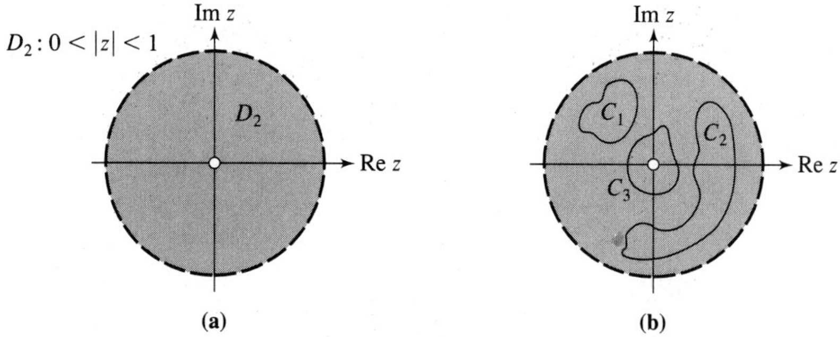

Figure 4.17

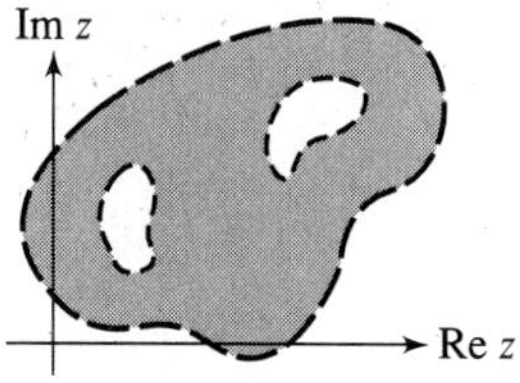

Figure 4.18

piecewise-smooth curves in D from A to B. There are two ways to ensure that a line integral is independent of path in a domain D.

1. Show that $P\hat{\imath} + Q\hat{\jmath}$ is the gradient of some function $\phi(x, y)$ at every point of D; $\nabla\phi = P\hat{\imath} + Q\hat{\jmath}$.
2. If D is simply connected, show that $\partial Q/\partial x = \partial P/\partial y$.

When a line integral is known to be independent of path and its value is required along some curve with initial point A and final point B, we can either replace the given curve with a simpler curve or take the difference in the values of the function $\phi(x, y)$ in (1) at B and A:

$$\int_C P(x,y)\,dx + Q(x,y)\,dy = \Big\{\phi(x,y)\Big\}_A^B = \phi(x_B, y_B) - \phi(x_A, y_A). \qquad (4.10)$$

EXAMPLE 4.7 ....

Evaluate $\int_C 2xy\,dx + (x^2 + y)\,dy$, where C is the curve $y = x^3 - 2x^2 + 1$ from $(-1, -2)$ to $(3, 10)$.

Solution Because $\nabla(x^2y+y^2/2) = 2xy\hat{\mathbf{i}}+(x^2+y)\hat{\mathbf{j}}$, the line integral is independent of path in the whole xy-plane, and

$$\int_C 2xy\,dx + (x^2+y)\,dy = \left\{x^2y + \frac{y^2}{2}\right\}_{(-1,-2)}^{(3,10)} = 140.$$

EXAMPLE 4.8

Evaluate the line integral $\displaystyle\int_C \frac{x}{x^2+y^2}\,dx + \frac{y}{x^2+y^2}\,dy$, where C is that part of the parabola $y = x^2 - 1$ from $(2,3)$ to $(-1,0)$ in Figure 4.19.

Solution Because

$$\nabla[(1/2)\ln(x^2+y^2)] = \frac{x}{x^2+y^2}\hat{\mathbf{i}} + \frac{y}{x^2+y^2}\hat{\mathbf{j}}$$

in the domain shown, the line integral is independent of path therein, and

$$\int_C \frac{x}{x^2+y^2}\,dx + \frac{y}{x^2+y^2}\,dy = \left\{\frac{1}{2}\ln(x^2+y^2)\right\}_{(2,3)}^{(-1,0)}$$
$$= \frac{1}{2}\ln(1) - \frac{1}{2}\ln(13) = -\frac{1}{2}\ln(13).$$

We could also have verified the existence of a function $\phi(x,y)$ such that

$$\nabla\phi = \frac{x}{x^2+y^2}\hat{\mathbf{i}} + \frac{y}{x^2+y^2}\hat{\mathbf{j}}$$

by showing that

$$\frac{\partial}{\partial y}\left(\frac{x}{x^2+y^2}\right) = \frac{\partial}{\partial x}\left(\frac{y}{x^2+y^2}\right).$$

The solution would then proceed as before.

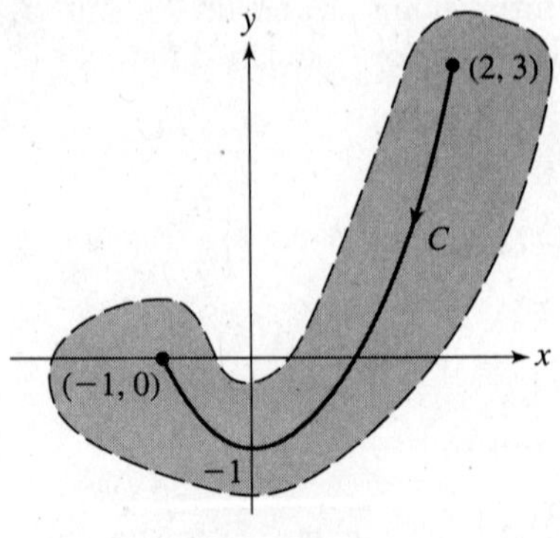

Figure 4.19

If a line integral is independent of path in a domain D and C is a closed, piecewise-smooth curve in D that contains only points of D in its interior, its value is zero:

$$\oint_C P\,dx + Q\,dy = 0. \tag{4.11}$$

The circle on the integral indicates that C is a closed curve.

EXAMPLE 4.9

Evaluate the line integral of Example 4.8 once counterclockwise around the circle $x^2 + y^2 = 1$.

Solution In this example it does no good to show that

$$\frac{\partial}{\partial y}\left(\frac{x}{x^2+y^2}\right) = \frac{\partial}{\partial x}\left(\frac{y}{x^2+y^2}\right)$$

because this cannot be shown to hold in a simply connected domain that contains the circle $x^2+y^2 = 1$. It fails to hold at $(0,0)$, and this point is interior to $x^2+y^2 = 1$. On the other hand, we need not follow this line of argument. Parametric equations $x = \cos t$, $y = \sin t$, $0 \le t \le 2\pi$, lead to an immediate solution:

$$\oint_C \frac{x}{x^2+y^2}\,dx + \frac{y}{x^2+y^2}\,dy = \int_0^{2\pi} \cos t(-\sin t\,dt) + \sin t(\cos t\,dt) = 0.$$

When a line integral is required around a closed curve, and the line integral is not independent of path, Green's theorem can sometimes be used to advantage. It is stated as follows.

■ **THEOREM 4.3** (Green's Theorem) Suppose $P(x, y)$ and $Q(x, y)$ have continuous first partial derivatives in a domain containing a simple, closed, piecewise-smooth curve C and its interior R (Figure 4.20). Then

$$\oint_C P\,dx + Q\,dy = \iint_R \left(\frac{\partial Q}{\partial x} - \frac{\partial P}{\partial y}\right) dA. \tag{4.12}$$

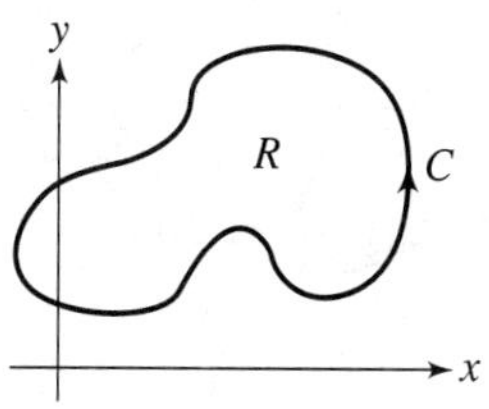

Figure 4.20

The arrowhead on the circle indicates that C is traversed in the counterclockwise direction. ■

EXAMPLE 4.10 ....

Use Green's theorem to evaluate $\oint_C y^2\,dx + x^2\,dy$, where C is the triangular curve in Figure 4.21.

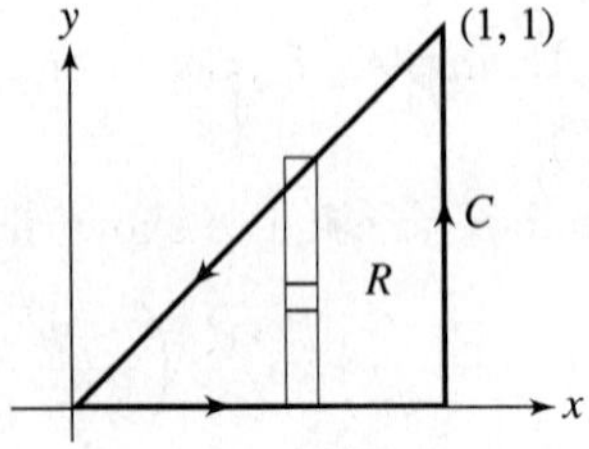

Figure 4.21

Solution Equation 4.12 gives

$$\oint_C y^2\,dx + x^2\,dy = \iint_R (2x - 2y)\,dA = 2\int_0^1 \int_0^x (x - y)\,dy\,dx$$
$$= 2\int_0^1 \left\{xy - \frac{y^2}{2}\right\}_0^x dx = \int_0^1 x^2\,dx = \left\{\frac{x^3}{3}\right\}_0^1 = \frac{1}{3}.$$

....

Exercises 4.2

In Exercises 1–10 find a parametric representation for the curve wherein the parameter increases along the curve. Find a tangent vector at each point on the curve.

1. $x + 2y = 6$ from $(1, 5/2)$ to $(-3, 9/2)$
2. $x^2 + y^2 = 9$ directed counterclockwise with initial point $(3, 0)$
3. $x^2 + y^2 = 9$ directed clockwise with initial point $(0, 3)$
4. $y = x^2 + 2x + 5$ from $(0, 5)$ to $(-2, 5)$
5. $x = y^3 + y$ from $(-2, -1)$ to $(10, 2)$
6. $x^2 + y^2 + 2x - 4y = 0$ directed counterclockwise with initial point $(\sqrt{5} - 1, 2)$
7. $4 - x^2 - 2y^2 = 0$, $y \geq 0$ from $(-2, 0)$ to $(2, 0)$
8. $x + y^3 + xy = 5y^2$ from $(2, 1)$ to $(4, 2)$
9. $x^2 - y^2 = 1$ from $(\sqrt{2}, 1)$ to $(\sqrt{5}, -2)$
10. the straight line segment from (x_1, y_1) to (x_2, y_2)

11. Is the curve $x = 3\cos t$, $y = 3\sin t$, $0 \le t \le 3\pi$ closed?

12. Find parametric equations for the curve in Example 4.2 for which $x'(t)$ and $y'(t)$ are continuous and do not vanish simultaneously at any point on C.

In Exercises 13–27 evaluate the line integral.

13. $\int_C x\,dx + x^2y\,dy$, where C is the curve $y = x^3$ from $(-1,1)$ to $(2,8)$

14. $\int_C xy^2\,dx + x^2y\,dy$, where C is the curve $y = x^2$ from $(0,0)$ to $(1,1)$

15. $\oint_C y^2\,dx + x^2\,dy$ once counterclockwise around $x^2 + y^2 = 1$

16. $\int_C x\,dx + (x+y)\,dy$, where C is the curve $x = 1 + y^2$ from $(2,1)$ to $(2,-1)$

17. $\int_C (3x^2 + y)\,dx + x\,dy$, where C is the line segment from $(2,1)$ to $(-3,2)$

18. $\oint_C (x^2 + 2y^2)\,dy$ once counterclockwise around $x^2 - 4x + y^2 + 3 = 0$

19. $\oint_C x^2y\,dx + (x-y)\,dy$ once counterclockwise around the curve enclosing the area bounded by $x = 1 - y^2$ and $y = x + 1$

20. $\oint_C y\cos x\,dx + \sin x\,dy$ once clockwise around $x^2 + y^2 - 2x + 4y = 7$

21. $\oint_C x^2e^y\,dx + (x+y)\,dy$ once clockwise around the square with vertices $(\pm 1, 1)$ and $(\pm 1, -1)$

22. $\int_C y^2\,dx + x^2\,dy$, where C is the semicircle $x = \sqrt{1 - y^2}$ from $(0,1)$ to $(0,-1)$

23. $\int_C 3x^2y^3\,dx + 3x^3y^2\,dy$, where C is the curve $y = e^x$ from $(0,1)$ to $(1,e)$

24. $\oint_C (x^2 + y^3)\,dx + (x^3 - y^3)\,dy$ once counterclockwise around the curve enclosing the area bounded by $x = y^2 - 1$, $x = 1 - y^2$

25. $\oint_C y^2\,dx + x^2\,dy$ once clockwise around the curve $|x| + |y| = 1$

26. $\oint_C (x^2 + 2y^2)\,dy$ twice clockwise around the circle $(x-2)^2 + y^2 = 1$

27. $\int_C -\frac{1}{x^2}\,\mathrm{Tan}^{-1}\,y\,dx + \frac{1}{x + xy^2}dy$, where C is the curve $x = y^2 + 1$ from $(2,-1)$ to $(10,3)$

28. Consider the line integral $\int_C xy\,dx + x^2\,dy$, where C is the quarter circle $x^2+y^2=9$ from $(3,0)$ to $(0,3)$. Show that for each of the following parametrizations of C the value of the line integral is the same.

(a) $x = 3\cos t,\ y = 3\sin t,\ 0 \le t \le \pi/2$

(b) $x = \sqrt{9-y^2},\ 0 \le y \le 3$

29. Evaluate the line integral $\int_C xy\,dx + x\,dy$ from $(-5,3)$ to $(4,0)$ along each of the following curves:

(a) the straight line segment joining the points

(b) $x = 4 - y^2$

(c) $3y = x^2 - 16$

30. Evaluate $\oint_C \dfrac{-y\,dx + x\,dy}{x^2+y^2}$

(a) Once counterclockwise around the circle $x^2+y^2=1$.

(b) Once counterclockwise around the circle $(x-2)^2+y^2=1$.

31. Evaluate $\int_C \dfrac{y}{x^2+y^2}\,dx - \dfrac{x}{x^2+y^2}\,dy$, where C is the set of line segments joining successively the points $(1,0)$, $(1,1)$, $(-1,1)$, and $(-1,0)$

32. Evaluate $\oint_C (2xye^{x^2y} + x^2y)\,dx + x^2e^{x^2y}\,dy$ once clockwise around the ellipse $x^2 + 4y^2 = 4$.

33. Evaluate $\oint_C (3x^2y^3 - x^2y)\,dx + (xy^2 + 3x^3y^2)\,dy$ once counterclockwise around the circle $x^2+y^2=9$.

34. Evaluate $\int_C (x-y)(dx+dy)$, where C is the semicircular part of $x^2+y^2=4$ above $y=x$ from $(-\sqrt{2},-\sqrt{2})$ to $(\sqrt{2},\sqrt{2})$.

35. (a) In what domains is the line integral $\int \dfrac{x\,dx + y\,dy}{x^2+y^2}$ independent of path?

(b) Evaluate the integral clockwise around the curve $x^2+y^2-2y=1$.

36. Find all possible values for the line integral $\oint_C \dfrac{-y\,dx + x\,dy}{x^2+y^2}$ for curves not passing through the origin.

37. Why is it possible to remove the scale factor t^2 in Example 4.2, but it is not possible to remove a t in Example 4.5?

4.3 Contour Integrals

In this section we define the integral of a complex function f defined in some domain D from a point z_0 in D to a point z_1 in D. Because f is a function of one variable, we might expect the resulting integral to be much like the definite integral of a real function of one variable in elementary calculus. It is not quite this simple, however, due to the fact that specification of z as a complex number requires the two-dimensional complex plane. Unlike the definite integral of $f(x)$ from $x = a$ to $x = b$ where there is only one way to go from from $x = a$ to $x = b$ (namely, along the x-axis), there are infinitely many ways in the complex plane to proceed from z_0 to z_1 (Figure 4.22). This is reminiscent of line integrals in the plane, which we reviewed in Section 4.2, and we therefore use these results to define the integral of a complex function f from one complex number z_0 to another complex number z_1.

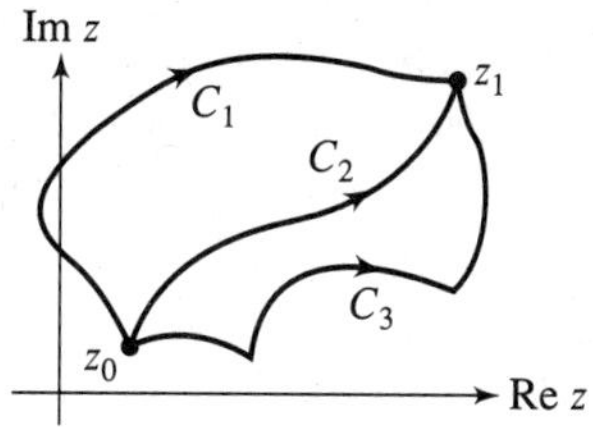

Figure 4.22

Suppose that C is a smooth curve in D with parametric representation

$$C: x = x(t),\ y = y(t), \qquad \alpha \le t \le \beta. \tag{4.13}$$

Using these, each complex number $z = x + yi$ on C can be expressed in terms of t; that is, we can write

$$C:\ z = z(t) = x(t) + y(t)\,i, \qquad \alpha \le t \le \beta. \tag{4.14}$$

If we differentiate this equation with respect to t, we obtain

$$\frac{dz}{dt} = \frac{dx}{dt} + \frac{dy}{dt} i. \tag{4.15}$$

Recalling that dx/dt and dy/dt are the components of a tangent vector to C, we have just shown that at any point on a curve C in the complex plane, the real and imaginary parts of dz/dt are the components of a tangent vector to C.

Suppose now that f is a continuous function on C with real and imaginary parts $u(x, y)$ and $v(x, y)$; that is, $f(z) = u(x, y) + v(x, y)i$. To define the integral of f along C, denoted by

$$\int_C f(z)\,dz,$$

we first set $f(z) = u + vi$ and $dz = dx + dy\,i$ (see 4.15):

$$\int_C f(z)\,dz = \int_C (u + vi)(dx + dy\,i) = \int_C (u\,dx - v\,dy) + (v\,dx + u\,dy)i. \tag{4.16}$$

We now separate this expression into real and imaginary parts:

$$\int_C f(z)\,dz = \int_C u\,dx - v\,dy + i \int_C v\,dx + u\,dy. \tag{4.17}$$

But on the right we have real line integrals along the curve C in the xy-plane. We use 4.17 as the definition of the integral on the left; that is, we define the **contour integral** of a continuous function f along a smooth curve C as the complex number on the right side of 4.17. Because these line integrals can be evaluated by expressing u, v, dx, and dy in terms of some parameter along C, we can say that to evaluate a contour integral, express $f(z)$ and dz in terms of any parametric representation of C and evaluate the resulting definite integrals,

$$\int_C f(z)\,dz = \int_\alpha^\beta f[z(t)]\,z'(t)\,dt. \tag{4.18}$$

When C is piecewise smooth rather than smooth, the value of the contour integral is defined as the sum of its values along the smooth parts, each of which can be evaluated as before. Let us illustrate with some examples.

EXAMPLE 4.11 ....

Evaluate $\displaystyle\int_C z^2\,dz$, where C is one quarter of the circle $|z| = 1$ joining 1 to i.

Solution With the parametric representation

$$C : x = \cos t,\ y = \sin t, \qquad 0 \le t \le \pi/2,$$

for the curve (Figure 4.23),

$$\begin{aligned}
\int_C z^2\,dz &= \int_C (x+yi)^2(dx+dy\,i)\\
&= \int_C (x^2-y^2+2xyi)(dx+dy\,i)\\
&= \int_0^{\pi/2} (\cos^2 t - \sin^2 t + 2\cos t\,\sin t\,i)(-\sin t + \cos t\,i)\,dt\\
&= \int_0^{\pi/2} \big[(-\cos^2 t\,\sin t + \sin^3 t - 2\cos^2 t\,\sin t)\\
&\qquad + (\cos^3 t - \sin^2 t\,\cos t - 2\sin^2 t\,\cos t)i\big]\,dt\\
&= \int_0^{\pi/2} \{[-3\cos^2 t\,\sin t + \sin t(1-\cos^2 t)]\\
&\qquad + [\cos t(1-\sin^2 t) - 3\sin^2 t\,\cos t]i\}\,dt\\
&= \int_0^{\pi/2} \big[(\sin t - 4\cos^2 t\,\sin t) + (\cos t - 4\sin^2 t\,\cos t)i\big]\,dt\\
&= \left\{\left(-\cos t + \frac{4\cos^3 t}{3}\right) + \left(\sin t - \frac{4\sin^3 t}{3}\right)i\right\}_0^{\pi/2}\\
&= -\frac{(1+i)}{3}.
\end{aligned}$$

Alternatively, because the modulus of z is constant along C, we could use the argument of z as parameter by writing $C : z = e^{\theta i}$, $0 \le \theta \le \pi/2$, in which case

$$\int_C z^2\,dz = \int_0^{\pi/2} e^{2\theta i} i e^{\theta i}\,d\theta = i\int_0^{\pi/2} e^{3\theta i}\,d\theta = i\left\{\frac{e^{3\theta i}}{3i}\right\}_0^{\pi/2} = -\frac{(1+i)}{3}.$$

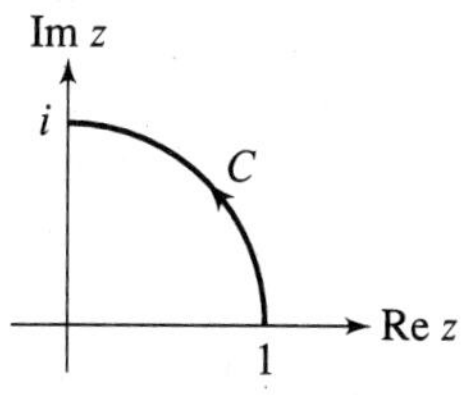

Figure 4.23

When the path of integration in a contour integral is a closed curve, we place a circle on the integral sign,

$$\oint_C f(z)\,dz \quad \text{or} \quad \oint_C f(z)\,dz.$$

The arrowheads indicate a clockwise direction (Figure 4.24(a)) or counterclockwise direction (Figure 4.24(b)) around the curve.

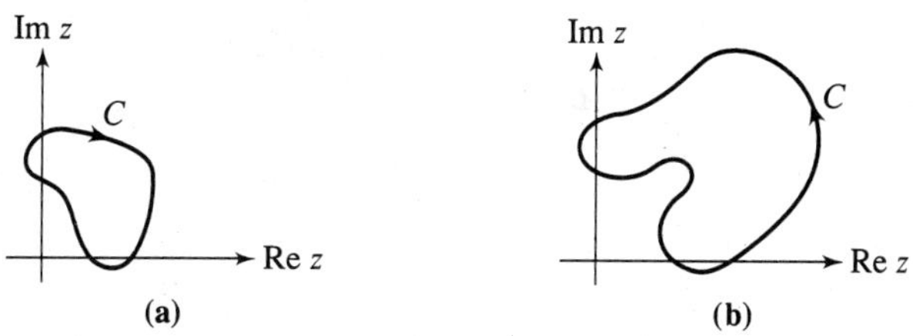

Figure 4.24

EXAMPLE 4.12

Evaluate $\oint_C e^z\,dz$, where $C = C_1 + C_2 + C_3$ is the closed curve in Figure 4.25.

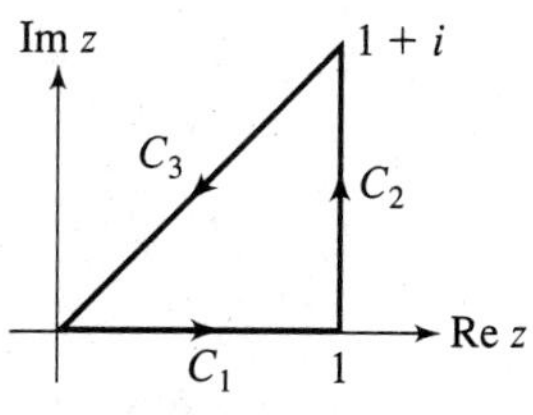

Figure 4.25

Solution $\oint_C e^z\,dz = \int_{C_1} e^z\,dz + \int_{C_2} e^z\,dz + \int_{C_3} e^z\,dz$, where

$$C_1 : y = 0, \qquad 0 \le x \le 1;$$
$$C_2 : x = 1, \qquad 0 \le y \le 1;$$
$$C_3 : x = t, y = t, \qquad 1 \ge t \ge 0.$$

Thus,

$$\begin{aligned}\oint_C e^z\,dz &= \int_0^1 e^x\,dx + \int_0^1 e^{1+yi}\,i\,dy + \int_1^0 e^{t+ti}(dt + dt\,i)\\ &= \left\{e^x\right\}_0^1 + \left\{e^{1+yi}\right\}_0^1 + \left\{e^{(1+i)t}\right\}_1^0\\ &= (e-1) + (e^{1+i} - e) + (1 - e^{1+i}) = 0.\end{aligned}$$

....

The following properties of contour integrals are immediate consequences of similar properties for line integrals. When $\int_C f(z)\,dz$ and $\int_C g(z)\,dz$ both exist,

$$\int_C [f(z) + g(z)]\,dz = \int_C f(z)\,dz + \int_C g(z)\,dz; \qquad \textbf{(4.19a)}$$

$$\int_C kf(z)\,dz = k\int_C f(z)\,dz, \qquad k \text{ a complex constant.} \qquad \textbf{(4.19b)}$$

Furthermore,

$$\int_C f(z)\,dz = \int_{C_1} f(z)\,dz + \int_{C_2} f(z)\,dz, \qquad \text{where } C = C_1 + C_2; \qquad \textbf{(4.20a)}$$

$$\int_{-C} f(z)\,dz = -\int_C f(z)\,dz, \qquad \text{where } -C \text{ is } C \text{ in the reverse direction.} \qquad \textbf{(4.20b)}$$

Expression 4.17 leads to the following two properties:

$$\text{Re}\int_C f(z)\,dz = \int_C \text{Re}[f(z)\,dz]; \qquad \textbf{(4.21a)}$$

$$\text{Im}\int_C f(z)\,dz = \int_C \text{Im}[f(z)\,dz]. \qquad \textbf{(4.21b)}$$

Integrals on the left sides of 4.21 are contour integrals in the complex z-plane; integrals on the right are real line integrals in the xy-plane.

A much less obvious property of contour integrals, but one that we use extensively, is

$$\left|\int_C f(z)\,dz\right| \le \int_C |f(z)|\,|dz| \le ML, \tag{4.22}$$

when M is the maximum value of $|f(z)|$ for z on C and L is the length of C. To verify the left inequality, suppose that r and θ are modulus and argument of the contour integral of $f(z)$ along C,

$$re^{\theta i} = \int_C f(z)\,dz.$$

Then,

$$r = e^{-\theta i}\int_C f(z)\,dz = \int_C e^{-\theta i} f(z)\,dz.$$

Because the real part of any complex number is less than or equal to its modulus, we can state that for any z on C and $dz = dx + dy\,i$,

$$\mathrm{Re}[e^{-\theta i} f(z)\,dz] \le |e^{-\theta i} f(z)\,dz| = |f(z)|\,|dz|.$$

Integration of this result along C gives

$$\begin{aligned}\int_C |f(z)|\,|dz| &\ge \int_C \mathrm{Re}[e^{-\theta i} f(z)\,dz]\\ &= \mathrm{Re}\int_C e^{-\theta i} f(z)\,dz \qquad \text{(property 4.21a)}\\ &= \mathrm{Re}(r) = r = \left|\int_C f(z)\,dz\right|.\end{aligned}$$

The right inequality in 4.22 is a property of real line integrals when we interpret $|dz| = ds = \sqrt{(dx)^2 + (dy)^2}$ as an element of length along C.

Definite integrals have many geometric and physical interpretations (area, volume, fluid pressure, etc.), as do line integrals (area, work, etc.). Complex contour integrals can also have physical interpretations (see Exercise 13). More important, many areas of applied mathematics make use of complex variable theory, and much of this theory is based on contour integration.

Exercises 4.3

In Exercises 1–8 evaluate the contour integral.

1. $\displaystyle\int_C z\,dz$, where C is the straight line from $z = 0$ to $z = 1 + 2i$

2. $\displaystyle\int_C (3z^2 + z)\,dz$, where C is the straight line from $z = 1$ to $z = i$

3. $\int_C e^z\,dz$, where C is the broken line from $z = -3i$ to $z = 0$ and then from $z = 0$ to $z = 2$

4. $\oint_C z^2\,dz$ once around the circle $C : |z| = 4$

5. $\oint_C \frac{1}{z^2}\,dz$ once around the circle $C : |z - 2| = 1$

6. $\int_C \sin z\,dz$, where C is the straight line from $z = 2i$ to $z = 2$

7. $\int_C \cosh z\,dz$, where C is the broken line from $z = -2$ to $z = 0$ and then from $z = 0$ to $z = i$

8. $\int_C \operatorname{Log} z\,dz$, where C is the straight line from $z = 1 + i$ to $z = 3 + 3i$

In Exercises 9–12 use inequalities 4.22, 1.32, and 1.33 to verify the result.

9. $\left|\oint_C z^{14}\,dz\right| \le 2^{15}\pi R^{15}$, where C is once around the circle $|z - R| = R$

10. $\left|\int_C (z^2 + 4z + 3)\,dz\right| \le 15\pi$, where C is the semicircle $|z - 1| = 1$, $\operatorname{Im} z \ge 0$ from $z = 0$ to $z = 2$

11. $\left|\oint_C \frac{1}{1 + z^4}\,dz\right| \le \frac{2\pi R}{R^4 - 1}$, where C is once around the circle $|z| = R > 1$

12. $\left|\int_C \frac{e^{zi}}{z^2 + 1}\,dz\right| \le \frac{\pi R}{R^2 - 1}$, where C is the semicircle $|z| = R > 1$, $\operatorname{Im} z \ge 0$ from $z = -R$ to $z = R$

13. Suppose that $f(z)$ is the complex potential associated with fluid flow in some domain D. Show that when ρ is taken as unity, circulation and flux for a curve C in D are real and imaginary parts of the contour integral of $\epsilon f'(z)$ along C; that is,

$$(\text{Circulation}) + (\text{Flux})i = \epsilon \int_C f'(z)\,dz.$$

4.4 Independence of Path

Because real line integrals can be independent of path and contour integrals are defined in terms of line integrals, it follows that contour integrals can also be independent of path. Formally we make the following definition.

Definition 4.3

A contour integral $\int_C f(z)\,dz$ is said to be independent of path in a domain D if for any pair of complex numbers z_0 and z_1 in D, the value of the contour integral is the same for all piecewise-smooth curves C in D from z_0 to z_1.

According to the following theorem, contour integrals are independent of path when their integrands have antiderivatives.

■ **THEOREM 4.4** The contour integral of a continuous function f is independent of path in a domain D if and only if f has an antiderivative in D.

Proof For the sufficiency part of the theorem, we suppose that f has an antiderivative F in D. Let z_0 and z_1 be any two points in D, and C be any smooth curve in D joining z_0 to z_1. By equation 4.17,

$$\int_C f(z)\,dz = \int_C u\,dx - v\,dy + i\int_C v\,dx + u\,dy.$$

If we set $F(z) = U(x,y) + V(x,y)i$, then

$$F'(z) = \frac{\partial U}{\partial x} + \frac{\partial V}{\partial x}i = u + vi = f(z).$$

Hence, $u = \partial U/\partial x$ and $v = \partial V/\partial x$, and

$$\int_C f(z)\,dz = \int_C \frac{\partial U}{\partial x}dx - \frac{\partial V}{\partial x}dy + i\int_C \frac{\partial V}{\partial x}dx + \frac{\partial U}{\partial x}dy.$$

Because F is analytic in D, it must satisfy the Cauchy-Riemann equations,

$$\frac{\partial U}{\partial x} = \frac{\partial V}{\partial y}, \qquad \frac{\partial V}{\partial x} = -\frac{\partial U}{\partial y},$$

and using these we can express the contour integral in the form

$$\int_C f(z)\,dz = \int_C \frac{\partial U}{\partial x}dx + \frac{\partial U}{\partial y}dy + i\int_C \frac{\partial V}{\partial x}dx + \frac{\partial V}{\partial y}dy.$$

Because each of these line integrals is independent of path in D,

$$\int_C f(z)\,dz = \Big\{U(x,y)\Big\}_{z_0}^{z_1} + i\Big\{V(x,y)\Big\}_{z_0}^{z_1} = \Big\{U + Vi\Big\}_{z_0}^{z_1} = F(z_1) - F(z_0).$$

If now C is any piecewise-smooth curve with discontinuities in its tangent vector at n points γ_j along C (Figure 4.26), then on each smooth subcurve C_j joining γ_{j-1} to γ_j,

$$\int_{C_j} f(z)\,dz = F(\gamma_j) - F(\gamma_{j-1}),$$

where $\gamma_0 = z_0$ and $\gamma_{n+1} = z_1$. Addition of these results gives

$$\int_C f(z)\,dz = \sum_{j=1}^{n+1} \int_{C_j} f(z)\,dz = \sum_{j=1}^{n+1} [F(\gamma_j) - F(\gamma_{j-1})] = F(z_1) - F(z_0).$$

We have shown then that the contour integral is independent of path in D.

For the necessity part of the theorem, we assume that the contour integral is independent of path in D. If z' is any fixed point in D, then the contour integral

$$F(z) = \int_C f(z)\,dz,$$

where C is any piecewise-smooth curve in D from z' to z, defines a single-valued function F in D. We claim that F is an antiderivative of f. To prove this, we show that $F'(z_0) = f(z_0)$ for any point z_0 in D; that is, we show that

$$\lim_{\Delta z \to 0} \left[\frac{F(z_0 + \Delta z) - F(z_0)}{\Delta z} - f(z_0) \right] = 0.$$

Because D is a domain, there exists a δ-neighborhood of z_0 that is completely contained in D. Suppose $z_0 + \Delta z$ is contained in this neighborhood and $C = C'' + C'$ is the curve joining z' to z_0, where C'' is any piecewise-smooth curve in D joining z' to $z_0 + \Delta z$ and C' is the straight-line segment joining $z_0 + \Delta z$ to z_0 (Figure 4.27).

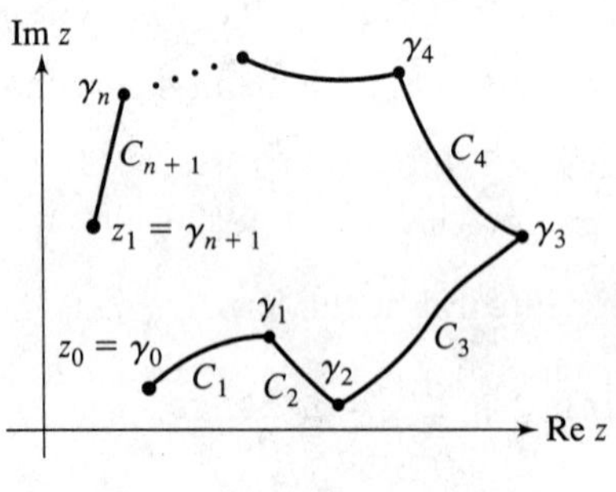

Figure 4.26

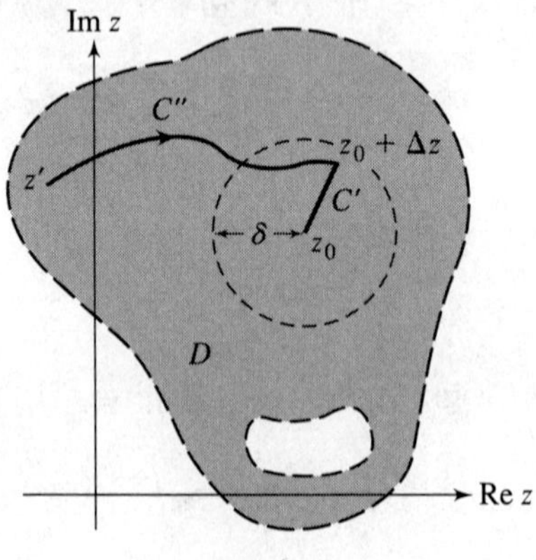

Figure 4.27

Now

$$\begin{aligned} F(z_0 + \Delta z) - F(z_0) &= \int_{C''} f(z)\,dz - \int_{C''+C'} f(z)\,dz \\ &= \int_{C''} f(z)\,dz - \int_{C''} f(z)\,dz - \int_{C'} f(z)\,dz \\ &= -\int_{C'} f(z)\,dz, \end{aligned}$$

and $\int_{C'} dz = -\Delta z$. Hence,

$$\frac{F(z_0+\Delta z)-F(z_0)}{\Delta z} - f(z_0) = -\frac{1}{\Delta z}\int_{C'} f(z)\,dz + \frac{f(z_0)}{\Delta z}\int_{C'} dz$$
$$= \frac{1}{\Delta z}\int_{C'} [f(z_0)-f(z)]\,dz.$$

Because f is continuous, given any $\epsilon > 0$, there exists a $\delta > 0$ such that $|f(z_0) - f(z)| < \epsilon$ whenever $|z - z_0| < \delta$. By inequality 4.22 then,

$$\left|\frac{F(z_0+\Delta z)-F(z_0)}{\Delta z} - f(z_0)\right| = \left|\frac{1}{\Delta z}\int_{C'} [f(z_0)-f(z)]\,dz\right| \le \frac{1}{|\Delta z|}\epsilon\,|\Delta z| = \epsilon.$$

This verifies that $F'(z_0) = f(z_0)$. ■

This theorem has yielded a very simple and important method for evaluating contour integrals.

■ **COROLLARY** If f is continuous in a domain D and has an antiderivative F in D, then for any piecewise-smooth curve C in D joining point z_0 to point z_1,

$$\int_C f(z)\,dz = F(z_1) - F(z_0). \qquad \textbf{(4.23)}$$ ■

Notice the dichotomy that we have here. Because of the two-dimensional nature of the complex plane, contour integrals, although integrals of a function of one variable, are analogous to line integrals. They can be evaluated by substituting from parametric equations of the curve along which integration is performed. But many contour integrals are independent of the path taken between initial point and final point, which is the case when the integrand has an antiderivative. In that case, the value of the contour integral is the difference in values of the antiderivative at final and initial points, which is how definite integrals of functions of one real variable are evaluated. What we are saying is that sometimes it is advantageous to think of contour integrals as similar to real definite integrals; sometimes it is better to think of them as analogues of real line integrals. Be prepared to think either way or both ways.

The following theorem is not particularly practical in determining whether a contour integral is independent of path, but it is very important in deriving practical tests for independence of path.

■ **THEOREM 4.5** The contour integral of a continuous function f is independent of path in a domain D if and only if the contour integral of f around every closed, piecewise-smooth curve C in D vanishes,

$$\oint_C f(z)\,dz = 0. \qquad \textbf{(4.24)}$$

Proof Suppose first of all that the value of the contour integral around every closed, piecewise-smooth curve in D vanishes. Let z_0 and z_1 be any two points in D, and C_1 and C_2 be any two piecewise-smooth curves in D from z_0 to z_1 (Figure 4.28). We can always produce a third curve C_3 from z_0 to z_1 that does not intersect C_1 or C_2. Because the contour integral around closed curves vanishes, we can say that

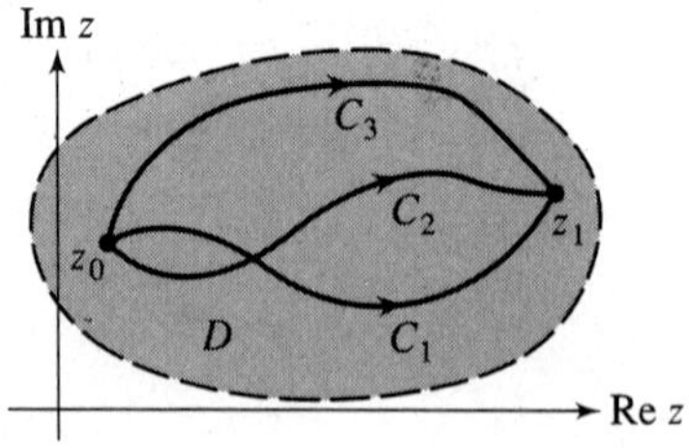

Figure 4.28

$$0 = \oint_{C_2 - C_3} f(z)\,dz = \int_{C_2} f(z)\,dz + \int_{-C_3} f(z)\,dz = \int_{C_2} f(z)\,dz - \int_{C_3} f(z)\,dz.$$

In other words,

$$\int_{C_3} f(z)\,dz = \int_{C_2} f(z)\,dz.$$

Similarly,

$$\int_{C_3} f(z)\,dz = \int_{C_1} f(z)\,dz.$$

Consequently,

$$\int_{C_1} f(z)\,dz = \int_{C_2} f(z)\,dz,$$

and the contour integral is independent of path in D.

Conversely, suppose that the contour integral of f is independent of path in D. Then f has an antiderivative F in D and the value of the contour integral along any curve C from z_0 to z_1 is $F(z_1) - F(z_0)$. When the curve is closed, z_0 and z_1 are identical and $F(z_1) - F(z_0) = 0$. ■

With Theorem 4.5 and the corollary to Theorem 4.4, Examples 4.11 and 4.12 are trivial. In Example 4.11, $z^3/3$ is an antiderivative of z^2 in the whole complex plane, and therefore

$$\int_C z^2\,dz = \left\{\frac{z^3}{3}\right\}_1^i = \frac{1}{3}(i^3 - 1) = -\frac{1+i}{3}.$$

Because e^z is its own antiderivative, the contour integral in Example 4.12 is independent of path. By Theorem 4.5, then, its value around the closed curve in Figure 4.25 vanishes. In fact, the integral of e^z around every closed curve vanishes.

EXAMPLE 4.13 ....

Evaluate $\displaystyle\int_C \frac{1}{z}\,dz$ along the paths in Figure 4.29(a) and (b).

Solution The integrand $1/z$ is continuous in the domain in Figure 4.29(a), and in this domain $d(\operatorname{Log} z)/dz = 1/z$ (the branch cut of $\operatorname{Log} z$ is the negative real axis). Hence,

$$\int_C \frac{1}{z}\,dz = \Big\{\operatorname{Log} z\Big\}_{-i}^{i} = \operatorname{Log}(i) - \operatorname{Log}(-i) = [\ln 1 + i(\pi/2)] - [\ln 1 + i(-\pi/2)] = \pi i.$$

Once again $1/z$ is continuous in the domain in Figure 4.29(b), and in this domain $d(\log_0 z)/dz = 1/z$ (the branch cut of $\log_0 z$ is the positive real axis). Thus,

$$\int_C \frac{1}{z}\,dz = \Big\{\log_0 z\Big\}_{-i}^{i} = \log_0(i) - \log_0(-i) = [\ln 1 + i(\pi/2)] - [\ln 1 + i(3\pi/2)] = -\pi i.$$

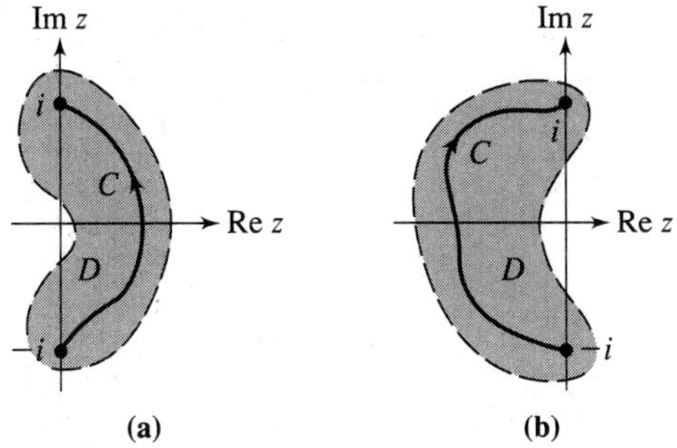

Figure 4.29

EXAMPLE 4.14

Evaluate $\oint_C \frac{1}{(z-z_0)^n}\,dz$, where n is an integer and $C : |z - z_0| = r$.

Solution If $n \neq 1$, then in the domain $|z - z_0| > 0$ (Figure 4.30), the integrand has antiderivative

$$\int \frac{1}{(z-z_0)^n}\,dz = \frac{1}{(1-n)(z-z_0)^{n-1}} + k.$$

Consequently, for $n \neq 1$,

$$\oint_C \frac{1}{(z-z_0)^n}\,dz = 0.$$

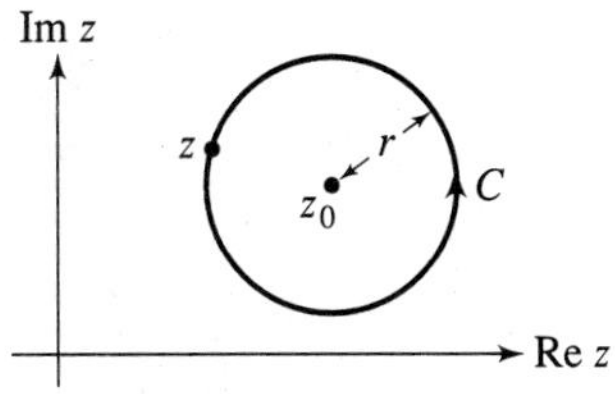

Figure 4.30

For $n = 1$, on the other hand, $1/(z - z_0)$ does not have an antiderivative in any domain that contains C. (Every branch of $\log(z - z_0)$ has a branch cut that intersects C.) If we set $z = z_0 + re^{\theta i}$, $-\pi < \theta \leq \pi$, on C, however,

$$\oint_C \frac{1}{z - z_0}\, dz = \int_{-\pi}^{\pi} \frac{1}{re^{\theta i}}\, rie^{\theta i}\, d\theta = \left\{\theta i\right\}_{-\pi}^{\pi} = 2\pi i.$$

Exercises 4.4

For Exercises 1–8 evaluate the contour integrals in Exercises 1–8 of Exercises 4.3 (page 155).

In Exercises 9–20 evaluate the contour integral.

9. $\int_C \frac{1}{(z-1)^3}\, dz$, where C is the semicircle $|z - 1| = 1$, $\operatorname{Im} z \geq 0$ from $z = 0$ to $z = 2$

10. $\int_C \frac{z}{z^4 + 2z^2 + 1}\, dz$, where C is the broken straight line joining successively $1 + i$, $2i$, and $-1 + i$

11. $\int_C e^z \sin z\, dz$, where C is the curve $\operatorname{Im} z = (\operatorname{Re} z)^2$ from $z = 0$ to $z = 1 + i$

12. $\int_C z \sinh 2z\, dz$, where C is the straight line from $z = -1$ to $z = i$

13. $\int_C \sin^2 z\, dz$, where C is the right half of the ellipse $|z - i| + |z + i| = 6$ from $z = -3i$ to $z = 3i$

14. $\int_C \frac{1}{z}\, dz$, where C is the quarter-circle $|z - i| = 2$ from $z = -i$ to $z = -2 + i$

15. $\int_C \frac{1}{z^2 - z}\, dz$, where C is any curve from $z = -2$ to $z = -3 + i$ for which $\operatorname{Re} z$ is always negative

16. $\int_C \frac{1}{\sqrt{z}}\, dz$, where C is the quarter-circle $|z| = 2$ from $z = \sqrt{2} - \sqrt{2}i$ to $z = \sqrt{2} + \sqrt{2}i$

17. $\int_C \frac{1}{z^2 + 4}\, dz$, where C is the straight line from $z = 1$ to $z = i$

18. $\int_C z^{1/2}\, dz$, where C is the straight line from $z = -2 - 2\sqrt{3}i$ to $z = -2 + 2\sqrt{3}i$ and $z^{1/2}$ is the branch of the square root function $z^{1/2} = e^{(1/2)\log_{-\pi/2} z}$

19. $\oint_C \operatorname{Log} z\, dz$, where C is the square with vertices ± 1, $\pm i$

20. $\oint_C \sqrt{z}\, dz$, where $C : |z| = 3$

4.5 The Cauchy-Goursat Theorem

The contour integral of a continuous function f is independent of path in a domain D if and only if f has an antiderivative in D or, equivalently, if and only if the contour integral of f around every closed, piecewise-smooth curve vanishes. In other words, we have two methods for determining whether the contour integral of a continuous function is independent of path:

1. Prove that f has an antiderivative in D.
2. Prove that the contour integral of f around every closed, piecewise-smooth curve in D is equal to zero.

This is not a completely satisfactory situation. Condition (2) clearly is not the easiest to put into practice. Condition (1) may be somewhat difficult also. For instance, if the integrand is given in terms of its real and imaginary parts, it would be necessary to find its definition $f(z)$ in terms of z. Even if f is defined in terms of z, say, $f(z) = \sin(1/z)$, how shall we know whether it has an antiderivative, or what the antiderivative is?

What we need then is a condition that guarantees existence of the antiderivative of a function. To find such a condition, suppose that C is a simple, closed, piecewise-smooth curve (Figure 4.31) and that f' is continuous in a domain containing C and its interior R.

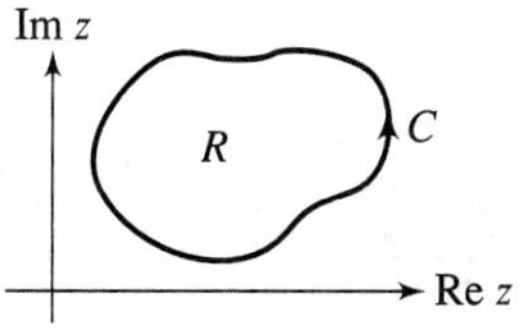

Figure 4.31

Because

$$f'(z) = \frac{\partial u}{\partial x} + \frac{\partial v}{\partial x} i = \frac{\partial v}{\partial y} - \frac{\partial u}{\partial y} i,$$

it follows that the first partial derivatives of u and v are continuous inside and on C. We may therefore apply Green's theorem (Theorem 4.3) to each of the real line integrals in 4.17:

$$\begin{aligned}\oint_C f(z)\,dz &= \oint_C u\,dx - v\,dy + i\oint_C v\,dx + u\,dy \\ &= \iint_R \left(-\frac{\partial v}{\partial x} - \frac{\partial u}{\partial y}\right) dA + i\iint_R \left(\frac{\partial u}{\partial x} - \frac{\partial v}{\partial y}\right) dA.\end{aligned}$$

But if f' is continuous, then certainly f is analytic (analyticity requiring only existence of f'), and hence f must satisfy the Cauchy-Riemann equations. Consequently, each integrand vanishes in R, and

$$\oint_C f(z)\,dz = 0.$$

We have just established the following result.

■ **THEOREM 4.6** If the derivative f' of a complex function f is continuous in a domain containing a simple, closed, piecewise-smooth curve C and its interior, then

$$\oint_C f(z)\,dz = 0. \qquad \blacksquare$$

If C is closed but not simple (Figure 4.32), the contour integral in Theorem 4.6 can be expressed as a sum of contour integrals over simple, closed curves, and hence the simple character of C is an unnecessary restriction. The theorem can be generalized even further, but not so trivially. The continuity of f' was required because we wished to use Green's theorem. It can be shown, by completely different methods, that this condition is unnecessary also. We state the result here and prove it at the end of this section.

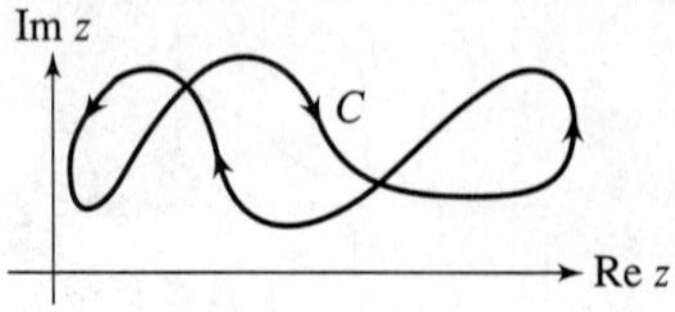

Figure 4.32

■ **THEOREM 4.7** (Cauchy-Goursat Theorem) If f is analytic inside and on a closed, piecewise-smooth curve C, then

$$\oint_C f(z)\,dz = 0. \qquad \blacksquare$$

We illustrate the Cauchy-Goursat theorem with a simple example before indicating its importance in complex variable theory.

EXAMPLE 4.15 ....

Evaluate $\oint_C \frac{e^z}{z^2 - 9}\, dz$ where $C : |z| = 2$.

Solution Singularities of $e^z/(z^2 - 9)$ occur at $z = \pm 3$ (Figure 4.33). The integrand is therefore analytic inside and on C, and the Cauchy-Goursat theorem implies that the value of the integral is zero.

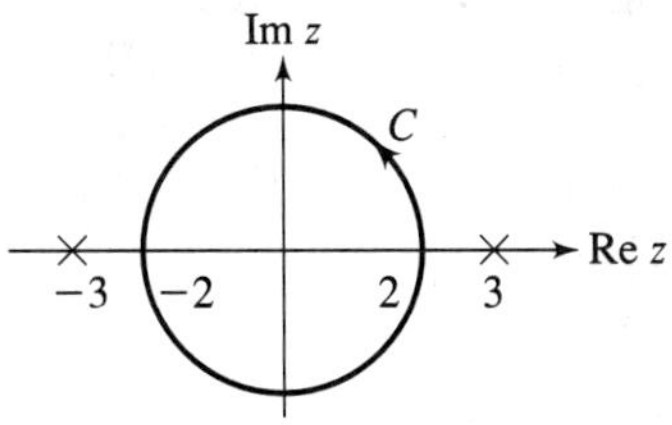

Figure 4.33

....

The consequences of the Cauchy-Goursat theorem can be developed in two directions. First, it has a straightforward extension that is sometimes useful in the evaluation of contour integrals around closed curves inside of which the integrand is not analytic. Second, the consequences of the Cauchy-Goursat theorem in the theory of analytic functions cannot be overstated. We discuss the extension of the Cauchy-Goursat theorem to other contour integrals before indicating its impact on analytic functions.

The Cauchy-Goursat theorem yields a value of zero for the contour integral of an analytic function around a closed curve that contains only points at which the function is analytic. If the curve encloses singularities of the function, we might be able to resort to the antiderivative (Theorem 4.4), but not the Cauchy-Goursat theorem. For instance, the integrand of

$$\oint_C \frac{1}{(z-1)^2}\, dz$$

has a singularity at $z = 1$ interior to the curve C in Figure 4.34. We cannot, therefore, use the Cauchy-Goursat theorem. However, $1/(z-1)^2$ has antiderivative $-1/(z-1)$ in the domain $|z - 1| > 0$. Consequently, by Theorem 4.4, the value of this contour integral is zero.

Neither Theorem 4.4 nor 4.7 is applicable to the integrals

$$\oint_C \frac{1}{z-1}\, dz, \qquad \oint_C \frac{e^z}{z^2-1}\, dz$$

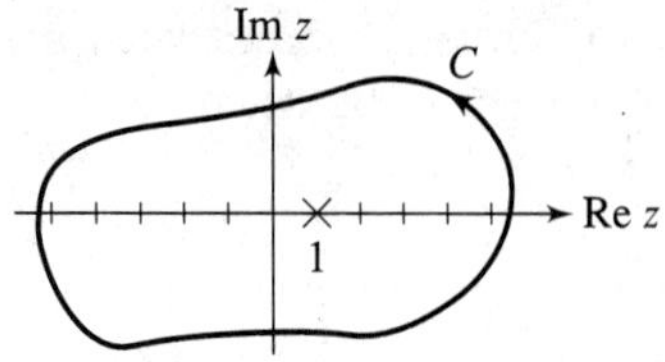

Figure 4.34

when C is the curve of Figure 4.34. The Cauchy-Goursat theorem cannot be used, because integrands are not analytic at all points interior to C. Theorem 4.4 cannot be applied to the first integral, because $1/(z-1)$ does not have an antiderivative in any domain that contains C (any domain that contains C also contains points on the branch cut of every branch of $\log(z-1)$). It is not at all clear whether $e^z/(z^2-1)$ has an antiderivative, and consequently Theorem 4.4 fails on the second integral as well. For such examples the following theorem permits us to simplify the problem by replacing the given curve with another, or others.

■ **THEOREM 4.8** Let C be a simple, closed, piecewise-smooth curve and $C_1, C_2, \ldots, C_n$ be disjoint, simple, closed, piecewise-smooth curves in the interior of C. If f is analytic at all points that are both inside or on C and outside or on each C_j, then

$$\oint_C f(z)\,dz = \sum_{j=1}^{n} \oint_{C_j} f(z)\,dz. \tag{4.25}$$

Proof Construct $n+1$ piecewise-smooth curves γ_j joining C_{j-1} to C_j (γ_1 joins C to C_1, and γ_{n+1} joins C_n to C), so that no two γ_j intersect (Figure 4.35). This divides the region inside C and outside the C_j into two parts. Each part is bounded by a closed, piecewise-smooth curve, say, L_1 and L_2, inside and on which f is analytic. Thus, by the Cauchy-Goursat theorem, the contour integrals of f around L_1 and L_2 vanish:

$$\oint_{L_1} f(z)\,dz = 0, \qquad \oint_{L_2} f(z)\,dz = 0.$$

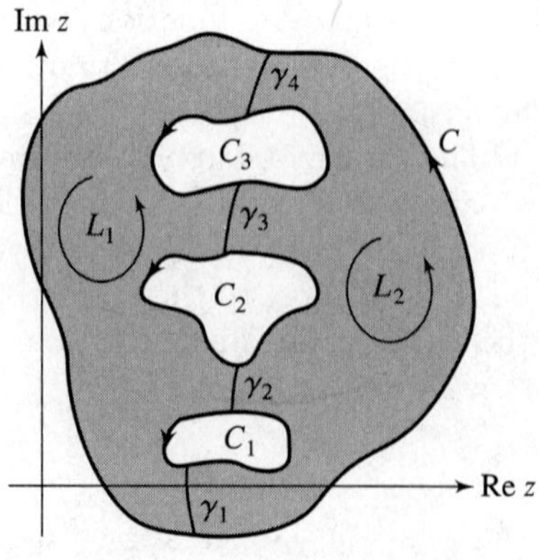

Figure 4.35

Addition of these equations gives

$$\oint_{L_1} f(z)\,dz + \oint_{L_2} f(z)\,dz = 0.$$

But if L_1 and L_2 are traversed in the directions indicated, Figure 4.35 clearly shows that this combination of contour integrals amounts to:

1. the contour integral around C in the direction shown; plus
2. the contour integrals around each C_j in the opposite direction to that shown; plus
3. the contour integrals over each γ_j, once in each direction.

Consequently,

$$\oint_C f(z)\,dz + \sum_{j=1}^{n} \oint_{-C_j} f(z)\,dz = 0,$$

or

$$\oint_C f(z)\,dz = \sum_{j=1}^{n} \oint_{C_j} f(z)\,dz. \qquad \blacksquare$$

EXAMPLE 4.16 ....

Evaluate $\displaystyle\oint_C \frac{1}{z-1}\,dz$, where C is the curve in Figure 4.34.

Solution The integrand $1/(z-1)$ fails to be analytic at the point $z = 1$ inside C, and hence the Cauchy-Goursat theorem is not applicable. With Theorem 4.8, we can replace integration around the somewhat ugly curve C with integration around the much simpler curve $C' : |z-1| = 1$. The reason for this is that $1/(z-1)$ is analytic at all points that are both inside and on C and outside or on C' (Figure 4.36). Hence,

$$\oint_C \frac{1}{z-1}\,dz = \oint_{C'} \frac{1}{z-1}\,dz = 2\pi i$$

(see Example 4.14).

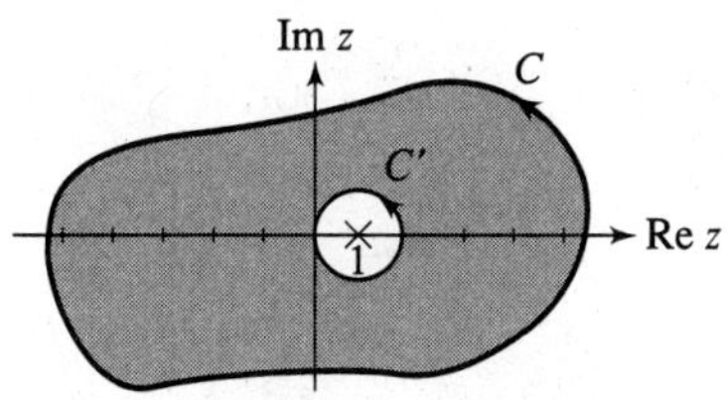

Figure 4.36

....

A slight generalization of this example occurs so often in future work that we state it as a theorem. It is proved in exactly the same way as Example 4.16.

■ **THEOREM 4.9** If C is a simple, closed, piecewise-smooth curve and z_0 is interior to C, then

$$\oint_C \frac{1}{z - z_0}\, dz = 2\pi i. \tag{4.26}$$ ■

In the next example, the integrand has two singularities inside the contour of integration.

EXAMPLE 4.17 ....

Evaluate $\oint_C \frac{1}{z^2 - 4} dz$, where C is the curve $|z| = 4$.

Solution Because $1/(z^2 - 4)$ is analytic at all points both inside and on C and outside and on the curves $C' : |z + 2| = 1$ and $C'' : |z - 2| = 1$ in Figure 4.37, Theorem 4.8 allows us to replace integration around C by integrations around C' and C'':

$$\oint_C \frac{1}{z^2 - 4} dz = \oint_{C'} \frac{1}{(z+2)(z-2)} dz + \oint_{C''} \frac{1}{(z+2)(z-2)} dz.$$

The partial fraction decomposition of $1/(z^2 - 4)$ is $\frac{1}{z^2 - 4} = \frac{-1/4}{z+2} + \frac{1/4}{z-2}$. It follows that

$$\oint_C \frac{1}{z^2 - 4} dz = \frac{1}{4}\oint_{C'} \left(\frac{1}{z-2} - \frac{1}{z+2}\right) dz + \frac{1}{4}\oint_{C''} \left(\frac{1}{z-2} - \frac{1}{z+2}\right) dz.$$

Because $1/(z - 2)$ is analytic inside and on C', its integral around C' is zero (by the Cauchy-Goursat theorem). Similarly, the contour integral of $1/(z + 2)$ around

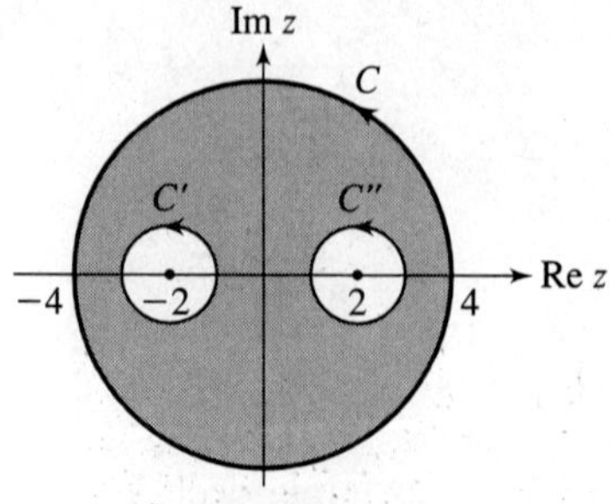

Figure 4.37

C'' vanishes. According to Theorem 4.9,

$$\oint_{C'} \frac{1}{z+2}\, dz = 2\pi i \quad \text{and} \quad \oint_{C''} \frac{1}{z-2}\, dz = 2\pi i.$$

Consequently,

$$\oint_{C} \frac{1}{z^2-4}\, dz = \frac{1}{4}(-2\pi i) + \frac{1}{4}(2\pi i) = 0.$$

By Theorem 4.8 we could replace the contour integral $\oint_C \frac{e^z}{z^2-9}\, dz$, where C is the curve in Figure 4.34, with the integrals

$$\oint_C \frac{e^z}{z^2-9}\, dz = \oint_{C'} \frac{e^z}{z^2-9}\, dz + \oint_{C''} \frac{e^z}{z^2-9}\, dz,$$

where C' and C'' are unit circles centered at $z = \pm 3$ (Figure 4.38). Each of these integrals is evaluated in the next section.

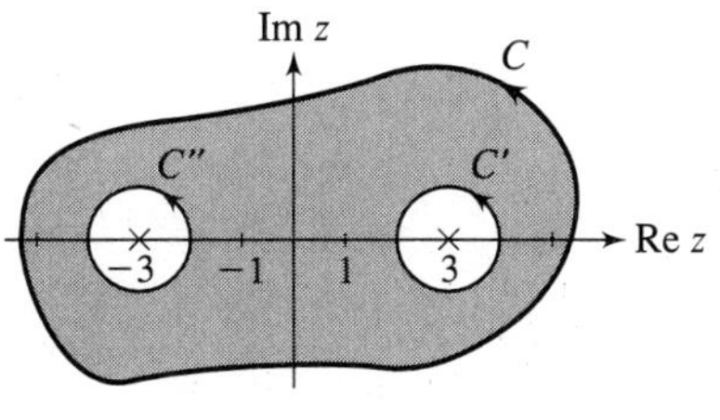

Figure 4.38

The Cauchy-Goursat theorem is not just a tool for evaluation of contour integrals; it and its consequences are foremost in the theory of analytic functions. Three immediate implications are discussed here; other consequences are discussed in Section 4.6. Although the three results mentioned here require very little proof, their importance justifies designating them as theorems rather than corollaries to the Cauchy-Goursat theorem.

According to Theorem 4.5, the contour integral of a complex function f is independent of path in a domain D if the contour integral of f around every closed, piecewise-smooth curve C in D vanishes:

$$\oint_C f(z)\, dz = 0.$$

According to the Cauchy-Goursat theorem, if f is analytic inside and on C, the contour integral of f around C vanishes. Can we put these two results together and say that when f is analytic in a domain D, its contour integral is independent of path in D? No, we could do this only if we could guarantee that every closed, piecewise-smooth curve in D contained points at which f is analytic, and this is not, in general, possible. Let us illustrate with the function $f(z) = 1/z$ and two different domains. First consider the domain $D^* : 0 < |z| < 1$, the unit circle with center $z = 0$ deleted. The contour integral of $1/z$ around any closed, piecewise-smooth curve in D^* that does not enclose $z = 0$ (Figure 4.39(a)) has value zero, because the curve contains only points of analyticity of $1/z$. However, the contour integral around curves that enclose $z = 0$ (Figure 4.39(b)) do not all have value zero. For instance, around $|z| = 1/2$ the value is $\pm 2\pi i$, depending on clockwise or counterclockwise orientation. Thus, the contour integral of $1/z$ is not independent of path in $0 < |z| < 1$.

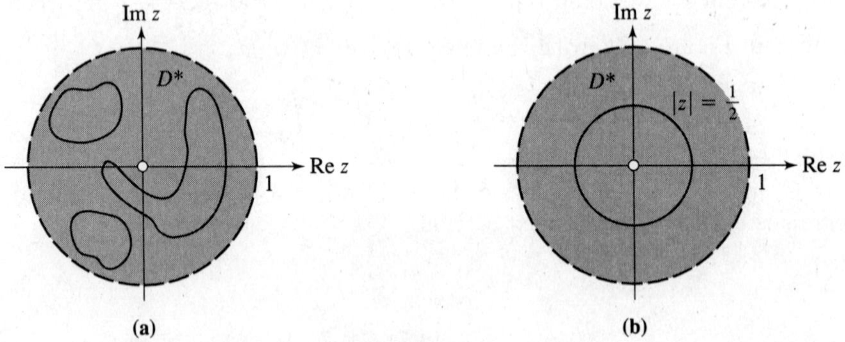

Figure 4.39

Consider now the domain D in Figure 4.40. Every closed, piecewise-smooth curve C in this domain contains only points at which $1/z$ is analytic. Hence, for any such curve

$$\oint_C \frac{1}{z}\, dz = 0,$$

and the contour integral of $1/z$ is independent of path in D.

Domains D and D^* are different. We can say that every closed curve in D contains only points of D; we cannot say this for D^*. This attribute of D was described in Section 4.2; D is a simply connected domain, D^* is not. The domains in Figures 4.41 and 4.42 are not simply connected; the first has two holes, and the second has three holes.

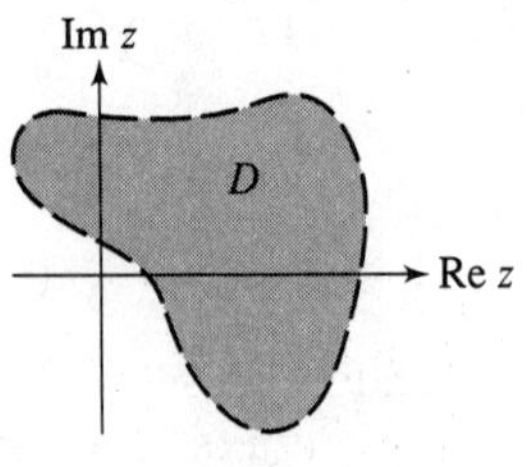

Figure 4.40

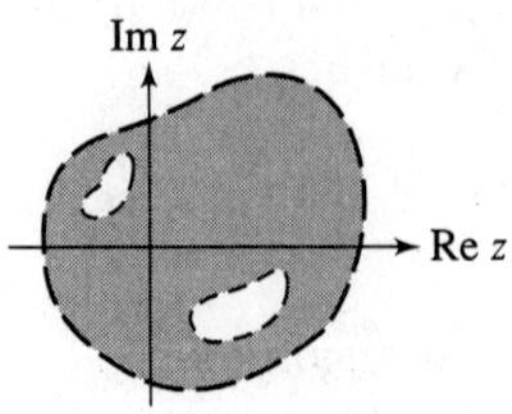

Figure 4.41

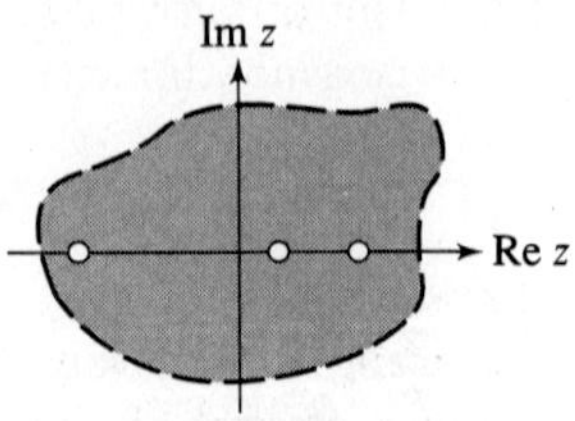

Figure 4.42

For simply connected domains, the Cauchy-Goursat theorem implies the following result.

■ **THEOREM 4.10** When f is analytic in a simply connected domain D, the contour integral of f around every closed, piecewise smooth curve in D vanishes. ■

Theorem 4.5 combines with this result to give the next theorem.

■ **THEOREM 4.11** When f is analytic in a simply connected domain D, the contour integral of f is independent of path in D. ■

Furthermore, Theorem 4.4 now implies that functions analytic in simply connected domains have antiderivatives therein.

■ **THEOREM 4.12** When f is analytic in a simply connected domain D, then f has an antiderivative in D. ■

More important consequences of the Cauchy-Goursat theorem are developed in the next section.

Appendix (Proof of Cauchy-Goursat Theorem)

The Cauchy-Goursat theorem states that when f is analytic inside and on a closed, piecewise-smooth curve C, then

$$\oint_C f(z)\,dz = 0.$$

Our discussion leading from Theorem 4.6 to the Cauchy-Goursat theorem indicated that there is no loss in generality in assuming that C is a simple curve (since curves that intersect themselves can be decomposed into simple ones). We therefore assume that C is simple, and we begin with a curve C that is a triangle (Figure 4.43(a)).

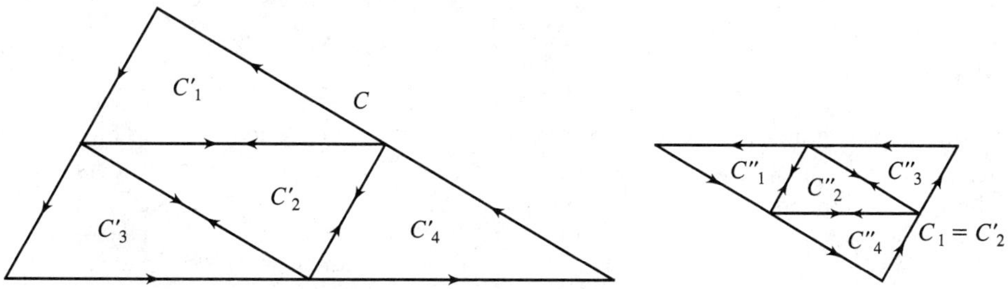

(a) (b)

Figure 4.43

To show that the contour integral of f around this curve vanishes, it is sufficient to show that its modulus vanishes,

$$\left|\oint_C f(z)\,dz\right| = 0,$$

and this is true if we can show that given any $\epsilon > 0$, no matter how small,

$$\left|\oint_C f(z)\,dz\right| < \epsilon.$$

Suppose we denote by C_1', C_2', C_3', and C_4' the edges of the four triangles obtained by joining midpoints of the sides of triangle C. Because of internal cancellations,

$$\oint_C f(z)\,dz = \oint_{C_1'} f(z)\,dz + \oint_{C_2'} f(z)\,dz + \oint_{C_3'} f(z)\,dz + \oint_{C_4'} f(z)\,dz.$$

If we take moduli, then inequality 1.32 gives

$$\left|\oint_C f(z)\,dz\right| \le \left|\oint_{C_1'} f(z)\,dz\right| + \left|\oint_{C_2'} f(z)\,dz\right| + \left|\oint_{C_3'} f(z)\,dz\right| + \left|\oint_{C_4'} f(z)\,dz\right|.$$

One of the contour integrals on the right is at least as large as each of the other three. If we denote this contour by C_1 (so that C_1 is one of C_1', C_2', C_3', or C_4', say C_2'), then

$$\left|\oint_C f(z)\,dz\right| \le 4\left|\oint_{C_1} f(z)\,dz\right|.$$

We now repeat this process on triangle C_1 (Figure 4.43(b)), yielding a contour C_2 for which

$$\left|\oint_C f(z)\,dz\right| \le 4^2\left|\oint_{C_2} f(z)\,dz\right|.$$

Continuation leads to a sequence of contours $\{C_n\}$ where the interior of C_n lies in the interior of C_{n-1} and

$$\left|\oint_C f(z)\,dz\right| \le 4^n\left|\oint_{C_n} f(z)\,dz\right|.$$

What seems geometrically clear but is not easily proved is the following fact: There is a unique point z_0 that belongs to each and every one of the triangles C_n. Assuming that this is indeed the case, f is analytic at z_0, and we can write that

$$f'(z_0) = \frac{f(z) - f(z_0)}{z - z_0} + g(z),$$

where g is continuous in some δ-neighborhood of z_0 and where $\lim_{z\to z_0} g(z) = 0$.

When this equation is solved for $f(z)$, the result is

$$f(z) = f(z_0) + f'(z_0)(z - z_0) - g(z)(z - z_0).$$

Suppose n is sufficiently large that C_n is contained in the δ-neighborhood of z_0 (Figure 4.44). Then

$$\begin{aligned}\oint_{C_n} f(z)\,dz &= \oint_{C_n} [f(z_0) + f'(z_0)(z - z_0) - g(z)(z - z_0)]\,dz \\ &= f(z_0)\oint_{C_n} dz + f'(z_0)\oint_{C_n} (z - z_0)\,dz - \oint_{C_n} g(z)(z - z_0)\,dz \\ &= -\oint_{C_n} g(z)(z - z_0)\,dz.\end{aligned}$$

We propose to use inequality 4.22 on the modulus of the contour integral on the right. This requires us to know the length of C_n and to have a maximum value for the modulus of the integrand. Clearly the length L_n of C_n is one-half the length of C_{n-1}, and therefore $L_n = L/2^n$, where L is the length of C. Because z_0 is inside or on C_n and z is on C_n, it follows that $|z - z_0|$, the length of the line joining z_0 and z, must be less than one-half the length of C_n; i.e., $|z - z_0| < L_n/2 = L/2^{n+1}$. Because $\lim_{z\to z_0} g(z) = 0$, δ can be chosen sufficiently small that for all z in the δ-neighborhood, $|g(z)| < 2\epsilon/L^2$. With these results, inequality 4.22 applied to the preceding contour integral gives

$$\left|\oint_{C_n} f(z)\,dz\right| = \left|\oint_{C_n} g(z)(z - z_0)\,dz\right| \le \left(\frac{2\epsilon}{L^2}\right)\left(\frac{L}{2^{n+1}}\right)\left(\frac{L}{2^n}\right) = \frac{\epsilon}{4^n}.$$

Consequently,

$$\left|\oint_C f(z)\,dz\right| \le 4^n \left|\oint_{C_n} f(z)\,dz\right| \le 4^n\left(\frac{\epsilon}{4^n}\right) = \epsilon.$$

As indicated earlier, because ϵ can be made arbitrarily small, this result implies that

$$\oint_C f(z)\,dz = 0.$$

Suppose now that C is a polygon rather than a triangle (Figure 4.45). Interior lines can be introduced so that the interior of C is divided into a finite number of triangles. The contour integral of f around each of these triangles is zero, and the sum of the contour integrals around the triangles is equal to the integral around C. Hence the contour integral of f around a polygonal path also vanishes.

The final step in the proof should be to show that any curved path can be approximated arbitrarily closely by a polygonal one. We omit this part of the proof.

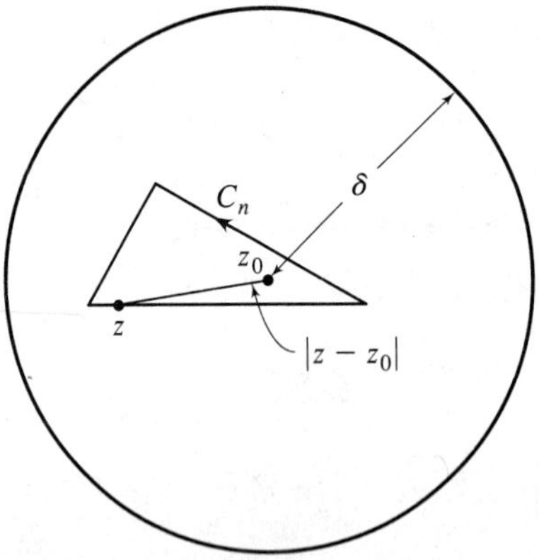

Figure 4.44

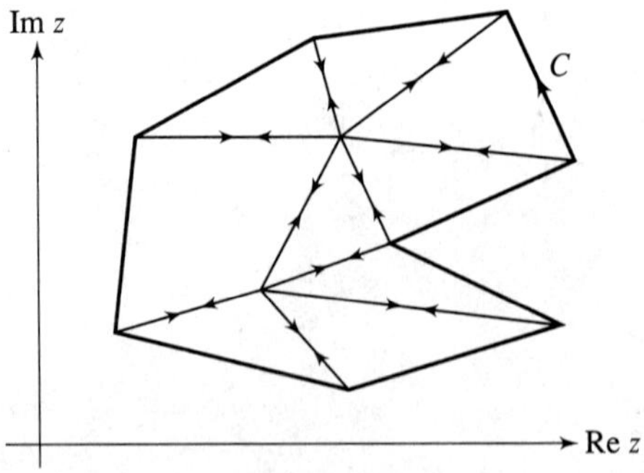

Figure 4.45

Exercises 4.5

For Exercises 1–10 evaluate the contour integral.

1. $\oint_C \frac{e^z}{z-5}\,dz$, where $C : |z| = 3$

2. $\oint_C \frac{\cosh 2z}{(z+2)^3}\,dz$, where $C : |z-2| = 2$

3. $\oint_C \frac{1}{z^2+4z+16}\,dz$, where $C : |z| = 1$

4. $\oint_C \frac{1}{1-z^4}\,dz$, where $C : |z+4i| = 1$

5. $\oint_C \frac{1}{z^2-1}\,dz$, where $C : |z| = 2$

6. $\oint_C \frac{1}{z^2+2z}\,dz$, where $C : |z+1| = 3$

7. $\oint_C \frac{z}{z^2+1}\,dz$, where C is the square with vertices $z = 0$, $z = \pm 1 + i$, $z = 2i$

8. $\oint_C \frac{1}{z^2 \sin z}\,dz$, where $C : |z-2-2i| = 1$

9. $\oint_C \frac{1}{z^2+4z+5}\,dz$, where $C : |z+2-2i| = 2$

10. $\oint_C \frac{1}{z^3+1}\,dz$, where $C : |z-i| = 1$

In Exercises 11–19 determine whether the domain is simply connected.

11. $1 < |z-3| < 2$

12. $|z+1| + |z-1| < 4$

13. $|\text{Re}\, z| < 3$

14. $|\text{Re}\, z| > 3$

15. $2\,\text{Re}\, z + 3\,\text{Im}\, z < 6$

16. $0 < \text{Arg}\, z < \pi/2$

17. $|z| < 5$, $z \neq 1, 2, 3$

18. $|z| > 5$

19. The domain of analyticity of $\text{Log}\, z$

20. Is $0 < |\text{Im}\, z| < 1$ a simply connected domain?

21. **(a)** Prove that when z_1 and z_2 are interior to a simple, closed, piecewise-smooth curve C, then

$$\oint_C \frac{1}{(z-z_1)(z-z_2)}\, dz = 0.$$

(b) If $z_1, \ldots, z_n$ are n points interior to C, is

$$\oint_C \frac{1}{(z-z_1)\cdots(z-z_n)}\, dz = 0?$$

22. Evaluate $\displaystyle\int_C z^i\, dz$, where $i \neq$ integer, z^i is the principal branch of the function, and C is any contour from $z = -1$ to $z = 1$ that, except for endpoints, lies in the half-plane $\text{Im}\, z > 0$.

23. Suppose that $P(z)$ is a polynomial of degree greater than 1 and C is a simple, closed, piecewise-smooth curve that contains all zeros of $P(z)$.

(a) Show that

$$\oint_C \frac{1}{P(z)}\, dz = \oint_{C_R} \frac{1}{P(z)}\, dz$$

if C_R is a circle centered at the origin with radius R that contains C.

(b) By writing $P(z)$ in factored form

$$P(z) = A(z-z_1)^{k_1}(z-z_2)^{k_2}\cdots(z-z_m)^{k_m}$$

where A is a constant, z_j are the zeros of $P(z)$ with multiplicities k_j, and $k_1 + k_2 + \cdots + k_m = n$, show that on C_R,

$$\frac{1}{|P(z)|} \leq \frac{1}{|A|(R-|z_1|)^{k_1}(R-|z_2|)^{k_2}\cdots(R-|z_m|)^{k_m}}.$$

(c) Use inequality 4.22 and parts (a) and (b) to show that

$$\oint_C \frac{1}{P(z)}\, dz = 0.$$

4.6 Cauchy Integral Formulas

In this section we develop the Cauchy integral formulas. They serve two purposes. They provide an additional method for evaluation of contour integrals, but, far more importantly, they lead to properties of analytic functions that are unparalleled in real calculus.

When f is analytic inside and on a simple, closed, piecewise-smooth curve C, the Cauchy-Goursat theorem states that

$$\oint_C f(z)\,dz = 0.$$

When z_0 is a point interior to C, the integrand of

$$\oint_C \frac{f(z)}{z-z_0}\,dz$$

for the same curve C and function f has an isolated singularity at z_0 (Figure 4.46).

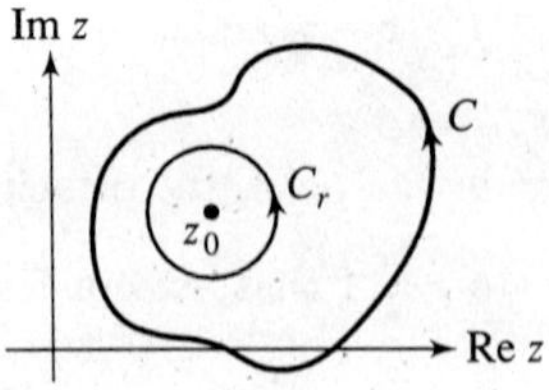

Figure 4.46

Theorem 4.8 (a consequence of the Cauchy-Goursat theorem) implies that when $C_r : |z - z_0| = r$ is a circle with center z_0 that lies in the interior of C, then the contour integral around C has the same value as the contour integral around C_r,

$$\oint_C \frac{f(z)}{z-z_0}\,dz = \oint_{C_r} \frac{f(z)}{z-z_0}\,dz.$$

We manipulate the integral on the right,

$$\begin{aligned}\oint_C \frac{f(z)}{z-z_0}\,dz = \oint_{C_r} \frac{f(z)}{z-z_0}\,dz &= \oint_{C_r} \frac{f(z_0)}{z-z_0}\,dz + \oint_{C_r} \frac{f(z)-f(z_0)}{z-z_0}\,dz \\ &= f(z_0)\oint_{C_r} \frac{1}{z-z_0}\,dz + \oint_{C_r} \frac{f(z)-f(z_0)}{z-z_0}\,dz \\ &= 2\pi i f(z_0) + \oint_{C_r} \frac{f(z)-f(z_0)}{z-z_0}\,dz,\end{aligned}$$

(see Theorem 4.9). Because the integral on the left side of this equation has a fixed value, as does $2\pi i f(z_0)$, we may state that for every circle C_r interior to C, the value of the contour integral

$$\oint_{C_r} \frac{f(z) - f(z_0)}{z - z_0}\, dz \tag{4.27}$$

is always the same. We now show that this value must be zero. Suppose M is the maximum value of $|f(z) - f(z_0)|$ on C_r. By property 4.22 for contour integrals,

$$\left| \oint_{C_r} \frac{f(z) - f(z_0)}{z - z_0}\, dz \right| \le \frac{M}{r}(2\pi r) = 2\pi M.$$

Now M is a function of r, say $M(r)$, and since f is continuous in D, $\lim_{r\to 0} M(r) = 0$. What have we shown? First, the value of contour integral 4.27 is the same for all circles $C_r : |z - z_0| = r$ inside C. Second, by choosing r sufficiently small, we can make the value of the integral, in modulus, arbitrarily close to zero. It follows that the value of this contour integral must be zero for all C_r, and consequently

$$\oint_C \frac{f(z)}{z - z_0}\, dz = 2\pi i f(z_0), \quad \text{or} \quad f(z_0) = \frac{1}{2\pi i} \oint_C \frac{f(z)}{z - z_0}\, dz.$$

This result is called Cauchy's integral formula. We restate it in the following theorem, replacing z_0 by z and z by ζ.

■ **THEOREM 4.13** (Cauchy Integral Formula) When f is analytic inside and on a simple, closed, piecewise-smooth curve C, its value at any point z interior to C is given by the contour integral

$$f(z) = \frac{1}{2\pi i} \oint_C \frac{f(\zeta)}{\zeta - z}\, d\zeta. \tag{4.28}$$ ■

This formula states that the values of an analytic function interior to a curve C are completely determined by its values on C. This result is amazingly different compared to the situation for real functions. When a real function $f(x)$ is differentiable on $a \le x \le b$, its values at $x = a$ and $x = b$ in no way dictate its values between $x = a$ and $x = b$. Likewise, when a function $f(x, y)$ is differentiable inside and on a simple, closed, piecewise-smooth curve C, its values on C do not determine its values inside C. Just the opposite is true for complex analytic functions. Values on a simple, closed, piecewise-smooth curve completely determine its values inside C. The following theorem enables us to extend the Cauchy integral formula to an even more startling result.

■ **THEOREM 4.14** Let f be continuous on a piecewise-smooth curve C (Figure 4.47). At each point z not on C, define the value of a function F by

$$F(z) = \int_C \frac{f(\zeta)}{\zeta - z}\, d\zeta. \tag{4.29a}$$

Then F is analytic in the domain consisting of all points not on C, and

$$F'(z) = \int_C \frac{f(\zeta)}{(\zeta - z)^2}\, d\zeta. \qquad \textbf{(4.29b)}$$

Proof By the definition of the derivative of a complex function,

$$\begin{aligned} F'(z) &= \lim_{\Delta z \to 0} \frac{F(z+\Delta z) - F(z)}{\Delta z} \\ &= \lim_{\Delta z \to 0} \frac{1}{\Delta z}\left[\int_C \frac{f(\zeta)}{\zeta - (z+\Delta z)}\, d\zeta - \int_C \frac{f(\zeta)}{\zeta - z}\, d\zeta\right] \\ &= \lim_{\Delta z \to 0} \int_C \frac{f(\zeta)}{(\zeta - z - \Delta z)(\zeta - z)}\, d\zeta. \end{aligned}$$

To verify 4.29(b), we show that

$$\begin{aligned} 0 &= \lim_{\Delta z \to 0}\left[\int_C \frac{f(\zeta)}{(\zeta - z - \Delta z)(\zeta - z)}\, dz - \int_C \frac{f(\zeta)}{(\zeta - z)^2}\, d\zeta\right] \\ &= \lim_{\Delta z \to 0} \int_C \frac{f(\zeta)\Delta z}{(\zeta - z - \Delta z)(\zeta - z)^2}\, d\zeta. \end{aligned}$$

To estimate the maximum value of the modulus of this integral, let M be the maximum value of $|f(\zeta)|$ for ζ on C and d be the minimum value of $|\zeta - z|$ for ζ on C (Figure 4.48). With inequality 1.33, we may write that $|\zeta - z - \Delta z| \geq |\zeta - z| - |\Delta z| > 0$, provided $|\Delta z|$ is chosen smaller than d.

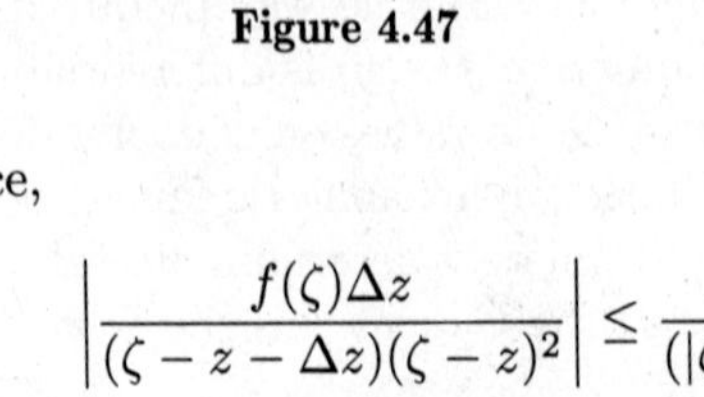

Figure 4.47

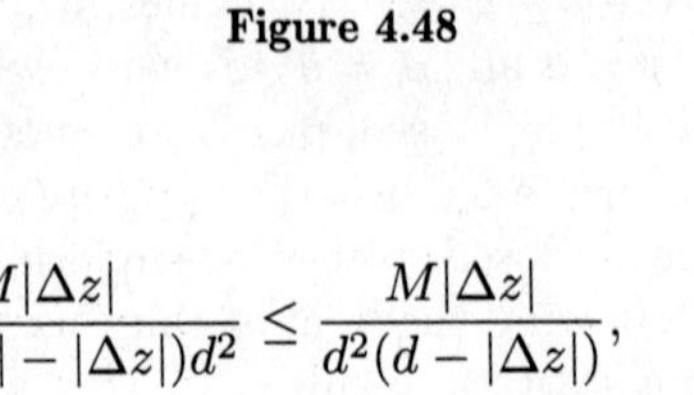

Figure 4.48

Hence,

$$\left|\frac{f(\zeta)\Delta z}{(\zeta - z - \Delta z)(\zeta - z)^2}\right| \leq \frac{M|\Delta z|}{(|\zeta - z| - |\Delta z|)d^2} \leq \frac{M|\Delta z|}{d^2(d - |\Delta z|)},$$

and therefore, using inequality 4.22,

$$\left|\int_C \frac{f(\zeta)\Delta z}{(\zeta - z - \Delta z)(\zeta - z)^2}\, d\zeta\right| \leq \frac{M|\Delta z|}{d^2(d - |\Delta z|)} L,$$

where L is the length of C. Clearly the limit of this expression as $\Delta z \to 0$ is zero, and the proof is complete. ■

This theorem can be extended in a similar way to show not only that F' exists, but that F has derivatives of all orders. In particular,

$$F^{(n)}(z) = n! \int_C \frac{f(\zeta)}{(\zeta - z)^{n+1}}\, d\zeta. \tag{4.29c}$$

What Theorem 4.14 and the extension in 4.29(c) show is that the operations of differentiation and contour integration can be interchanged. When we combine 4.29(c) with Cauchy's integral formula, we obtain a most startling result.

■ **THEOREM 4.15** (Generalized Cauchy Integral Formula) Let f be analytic inside and on a simple, closed, piecewise-smooth curve C. Then for any point z interior to C and $n = 0, 1, 2, \ldots,$

$$f^{(n)}(z) = \frac{n!}{2\pi i} \oint_C \frac{f(\zeta)}{(\zeta - z)^{n+1}}\, d\zeta. \tag{4.30}$$ ■

The generalized Cauchy integral formula has the most profound consequences. In particular, we can immediately prove that an analytic function has derivatives of all orders.

■ **THEOREM 4.16** If f is analytic in a domain D, then its derivatives $f', f'', \ldots$ are analytic in D.

Proof Given any point z in D, there exists a circle C with center z that is contained in D and that contains in its interior only points of D. By the generalized Cauchy integral formula

$$f^{(n)}(z) = \frac{n!}{2\pi i} \oint_C \frac{f(\zeta)}{(\zeta - z)^{n+1}}\, d\zeta, \qquad n = 0, 1, \ldots,$$

and all derivatives of f exist at z. ■

This situation is totally different than that for real functions. The existence of the first derivative of a real function implies nothing about higher-order derivatives. For example, the function $x^{5/3}$ has a first derivative for every value of x including $x = 0$, but it does not have a second derivative at $x = 0$. The function $(x + 1)^{8/3}$ has a first and second derivative for all x but no third derivative at $x = -1$. In fact, the existence of n derivatives of a function $f(x)$ does not guarantee the $(n+1)$ derivative. For complex functions, on the other hand, existence of the first derivative in some domain D implies existence of derivatives of all orders at all points in D.

Existence of all derivatives of a complex function f implies existence of all partial derivatives of its real and imaginary parts. This is proved in the following corollary.

■ **COROLLARY** If $f(z) = u + vi$ is analytic in a domain D, then all partial derivatives of u and v exist and are continuous in D.

Proof Theorem 4.16 implies that when f is analytic in D, so also is

$$f'(z) = \frac{\partial u}{\partial x} + \frac{\partial v}{\partial x}i = \frac{\partial v}{\partial y} - \frac{\partial u}{\partial y}i.$$

But analyticity of f' implies continuity of f'; hence each of the first partial derivatives of u and v is continuous in D.

Because f' is analytic in D, we can write (by the corollary to Theorem 2.6)

$$f''(z) = \frac{\partial^2 u}{\partial x^2} + \frac{\partial^2 v}{\partial x^2}i = \frac{\partial^2 v}{\partial x \partial y} - \frac{\partial^2 u}{\partial x \partial y}i = \frac{\partial^2 v}{\partial y \partial x} - \frac{\partial^2 u}{\partial y \partial x}i = -\frac{\partial^2 u}{\partial y^2} - \frac{\partial^2 v}{\partial y^2}i.$$

Again, analyticity of f'' implies continuity of f'', and we therefore obtain continuity of all second partial derivatives of u and v. This process may be continued indefinitely to give the corollary. ■

Theorem 4.10 states that when a function f is analytic in a simply connected domain, the contour integral of f around every closed, piecewise-smooth curve in the domain vanishes. According to the next theorem, the converse is true even when the domain is not simply connected.

■ **THEOREM 4.17** (Morera's Theorem) If f is continuous in a domain D and if the contour integral of f is zero around every closed, piecewise-smooth curve in D, then f is analytic in D.

Proof If the contour integral of f vanishes around every closed, piecewise-smooth curve in D, then by Theorem 4.5, the contour integral is independent of path in D. Consequently, by Theorem 4.4, f has an antiderivative in D, say, F. Hence F is analytic in D, and by Theorem 4.16, its derivative $F' = f$ is analytic in D also. ■

Before discussing further implications of analyticity, we indicate how the generalized Cauchy integral formula can be used to evaluate contour integrals around closed curves. The formula implies that when f is analytic inside and on a closed, piecewise-smooth curve C and n is a nonnegative integer, the value of the contour integral is

$$\oint_C \frac{f(z)}{(z-z_0)^{n+1}}\,dz = \begin{cases} 0, & z_0 \text{ outside } C \\ \dfrac{2\pi i}{n!} f^{(n)}(z_0), & z_0 \text{ inside } C. \end{cases} \qquad \textbf{(4.31)}$$

Be clear in your mind that certain conditions must be met before using 4.31. First, C must be a closed, piecewise smooth curve. Second, the integrand must be of the form $f(z)/(z-z_0)^{n+1}$, where f is analytic inside and on C and z_0 is interior to C.

EXAMPLE 4.18 ....

Evaluate:

(a) $\oint_C \frac{e^z \sin z}{z-5}\, dz$, where $C : |z| = 6$

(b) $\oint_C \frac{z^2}{(z-1)^n}\, dz$, where $C : |z| = 2$, and n is a positive integer.

Solution **(a)** Because $e^z \sin z$ is an entire function and $z = 5$ is interior to C (Figure 4.49), the Cauchy integral formula gives

$$\oint_C \frac{e^z \sin z}{z-5}\, dz = 2\pi i \left\{e^z \sin z\right\}_{|z=5} = 2\pi i(e^5 \sin 5).$$

(b) Because z^2 is entire and $z = 1$ is interior to C (Figure 4.50), we have by the generalized Cauchy integral formula,

$$\oint_C \frac{z^2}{(z-1)^n}\, dz = \frac{2\pi i}{(n-1)!} \frac{d^{n-1}}{dz^{n-1}} \left\{z^2\right\}_{|z=1} = \begin{cases} 2\pi i, & n = 1 \\ 4\pi i, & n = 2 \\ 2\pi i, & n = 3 \\ 0, & n > 3. \end{cases}$$

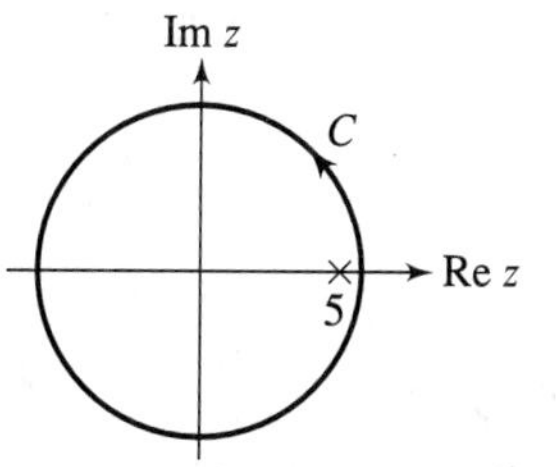

Figure 4.49

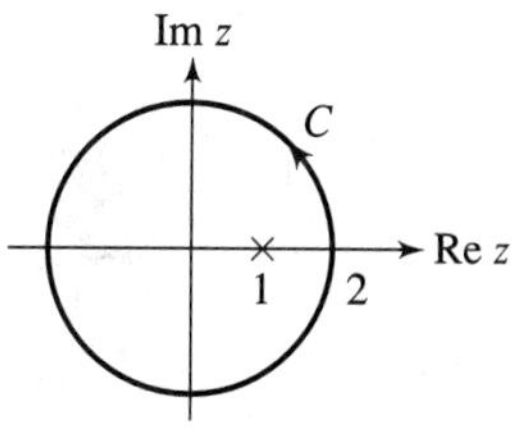

Figure 4.50

....

EXAMPLE 4.19 ....

Evaluate $\oint_C \frac{z^2}{2z^2+9}\, dz$, where C is the rectangle in Figure 4.51.

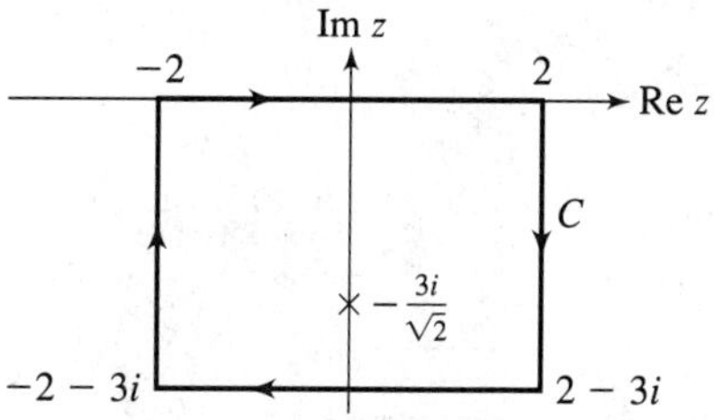

Figure 4.51

Solution Factoring the denominator of the integrand gives

$$\frac{z^2}{2z^2+9} = \frac{z^2}{2(z+3i/\sqrt{2})(z-3i/\sqrt{2})}.$$

Because $z^2/[2(z-3i/\sqrt{2})]$ is analytic inside and on C, the Cauchy integral formula gives

$$\oint_C \frac{z^2}{2z^2+9}\,dz = -\oint_{-C} \frac{z^2}{2z^2+9}\,dz = -\oint_{-C} \frac{\dfrac{z^2}{2(z-3i/\sqrt{2})}}{z+3i/\sqrt{2}}\,dz$$

$$= -2\pi i \left\{ \frac{z^2}{2(z-3i/\sqrt{2})} \right\}_{|z=-3i/\sqrt{2}} = -2\pi i \left[\frac{-9/2}{2(-6i/\sqrt{2})} \right] = \frac{-3\sqrt{2}\pi}{4}.$$

EXAMPLE 4.20

Evaluate $\oint_C \frac{2z+1}{z(z-1)^2}\,dz$, where C is the curve in Figure 4.52.

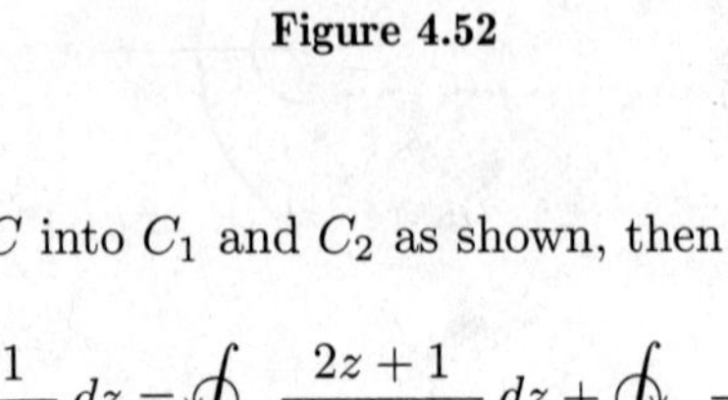

Figure 4.52

Solution If we divide C into C_1 and C_2 as shown, then

$$\oint_C \frac{2z+1}{z(z-1)^2}\,dz = \oint_{C_1} \frac{2z+1}{z(z-1)^2}\,dz + \oint_{C_2} \frac{2z+1}{z(z-1)^2}\,dz.$$

Because $(2z+1)/(z-1)^2$ is analytic inside and on C_2,

$$\oint_{C_2} \frac{2z+1}{z(z-1)^2}\,dz = -\oint_{-C_2} \frac{\dfrac{2z+1}{(z-1)^2}}{z}\,dz = -2\pi i \left\{ \frac{2z+1}{(z-1)^2} \right\}_{|z=0} = -2\pi i.$$

Because $(2z+1)/z$ is analytic inside and on C_1,

$$\oint_{C_1} \frac{2z+1}{z(z-1)^2}\,dz = \oint_{C_1} \frac{\frac{2z+1}{z}}{(z-1)^2}\,dz = 2\pi i \frac{d}{dz}\left\{\frac{2z+1}{z}\right\}_{|z=1}$$

$$= 2\pi i \left[\frac{z(2)-(2z+1)(1)}{z^2}\right]_{|z=1} = -2\pi i.$$

Finally,

$$\oint_C \frac{2z+1}{z(z-1)^2}\,dz = -2\pi i - 2\pi i = -4\pi i.$$

EXAMPLE 4.21

Redo Example 4.17 using equation 4.31.

Solution The integrand $1/(z^2-4)$ has two singularities $z = \pm 2$ inside $C : |z| = 4$. According to Theorem 4.8, however, we may replace the integral around C with integrals around the curves $C' : |z+2| = 1$ and $C'' : |z-2| = 1$ (Figure 4.53),

$$\oint_C \frac{1}{z^2-4}\,dz = \oint_{C'} \frac{1}{(z+2)(z-2)}\,dz + \oint_{C''} \frac{1}{(z+2)(z-2)}\,dz.$$

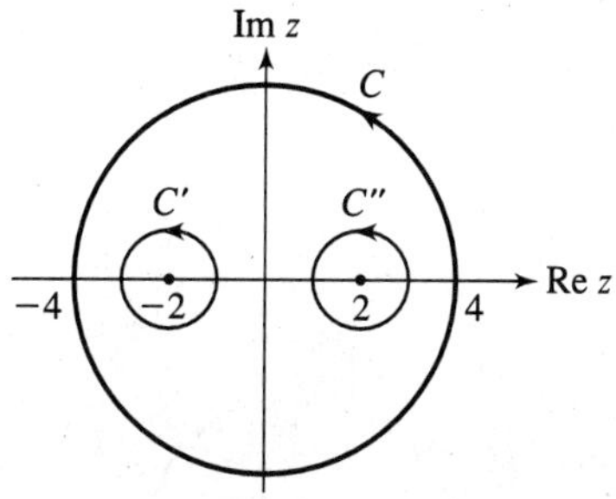

Figure 4.53

Because $1/(z+2)$ is analytic inside and on C'' and $z = 2$ is interior to C'', 4.31 gives

$$\oint_{C''} \frac{1}{(z+2)(z-2)}\,dz = \oint_{C''} \frac{\frac{1}{z+2}}{z-2}\,dz = 2\pi i \left\{\frac{1}{z+2}\right\}_{|z=2} = \frac{\pi i}{2}.$$

Similarly,

$$\oint_{C'} \frac{1}{(z+2)(z-2)}\, dz = \oint_{C'} \frac{\dfrac{1}{z-2}}{z+2}\, dz = 2\pi i \left\{ \frac{1}{z-2} \right\}_{|z=-2} = -\frac{\pi i}{2}.$$

Thus,

$$\oint_C \frac{1}{z^2-4}\, dz = \frac{\pi i}{2} - \frac{\pi i}{2} = 0.$$

EXAMPLE 4.22

Show that if C is a closed, piecewise-smooth curve not passing through $z = 0$, the contour integral

$$I = \frac{1}{2\pi i} \oint_C \frac{1}{z}\, dz$$

is equal to the number of times that C encircles $z = 0$ in the counterclockwise direction.

Solution If C has a loop that does not encircle $z = 0$ (C_1 in Figure 4.54), then

$$\oint_{C_1} \frac{1}{z}\, dz = 0$$

(because $1/z$ is analytic inside and on C_1). Consequently, any loops of C that do not encircle $z = 0$ may be removed from C as making no contribution to I. What remains is either no curve at all (Figure 4.55) or a curve C' that encircles $z = 0$ (Figure 4.56).

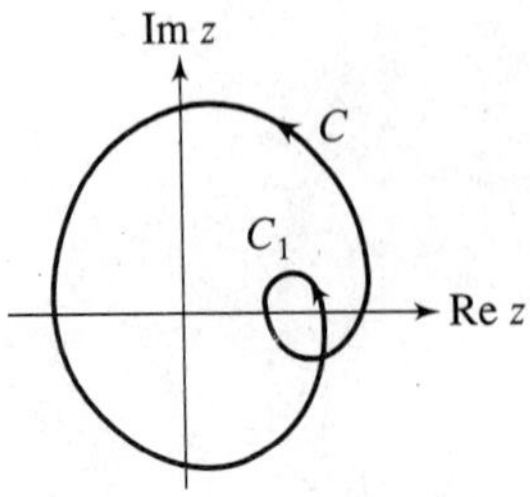

Figure 4.54

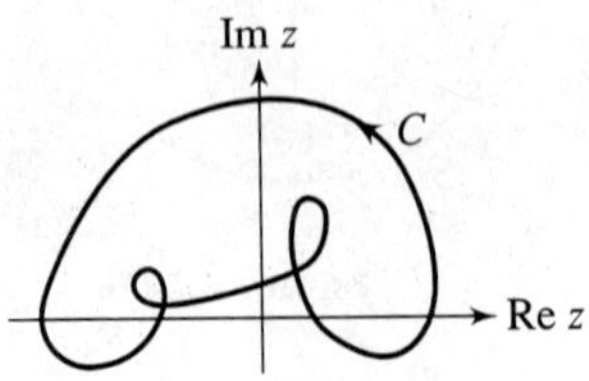

Figure 4.55

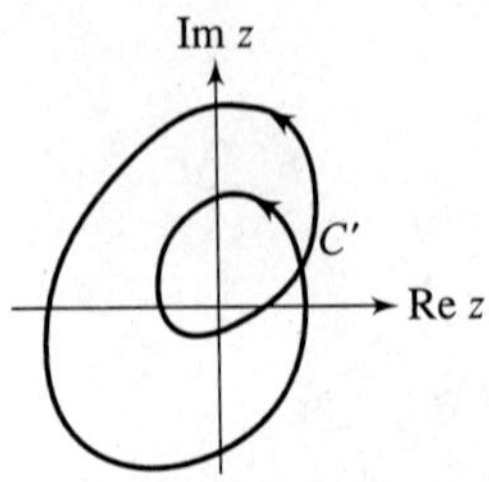

Figure 4.56

When no curve remains, I is equal to zero, and this is also the number of times that C encircles $z = 0$.

When C' does encircle $z = 0$, let n represent the number of times ($n = 2$ in Figure 4.56). Then C' can be subdivided into n loops C_j, each of which encircles

$z = 0$ once. But according to Theorem 4.9, the contour integral around each C_j is equal to $2\pi i$. Thus, if C' encircles $z = 0$ n times in the counterclockwise direction, then $I = n$. If C' encircles $z = 0$ n times in the clockwise direction, $I = -n$.

We now prove an extension to Cauchy's integral formula 4.28 that is useful in taking inverse Laplace transforms (Section 8.3).

■ **THEOREM 4.18** Suppose that f is analytic in the half-plane $x > \delta > 0$ and that in this half-plane, there exist constants M, R, and $k > 1$ such that

$$|f(z)z^k| < M \quad \text{for} \quad |z| > R.$$

If z is a point with $\operatorname{Re} z > \gamma > \delta$, then

$$f(z) = -\frac{1}{2\pi i} \lim_{\beta \to \infty} \int_{\gamma - \beta i}^{\gamma + \beta i} \frac{f(\zeta)}{\zeta - z}\, d\zeta. \tag{4.32}$$

Proof For the given z it is always possible to draw a circular arc centered at $z = 0$ with radius r, which, along with the line segment $\operatorname{Re} z = \gamma$ from $z = \gamma - \beta i$ to $z = \gamma + \beta i$, encloses z (Figure 4.57). Because f is analytic inside and on the curve composed of the straight-line segment C and the circular arc Γ, Cauchy's integral formula gives

$$f(z) = \frac{1}{2\pi i} \int_C \frac{f(\zeta)}{\zeta - z}\, d\zeta + \frac{1}{2\pi i} \int_\Gamma \frac{f(\zeta)}{\zeta - z}\, d\zeta.$$

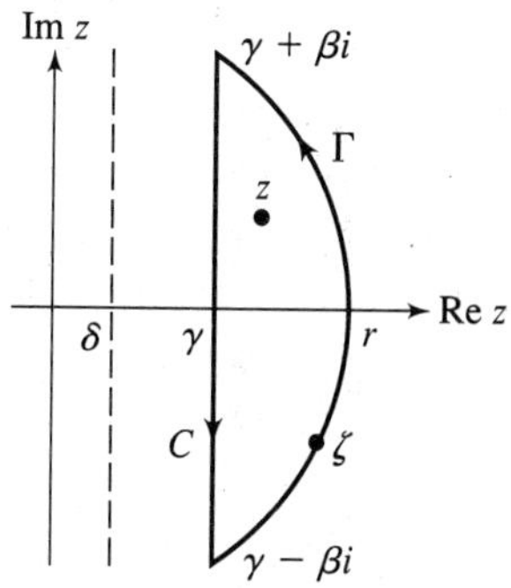

Figure 4.57

If r is chosen larger than R, then on Γ,

$$\left|\frac{f(\zeta)}{\zeta - z}\right| \le \frac{M/|\zeta|^k}{|\zeta - z|} \le \frac{M}{r^k(r - |z|)}.$$

According to inequality 4.22,

$$\left|\int_\Gamma \frac{f(\zeta)}{\zeta - z}\, d\zeta\right| < \frac{M}{r^k(r - |z|)}(\pi r) = \frac{\pi M}{r^{k-1}(r - |z|)},$$

and this quantity approaches zero as $r \to \infty$. Consequently, we may write that

$$f(z) = \frac{1}{2\pi i} \lim_{\beta \to \infty} \int_C \frac{f(\zeta)}{\zeta - z}\, d\zeta = -\frac{1}{2\pi i} \lim_{\beta \to \infty} \int_{\gamma - \beta i}^{\gamma + \beta i} \frac{f(\zeta)}{\zeta - z}\, d\zeta.$$

■

Exercises 4.6

In Exercises 1–13 evaluate the contour integral.

1. $\oint_C \frac{e^z}{z-2}\,dz$, where $C: |z| = 3$

2. $\oint_C \frac{z^3}{z+\pi i/2}\,dz$, where $C: |z+1| = 5$

3. $\oint_C \frac{\sin z}{(z+2)^3}\,dz$, where $C: |z-2| + |z+2| = 10$

4. $\oint_C \frac{z^2}{(z+5i)^4}\,dz$, where $C: |z| = 6$

5. $\oint_C \frac{z}{(z^2+4)^2}\,dz$, where $C: |z-i| = 2$

6. $\oint_C \frac{1}{z^2+4z+16}\,dz$, where $C: |z+2-3i| = 3$

7. $\oint_C \frac{\cos z}{(z^2-4)^2}\,dz$, where $C: |z-2| = 2$

8. $\oint_C \frac{1}{z^2+z-2}\,dz$, where: (a) $C: |z-1| = 2$, (b) $C: |z+2| = 1$, and (c) $C: |z| = 3$

9. $\oint_C \frac{3z-2}{z^2-z}\,dz$, where C is the square with vertices $2 \pm 2i$, $-2 \pm 2i$

10. $\oint_C \frac{\operatorname{Log} z}{z-2}\,dz$, where $C: |z-2| = 1$

11. $\oint_C \frac{\sqrt{z}}{(z-4i)^2}\,dz$, where $C: |z-5i| + |z-3i| = 6$

12. $\oint_C \frac{z \log_{\pi/2} z}{(z-2)^2}\,dz$, where C is the square with vertices 1, $2+i$, 3, and $2-i$

13. $\oint_C \frac{2z^2-z+1}{(z-1)^2(z+1)}\,dz$, where C is the curve of Figure 4.52

14. Functions f and g are analytic inside and on a simple, closed, piecewise-smooth curve C. If $f(z) = g(z)$ on C, is it necessary that $f(z) = g(z)$ inside C?

15. Prove that all derivatives of real and imaginary parts of a function f analytic in a domain D are harmonic in D.

16. The Legendre polynomials for complex arguments are defined by

$$P_m(z) = \frac{1}{2^m m!}\frac{d^m}{dz^m}(z^2-1)^m.$$

(a) Show that when C is a simple, closed, piecewise-smooth curve and z_0 is interior to C,

$$P_m(z_0) = \frac{-i}{2^{m+1}\pi} \oint_C \frac{(z^2-1)^m}{(z-z_0)^{m+1}}\, dz,$$

called *Schlaefli's formula.*

(b) Use this result to prove that $P_m(1) = 1$ and $P_m(-1) = (-1)^m$.

(c) By setting $\zeta = z + \sqrt{z^2-1}\, e^{\theta i}$, $-\pi < \theta \le \pi$, in Schlaefli's formula, show that

$$P_m(z) = \frac{1}{\pi} \int_0^{\pi} [z + \sqrt{z^2-1}\, \cos\theta]^m \, d\theta.$$

17. Evaluate $\oint_C \frac{z}{z^4-1}\, dz$, where C is the circle $|z-R| = R$, with $R \ne \frac{1}{2}$.

4.7 Bounds for Moduli of Analytic Functions

We have seen that an adequate pictorial representation of complex functions does not exist; they are often regarded as mappings from one plane to another. In some applications, however, it is only the modulus $|f(z)|$ of a function $f(z)$ that is important, and this can be illustrated geometrically by setting $z = x + yi$ and drawing the surface $|w| = |f(x+yi)|$. Three examples are shown in Figure 4.58. Cauchy's generalized integral formulas lead to properties of moduli of analytic functions, properties that are reflected in surfaces like those in Figure 4.58. They also lead to important facts about harmonic functions.

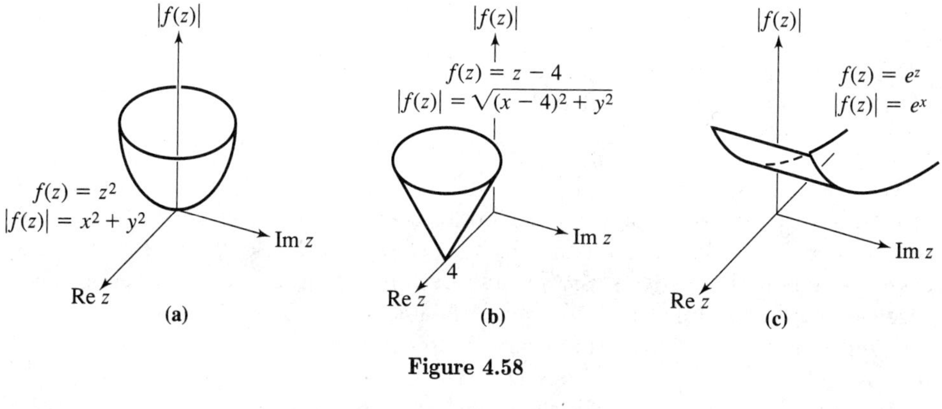

Figure 4.58

Cauchy's generalized integral formula indicates that when the values of an analytic function are known on a closed curve, its values — and those of its derivatives — are known everywhere inside the curve. The following somewhat weaker result indicates that when the maximum value of the modulus of an analytic function is known on a circle, maximum values for moduli of the derivatives of the function at the center of the circle are known.

■ **THEOREM 4.19** Let f be analytic inside and on a circle C_r of radius r centered at z_0. If $|f(z)| \leq M$ for all z on C_r, then

$$|f^{(n)}(z_0)| \leq \frac{n!M}{r^n}, \qquad n = 1, 2, \ldots. \tag{4.33}$$

Proof By the generalized Cauchy integral formula,

$$f^{(n)}(z_0) = \frac{n!}{2\pi i} \oint_{C_r} \frac{f(z)}{(z - z_0)^{n+1}} \, dz,$$

and hence using property 4.22,

$$|f^{(n)}(z_0)| \leq \frac{n!}{2\pi} \frac{M}{r^{n+1}} (2\pi r) = \frac{n! \, M}{r^n}.$$ ■

We now use this result to prove that the only functions that are entire and bounded are constant functions. A function f is said to be **bounded** in a domain D if there exists a (real) number $M > 0$ such that $|f(z)| \leq M$ for all z in D.

■ **THEOREM 4.20** (Liouville's Theorem) If an entire function is bounded, then it is constant.

Proof Suppose f is entire and bounded. Then there exists a constant M such that $|f(z)| \leq M$ for all z. If z_0 is any point in the complex plane and $C_r : |z - z_0| = r$ is a circle centered at z_0 with radius r, then by Theorem 4.19,

$$|f'(z_0)| \leq \frac{M}{r}.$$

Because r can be made arbitrarily large — and, therefore, $|f'(z_0)|$ can be arbitrarily small — it follows that $f'(z_0) = 0$, and $f(z)$ is constant. ■

We saw instances of Liouville's theorem in Chapter 3, but each situation was investigated separately. For example, because $e^z = e^x(\cos y + \sin y \, i)$, it is clear that $|e^z| = e^x$, which is unbounded as $x \to \infty$. Moduli of functions $\sin z$ and $\cos z$ are unbounded as $y \to \pm\infty$ for any fixed x (see Exercise 18 in Section 3.3). Likewise, moduli of $\sinh z$ and $\cosh z$ are unbounded (see Exercise 14 in Section 3.4). Liouville's theorem indicates that every one of these functions must be unbounded, because each is entire and not constant.

The modulus of a complex function can be visualized geometrically as a surface, as in Figure 4.58. Liouville's theorem indicates that such surfaces for nonconstant, entire functions must be unbounded in some direction(s). These figures suggest other properties of the modulus of a complex function. In particular, suppose that one of the surfaces in Figure 4.58 is restricted to lie above some domain D. No matter what this domain is, the surface has no maximum point therein; in other words, the modulus $|f(z)|$ does not take on a maximum value in D. Furthermore, when $|f(z)|$ is defined on the region consisting of D and its boundary, its maximum value occurs on the boundary. That these facts are always true is proved in the following theorem.

■ **THEOREM 4.21** (Maximum Modulus Principle) If f is analytic and not constant in a domain D, then $|f(z)|$ cannot take on a maximum value in D.

Proof Suppose to the contrary that there exists a point z_0 in D such that $|f(z)| \le |f(z_0)|$ for all z in D. Cauchy's integral formula states that when $C_r : |z - z_0| = r$ is a circle that contains only points of D (Figure 4.59),

$$f(z_0) = \frac{1}{2\pi i} \oint_{C_r} \frac{f(z)}{z - z_0}\, dz.$$

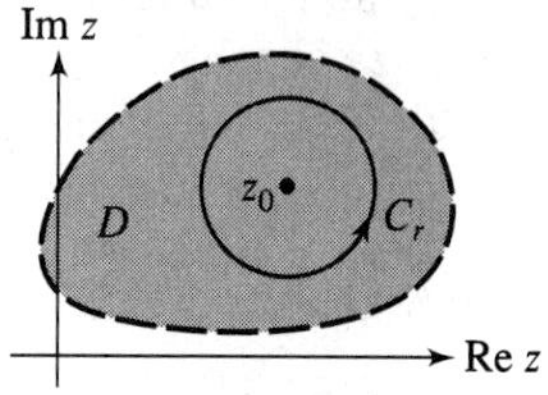

Figure 4.59

If C_r is represented parametrically by $z = z_0 + re^{\theta i}$, $-\pi \le \theta \le \pi$, then

$$f(z_0) = \frac{1}{2\pi i} \int_{-\pi}^{\pi} \frac{f(z_0 + re^{\theta i}) rie^{\theta i}}{re^{\theta i}}\, d\theta = \frac{1}{2\pi} \int_{-\pi}^{\pi} f(z_0 + re^{\theta i})\, d\theta.$$

According to inequality 4.22, we may write

$$|f(z_0)| \le \frac{1}{2\pi} \int_{-\pi}^{\pi} |f(z_0 + re^{\theta i})|\, d\theta,$$

and notice that the integral on the right is a definite integral, not a contour integral. Suppose at some point on C_r, $|f(z)| < |f(z_0)|$. Then due to the continuity of $f(z)$, there exists an arc of C_r along which $|f(z)| < |f(z_0)|$. This means that there is an interval $-\pi \le \theta_1 \le \theta \le \theta_2 \le \pi$ in which $|f(z_0 + re^{\theta i})| < |f(z_0)|$, and hence

$$|f(z_0)| \le \frac{1}{2\pi} \int_{-\pi}^{\pi} |f(z_0 + re^{\theta i})|\, d\theta < \frac{1}{2\pi} |f(z_0)|(2\pi) = |f(z_0)|,$$

a contradiction. It follows therefore that at every point on C_r, $|f(z)| = |f(z_0)|$. But the radius of C_r is arbitrary (except that C_r be in D). Hence $|f(z)| = |f(z_0)|$ for all points in and on C_r.

Let S be the set of all points in D for which $|f(z)| = |f(z_0)|$. If z' is any point in S, then by the foregoing proof, we can construct a circle around z' in which $|f(z)| = |f(z')| = |f(z_0)|$. In other words, S is an open set. If z'' is in $D - S$, then $|f(z'')| \ne |f(z_0)|$. Because $f(z)$ is continuous, there exists a circle centered at z'' in which $|f(z)| \ne |f(z_0)|$; that is, $D - S$ is open also. But the only way for an open subset S of a domain D (in particular a connected set) to have an open complement $D - S$ is for the complement to be empty (see Exercise 13 in Section 2.1). Consequently, $S = D$ and $|f(z)|$ is constant in D. By Theorem 2.8, $f(z)$ is constant in D, a contradiction. No such point z_0 for which $|f(z)| \le |f(z_0)|$ can therefore exist. ■

■ **COROLLARY** If f is analytic in a bounded domain D and continuous on the region consisting of D and its boundary $\beta(D)$, then $|f(z)|$ attains its maximum on $\beta(D)$.

Proof Because a continuous (real) function on a closed and bounded set must attain a maximum at some point in that set, $|f(z)|$ must attain a maximum in D or on $\beta(D)$. If $f(z) \neq$ constant in D, it cannot attain this maximum in D (by the maximum modulus principle) and must therefore attain it on $\beta(D)$. If $f(z)$ is constant in D, then $|f(z)|$ attains its maximum at every point of D and $\beta(D)$. Thus, in either case, $|f(z)|$ attains its maximum on $\beta(D)$. ■

EXAMPLE 4.23 ....

If $f(z) = z/(z^2+2)$, find the maximum value of $|f(z)|$ for $|z| \leq 1$.

Solution Since f is analytic for $|z| < 1$ and continuous for $|z| \leq 1$, $|f(z)|$ attains its maximum value on the boundary $|z| = 1$. For $z = e^{\theta i}$, $-\pi \leq \theta \leq \pi$, on $|z| = 1$,

$$|f(z)| = \left|\frac{e^{\theta i}}{e^{2\theta i}+2}\right| = \frac{1}{|e^{2\theta i}+2|} = \frac{1}{\sqrt{(2+\cos 2\theta)^2+\sin^2 2\theta}} = \frac{1}{\sqrt{5+4\cos 2\theta}}.$$

This expression is a maximum when $5+4\cos 2\theta$ is a minimum. But clearly the minimum value is 1, occuring when $\theta = \pm\pi/2$. The maximum value of $|f(z)|$ is, therefore, 1. ....

There is also a minimum modulus principle. Consider again the surfaces in Figure 4.58. If D is a domain in the complex plane in Figure 4.58(b) and D does not contain the zero $z = 4$ of $f(z) = z - 4$, then the surface has no relative minimum in D. On the other hand, if the domain contains $z = 4$, then the surface has a minimum at $z = 4$. Similarly, the surface in Figure 4.58(a) has a relative minimum only when $z = 0$, the zero of $f(z) = z^2$, is contained in D. The surface in Figure 4.58(c) does not have a relative minimum for any domain, and e^z has no zeros. These are instances of the minimum modulus principle.

■ **THEOREM 4.22** (Minimum Modulus Principle) Let f be analytic and not constant in a domain D. If f has no zeros in D, then $|f(z)|$ cannot take on a minimum value in D.

Proof When f is analytic and never zero in D, then $1/f$ is also analytic in D. By the maximum modulus principle, $|1/f(z)| = 1/|f(z)|$ cannot take on a maximum value in D; that is, $|f(z)|$ cannot take on a minimum value in D. ■

Further, we can prove the following analogue to the corollary of the maximum modulus principle.

■ **COROLLARY** Let f be analytic in a bounded domain D and be continuous on the region consisting of D and its boundary $\beta(D)$. If f does not vanish in D, then $|f(z)|$ attains its minimum on $\beta(D)$. ■

Exercises 4.7

1. Use Liouville's theorem to prove the fundamental theorem of algebra: Every polynomial of degree greater than zero with complex coefficients has at least one zero. In Exercise 40 of Section 5.4, this result is extended to show that every polynomial of degree n has exactly n zeros.
2. Find all functions f analytic in the domain $D : |z| < R$ that satisfy $f(0) = i$ and $|f(z)| \leq 1$ for all z in D.
3. A function f has period T if $f(z+T) = f(z)$ for all z. Prove that if an entire function has both a real and an imaginary period, then it must be a constant function.
4. What is the maximum value of $|z^2/(z^3 - 15)|$ for $|z| \leq 2$?
5. What is the maximum value of $|\sin z|$ on the rectangle $0 \leq \text{Re}\, z \leq \pi/2$, $0 \leq \text{Im}\, z \leq a$?
6. What is the maximum value of $|\cos 2z|$ on the rectangle $0 \leq \text{Re}\, z \leq \pi/2$, $a \leq \text{Im}\, z \leq b$?
7. What is the minimum value of $|e^{1-2z}|$ on the region $|\text{Re}\, z| + |\text{Im}\, z| \leq 4$?
8. What is the minimum value of $|\cos 2z|$ on $|z| \leq 1$?
9. What is the maximum value of $|\text{Log}\, z|$ on $|z| \leq 2$, $\text{Re}\, z \geq 1$?
10. What is the minimum value of $|z/(z^2 + 9)|$ on $1 \leq |z| \leq 2$?
11. Suppose a nonconstant function f is analytic inside and on a simple, closed, piecewise-smooth curve C. If $|f(z) - 1| < 1$ on C, show that $f(z)$ is never zero inside C.
12. A function f is entire. Is it possible for $w = f(z)$ to map the z-plane into the circle $|w| < 1$? Explain.

4.8 Applications to Harmonic Functions

In Chapter 2 we showed that the real and imaginary parts of an analytic function are harmonic functions, subject to the provision that these functions have continuous second partial derivatives. This restriction has been removed by the corollary to Theorem 4.16, because the real and imaginary parts of an analytic function have continuous partial derivatives of all orders. We also stated that given a harmonic function $u(x, y)$ defined in some special kind of domain D, it is possible to find a harmonic function $v(x, y)$, called the harmonic conjugate of $u(x, y)$, such that the function $u(x, y) + v(x, y)i$ is an analytic function f in D. That this cannot always be done for arbitrary domains is illustrated by the function $u(x, y) = (1/2)\ln(x^2 + y^2)$, which is harmonic in the domain consisting of the entire xy-plane with the origin deleted. It is not possible to find a function $v(x, y)$ such that $u + vi$ is analytic in this domain. Indeed, we have already seen that $u(x, y)$ is the real part of the function $\log z = \ln|z| + (\arg z)i$, and $\log z$ cannot be made analytic in any domain that completely surrounds the origin.

The following theorem proves that harmonic conjugates can always be found in simply connected domains.

■ **THEOREM 4.23** If $u(x, y)$ is harmonic in a simply connected domain D, there exists a harmonic conjugate $v(x, y)$ in D.

Proof If we define a function $F(z) = \partial u/\partial x - i\,\partial u/\partial y$, then the real and imaginary parts of $F(z)$ satisfy the Cauchy-Riemann equations at each point of D:

$$\frac{\partial}{\partial x}\left(\frac{\partial u}{\partial x}\right) = \frac{\partial}{\partial y}\left(-\frac{\partial u}{\partial y}\right) \qquad \text{(because } u \text{ satisfies Laplace's equation)},$$

$$\frac{\partial}{\partial x}\left(-\frac{\partial u}{\partial y}\right) = -\frac{\partial}{\partial y}\left(\frac{\partial u}{\partial x}\right) \qquad \text{(because second partial derivatives of } u \text{ are continuous)}.$$

Consequently, F is analytic in D, and by Theorem 4.12, F has an antiderivative in D. If we denote this antiderivative by $f(z) = \phi + vi$, then

$$f'(z) = \frac{\partial \phi}{\partial x} - \frac{\partial \phi}{\partial y} i = F(z) = \frac{\partial u}{\partial x} - \frac{\partial u}{\partial y} i.$$

But this equation implies that $\phi = u + k$, where k is a complex constant. If we choose $k = 0$, then $f(z) = u + vi$; that is, we have found a function f analytic in D that has $u(x, y)$ as its real part (and, therefore, $v(x, y)$ as the harmonic conjugate of $u(x, y)$). ■

The maximum and minimum modulus principles for analytic functions in Theorems 4.21 and 4.22 and their corollaries imply similar principles about harmonic functions.

■ **THEOREM 4.24** If $u(x, y)$ is harmonic and not constant in a domain D, then $u(x, y)$ cannot take on a maximum or minimum value in D.

Proof We shall prove only the maximum case (the minimum case is similar, see Exercise 1). First, suppose that D is simply connected. Then, by Theorem 4.23, there exists a function $f(z) = u + vi$, analytic in D, which has $u(x, y)$ as its real part. Now the function $e^{f(z)}$ is also analytic in D, and by the maximum modulus principle $|e^{f(z)}|$ cannot take on a maximum value in D. But $|e^{f(z)}| = |e^{u+vi}| = e^u$, and therefore e^u cannot take on a maximum value in D. Because e^u is maximized when u is maximized, it follows that $u(x, y)$ cannot take on a maximum value in D.

Take now the case that D is just a domain (and could therefore be multiply connected). Suppose to the contrary of the theorem that $u(x, y)$ takes on a maximum value at some point (x_0, y_0) in D (Figure 4.60). If we draw an open

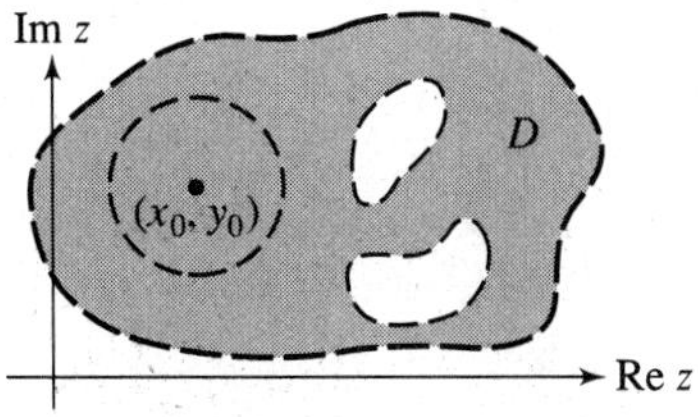

Figure 4.60

circle around (x_0, y_0) that contains only points of D, then in this simply connected domain, $u(x, y)$ takes on a maximum value, contradicting the previous paragraph. Consequently, $u(x, y)$ cannot take on a maximum value in D. ■

Corollaries corresponding to those of Theorems 4.21 and 4.22 can be combined into one.

■ **COROLLARY 1** If $u(x, y)$ is harmonic in a bounded domain D and continuous on the region consisting of D and its boundary $\beta(D)$, then $u(x, y)$ attains its maximum and minimum values on $\beta(D)$. ■

In addition, we can prove the following result.

■ **COROLLARY 2** If $u(x, y)$ is harmonic in a bounded domain D and continuous on the region consisting of D and its boundary $\beta(D)$, and if $u(x, y)$ is constant on $\beta(D)$, then $u(x, y)$ is constant in D.

Proof If k is the constant value of $u(x, y)$ on $\beta(D)$, then according to Corollary 1, k must be both the maximum and minimum value of $u(x, y)$ on D and $\beta(D)$. Therefore, $u(x, y)$ must be equal to k everywhere in D. ■

In Sections 2.6–2.8, we saw how important solutions to Laplace's equation are in the fields of heat and fluid flow and electrostatics. An important class of problems in these fields, and one that we deal with extensively in Chapter 7, is to find a solution to Laplace's equation in some domain D that also takes on prescribed values on the boundary $\beta(D)$ of D. The following corollary indicates that there cannot be more than one solution to the problem; that is, if there is a solution, it is unique.

■ **COROLLARY 3** There cannot be more than one function $u(x, y)$ that is harmonic in a bounded domain D, is continuous on the region consisting of D and its boundary $\beta(D)$, and takes on prescribed values on $\beta(D)$.

Proof Suppose to the contrary that $u_1(x,y)$ and $u_2(x,y)$ are two solutions of Laplace's equation in D that are continuous on the region consisting of D and its boundary and have the same values at each point on $\beta(D)$. Then the function $u(x,y) = u_1(x,y) - u_2(x,y)$ satisfies Laplace's equation in D, is continuous on the region consisting of D and $\beta(D)$, and has value zero on $\beta(D)$. By Corollary 2, $u(x,y)$ must vanish everywhere in D, and hence $u_1(x,y) = u_2(x,y)$ in D. ■

Theorem 4.24 and its corollaries provide direct implications about the physical phenomena of heat and fluid flow and electrostatics. We shall see them in the detailed discussions of these applications in Sections 7.7–7.9.

Poisson's Integral Formulas

When a function f is analytic inside and on a simple, closed, piecewise-smooth curve C, the value of f at any point z interior to C can, according to Cauchy's integral formula, be expressed in terms of its values on C,

$$f(z) = \frac{1}{2\pi i}\oint_C \frac{f(\zeta)}{\zeta - z}\,d\zeta.$$

If we set $f(z) = u + vi$, then it is possible (given parametric equations for C) to take real and imaginary parts of both sides of this equation and thereby have line integrals describing the values of u and v at points interior to C in terms of their values on C. The difficulty would be that each integral would contain the values of both u and v on C. We ask now whether it is possible to develop a line integral that describes u at a point interior to C in terms of only the values of u on C. In general this is a very difficult problem, and we therefore consider it with a very simple curve.

In particular, suppose that f is analytic inside and on the circle $C : |z| = R$ (Figure 4.61).

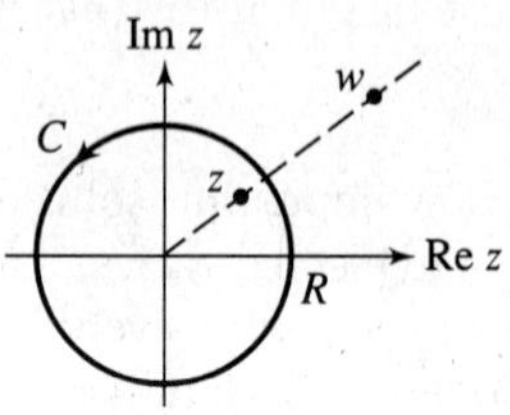

Figure 4.61

When z is interior to C,

$$f(z) = \frac{1}{2\pi i}\oint_C \frac{f(\zeta)}{\zeta - z}\,d\zeta.$$

If we write $z = re^{\theta i}$, then the point $w = (R^2/r)e^{\theta i}$ is clearly exterior to C. As a result, the function $f(z)/(z-w)$ is analytic in a domain containing C, and by the Cauchy-Goursat theorem

$$0 = \frac{1}{2\pi i}\oint_C \frac{f(\zeta)}{\zeta - w}\,d\zeta.$$

If we subtract the second of these equations from the first,

$$f(z) = \frac{1}{2\pi i}\oint_C f(\zeta)\left(\frac{1}{\zeta - z} - \frac{1}{\zeta - w}\right)d\zeta = \frac{1}{2\pi i}\oint_C \frac{f(\zeta)(z-w)}{(\zeta - z)(\zeta - w)}\,d\zeta.$$

On C we set $\zeta = Re^{\phi i}$, $-\pi \le \phi \le \pi$, in which case

$$\begin{aligned}
f(re^{\theta i}) &= \frac{1}{2\pi i}\int_{-\pi}^{\pi} \frac{f(Re^{\phi i})(re^{\theta i} - R^2 e^{\theta i}/r)}{(Re^{\phi i} - re^{\theta i})(Re^{\phi i} - R^2 e^{\theta i}/r)} Rie^{\phi i}\,d\phi \\
&= \frac{1}{2\pi}\int_{-\pi}^{\pi} \frac{f(Re^{\phi i})(r^2 - R^2)e^{(\theta+\phi)i}}{(Re^{\phi i} - re^{\theta i})(re^{\phi i} - Re^{\theta i})}\,d\phi \\
&= \frac{1}{2\pi}\int_{-\pi}^{\pi} \frac{(R^2 - r^2)f(Re^{\phi i})}{(Re^{\phi i} - re^{\theta i})(Re^{-\phi i} - re^{-\theta i})}\,d\phi \\
&= \frac{1}{2\pi}\int_{-\pi}^{\pi} \frac{(R^2 - r^2)f(Re^{\phi i})}{R^2 + r^2 - rR[e^{(\theta-\phi)i} + e^{-(\theta-\phi)i}]}\,d\phi \\
&= \frac{1}{2\pi}\int_{-\pi}^{\pi} \frac{(R^2 - r^2)f(Re^{\phi i})}{R^2 + r^2 - 2rR\cos(\theta - \phi)}\,d\phi.
\end{aligned}$$

This equation

$$f(re^{\theta i}) = \frac{R^2 - r^2}{2\pi}\int_{-\pi}^{\pi} \frac{f(Re^{\phi i})}{R^2 + r^2 - 2rR\cos(\theta - \phi)}\,d\phi,$$

like the Cauchy integral formula, expresses $f(z)$ at points interior to C in terms of its values on C. It does, however, divide very conveniently into real and imaginary parts. Indeed, if we set $f(z) = f(re^{\theta i}) = u(r,\theta) + v(r,\theta)i$, then

$$u(r,\theta) + v(r,\theta)i = \frac{R^2 - r^2}{2\pi}\int_{-\pi}^{\pi} \frac{u(R,\phi) + v(R,\phi)i}{R^2 + r^2 - 2rR\cos(\theta - \phi)}\,d\phi$$

and real and imaginary parts give

$$u(r,\theta) = \frac{R^2 - r^2}{2\pi}\int_{-\pi}^{\pi} \frac{u(R,\phi)}{R^2 + r^2 - 2rR\cos(\theta - \phi)}\,d\phi, \tag{4.34a}$$

$$v(r,\theta) = \frac{R^2 - r^2}{2\pi}\int_{-\pi}^{\pi} \frac{v(R,\phi)}{R^2 + r^2 - 2rR\cos(\theta - \phi)}\,d\phi. \tag{4.34b}$$

Each of these formulas (and they are the same) is called the **Poisson integral formula** for a circle. It expresses the value of a harmonic function, say, $u(r,\theta)$, at points interior to a circle of radius R centered at the origin, in terms of its values on the circumference of the circle.

Our development of 4.34(a) was based on the assumption that f is analytic inside and on the circle $|z| < R$. The results are valid with somewhat weaker restrictions, but the proof is correspondingly more difficult. We simply state the result.

■ **THEOREM 4.25** (Poisson Integral Formula) When U is a piecewise-continuous function on the circle $C : r = R$, the function

$$u(r,\theta) = \frac{R^2 - r^2}{2\pi} \int_{-\pi}^{\pi} \frac{U(R,\phi)}{R^2 + r^2 - 2rR\cos(\theta - \phi)}\, d\phi \tag{4.35}$$

is harmonic inside C. Furthermore, as (r,θ) approaches any point on C where U is continuous, $u(r,\theta)$ approaches the value of U at that point. In particular, if U is continuous on C, so also is $u(r,\theta)$. ■

Poisson's integral formula defines values for the solution $u(r,\theta)$ of Laplace's equation in $r < R$ in terms of its values $U(R,\phi)$ on $r = R$. Unfortunately, the difficulty in finding an antiderivative for the integrand, even for the simplest of functions $U(R,\phi)$, renders the formula of limited practical use. For example, if

$$U(R,\phi) = \begin{cases} 1, & 0 < \phi < \pi \\ 0, & \pi < \phi < 2\pi, \end{cases}$$

then 4.35 becomes

$$u(r,\theta) = \frac{R^2 - r^2}{2\pi} \int_{0}^{\pi} \frac{1}{R^2 + r^2 - 2rR\cos(\theta - \phi)}\, d\phi,$$

not a pleasant integration problem. Notice, however, that the solution of Laplace's equation at the center of the circle is

$$u(0,\theta) = \frac{R^2}{2\pi} \int_0^{\pi} \frac{1}{R^2}\, d\phi = \frac{1}{2}.$$

This value $\frac{1}{2}$ also happens to be the average value of $U(R,\phi)$ on the circumference of the circle $r = R$. This is not a peculiarity of this example, but as we now show is a property of all solutions of Laplace's equation in the circle $r = R$. According to Poisson's integral formula 4.35, the value of $u(r,\theta)$ at the center of the circle $r = R$ is

$$u(0,\theta) = \frac{1}{2\pi} \int_{-\pi}^{\pi} U(R,\phi)\, d\phi = \frac{1}{2\pi R} \int_{-\pi}^{\pi} U(R,\phi)\, R\, d\phi; \tag{4.36a}$$

This expression is the average value of $U(R,\phi)$ on $r = R$.

In Exercise 12 you are asked to show that the average value of $u(r,\theta)$ around every circle centered at the origin (not just $r=R$) is equal to its value $u(0,\theta)$ at the center of the circle. In other words, we can write that

$$u(0,\theta)=\frac{1}{2\pi r}\int_{-\pi}^{\pi}u(r,\phi)r\,d\phi$$

for every r in $0<r<R$. If we multiply this equation by r and integrate with respect to r from $r=0$ to $r=R$,

$$\int_0^R u(0,\theta)r\,dr=\int_0^R\frac{1}{2\pi}\left(\int_{-\pi}^{\pi}u(r,\phi)r\,d\phi\right)dr,$$

or,

$$\frac{R^2}{2}u(0,\theta)=\frac{1}{2\pi}\int_0^R\int_{-\pi}^{\pi}u(r,\phi)r\,d\phi\,dr.$$

Multiplication by $2/R^2$ gives

$$u(0,\theta)=\frac{1}{\pi R^2}\int_{-\pi}^{\pi}\int_0^R u(r,\phi)r\,dr\,d\phi. \tag{4.36b}$$

Thus, not only is the value of $u(r,\theta)$ at the center of the circle $r=R$ equal to its average value on the circle $r=R$, it is also equal to its average over the interior of the circle.

A second integral formula for a half-space proves even more useful. To develop this formula, we suppose that f is an analytic function at all points in the upper half-plane $\operatorname{Im} z\geq 0$ and that f is bounded therein. Let C be the curve consisting of the semicircle $\Gamma:|z|=R$ ($\operatorname{Im} z\geq 0$) and that part of the real axis $-R\leq\operatorname{Re} z\leq R$ (Figure 4.62).

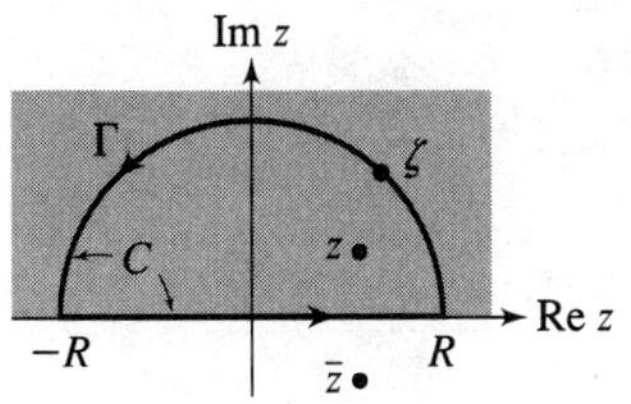

Figure 4.62

When z is interior to C, Cauchy's integral formula gives

$$f(z)=\frac{1}{2\pi i}\oint_C\frac{f(\zeta)}{\zeta-z}\,d\zeta.$$

Because $\overline{z}$ is clearly exterior to C, it follows that

$$0 = \frac{1}{2\pi i}\oint_C \frac{f(\zeta)}{\zeta - \overline{z}}\,d\zeta.$$

If we subtract the second of these from the first,

$$f(z) = \frac{1}{2\pi i}\oint_C \frac{f(\zeta)}{\zeta - z}\,d\zeta - \frac{1}{2\pi i}\oint_C \frac{f(\zeta)}{\zeta - \overline{z}}\,d\zeta = \frac{1}{2\pi i}\oint_C \frac{(z-\overline{z})f(\zeta)}{(\zeta - z)(\zeta - \overline{z})}\,d\zeta.$$

We now divide this contour integral into two integrals, one along the real axis and one along Γ. We set $z = x + yi$ and $f(z) = u(x,y) + v(x,y)i$, and on the real axis we set $\zeta = X$:

$$\begin{aligned} f(z) &= \frac{1}{2\pi i}\int_{-R}^{R} \frac{2yi[u(X,0) + v(X,0)i]}{(X - x - yi)(X - x + yi)}\,dX + \frac{1}{2\pi i}\int_\Gamma \frac{2yif(\zeta)}{(\zeta - z)(\zeta - \overline{z})}\,d\zeta \\ &= \frac{y}{\pi}\int_{-R}^{R} \frac{u(X,0) + v(X,0)i}{(X-x)^2 + y^2}\,dX + \frac{y}{\pi}\int_\Gamma \frac{f(\zeta)}{(\zeta - z)(\zeta - \overline{z})}\,d\zeta. \end{aligned} \tag{4.37}$$

Because f is bounded for $\operatorname{Im} z \geq 0$, there exists a constant M such that $|f(z)| \leq M$ for all such z. As a result, the modulus of the integrand for the contour integral along Γ is

$$\left|\frac{f(\zeta)}{(\zeta - z)(\zeta - \overline{z})}\right| \leq \frac{M}{|\zeta - z||\zeta - \overline{z}|}.$$

Using inequality 1.33, we can write

$$\left|\frac{f(\zeta)}{(\zeta - z)(\zeta - \overline{z})}\right| \leq \frac{M}{(|\zeta| - |z|)(|\zeta| - |\overline{z}|)} = \frac{M}{(R - |z|)(R - |z|)} = \frac{M}{(R - |z|)^2}.$$

Inequality 4.22 now gives

$$\left|\int_\Gamma \frac{f(\zeta)}{(\zeta - z)(\zeta - \overline{z})}\,d\zeta\right| \leq \frac{M}{(R - |z|)^2}(\pi R),$$

and this expression approaches zero as $R \to \infty$. We use this result by taking limits as $R \to \infty$ in 4.37:

$$\begin{aligned} f(z) &= \lim_{R\to\infty}\left[\frac{y}{\pi}\int_{-R}^{R} \frac{u(X,0) + v(X,0)i}{(X-x)^2 + y^2}\,dX + \frac{y}{\pi}\int_\Gamma \frac{f(\zeta)}{(\zeta - z)(\zeta - \overline{z})}\,d\zeta\right] \\ &= \frac{y}{\pi}\int_{-\infty}^{\infty} \frac{u(X,0) + v(X,0)i}{(X-x)^2 + y^2}\,dX. \end{aligned}$$

When we take real parts (the imaginary parts give a similar formula),

$$u(x,y) = \frac{y}{\pi}\int_{-\infty}^{\infty} \frac{u(X,0)}{(X-x)^2 + y^2}\,dX. \tag{4.38}$$

This is Poisson's integral formula for the half-space $y \geq 0$. It expresses the values of a function $u(x, y)$ harmonic in the upper half-plane in terms of its values on the x-axis.

EXAMPLE 4.24 ....

Find the solution of Laplace's equation in the half-plane $y > 0$ that has value 0 on the negative x-axis and value 1 on the positive x-axis.

Solution By Poisson's integral formula 4.38,

$$u(x,y) = \frac{y}{\pi}\int_0^{\infty} \frac{1}{(X-x)^2+y^2}\,dX = \frac{y}{\pi}\left\{\frac{1}{y}\operatorname{Tan}^{-1}\left(\frac{X-x}{y}\right)\right\}_0^{\infty}$$

$$= \frac{1}{\pi}\left[\frac{\pi}{2} - \operatorname{Tan}^{-1}\left(-\frac{x}{y}\right)\right] = \frac{1}{2} + \frac{1}{\pi}\operatorname{Tan}^{-1}\left(\frac{x}{y}\right).$$

It is interesting to draw curves along which $u(x, y)$ is constant (see Exercise 5).

....

Exercises 4.8

1. Prove the minimum case for Theorem 4.24.
2. Use the result of Exercise 15 in Section 4.6 to prove that when a function is harmonic in a simply connected domain, all its derivatives are harmonic therein also.
3. Prove that

$$\int_0^{2\pi} \frac{1}{R^2 + r^2 - 2rR\cos(\theta - \phi)}\,d\phi = \frac{2\pi}{R^2 - r^2}.$$

4. Prove that if an entire function has either a bounded real part or a bounded imaginary part, then it is a constant function.
5. **(a)** Show that curves along which $u(x, y) = C$ in Example 4.24 are radial lines eminating from the origin.
   (b) What is the limit of $u(x, y)$ as (x, y) approaches a point on the positive x-axis, the negative x-axis, and the origin?

In Exercises 6–10 find a function $u(x, y)$ harmonic for $y > 0$ that has the given values on the x-axis.

6. $u(x, 0) = 1$
7. $u(x,0) = \begin{cases} 1, & |x| < 1 \\ 0, & |x| > 1 \end{cases}$
8. $u(x,0) = \begin{cases} -1, & x < 0 \\ 1, & x > 0 \end{cases}$

9. $u(x,0) = \begin{cases} -1, & |x| > c \\ 1, & |x| < c \end{cases}$ $c > 0$ a constant

10. $u(x,0) = \begin{cases} 0, & |x| < 1 \\ x^{-1}, & |x| > 1 \end{cases}$

11. Suppose $u(x,y)$ is a continuous steady-state temperature distribution in the circle $x^2 + y^2 \le R^2$. Can $u(x,y)$ have a relative maximum or minimum at any point in $x^2 + y^2 < R^2$?

12. Suppose $u(r,\theta)$ is the solution of Laplace's equation in $r < R$ taking prescribed values on $r = R$. Show that the average value of $u(r,\theta)$ around every circle $r = R_1 < R$ is the same. What is this value?

13. Verify by direct substitution that $u(r,\theta)$ as defined by 4.35 satisfies Laplace's equation:

$$\frac{\partial^2 u}{\partial r^2} + \frac{1}{r}\frac{\partial u}{\partial r} + \frac{1}{r^2}\frac{\partial^2 u}{\partial \theta^2} = 0.$$

14. Show that when $u(r,\theta)$ is harmonic in $|z| < R$, the complex conjugate $v(r,\theta)$ of $u(r,\theta)$ can be expressed in the form

$$v(r,\theta) = C + \frac{rR}{\pi}\int_{-\pi}^{\pi} \frac{\sin(\theta-\phi)u(R,\phi)}{R^2 + r^2 - 2rR\cos(\theta-\phi)}\,d\phi,$$

where C is a constant and $u(R,\phi)$ are the values of $u(r,\theta)$ on $|z| = R$.

15. Use 4.35 and the result of Exercise 14 to show that when $u(r,\theta)$ is harmonic in $|z| < R$ and $u(R,\phi)$ are its values on $|z| = R$, then the function $f(z)$ that is analytic in $|z| < R$ and has $u(r,\theta)$ as its real part can be expressed in the form

$$f(z) = Ci + \frac{1}{2\pi}\int_{-\pi}^{\pi} u(R,\phi)\left(\frac{\zeta + z}{\zeta - z}\right)d\phi,$$

where $\zeta = Re^{\phi i}$ and C is a real constant.

16. **(a)** Show that $u_1(x,y) = y + 1$ and $u_2(x,y) = 1$ both satisfy Laplace's equation in the half-plane $y > 0$ and have value 1 on the x-axis.

(b) Do the results in part (a) contradict uniqueness of solutions of Laplace's equation in Corollary 3 to Theorem 4.24?

(c) Which, if either, of these solutions is given by Poisson's integral formula 4.38 (see Exercise 6)? Explain.

CHAPTER 5

Power Series and Laurent Series

In Chapter 2 we developed properties of analytic functions using derivatives. Contour integrals opened a floodgate of properties in Chapter 4. In this chapter infinite series uncover even more properties of analytic functions. We begin in Sections 5.1 and 5.2 with sequences and series of (complex) constants. These pave the way for power series and Laurent series in Sections 5.3 and 5.4. Laurent series, which have no counterpart in real calculus, provide a classification for singularities of complex functions.

5.1 Infinite Sequences of Constants

An infinite sequence of constants is a complex-valued function f whose domain is the set of positive integers. It assigns a complex number $f(n)$ to each positive integer n. Quite often we list these numbers as follows:

$$f(1), f(2), \ldots, f(n), \ldots \tag{5.1}$$

and regard the list as the sequence rather than the rule by which it is formed. We call $f(1)$ the first term of the sequence, $f(2)$ the second term, and $f(n)$ the nth term of the sequence. It is customary to replace $f(1)$ by c_1, $f(2)$ by c_2, etc., so that the sequence is

$$c_1, c_2, \ldots, c_n, \ldots. \tag{5.2}$$

To represent this even more compactly, we often write c_n in braces:

$$\{c_n\} = c_1, c_2, \ldots, c_n, \ldots. \tag{5.3}$$

EXAMPLE 5.1

Write out the first six terms of the following sequences:

$$\text{(a)}\ \left\{\frac{i^{n+1} + n^2 i}{n^2}\right\} \qquad \text{(b)}\ \left\{\frac{(1+i)^n}{n!}\right\}$$

Solution The first six terms of these sequences are as follows:

$$\text{(a)}\ -1+i,\quad \frac{3i}{4},\quad \frac{1+9i}{9},\quad \frac{17i}{16},\quad \frac{-1+25i}{25},\quad \frac{35i}{36}$$

$$\text{(b)}\ 1+i,\quad i,\quad \frac{-1+i}{3},\quad -\frac{1}{6},\quad -\frac{1+i}{30},\quad -\frac{i}{90}$$

Most often we are interested in whether a sequence has a limit. Does there exist a number to which terms of the sequence get arbitrarily close, and stay close, as n gets very large? This is stated precisely in the following definition.

Definition 5.1
A sequence $\{c_n\}$ has limit L if, given any $\epsilon > 0$, there exists an integer N such that for all $n > N$,

$$|c_n - L| < \epsilon.$$

The definition for the limit of a real sequence is identical to Definition 5.1, except that in the real case vertical lines in $|c_n - L|$ are interpreted as absolute values instead of moduli.

When L is the limit of a sequence $\{c_n\}$, we write

$$\lim_{n\to\infty} c_n = L, \tag{5.4}$$

and we say that the sequence converges to L. For instance, the first sequence in Example 5.1 converges:

$$\lim_{n\to\infty} \frac{i^{n+1} + n^2 i}{n^2} = \lim_{n\to\infty}\left(\frac{i^{n+1}}{n^2} + i\right) = 0 + i = i.$$

Convergent sequences can be added, subtracted, multiplied, and divided, term by term.

■ **THEOREM 5.1** If sequences $\{c_n\}$ and $\{d_n\}$ have limits C and D, then

(i) $\{kc_n\}$ has limit kC;
(ii) $\{c_n \pm d_n\}$ have limits $C \pm D$;
(iii) $\{c_n d_n\}$ has limit CD;
(iv) $\{c_n/d_n\}$ has limit C/D, provided $D \neq 0$ and none of the $d_n = 0$. ■

For example, because

$$\lim_{n\to\infty}\left(1 + \frac{2i}{n}\right) = 1 \quad\text{and}\quad \lim_{n\to\infty} \frac{n}{\sin{(ni)}} = \lim_{n\to\infty} \frac{n}{i\sinh n} = 0,$$

we can say that the sequence $\left\{\left(1 + \frac{2i}{n}\right)\left(\frac{n}{\sin{(ni)}}\right)\right\}$ has limit $1 \cdot 0 = 0$.

The following theorem indicates that, if it is convenient to do so, discussion of the convergence of a sequence $\{c_n\}$ can be separated into discussions of convergence of the real and imaginary parts of c_n.

■ **THEOREM 5.2** Suppose $c_n = x_n + y_n i$ and $L = X + Yi$. Then $\lim_{n\to\infty} c_n = L$ if and only if $\lim_{n\to\infty} x_n = X$ and $\lim_{n\to\infty} y_n = Y$.

Proof Suppose first of all that $\lim_{n\to\infty} x_n = X$ and $\lim_{n\to\infty} y_n = Y$. Then given any $\epsilon > 0$, there exists an integer N_1 such that when $n > N_1$, $|x_n - X| < \epsilon/2$. Similarly, there exists an integer N_2 such that when $n > N_2$, $|y_n - Y| < \epsilon/2$. Consequently, when n is greater than the larger of N_1 and N_2,

$$\begin{aligned}|c_n - L| &= |(x_n + y_n i) - (X + Yi)| = |(x_n - X) + (y_n - Y)i| \\ &\leq |x_n - X| + |y_n - Y| < \frac{\epsilon}{2} + \frac{\epsilon}{2} = \epsilon,\end{aligned}$$

and this proves that $\lim_{n\to\infty} c_n = L$.

Conversely, if $\lim_{n\to\infty} c_n = L$, then given any $\epsilon > 0$, there exists an integer N such that for $n > N$,

$$\begin{aligned}\epsilon > |c_n - L| &= |(x_n + y_n i) - (X + Yi)| = |(x_n - X) + (y_n - Y)i| \\ &= \sqrt{(x_n - X)^2 + (y_n - Y)^2}.\end{aligned}$$

It follows from this that for $n > N$,

$$|x_n - X| < \epsilon \quad \text{and} \quad |y_n - Y| < \epsilon.$$

Hence, $\lim_{n\to\infty} x_n = X$ and $\lim_{n\to\infty} y_n = Y$. ■

EXAMPLE 5.2 ....

Does the sequence $\left\{\left(1 + \frac{1}{n}\right)^n - \left(1 + \frac{1}{n}\right)^3 i\right\}$ have a limit?

Solution Because

$$\lim_{n\to\infty}\left(1 + \frac{1}{n}\right)^n = e \quad \text{and} \quad \lim_{n\to\infty}\left[-\left(1 + \frac{1}{n}\right)^3\right] = -1,$$

it follows by Theorem 5.2 that $\lim_{n\to\infty}\left[\left(1 + \frac{1}{n}\right)^n - \left(1 + \frac{1}{n}\right)^3 i\right] = e - i.$

....

The analogous result to Theorem 5.2 involving modulus and argument is not quite as strong (see Exercise 15).

■ **THEOREM 5.3** Suppose $c_n = r_n e^{\theta_n i}$ and $L = Re^{\Theta i}$, where θ_n and Θ are principal values of arguments. Then, $\lim_{n\to\infty} c_n = L$ if $\lim_{n\to\infty} r_n = R$ and $\lim_{n\to\infty} \theta_n = \Theta$. ■

The following result is an immediate consequence of Theorem 5.3.

■ **THEOREM 5.4** $\lim_{n\to\infty} c_n = 0$ if and only if $\lim_{n\to\infty} |c_n| = 0$. ■

For instance, we can write for the second sequence in Example 5.1,

$$\lim_{n\to\infty} \left|\frac{(1+i)^n}{n!}\right| = \lim_{n\to\infty} \frac{|1+i|^n}{n!} = \lim_{n\to\infty} \frac{(\sqrt{2})^n}{n!} = 0,$$

and therefore by Theorem 5.4, $\lim_{n\to\infty} \frac{(1+i)^n}{n!} = 0$.

EXAMPLE 5.3 ....

Show that $\lim_{n\to\infty} z^n = 0$ if $|z| < 1$ and that the limit does not exist when $|z| > 1$. What can you say if $|z| = 1$?

Solution If $|z| < 1$,

$$\lim_{n\to\infty} |z^n| = \lim_{n\to\infty} |z|^n = 0;$$

therefore, by Theorem 5.4, $\lim_{n\to\infty} z^n = 0$. On the other hand, if $|z| > 1$, then

$$\lim_{n\to\infty} |z^n| = \lim_{n\to\infty} |z|^n = \infty,$$

and therefore $\lim_{n\to\infty} z^n$ does not exist.

If $z = 1$, then certainly $\lim_{n\to\infty} z^n = 1$. To discuss other points on the circle $|z| = 1$, we use the fact that if a sequence $\{c_n\}$ converges, then $\lim_{n\to\infty} (c_n - c_{n-1}) = 0$, (part (ii) of Theorem 5.1), and Theorem 5.4 then implies that $\lim_{n\to\infty} |c_n - c_{n-1}| = 0$. It follows that if $\lim_{n\to\infty} |c_n - c_{n-1}| \neq 0$ for a sequence $\{c_n\}$, then the sequence cannot converge. This is the case for $\{z^n\}$ when $|z| = 1$ and $z \neq 1$:

$$\lim_{n\to\infty} |z^n - z^{n-1}| = \lim_{n\to\infty} |z^{n-1}|\,|z-1| = \lim_{n\to\infty} |z-1| = |z-1| \neq 0.$$

Hence, $\{z^n\}$ diverges for all z on $|z| = 1$ except $z = 1$.

....

Exercises 5.1

In Exercises 1–10 find the limit of the sequence, if it exists.

1. $\left\{\frac{1-2^{n-1}}{2^n} + \frac{i}{3^n - 1}\right\}$

2. $\{(1+i)^n\}$

3. $\left\{\dfrac{(-1)^n + i^n}{n}\right\}$

4. $\left\{\sin\left(\dfrac{ni}{n-1}\right)\right\}$

5. $\left\{\sinh\left(\dfrac{n^2+i}{1-n^2 i}\right)\right\}$

6. $\{\tan^n(i/n)\}$

7. $\left\{\sqrt[3]{1-2e^{(-1+ni)/n}}\right\}$

8. $\left\{i\,\mathrm{Log}\left(\dfrac{3i^n - 2n}{n}\right)\right\}$

9. $\left\{\mathrm{Tan}^{-1}(ni+i)\right\}$

10. $\{\sqrt[n]{i}\}$

11. If $\lim_{n\to\infty} c_n = L$, show that $\lim_{n\to\infty} \overline{c_n} = \overline{L}$.

12. Use Theorem 5.3 to prove that $\displaystyle\lim_{n\to\infty}\left(1+\frac{i}{n}\right)^n = e^i$.

13. Prove that $\displaystyle\lim_{n\to\infty}\left(1+\frac{z}{n}\right)^n = e^z$.

14. Do the following sequences have limits?

(a) $\left\{\mathrm{Log}\left(\dfrac{n+i}{2.5+n}\right)\right\}$ (b) $\left\{\mathrm{Log}\left(\dfrac{n+i}{2.5-n}\right)\right\}$ (c) $\left\{\mathrm{Log}\left(\dfrac{n+(-1)^n i}{2.5-n}\right)\right\}$

15. Verify Theorem 5.3. Is the converse true; that is, if $\lim_{n\to\infty} c_n = L$, is $\lim_{n\to\infty} r_n = R$ and $\lim_{n\to\infty} \theta_n = \Theta$?

5.2 Infinite Series of Constants

When the terms of an infinite sequence are added together, we obtain what is called an **infinite series**:

$$\sum_{n=1}^{\infty} c_n = c_1 + c_2 + \cdots + c_n + \cdots. \tag{5.5}$$

If the sequence of partial sums $\{S_n\}$, where

$$S_n = c_1 + c_2 + \cdots + c_n,$$

has limit S, we say that the series has sum S or that the series converges to S, and we write

$$\sum_{n=1}^{\infty} c_n = S. \tag{5.6}$$

If $\lim_{n\to\infty} S_n$ does not exist, we say that the series does not have a sum or that it diverges.

For example, the nth partial sum of the **geometric series**

$$a + az + az^2 + \cdots + az^{n-1} + \cdots$$

is

$$S_n = a + az + az^2 + \cdots + az^{n-1}.$$

If we multiply this equation by z,

$$zS_n = az + az^2 + az^3 + \cdots + az^n,$$

and then subtract these equations,

$$S_n - zS_n = a - az^n.$$

This equation can be solved for S_n:

$$S_n = \frac{a(1 - z^n)}{1 - z},$$

provided, of course, $z \neq 1$. According to Example 5.2, $\lim_{n\to\infty} z^n = 0$ when $|z| < 1$. Hence, $\lim_{n\to\infty} S_n = a/(1 - z)$ when $|z| < 1$. Furthermore, when $|z| > 1$ or $|z| = 1$ (but $z \neq 1$), Example 5.2 also indicates that $\lim_{n\to\infty} z^n$ does not exist. Thus, for all z such that $|z| \geq 1$ and $z \neq 1$, $\lim_{n\to\infty} S_n$ does not exist. Finally when $z = 1$, $S_n = a + a + \cdots + a = na$, and certainly $\lim_{n\to\infty} S_n$ does not exist (unless trivially $a = 0$). To summarize,

$$\lim_{n\to\infty} S_n = \begin{cases} \dfrac{a}{1-z}, & |z| < 1 \\ \text{does not exist}, & |z| \geq 1. \end{cases}$$

Thus,

$$\sum_{n=1}^{\infty} az^{n-1} = a + az + az^2 + \cdots = \begin{cases} \dfrac{a}{1-z}, & |z| < 1 \\ \text{diverges}, & |z| \geq 1. \end{cases} \tag{5.7}$$

The separation of terms in a sequence into real and imaginary parts in Theorem 5.1 can be extended to series.

■ **THEOREM 5.5** A complex series converges if and only if the series of real and imaginary parts converge. To be precise, suppose $c_n = x_n + y_n i$ and $S = X + Yi$. Then

$$\sum_{n=1}^{\infty} c_n = S \quad \text{if and only if} \quad \sum_{n=1}^{\infty} x_n = X \quad \text{and} \quad \sum_{n=1}^{\infty} y_n = Y. \tag{5.8}$$ ■

EXAMPLE 5.4 ....

Show that $\displaystyle\sum_{n=1}^{\infty} \frac{1 + \sinh(i/n)}{2^n}$ converges.

Solution It is straightforward to identify real and imaginary parts of the nth term of the series

$$\frac{1 + \sinh(i/n)}{2^n} = \frac{1}{2^n} + \frac{\sin(1/n)}{2^n} i.$$

Because $\sum_{n=1}^{\infty} 1/2^n$ is a convergent (real) geometric series, and $\sum_{n=1}^{\infty} \sin(1/n)/2^n$ also converges (being term by term less than $\sum_{n=1}^{\infty} 1/2^n$), it follows that $\sum_{n=1}^{\infty} [1+\sinh(i/n)]/2^n$ converges. ....

To discuss convergence of a series $\sum_{n=1}^{\infty} c_n$ in terms of its sequence $\{S_n\}$ of partial sums is usually impractical; seldom is it possible to find a simple formula for S_n. Consequently, we must develop alternative ways to decide on the convergence of a series. Before doing this, however, we mention some fairly simple but important results concerning convergence of series. First, if a series $\sum_{n=1}^{\infty} c_n$ has a finite number of terms altered in any fashion whatsoever, the new series converges if and only if the original series converges. The new series may converge to a different sum, but it converges if the original series converges. The following theorem indicates that convergent series can be added, subtracted, and multiplied by constants. These properties can be verified using the definition of the sum of a series as the limit of the sequence of its partial sums.

■ **THEOREM 5.6** If $\sum_{n=1}^{\infty} c_n = C$ and $\sum_{n=1}^{\infty} d_n = D$, then

$$\sum_{n=1}^{\infty} (c_n \pm d_n) = C \pm D, \tag{5.9a}$$

$$\sum_{n=1}^{\infty} kc_n = kC \qquad \text{(when } k \text{ is a constant).} \tag{5.9b}$$ ■

The following result gives rise to the nth-term test for divergence of a series.

■ **THEOREM 5.7** If $\sum_{n=1}^{\infty} c_n$ converges, then $\lim_{n\to\infty} c_n = 0$.

Proof If $\sum_{n=1}^{\infty} c_n$ converges to S, its sequence of partial sums

$$S_1, S_2, \ldots, S_n, \ldots$$

has limit S. But then the sequence

$$0, S_1, S_2, \ldots, S_{n-1}, \ldots$$

also converges to S. When we subtract these sequences term by term, the sequence

$$S_1 - 0, S_2 - S_1, \ldots, S_n - S_{n-1}, \ldots$$

has limit $S - S = 0$ (according to Theorem 5.1). Because $S_n - S_{n-1} = c_n$, we have shown that $\lim_{n\to\infty} c_n = 0$. ■

The contrapositive of this result is the nth-term test.

■ **COROLLARY** (*n*th-Term Test) If $\lim_{n\to\infty} c_n \neq 0$ or does not exist, then $\sum_{n=1}^{\infty} c_n$ diverges. ■

The nth-term test is a test for divergence of a series, not convergence. It states that if $\lim_{n\to\infty} c_n$ exists and is equal to anything but zero or the limit does not exist, then the series $\sum_{n=1}^{\infty} c_n$ diverges. The nth-term test never indicates that a series converges. Even if $\lim_{n\to\infty} c_n = 0$, we can conclude nothing about the convergence or divergence of $\sum_{n=1}^{\infty} c_n$; it may converge or diverge.

The concept of convergence of an infinite series is the same whether the terms of the series are real or constant. In other words, everything that we have said in this section (with the exception of Theorem 5.5) is exactly as we would have said had series 5.5 been real instead of complex.

In real-variable theory, two types of series are discussed, those with only nonnegative terms and those with an infinite number of positive and negative terms. To determine whether a series with only nonnegative terms converges or diverges, various tests are developed, such as comparison, integral, ratio, and root. To discuss convergence of series with both positive and negative terms, the ideas of absolute and conditional convergence are introduced. The situation for complex series is different but similar. It is different because complex numbers are neither positive nor negative. The situation is similar in that we use absolute convergence as the vehicle to discuss convergence and divergence of most series.

Definition 5.2
A complex series $\sum_{n=1}^{\infty} c_n$ is said to be **absolutely convergent** (or to converge absolutely) if $\sum_{n=1}^{\infty} |c_n|$ converges.

It might seem more appropriate to say *converges in modulus* rather than *converges absolutely* because vertical bars indicate moduli rather than absolute values, but absolute convergence has become the accepted terminology. The importance of absolute convergence is contained in the following theorem.

■ **THEOREM 5.8** If a complex series converges absolutely, then it converges.

Proof Suppose the series $\sum_{n=1}^{\infty} c_n$ converges absolutely, where $c_n = a_n + b_n i$. This convergence means that $\sum_{n=1}^{\infty} |c_n| = \sum_{n=1}^{\infty} \sqrt{a_n^2 + b_n^2}$ has a sum, call it S. Now each term in the series $\sum_{n=1}^{\infty} |a_n|$ is nonnegative and is less than the corresponding term in $\sum_{n=1}^{\infty} \sqrt{a_n^2 + b_n^2}$. It follows that $\sum_{n=1}^{\infty} |a_n|$ converges; that is, the real series $\sum_{n=1}^{\infty} a_n$ of real parts of $\sum_{n=1}^{\infty} c_n$ converges absolutely. Similarly, the series of imaginary parts $\sum_{n=1}^{\infty} b_n$ converges absolutely, and by Theorem 5.5, $\sum_{n=1}^{\infty} c_n$ therefore converges. ■

According to Theorem 5.8, a complex series $\sum_{n=1}^{\infty} c_n$ converges when its series of moduli $\sum_{n=1}^{\infty} |c_n|$ converges. Because the latter series is a series of nonnegative reals, we can utilize the wealth of tests for such series in real variable theory. We describe five of these tests next but phrase them in the context of complex series and absolute convergence.

■ **THEOREM 5.9** (Comparison Test) If $|c_n| \leq a_n$ for all $n \geq N$ and $\sum_{n=N}^{\infty} a_n$ converges, then $\sum_{n=1}^{\infty} c_n$ converges absolutely. ■

■ **THEOREM 5.10** (Limit Comparison Test) If $\sum_{n=1}^{\infty} b_n$ is a convergent series of positive real numbers and

$$\lim_{n\to\infty} \frac{|c_n|}{b_n} = l, \qquad 0 \leq l < \infty,$$

then $\sum_{n=1}^{\infty} c_n$ converges absolutely. ■

■ **THEOREM 5.11** (Integral Test) If $|c_n| = f(n)$ and $f(x)$ is a continuous, decreasing function for $x \geq N$ such that

$$\int_N^{\infty} f(x)\,dx < \infty,$$

then $\sum_{n=1}^{\infty} c_n$ converges absolutely. ■

■ **THEOREM 5.12** (Limit Ratio Test) Suppose $c_n \neq 0$ and $\displaystyle\lim_{n\to\infty} \left|\frac{c_{n+1}}{c_n}\right| = L$. Then

(i) $\sum_{n=1}^{\infty} c_n$ converges absolutely if $L < 1$;
(ii) $\sum_{n=1}^{\infty} c_n$ diverges if $L > 1$ or if $\lim_{n\to\infty} |c_{n+1}/c_n| = \infty$;
(iii) $\sum_{n=1}^{\infty} c_n$ may converge or diverge if $L = 1$. ■

■ **THEOREM 5.13** (Limit Root Test) Suppose $\displaystyle\lim_{n\to\infty} \sqrt[n]{|c_n|} = \mathcal{L}$. Then,

(i) $\sum_{n=1}^{\infty} c_n$ converges absolutely if $\mathcal{L} < 1$;
(ii) $\sum_{n=1}^{\infty} c_n$ diverges if $\mathcal{L} > 1$ or if $\lim_{n\to\infty} \sqrt[n]{|c_n|} = \infty$;
(iii) $\sum_{n=1}^{\infty} c_n$ may converge or diverge if $\mathcal{L} = 1$. ■

The following examples illustrate the application of these tests.

EXAMPLE 5.5 ....

Determine whether the following series converge or diverge:

(a) $\displaystyle\sum_{n=1}^{\infty} \frac{(n+i)^2}{n^4}$ **(b)** $\displaystyle\sum_{n=1}^{\infty} \frac{(2+i)^n}{n!}$

(c) $\displaystyle\sum_{n=1}^{\infty} \left(2+\frac{i}{n^2}\right)^n$ **(d)** $\displaystyle\sum_{n=1}^{\infty} \frac{1}{n \operatorname{Log}^2 ni}$

Solution **(a)** Because

$$l = \lim_{n\to\infty} \frac{\left|\dfrac{(n+i)^2}{n^4}\right|}{\dfrac{1}{n^2}} = \lim_{n\to\infty} \frac{n^2(n^2+1)}{n^4} = 1$$

and $\sum_{n=1}^{\infty} 1/n^2$ converges, it follows that $\sum_{n=1}^{\infty} (n+i)^2/n^4$ converges absolutely by the limit comparison test.

(b) Because

$$L = \lim_{n\to\infty} \left|\frac{\dfrac{(2+i)^{n+1}}{(n+1)!}}{\dfrac{(2+i)^n}{n!}}\right| = \lim_{n\to\infty} \frac{|2+i|}{n+1} = 0,$$

the series $\sum_{n=1}^{\infty} (2+i)^n/n!$ converges absolutely by the limit ratio test.

(c) Because

$$\mathcal{L} = \lim_{n\to\infty} \sqrt[n]{\left|\left(2+\frac{i}{n^2}\right)^n\right|} = \lim_{n\to\infty} \left|2+\frac{i}{n^2}\right| = \lim_{n\to\infty} \sqrt{4+1/n^4} = 2,$$

the series diverges by the limit root test.

(d) First we note that

$$\left|\frac{1}{n\,\mathrm{Log}^2\, ni}\right| = \frac{1}{n|\ln n + \pi i/2|^2} = \frac{1}{n[(\ln n)^2 + \pi^2/4]}.$$

If we set $f(x) = 1/\{x[(\ln x)^2 + \pi^2/4]\}$, then for $x \geq 2$, $f(x) < 1/[x(\ln x)^2]$, and

$$\int_2^{\infty} f(x)\,dx < \int_2^{\infty} \frac{1}{x(\ln x)^2}\,dx = \left\{\frac{-1}{\ln x}\right\}_2^{\infty} = \frac{1}{\ln 2}.$$

Thus, the series converges absolutely by the integral test.

Exercises 5.2

In Exercises 1–20 determine whether the series converges or diverges.

1. $\displaystyle\sum_{n=1}^{\infty} i^n$

2. $\displaystyle\sum_{n=1}^{\infty} \frac{1+3^n i}{5^n}$

3. $\displaystyle\sum_{n=1}^{\infty} \frac{n+2i}{ni-3}$

4. $\displaystyle\sum_{n=1}^{\infty} \frac{1}{n^2+i}$

5. $\displaystyle\sum_{n=1}^{\infty} (1+i)^n$

6. $\displaystyle\sum_{n=2}^{\infty} \left[1-\left(\frac{3}{2+4i}\right)^n\right]$

7. $\sum_{n=2}^{\infty}\left(\frac{1}{2+3i}\right)^n$

8. $\sum_{n=1}^{\infty}\left(\frac{2-i}{3+i}\right)^{2n}$

9. $\sum_{n=1}^{\infty}\frac{2ni+3}{n^2+4i}$

10. $\sum_{n=1}^{\infty}\frac{e^{ni}}{n^2-4i}$

11. $\sum_{n=1}^{\infty}\frac{2^n}{(n+1)i^n}$

12. $\sum_{n=1}^{\infty}\frac{3-(2i)^n}{\cos ni}$

13. $\sum_{n=1}^{\infty}\frac{\sin ni}{(1+i)^n}$

14. $\sum_{n=1}^{\infty}\frac{\sqrt{n+i}}{n^2}$

15. $\sum_{n=1}^{\infty}\left(-2+\frac{i}{n}\right)^n$

16. $\sum_{n=1}^{\infty}\left(\frac{ni}{n+i}\right)^{n^2}$

17. $\sum_{n=1}^{\infty}\frac{(4-3i)^{2n}}{n!}$

18. $\sum_{n=1}^{\infty}\frac{(1+i)(2+i)\cdots(n+i)}{n^n}$

19. $\sum_{n=1}^{\infty}\frac{\operatorname{Log}(n+i)}{n^3}$

20. $\sum_{n=1}^{\infty}\frac{\tan(ni)}{n^2+1}$

21. The comparison test for real series states that when $c_n \geq d_n \geq 0$ for all n and $\sum_{n=1}^{\infty} d_n$ diverges, so also does $\sum_{n=1}^{\infty} c_n$. Decide if the following complex analogue is valid: If $|c_n| \geq d_n \geq 0$ for all n and $\sum_{n=1}^{\infty} d_n$ diverges, so also does $\sum_{n=1}^{\infty} c_n$.

22. For what values of z does the series $\sum_{n=1}^{\infty}\frac{e^{nz}}{(n+1)^2}$ converge?

23. Evaluate $\sum_{n=1}^{\infty}\frac{1}{(n+i)(n+i-1)}$. *Hint:* Use partial fractions.

In Exercises 24–28 verify the identity.

24. $1+\cos\theta+\cos 2\theta+\cdots+\cos n\theta = \dfrac{\sin\dfrac{(n+1)\theta}{2}\cos\dfrac{n\theta}{2}}{\sin\dfrac{\theta}{2}}$

25. $\sin\theta+\sin 2\theta+\cdots+\sin n\theta = \dfrac{\sin\dfrac{(n+1)\theta}{2}\sin\dfrac{n\theta}{2}}{\sin\dfrac{\theta}{2}}$

26. $\cos\theta+\cos 3\theta+\cdots+\cos(2n-1)\theta = \dfrac{\sin 2n\theta}{2\sin\theta}$

27. $\sin\theta+\sin 3\theta+\cdots+\sin(2n-1)\theta = \dfrac{\sin^2 n\theta}{\sin\theta}$

28. $\sin\theta - \sin 2\theta + \cdots + (-1)^{n-1}\sin n\theta = \begin{cases} \dfrac{\sin\dfrac{(n+1)\theta}{2}\cos\dfrac{n\theta}{2}}{\cos\dfrac{\theta}{2}}, & n \text{ odd} \\ -\dfrac{\cos\dfrac{(n+1)\theta}{2}\sin\dfrac{n\theta}{2}}{\cos\dfrac{\theta}{2}}, & n \text{ even} \end{cases}$

5.3 Complex Power Series

A **complex power series** is a series of the form

$$\sum_{n=0}^{\infty} a_n(z-z_0)^n = a_0 + a_1(z-z_0) + a_2(z-z_0)^2 + \cdots \tag{5.10}$$

where z_0 is a point in the complex plane and the a_n are complex constants. Such series are called power series in $z - z_0$ or power series about the point z_0. When a complex number z is substituted into 5.10, a series of complex constants results; this series may converge or diverge. It is clear that a complex power series always converges for $z = z_0$, but does it converge for other values of z? We now show that there are exactly three possibilities as far as values of z for which a power series yields a convergent series of constants:

1. The power series converges only for $z = z_0$.
2. The power series converges absolutely for all z.
3. There exists a number $R > 0$ such that the power series converges absolutely for $|z - z_0| < R$, diverges for $|z - z_0| > R$, and may or may not converge for $|z - z_0| = R$.

It is helpful to visualize (3) geometrically. Because $|z - z_0| = R$ is a circle centered at z_0 with radius R (Figure 5.1), situation (3) says that the power series converges absolutely for all points inside the circle, diverges for points outside the circle, and may or may not converge for points on the circle.

The power series $\sum_{n=0}^{\infty} n!(z-z_0)^n$ and $\sum_{n=0}^{\infty}(z-z_0)^n/n!$ establish (1) and (2). For any $z \neq z_0$, the limit ratio test on the first series gives

$$L = \lim_{n\to\infty}\left|\frac{(n+1)!(z-z_0)^{n+1}}{n!(z-z_0)^n}\right| = \lim_{n\to\infty}(n+1)|z-z_0| = \infty.$$

Hence, $\sum_{n=0}^{\infty} n!(z-z_0)^n$ diverges for all $z \neq z_0$. For the second series we calculate that

$$L = \lim_{n\to\infty}\left|\frac{(z-z_0)^{n+1}/(n+1)!}{(z-z_0)^n/n!}\right| = \lim_{n\to\infty}\frac{|z-z_0|}{n+1} = 0,$$

and therefore $\sum_{n=0}^{\infty} (z - z_0)^n/n!$ converges absolutely for all z. To establish that (3) is the only other possibility, we first prove that if a power series converges for z_1, then it converges absolutely for all points closer to z_0 than z_1 (Figure 5.2).

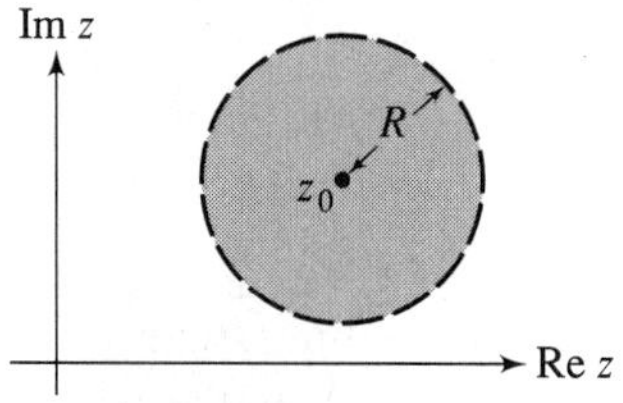

Figure 5.1

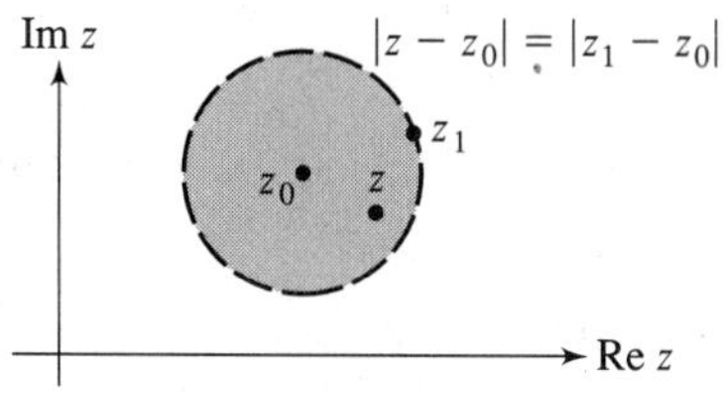

Figure 5.2

■ **THEOREM 5.14** If $\sum_{n=0}^{\infty} a_n(z - z_0)^n$ converges for z_1, then it converges for all z in the circle $|z - z_0| < |z_1 - z_0|$.

Proof If $\sum_{n=0}^{\infty} a_n(z_1 - z_0)^n$ converges, then the terms of the series must be bounded; that is, there exists a number $M > 0$ such that $|a_n(z_1 - z_0)^n| \le M$ for all $n \ge 0$. For any z in the circle $|z - z_0| < |z_1 - z_0|$,

$$|a_n(z - z_0)^n| = |a_n|\,|z - z_0|^n = |a_n| \left|\frac{z - z_0}{z_1 - z_0}\right|^n |z_1 - z_0|^n \le M \left|\frac{z - z_0}{z_1 - z_0}\right|^n .$$

Because $\sum_{n=0}^{\infty} M|(z - z_0)/(z_1 - z_0)|^n$ is a geometric series with common ratio $|(z - z_0)/(z_1 - z_0)| < 1$, it follows that $\sum_{n=0}^{\infty} a_n(z - z_0)^n$ converges absolutely in the circle $|z - z_0| < |z_1 - z_0|$ (by the comparison test). ■

We can now establish that (3) is the only possibility besides (1) and (2). Suppose that S is the set of points for which $\sum_{n=0}^{\infty} a_n(z - z_0)^n$ converges and that S contains more than the point z_0, but S is not the whole complex plane. If S contains a complex number z_1 that is farthest from z_0, then the power series converges absolutely for all z such that $|z - z_0| < |z_1 - z_0|$ and diverges for $|z - z_0| > |z_1 - z_0|$; that is, R in (3) is equal to $|z_1 - z_0|$. If S does not contain a complex number farthest from z_0, consider the set of all real numbers $d = |z - z_0|$, where z is in S. This set forms an interval $0 \le d < R$ on the real line. According to Theorem 5.14, the series converges absolutely for all z satisfying $|z - z_0| < R$. Furthermore, it does not converge for any z satisfying $|z - z_0| > R$. Thus, whether or not S has a complex number farthest from z_0, (3) is established.

In (3) we call R the **radius of convergence** of the power series. In (1) and (2) we say that $R = 0$ and $R = \infty$. Every power series then has a radius of convergence. If $R = 0$, the series converges only for $z = z_0$; if $R = \infty$, the series converges absolutely for all z; and if $0 < R < \infty$, the series converges absolutely for all points in the **circle of convergence** $|z - z_0| < R$, diverges for points outside the circle, and may or may not converge for points on the circle. For many power series, the radius of convergence can be calculated according to

$$R = \lim_{n\to\infty} \left| \frac{a_n}{a_{n+1}} \right| \tag{5.11a}$$

or

$$R = \lim_{n\to\infty} \frac{1}{|a_n|^{1/n}}, \tag{5.11b}$$

provided either limit exists or is equal to infinity. These limits can be established using the limit ratio and limit root tests in Section 5.2. Formulas 5.11 yield radii of convergence for real power series $\sum_{n=0}^{\infty} a_n(x-x_0)^n$, but in the context of real power series, R represents half the length of the interval of convergence. We now see the reason for calling R the radius of convergence of a power series.

An integral part of the discussion of the interval of convergence of a real power series is consideration of the endpoints of this interval. The counterpart of these endpoints for complex series is the circle $|z-z_0| = R$. We do not discuss whether complex power series converge or diverge on $|z-z_0| = R$ in these notes except for the results in Exercises 44–48 at the end of this section.

EXAMPLE 5.6 ....

Find circles of convergence for the following power series:

(a) $\displaystyle\sum_{n=0}^{\infty} \frac{(n+1)}{(3i)^n}(z-1)^n$

(b) $\displaystyle\sum_{n=0}^{\infty} \left(\frac{n+i}{n-2i}\right)^n (z-i)^n$

(c) $\displaystyle\sum_{n=1}^{\infty} \frac{1}{n2^n}(z-2+i)^{2n}$

Solution **(a)** Because

$$\lim_{n\to\infty} \left| \frac{\dfrac{n+1}{(3i)^n}}{\dfrac{n+2}{(3i)^{n+1}}} \right| = \lim_{n\to\infty} \frac{3(n+1)}{n+2} = 3,$$

the radius of convergence of the power series is $R = 3$. Its circle of convergence is $|z-1| < 3$ (Figure 5.3). The power series converges for all points inside this circle, diverges for points outside the circle, and may or may not converge for points on the circle.

(b) Because

$$\lim_{n\to\infty} \frac{1}{\left[\left|\left(\dfrac{n+i}{n-2i}\right)^n\right|\right]^{1/n}} = \lim_{n\to\infty}\left|\frac{n-2i}{n+i}\right| = 1,$$

the radius of convergence of the power series is $R = 1$. Its circle of convergence is $|z - i| < 1$ (Figure 5.4).

(c) Because the series contains only even powers of $z - 2 + i$, we first set $w = (z - 2 + i)^2$. Then

$$\sum_{n=1}^{\infty} \frac{1}{n2^n}(z - 2 + i)^{2n} = \sum_{n=1}^{\infty} \frac{1}{n2^n} w^n.$$

The radius of convergence of this series is

$$R_w = \lim_{n\to\infty} \frac{\dfrac{1}{n2^n}}{\dfrac{1}{(n+1)2^{n+1}}} = \lim_{n\to\infty} \frac{2(n+1)}{n} = 2;$$

that is, $\sum_{n=1}^{\infty} [1/(n2^n)]w^n$ converges absolutely for $|w| < 2$. It follows that $\sum_{n=1}^{\infty} [1/(n2^n)](z - 2 + i)^{2n}$ converges absolutely for $|z - 2 + i| < \sqrt{2}$ (Figure 5.5).

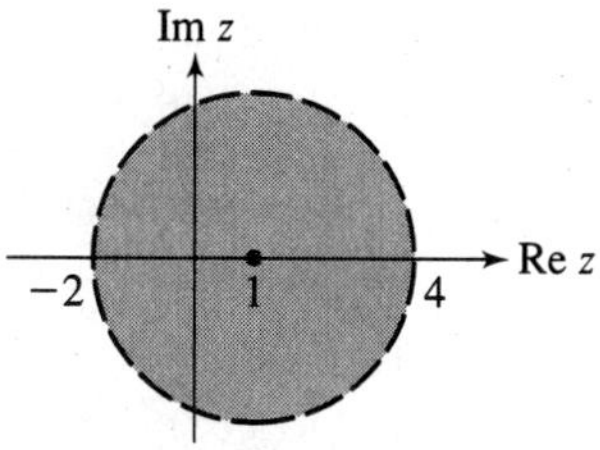

Figure 5.3

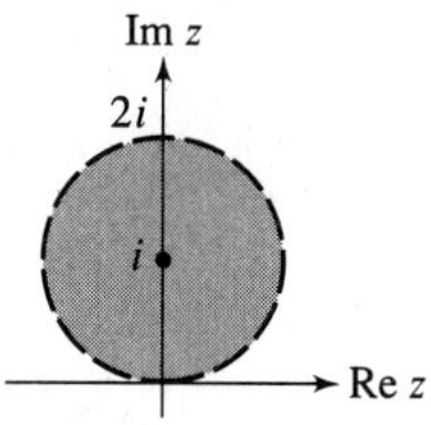

Figure 5.4

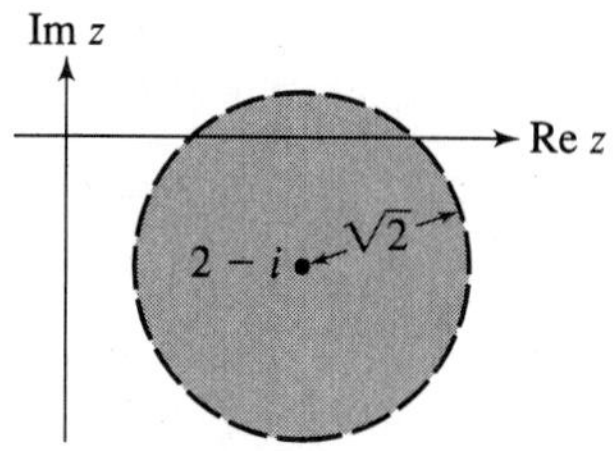

Figure 5.5

When a function f has the same value $f(z)$ as the sum of a series

$$\sum_{n=0}^{\infty} a_n (z - z_0)^n$$

at every point interior to its circle of convergence, we call f the sum of the series and write

$$f(z) = \sum_{n=0}^{\infty} a_n (z - z_0)^n, \qquad |z - z_0| < R. \tag{5.12}$$

For example, the geometric series $1 + z + z^2 + \cdots$ has sum $1/(1-z)$ provided z is in modulus less than 1, and we express this by writing

$$\frac{1}{1-z} = 1 + z + z^2 + \cdots, \qquad |z| < 1. \tag{5.13}$$

EXAMPLE 5.7 ....

What is the sum of the power series $\displaystyle\sum_{n=2}^{\infty} \frac{2^n}{3^{2n+1}}(z-1-2i)^n$?

Solution Because $\displaystyle\sum_{n=2}^{\infty} \frac{2^n}{3^{2n+1}}(z-1-2i)^n = \frac{1}{3}\sum_{n=2}^{\infty}\left[\frac{2}{9}(z-1-2i)\right]^n$, the power series is geometric with sum

$$\frac{1}{3}\frac{[\frac{2}{9}(z-1-2i)]^2}{1-\frac{2}{9}(z-1-2i)} = \frac{4(z-1-2i)^2}{27(11+4i-2z)},$$

provided $\left|\frac{2}{9}(z-1-2i)\right| < 1$, or $|z-1-2i| < \frac{9}{2}$. ....

Two complex power series about the same point z_0 can be added and subtracted term by term within their common circle of convergence; that is, if $f(z) = \sum_{n=0}^{\infty} a_n(z-z_0)^n$ for $|z-z_0| < R_1$ and $g(z) = \sum_{n=0}^{\infty} b_n(z-z_0)^n$ for $|z-z_0| < R_2$, then

$$f(z) \pm g(z) = \sum_{n=0}^{\infty}(a_n \pm b_n)(z-z_0)^n, \tag{5.14}$$

valid for $|z - z_0| <$ smaller of R_1 and R_2. It is not reasonable to multiply power series term by term. Indeed, to multiply the power series for $f(z)$ and $g(z)$,

$$[a_0 + a_1(z-z_0) + a_2(z-z_0)^2 + \cdots][b_0 + b_1(z-z_0) + b_2(z-z_0)^2 + \cdots],$$

the rules of algebra would suggest that every term in the first series should be multiplied by every term in the second series. When this is done and like terms are accumulated, the result is

$$a_0b_0 + (a_0b_1 + a_1b_0)(z-z_0) + (a_0b_2 + a_1b_1 + a_2b_0)(z-z_0)^2 + \cdots.$$

This series can be expressed in sigma notation as follows:

$$\sum_{n=0}^{\infty} c_n(z-z_0)^n \quad \text{where} \quad c_n = \sum_{j=0}^{n} a_j b_{n-j}.$$

This series converges to $f(z)g(z)$ in the common circle of convergence for the series for $f(z)$ and $g(z)$; that is,

$$f(z)g(z) = \sum_{n=0}^{\infty} c_n(z - z_0)^n, \qquad c_n = \sum_{j=0}^{n} a_j b_{n-j}, \tag{5.15}$$

valid for $|z - z_0| <$ smaller of R_1 and R_2.

A natural question at this point is, If f is the sum of a power series

$$\sum_{n=0}^{\infty} a_n(z - z_0)^n,$$

what properties does f possess? The following theorem indicates that every convergent power series converges to an analytic function in its circle of convergence.

■ **THEOREM 5.15** The sum f of a complex power series $\sum_{n=0}^{\infty} a_n(z - z_0)^n$ is an analytic function in its circle of convergence. Its derivative can be obtained by term-by-term differentiation of the power series

$$f'(z) = \sum_{n=0}^{\infty} n a_n (z - z_0)^{n-1}, \tag{5.16}$$

valid in the same circle of convergence. Furthermore, if C is any piecewise-smooth curve in the circle of convergence joining point A to point B, then

$$\int_C f(z)\,dz = \sum_{n=0}^{\infty} \int_C a_n(z - z_0)^n\,dz = \sum_{n=0}^{\infty} a_n \left\{ \frac{(z - z_0)^{n+1}}{n+1} \right\}_A^B. \tag{5.17}$$ ■

The proof of this result is given in the appendix at the end of the section. It can be omitted by those who would prefer not to interrupt the flow of the discussion.

EXAMPLE 5.8 ....

Use Theorem 5.15 to find a power series about $z = 0$ for $1/(1 - 2z)^2$.

Solution Because $1/(1 - 2z)$ can be regarded as the sum of a geometric series,

$$\frac{1}{1 - 2z} = 1 + 2z + 4z^2 + \cdots = \sum_{n=0}^{\infty} 2^n z^n,$$

valid for $|2z| < 1$ or $|z| < \frac{1}{2}$. Term-by-term differentiation gives

$$\frac{2}{(1 - 2z)^2} = \sum_{n=0}^{\infty} 2^n n z^{n-1} = \sum_{n=1}^{\infty} n 2^n z^{n-1},$$

or

$$\frac{1}{(1-2z)^2} = \sum_{n=1}^{\infty} n2^{n-1}z^{n-1} = \sum_{n=0}^{\infty} (n+1)2^n z^n, \qquad |z| < \frac{1}{2}.$$

EXAMPLE 5.9

Find the sum of the series $\sum_{n=1}^{\infty} \frac{1}{n}(z-2)^n$.

Solution If we set $f(z) = \sum_{n=1}^{\infty} \frac{1}{n}(z-2)^n$, $|z-2| < 1$, then term-by-term differentiation gives

$$f'(z) = \sum_{n=1}^{\infty} (z-2)^{n-1} = \frac{1}{1-(z-2)} = \frac{1}{3-z}.$$

Antidifferentiation now gives

$$f(z) = -\log_{\phi}(3-z) + C,$$

where the branch cut must lie in the half-plane $\operatorname{Re} z \geq 3$ (Figure 5.6). This result is true, because $f(z)$ is analytic in $|z-2| < 1$. Because $f(2) = 0$, it follows that

$$0 = -\log_{\phi}(1) + C,$$

and therefore

$$f(z) = \log_{\phi} 1 - \log_{\phi}(3-z).$$

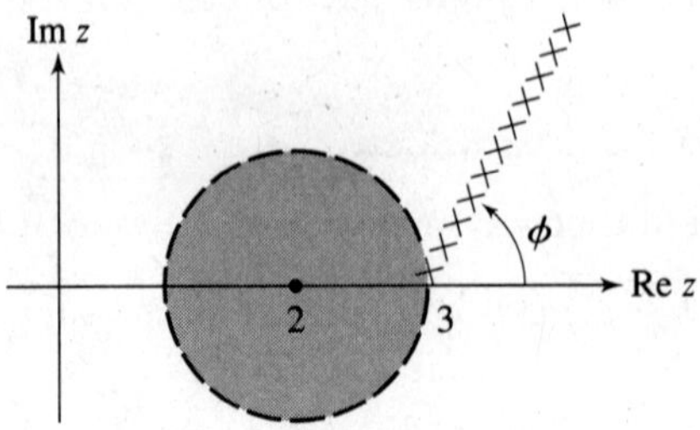

Figure 5.6

We complete this section with a proof of Theorem 5.15, but before doing so, it is important to point out that our discussions are leading to a new characterization of analytic functions. Theorem 5.15 states that the sum of a convergent power series is an analytic function. The converse result, that analytic functions can be expressed as convergent power series, is also valid (and is proved in Section 5.4). What this means is that the study of analytic functions is equivalent to the study of convergent power series. Properties of convergent power series are properties of analytic functions.

Appendix

To give a rigorous proof of Theorem 5.15, it is necessary to introduce the idea of uniform convergence. We do so in a more general setting than that of power series. To say that a series of functions $\sum_{n=0}^{\infty} f_n(z)$ has sum f in some domain D is to say that its sequence of partial sums has limit f; that is, given any $\epsilon > 0$, there exists an $\overline{N}$ such that when $N > \overline{N}$,

$$\left| f(z) - \sum_{n=0}^{N} f_n(z) \right| < \epsilon.$$

Now $\overline{N}$ is certainly a function of ϵ; the smaller ϵ is, the larger $\overline{N}$ is. In addition, $\overline{N}$ normally depends on z. For certain values of z (and fixed ϵ), the value of $\overline{N}$ may be larger than it is for other values of z. When $\overline{N}$ can be chosen independent of z, we say that the series converges uniformly to f. This is stated in the following definition.

Definition 5.3

A series $\sum_{n=0}^{\infty} f_n(z)$ is said to **converge uniformly** to f in a domain D if given any $\epsilon > 0$, there exists an $\overline{N}$ such that for $N > \overline{N}$ and all z in D,

$$\left| f(z) - \sum_{n=0}^{N} f_n(z) \right| < \epsilon.$$

The most common way to verify uniform convergence is with the Weierstrass M-test.

■ **THEOREM 5.16** (Weierstrass M-Test) A series $\sum_{n=0}^{\infty} f_n(z)$ is uniformly convergent in a domain D if there exists a convergent series of constants $\sum_{n=0}^{\infty} M_n$ such that $|f_n(z)| \leq M_n$ for all $n \geq 0$ and all z in D.

Proof Because $|f_n(z)| \leq M_n$ for any fixed z in D, it follows by the comparison test that the series $\sum_{n=0}^{\infty} f_n(z)$ converges. Let its sum be f; that is, let $f(z) = \sum_{n=0}^{\infty} f_n(z)$. Since $\sum_{n=0}^{\infty} M_n$ converges, given any $\epsilon > 0$, there exists an $\overline{N}$ such that for $N > \overline{N}$,

$$\sum_{n=N+1}^{\infty} M_n < \epsilon.$$

Because $|f_n(z)| \le M_n$, it follows that $\sum_{n=N+1}^{\infty} |f_n(z)| \le \sum_{n=1}^{\infty} M_n < \epsilon$. But this means that for $N > \overline{N}$ and all z in D,

$$\left| f(z) - \sum_{n=0}^{N} f_n(z) \right| = \left| \sum_{n=N+1}^{\infty} f_n(z) \right| \le \sum_{n=N+1}^{\infty} M_n < \epsilon.$$

Thus, $\sum_{n=0}^{\infty} f_n(z)$ converges uniformly to f in D. ■

For example, the power series $\sum_{n=0}^{\infty} z^n$ converges uniformly in the domain $|z| \le R < 1$, because therein $|z|^n \le R^n$ and $\sum_{n=0}^{\infty} R^n$ converges.

The importance of knowing that a series of functions converges uniformly is contained in the following theorem.

■ **THEOREM 5.17** When a series $\sum_{n=0}^{\infty} f_n(z)$ of analytic functions converges uniformly in a domain D, its sum f is analytic in D. The derivative of f in D can be obtained by term-by-term differentiation of the series

$$f'(z) = \sum_{n=0}^{\infty} f_n'(z). \tag{5.18}$$

In addition, if C is any piecewise-smooth curve in D, then

$$\int_C f(z)\,dz = \sum_{n=0}^{\infty} \int_C f_n(z)\,dz. \tag{5.19}$$

Proof First we verify that f is continuous in D. If z and z_0 are points in D, then

$$f(z) - f(z_0) = \left[f(z) - \sum_{n=0}^{N} f_n(z) \right] + \left[\sum_{n=0}^{N} f_n(z) - \sum_{n=0}^{N} f_n(z_0) \right] + \left[\sum_{n=0}^{N} f_n(z_0) - f(z_0) \right].$$

When we take moduli and use triangle inequality 1.32,

$$|f(z) - f(z_0)| \le \left| f(z) - \sum_{n=0}^{N} f_n(z) \right| + \left| \sum_{n=0}^{N} f_n(z) - \sum_{n=0}^{N} f_n(z_0) \right| + \left| f(z_0) - \sum_{n=0}^{N} f_n(z_0) \right|.$$

Because $\sum_{n=0}^{\infty} f_n(z)$ converges uniformly to f, given any $\epsilon > 0$, there exists an $\overline{N}$ such that for $N > \overline{N}$ and all z in D (including z_0),

$$\left| f(z) - \sum_{n=0}^{N} f_n(z) \right| < \frac{\epsilon}{3}.$$

Because $\sum_{n=0}^{N} f_n(z)$ is a finite sum of continuous functions, it is continuous. It follows that for fixed N and z_0, there exists a $\delta > 0$ such that when $|z - z_0| < \delta$ and z is in D,

$$\left| \sum_{n=0}^{N} f_n(z) - \sum_{n=0}^{N} f_n(z_0) \right| < \frac{\epsilon}{3}.$$

Hence, for any given $\epsilon > 0$, there exists a δ such that when $|z - z_0| < \delta$ and z is in D,

$$|f(z) - f(z_0)| < \frac{\epsilon}{3} + \frac{\epsilon}{3} + \frac{\epsilon}{3} = \epsilon;$$

that is, f is continuous at z_0.

We now use Morera's theorem to verify that f is analytic in D. Let z_0 be any point in D and D' be a neighborhood of z_0 that contains only points of D. Because f_n is analytic in D', if C is any simple, closed, piecewise-smooth curve in D' (a simply connected domain), then by the Cauchy-Goursat theorem,

$$\oint_C f_n(z)\, dz = 0, \qquad n = 0, 1, \ldots.$$

Consequently,

$$\left| \oint_C f(z)\, dz \right| = \left| \oint_C f(z)\, dz - \oint_C \sum_{n=0}^{N} f_n(z)\, dz \right| = \left| \oint_C \left[f(z) - \sum_{n=0}^{N} f_n(z) \right] dz \right|.$$

But for given $\epsilon > 0$, there exists $\overline{N}$ such that for $N > \overline{N}$ and all z in D',

$$\left| f(z) - \sum_{n=0}^{N} f_n(z) \right| < \frac{\epsilon}{L}$$

where L is the length of C. Hence, for $N > \overline{N}$, we can use inequality 4.22 to write

$$\left| \oint_C f(z)\, dz \right| < \frac{\epsilon}{L}(L) = \epsilon.$$

Because ϵ can be made arbitrarily small, it follows that

$$\left| \oint_C f(z)\, dz \right| = 0.$$

When C is not simple, it may be decomposed into simple curves so that for any closed piecewise-smooth curve in D',

$$\oint_C f(z)\, dz = 0.$$

Morera's theorem implies that f is analytic in D' and, therefore, at z_0. Hence f is analytic in D.

To prove 5.18 we show that for any point z_0 in D, given any $\epsilon > 0$, there exists an $\overline{N}$ such that for $N > \overline{N}$,

$$\left| f'(z_0) - \sum_{n=0}^{N} f_n'(z_0) \right| < \epsilon.$$

Because f and $\sum_{n=0}^{N} f_n(z)$ are both analytic in a neighborhood D' of z_0 that is contained in D, Cauchy's generalized integral formula gives

$$f'(z_0) = \frac{1}{2\pi i} \oint_{C_r} \frac{f(z)}{(z - z_0)^2}\, dz \quad \text{and} \quad \sum_{n=0}^{N} f_n'(z_0) = \frac{1}{2\pi i} \oint_{C_r} \frac{\sum_{n=0}^{N} f_n(z)}{(z - z_0)^2}\, dz$$

where $C_r : |z - z_0| = r$ is a circle of radius r centered at z_0 that is contained in D'. Consequently,

$$f'(z_0) - \sum_{n=0}^{N} f_n'(z_0) = \frac{1}{2\pi i} \oint_{C_r} \frac{f(z) - \sum_{n=0}^{N} f_n(z)}{(z - z_0)^2}\, dz.$$

Because $\sum_{n=0}^{\infty} f_n(z)$ converges uniformly to f in D', there exists an $\overline{N}$ such that for all z in D' and $N > \overline{N}$,

$$\left| f(z) - \sum_{n=0}^{N} f_n(z) \right| < \epsilon r.$$

Thus, for $N > \overline{N}$,

$$\left| f'(z_0) - \sum_{n=0}^{N} f_n'(z_0) \right| = \frac{1}{2\pi} \left| \oint_{C_r} \frac{f(z) - \sum_{n=0}^{N} f_n(z)}{(z - z_0)^2}\, dz \right| < \frac{1}{2\pi} \left(\frac{\epsilon r}{r^2} \right) (2\pi r) = \epsilon.$$

To verify 5.19 we show that given any $\epsilon > 0$, there exists an $\overline{N}$ such that for $N > \overline{N}$,

$$\left| \int_C f(z)\, dz - \sum_{n=0}^{N} \int_C f_n(z)\, dz \right| < \epsilon.$$

Because $\sum_{n=0}^{\infty} f_n(z)$ converges uniformly to f in D, there exists an $\overline{N}$ such that for all z in D and $N > \overline{N}$,

$$\left| f(z) - \sum_{n=0}^{N} f_n(z) \right| < \frac{\epsilon}{L}$$

where L is the length of C. Then, for $N > \overline{N}$,

$$\left| \int_C f(z)\, dz - \sum_{n=0}^{N} \int_C f_n(z)\, dz \right| = \left| \int_C \left[f(z) - \sum_{n=0}^{N} f_n(z) \right] dz \right| < \frac{\epsilon}{L}(L) = \epsilon.$$

This completes the proof of Theorem 5.17. ■

We now use the results of Theorem 5.17 to verify Theorem 5.15.

■ **Proof of Theorem 5.15** Let z^* be any point in the circle of convergence $|z - z_0| < R$ of the power series $\sum_{n=0}^{\infty} a_n(z - z_0)^n$ (Figure 5.7). It is always possible to draw a circle $|z - z_0| = r < R$ centered at z_0 that contains z^*. Because the power series converges absolutely for all points on $|z - z_0| = r$, we can say that $\sum_{n=0}^{\infty} |a_n| r^n$ is a convergent series of constants.

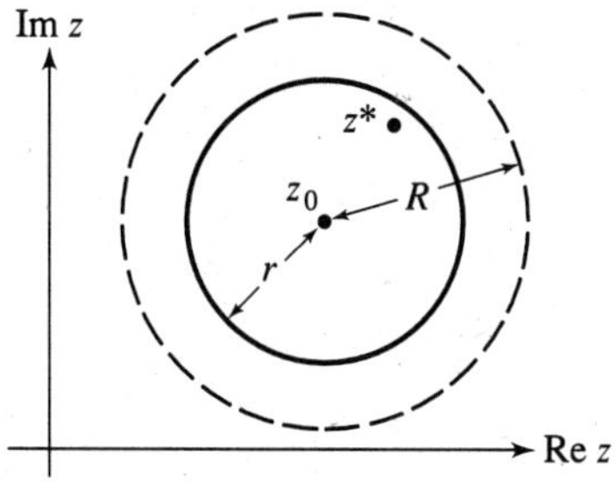

Figure 5.7

Because $|a_n(z - z_0)^n| < |a_n| r^n$ for every point z in $|z - z_0| < r$, the Weierstrass M-test implies that the power series converges uniformly in $|z - z_0| < r$. According to Theorem 5.17, the series is analytic in $|z - z_0| < r$ and, in particular, at z^*. In addition,

$$f'(z^*) = \sum_{n=0}^{\infty} n a_n (z^* - z_0)^{n-1}.$$

Because z^* is an arbitrary point in the circle of convergence $|z - z_0| < R$, this verifies 5.16. Property 5.17 follows immediately from 5.19. ■

Exercises 5.3

In Exercises 1–20 find the circle of convergence for the power series.

1. $\sum_{n=1}^{\infty} \frac{1}{n} z^n$

2. $\sum_{n=1}^{\infty} n^2 (z-2)^n$

3. $\sum_{n=1}^{\infty} n^2 i^n (z+i)^n$

4. $\sum_{n=0}^{\infty} \frac{1}{(n+i)^3} z^n$

5. $\sum_{n=1}^{\infty} \frac{1}{n^n} (z+3i)^n$

6. $\sum_{n=0}^{\infty} (-1)^n (n+1)^2 (z+4+i)^n$

7. $\sum_{n=0}^{\infty} \left(\frac{n-1}{n+1}\right) (2z)^n$

8. $\sum_{n=0}^{\infty} 2^n \left(\frac{n-i}{n+2i}\right) (z-4)^n$

9. $\sum_{n=0}^{\infty}(-i)^n(z+4)^{2n}$

10. $\sum_{n=0}^{\infty}(1+i)^n(z-4)^{2n}$

11. $\sum_{n=0}^{\infty}\frac{(n!)^3}{(3n)!}(z+1-i)^n$

12. $\sum_{n=1}^{\infty}\left(1+\frac{4}{n}\right)^{n^2}(z-2+i)^{2n}$

13. $\sum_{n=0}^{\infty}\frac{1}{(2+i)^n}(z-i)^{2n+1}$

14. $\sum_{n=0}^{\infty}\frac{(-1)^n}{3^n}(z+1)^{3n}$

15. $\sum_{n=1}^{\infty}\frac{2\cdot 4\cdot 6\cdots(2n)}{3\cdot 5\cdot 7\cdots(2n+1)}z^n$

16. $\sum_{n=1}^{\infty}\frac{[1\cdot 3\cdot 5\cdots(2n+1)]^2}{2^{2n}(2n)!}(z+1)^n$

17. $\sum_{n=1}^{\infty}\frac{1}{\sqrt{ni}}(z+2i-1)^n$

18. $\sum_{n=0}^{\infty}\frac{1}{\sqrt{n+i}}(z-2)^n$

19. $\sum_{n=1}^{\infty}\frac{1}{\operatorname{Log} n}z^n$

20. $\sum_{n=1}^{\infty}\frac{1}{\operatorname{Log} ni}z^n$

In Exercises 21–30 find the sum of the power series.

21. $\sum_{n=0}^{\infty}\frac{1}{2^n}z^n$

22. $\sum_{n=2}^{\infty}i^n(z+1)^n$

23. $\sum_{n=0}^{\infty}\frac{(-1)^{n+1}3^n}{5^{n+2}}(z+i)^{2n}$

24. $\sum_{n=1}^{\infty}\left(\frac{1+i}{1-i}\right)^n(z-i)^{3n}$

25. $\sum_{n=1}^{\infty}n(z+2i)^{n-1}$

26. $\sum_{n=2}^{\infty}n(n-1)(z-1-i)^{n-2}$

27. $\sum_{n=1}^{\infty}(n+1)z^{n-1}$

28. $\sum_{n=1}^{\infty}n^2z^{n-1}$

29. $\sum_{n=2}^{\infty}n3^nz^{2n}$

30. $\sum_{n=1}^{\infty}(n^2+2n)z^n$

31. Find the circle of convergence for the hypergeometric series

$$1+\frac{\alpha\beta}{\gamma}z+\frac{\alpha(\alpha+1)\beta(\beta+1)}{\gamma(\gamma+1)2!}z^2+\frac{\alpha(\alpha+1)(\alpha+2)\beta(\beta+1)(\beta+2)}{\gamma(\gamma+1)(\gamma+2)3!}z^3+\cdots,$$

where α, β, and γ are constant and γ is not zero or a negative integer.

32. When m is a nonnegative integer, the Bessel function of the first kind of order m is defined by

$$J_m(z)=\sum_{n=0}^{\infty}\frac{(-1)^n}{2^{2n+m}n!(n+m)!}z^{2n+m}.$$

What is its circle of convergence?

In Exercises 33–38 use formula 5.13 to find the required power series of the function.

33. $f(z) = \dfrac{1}{1+3z}$ about $z_0 = 0$

34. $f(z) = \dfrac{z^2}{2-z}$ about $z_0 = 0$

35. $f(z) = \dfrac{1}{1+z}$ about $z_0 = 1$

36. $f(z) = \dfrac{1}{1+z}$ about $z_0 = i$

37. $f(z) = \dfrac{z}{1+z}$ about $z = 1+i$

38. $f(z) = \dfrac{z}{(1-3z)^2}$ about $z = 0$

The results in Exercises 39–41 are useful in potential theory. Prove them for real x and θ.

39. $\displaystyle\sum_{n=1}^{\infty} x^n \cos n\theta = \frac{x\cos\theta - x^2}{1+x^2-2x\cos\theta}, \quad |x| < 1$

40. $\displaystyle\sum_{n=1}^{\infty} x^n \sin n\theta = \frac{x\sin\theta}{1+x^2-2x\cos\theta}, \quad |x| < 1$

41. $\displaystyle\sum_{n=1}^{\infty} \frac{x^n \cos n\theta}{n} = -\frac{1}{2}\ln\left(1+x^2-2x\cos\theta\right), \quad |x| < 1$

In Exercises 42–43 find the circle of convergence by separating the series into two series, each of which has an easily determined radius of convergence.

42. $1 - z + \dfrac{z^2}{4} - \dfrac{z^3}{3!} + \dfrac{z^4}{8} - \dfrac{z^5}{5!} + \dfrac{z^6}{16} + \cdots$

43. $\displaystyle\sum_{n=0}^{\infty} [3-(-1)^n]^n z^n$

In Exercises 44–48 we discuss convergence of power series for points on the circle of convergence using the following result attributed to Weierstrass: Suppose the coefficients a_n of a power series $\sum_{n=0}^{\infty} a_n z^n$ are such that for $n \geq N$, N some integer,

$$\frac{a_{n+1}}{a_n} = 1 - \frac{\alpha}{n} + C_n,$$

where $\sum_{n=N}^{\infty} C_n$ converges absolutely and α is a constant. Then the power series has radius of convergence equal to 1, and

(i) Converges absolutely for all $|z| = 1$ if $\operatorname{Re}\alpha > 1$;
(ii) Converges (but not absolutely) for all $|z| = 1$ if $0 < \operatorname{Re}\alpha \leq 1$, except for $z = 1$ where it diverges;
(iii) Diverges for all $|z| = 1$ if $\operatorname{Re}\alpha \leq 0$.

Use this result to examine convergence of the series on the circle of convergence in Exercises 44–46.

44. $\displaystyle\sum_{n=0}^{\infty} z^n$

45. $\displaystyle\sum_{n=1}^{\infty} \frac{z^n}{n}$

46. $\displaystyle\sum_{n=1}^{\infty} \frac{z^n}{n^2}$

To discuss convergence of power series $\sum_{n=0}^{\infty} a_n z^n$ on the circle of convergence $|z| = R$ when $R \neq 1$, we set $\zeta = z/R$ and apply the preceding result to the series $\sum_{n=0}^{\infty} a_n R^n \zeta^n$. Illustrate this by examining the series for convergence on the circle of convergence in Exercises 47 and 48.

47. $\displaystyle\sum_{n=0}^{\infty} \frac{z^n}{2^n}$

48. $\displaystyle\sum_{n=1}^{\infty} n^2 (3z)^{2n}$

5.4 Taylor and Maclaurin Series

In Section 5.3 we established the fact that every convergent power series sums to an analytic function in its circle of convergence. The following theorem provides the converse result — analytic functions can be expanded in power series.

■ **THEOREM 5.18** (Taylor Expansion Theorem) Let f be analytic in a domain D and z_0 be a point in D. Then f can be expanded in a power series:

$$f(z) = f(z_0) + f'(z_0)(z - z_0) + \frac{f''(z_0)}{2!}(z - z_0)^2 + \cdots, \qquad \text{(5.20)}$$

valid in all circles $|z - z_0| < r$ containing only points of D.

Proof When C_r is a circle with center z_0 and radius r that contains only points of D (Figure 5.8), Cauchy's integral formula states that for any point z inside C_r,

$$\begin{aligned}
f(z) &= \frac{1}{2\pi i} \oint_{C_r} \frac{f(\zeta)}{\zeta - z}\, d\zeta \\
&= \frac{1}{2\pi i} \oint_{C_r} \frac{f(\zeta)}{(\zeta - z_0) - (z - z_0)}\, d\zeta \\
&= \frac{1}{2\pi i} \oint_{C_r} \frac{f(\zeta)}{\zeta - z_0} \left(\frac{1}{1 - \dfrac{z - z_0}{\zeta - z_0}} \right) d\zeta
\end{aligned}$$

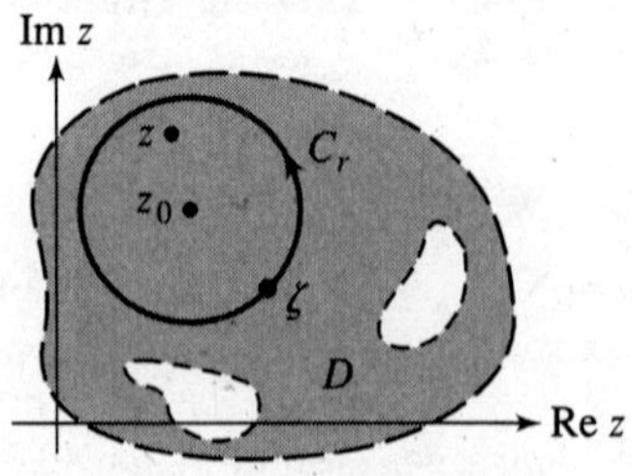

Figure 5.8

$$= \frac{1}{2\pi i} \oint_{C_r} \frac{f(\zeta)}{\zeta - z_0} \left[1 + \left(\frac{z - z_0}{\zeta - z_0}\right) + \left(\frac{z - z_0}{\zeta - z_0}\right)^2 + \cdots \right.$$

$$\left. + \left(\frac{z - z_0}{\zeta - z_0}\right)^{n-1} + \frac{\left(\dfrac{z - z_0}{\zeta - z_0}\right)^n}{1 - \left(\dfrac{z - z_0}{\zeta - z_0}\right)} \right] d\zeta.$$

When Cauchy's generalized integral formula is applied to the first n terms, the result is

$$f(z) = f(z_0)+f'(z_0)(z-z_0)+\frac{f''(z_0)}{2!}(z-z_0)^2+\cdots+\frac{f^{(n-1)}(z_0)}{(n-1)!}(z-z_0)^{n-1}+R_n,$$

where

$$R_n = \frac{(z-z_0)^n}{2\pi i}\oint_{C_r} \frac{f(\zeta)}{(\zeta - z)(\zeta - z_0)^n}\, d\zeta.$$

If M denotes the maximum value of $|f(z)|$ for ζ on C_r and $|z - z_0| = \rho$, then for all ζ on C_r,

$$\left|\frac{f(\zeta)}{(\zeta - z)(\zeta - z_0)^n}\right| \leq \frac{M}{(r-\rho)r^n}.$$

By inequality 4.22

$$|R_n| \leq \frac{\rho^n}{2\pi}\frac{M}{(r-\rho)r^n}(2\pi r) = \frac{rM}{r-\rho}\left(\frac{\rho}{r}\right)^n.$$

Because $\rho < r$, it follows that $R_n \to 0$ as $n \to \infty$, and we can write for any z inside C_r,

$$f(z) = f(z_0) + f'(z_0)(z - z_0) + \frac{f''(z_0)}{2!}(z - z_0)^2 + \cdots. \qquad \blacksquare$$

The expansion

$$f(z) = \sum_{n=0}^{\infty} \frac{f^{(n)}(z_0)}{n!}(z - z_0)^n \tag{5.21}$$

is called the **Taylor series** of f about z_0. The special case in which $z_0 = 0$,

$$f(z) = \sum_{n=0}^{\infty} \frac{f^{(n)}(0)}{n!} z^n, \tag{5.22}$$

is called the **Maclaurin series** of f.

Notice the simplicity of Theorem 5.18 compared to its real counterpart. The Taylor series

$$f(a) + f'(a)(x - a) + \frac{f''(a)}{2!}(x - a)^2 + \cdots$$

for a function $f(x)$ converges to $f(x)$ if all derivatives $f^{(n)}(a)$ exist and if Taylor remainders approach zero as $n \to \infty$. For complex analytic functions, there is no necessity to consider remainders; analyticity guarantees that remainders approach zero and that the Taylor series of $f(z)$ converges to $f(z)$.

It is straightforward to establish uniqueness of power series expansions.

■ **THEOREM 5.19** If f has a power series expansion about a point z_0 with nonzero radius of convergence, it must be the Taylor series about z_0.

Proof Suppose f has a power series expansion about a point z_0,

$$f(z) = a_0 + a_1(z - z_0) + a_2(z - z_0)^2 + \cdots,$$

valid in some circle of convergence $|z - z_0| < R$ with nonzero radius of convergence R. Substitution of $z = z_0$ gives $f(z_0) = a_0$. Differentiation with respect to z (see Theorem 5.15) yields

$$f'(z) = a_1 + 2a_2(z - z_0) + 3a_3(z - z_0)^2 + \cdots,$$

and when we set $z = z_0$, we obtain $f'(z_0) = a_1$. Another differentiation gives

$$f''(z) = 2a_2 + 3 \cdot 2a_3(z - z_0) + 4 \cdot 3a_4(z - z_0)^2 + \cdots,$$

and for $z = z_0$,

$$f''(z_0) = 2a_2, \quad \text{or} \quad a_2 = \frac{f''(z_0)}{2!}.$$

Continued differentiation and substitution of $z = z_0$ gives $a_n = f^{(n)}(z_0)/n!$. ■

What this theorem establishes is that there is only one power series expansion of a given function f about a given point z_0, its Taylor series. It follows that if we are required to find the Taylor series of f about z_0, we do not necessarily have to use formula 5.21. Instead, if by any means whatsoever, we can find a power series of the form $\sum_{n=0}^{\infty} a_n(z - z_0)^n$ for f (with nonzero radius of convergence), then this series must be the Taylor series for f. This point is illustrated in the next example.

EXAMPLE 5.10 ....

Find the Maclaurin series for $f(z) = z/(1 - 4z)$.

Solution We can use formula 5.22 by calculating

$$f(0) = 0,$$

$$f'(0) = \frac{(1 - 4z) - z(-4)}{(1 - 4z)^2}\bigg|_{z=0} = \frac{1}{(1 - 4z)^2}\bigg|_{z=0} = 1,$$

$$f''(0) = \frac{2(4)}{(1 - 4z)^3}\bigg|_{z=0} = 2(4),$$

$$f'''(0) = \frac{2(3)4^2}{(1 - 4z)^4}\bigg|_{z=0} = 3!\,4^2.$$

The pattern emerging is $f^{(n)}(0) = n!\,4^{n-1}$, and therefore

$$\frac{z}{1 - 4z} = \sum_{n=1}^{\infty} 4^{n-1} z^n,$$

with radius of convergence $R = \lim_{n\to\infty} \left| \frac{4^{n-1}}{4^n} \right| = \frac{1}{4}$.

Alternatively, because $1/(1 - 4z)$ may be regarded as the sum of an infinite geometric series with first term equal to 1 and common ratio $4z$,

$$\frac{z}{1-4z} = z\left[1 + (4z) + (4z)^2 + \cdots\right] = \sum_{n=1}^{\infty} 4^{n-1} z^n,$$

where $|4z| < 1$ or $|z| < 1/4$. ....

EXAMPLE 5.11 ....

Find the Taylor series for $\operatorname{Log} z$ about $z = 1$.

Solution Term-by-term integration of

$$\frac{1}{z} = \frac{1}{(z-1)+1} = \sum_{n=0}^{\infty} (-1)^n (z-1)^n, \qquad |z-1| < 1,$$

gives

$$\operatorname{Log} z = \sum_{n=0}^{\infty} \frac{(-1)^n}{n+1}(z-1)^{n+1} + C, \qquad |z-1| < 1.$$

Because $\operatorname{Log}(1) = 0$, we must choose $C = 0$, in which case

$$\operatorname{Log} z = \sum_{n=0}^{\infty} \frac{(-1)^n}{n+1}(z-1)^{n+1} = (z-1) - \frac{1}{2}(z-1)^2 + \frac{1}{3}(z-1)^3 + \cdots,$$

valid for $|z-1| < 1$. ....

The Taylor series in Example 5.11 is a representation for other branches of the logarithm function as well. In particular, it represents any branch of $\log z$ for which $\log 1 = 0$ and that does not have a branch cut passing through the interior of the circle of convergence. In other words, we may write

$$\log_\phi z = \sum_{n=1}^{\infty} \frac{(-1)^{n-1}}{n}(z-1)^n, \qquad |z-1| < 1,$$

provided $-3\pi/2 \le \phi \le -\pi/2$ (Figure 5.9).

This section and Section 5.3 have established the following equivalence between analytic functions and convergent power series: When a function f is analytic in a domain D and z_0 is a point of D, then f can be expressed in a Taylor series about z_0 that converges in every circle $|z - z_0| < r$ interior to D. Conversely, every power

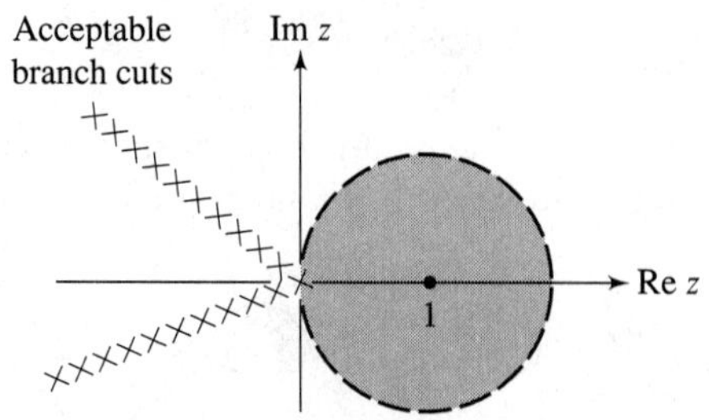

Figure 5.9

series $\sum_{n=0}^{\infty} a_n(z-z_0)^n$ with nonzero radius of convergence R sums to an analytic function in $|z-z_0|<R$ (and is, in fact, the Taylor series of that function).

In Chapter 2 we mentioned the importance of zeros of analytic functions. The next theorem gives an alternative method for determining the order of a zero.

■ **THEOREM 5.20** A function f has a zero of order m at z_0 if and only if it can be written in the form

$$f(z) = (z-z_0)^m g(z), \tag{5.23}$$

where g is analytic at z_0 and $g(z_0) \neq 0$.

Proof If f has a zero of order m at z_0, it is analytic in a neighborhood of z_0. In this neighborhood, it has a convergent Taylor series,

$$f(z) = f(z_0) + f'(z_0)(z-z_0) + \frac{f''(z_0)}{2}(z-z_0)^2 + \cdots.$$

Because $f(z_0) = f'(z_0) = \cdots = f^{(m-1)}(z_0) = 0$, and $f^{(m)}(z_0) \neq 0$, the Taylor series reduces to

$$\begin{aligned} f(z) &= \frac{f^{(m)}(z_0)}{m!}(z-z_0)^m + \frac{f^{(m+1)}(z_0)}{(m+1)!}(z-z_0)^{m+1} + \cdots \\ &= (z-z_0)^m \left[\frac{f^{(m)}(z_0)}{m!} + \frac{f^{(m+1)}(z_0)}{(m+1)!}(z-z_0) + \cdots\right]. \end{aligned}$$

If we define

$$g(z) = \frac{f^{(m)}(z_0)}{m!} + \frac{f^{(m+1)}(z_0)}{(m+1)!}(z-z_0) + \cdots,$$

then $f(z) = (z-z_0)^m g(z)$, where clearly $g(z_0) = f^{(m)}(z_0)/m! \neq 0$. Further, the series for g converges wherever the series for f converges, and hence g is analytic at z_0.

Conversely, if $f(z)$ can be expressed in the form $f(z) = (z - z_0)^m g(z)$, where g is analytic at z_0 and $g(z_0) \neq 0$, then

$$\begin{aligned} f(z_0) &= 0; \\ f'(z_0) &= [(z - z_0)^m g'(z) + m(z - z_0)^{m-1} g(z)]_{|z=z_0} = 0; \\ f''(z_0) &= [(z - z_0)^m g''(z) + 2m(z - z_0)^{m-1} g'(z) \\ &\qquad + m(m-1)(z - z_0)^{m-2} g(z)]_{|z=z_0} = 0. \end{aligned}$$

Continuation leads to vanishing derivatives of orders up to and including $m-1$. Leibnitz's rule for the mth derivative of a product gives

$$\begin{aligned} f^{(m)}(z_0) &= \left[(z - z_0)^m g^{(m)}(z) + \binom{m}{1} m(z - z_0)^{m-1} g^{(m-1)}(z) + \cdots \right. \\ &\qquad \left. + \, m(m-1)\cdots(2)(1)g(z)\right]_{|z=z_0} \\ &= m!\, g(z_0) \neq 0. \end{aligned}$$

Thus, z_0 is zero of order m. ■

An important consequence of Theorem 5.20 is the following corollary.

■ **COROLLARY** The zeros of a nonconstant analytic function are isolated; that is, every zero has a neighborhood inside of which it is the only zero.

Proof If z_0 is a zero of f of order m, then $f(z)$ can be written in form 5.23, where g is analytic at z_0 and $g(z_0) \neq 0$. Because g is therefore continuous at z_0, it follows that there exists a neighborhood of z_0 in which $g(z) \neq 0$. In this neighborhood, $f(z) \neq 0$ except at $z = z_0$. ■

We can also use Theorem 5.20 to establish L'Hôpital's rule for complex functions.

■ **THEOREM 5.21** (L'Hôpital's Rule) Let f and g be functions that are analytic in a domain D but not identically equal to zero. If $f(z_0) = g(z_0) = 0$ at a point z_0 in D, then

$$\lim_{z \to z_0} \frac{f(z)}{g(z)} = \lim_{z \to z_0} \frac{f'(z)}{g'(z)}, \tag{5.24}$$

provided the latter limit exists.

Proof If the orders of the zeros of f and g at z_0 are m and n, respectively, then in some neighborhood of z_0 we may write

$$f(z) = (z - z_0)^m F(z) \quad \text{and} \quad g(z) = (z - z_0)^n G(z),$$

where $F(z_0)G(z_0) \neq 0$. With these,

$$\frac{f(z)}{g(z)} = (z - z_0)^{m-n}\frac{F(z)}{G(z)},$$

and

$$\frac{f'(z)}{g'(z)} = \frac{m(z-z_0)^{m-1}F(z) + (z-z_0)^m F'(z)}{n(z-z_0)^{n-1}G(z) + (z-z_0)^n G'(z)}$$
$$= (z-z_0)^{m-n}\frac{mF(z) + (z-z_0)F'(z)}{nG(z) + (z-z_0)G'(z)}.$$

When $m = n$, the limit of each of these expressions for $f(z)/g(z)$ and $f'(z)/g'(z)$ as $z \to z_0$ is $F(z_0)/G(z_0)$. When $m > n$, both expressions have limit zero, and when $m < n$, neither limit exists. ■

Exercises 5.4

In Exercises 1–22 find the Taylor series for the function about the given point. In each case determine values of z for which the series converges to the function.

1. $\dfrac{z}{1+z^2}$ about $z_0 = 0$
2. $\dfrac{3}{1+z}$ about $z_0 = 1$
3. $\dfrac{z}{(1-z)^2}$ about $z_0 = 0$
4. e^z about $z_0 = 0$
5. $\sin z$ about $z_0 = 0$
6. $\cos z$ about $z_0 = 0$
7. $\sinh z$ about $z_0 = 0$
8. $\cosh z$ about $z_0 = 0$
9. e^{2z} about $z_0 = 1$
10. $\sin^2 z$ about $z_0 = 0$
11. $(z+1)\cosh 3z$ about $z_0 = 0$
12. $\sin z$ about $z_0 = \pi/4$
13. $\cosh z$ about $z_0 = i$
14. $\dfrac{\cos 2z}{z^2+1}$ about $z_0 = 0$
15. $\dfrac{1}{1+z}$ about $z_0 = 1$
16. $\dfrac{1}{1+z}$ about $z_0 = i$
17. $\dfrac{z}{(1-3z)^2}$ about $z_0 = 0$
18. $\dfrac{z}{1+z}$ about $z_0 = 1+i$
19. $\sin z$ about $z_0 = i$
20. e^z about $z_0 = 1+i$
21. $\sinh 4z$ about $z_0 = -i$
22. $\operatorname{Log} z$ about $z_0 = i$

In Exercises 23–28 identify all zeros and their orders for the function.

23. $\dfrac{z^2+1}{z^2(z+1)}$
24. $z^3 e^{z+2}$
25. $\dfrac{\sin 3z}{z^5}$
26. $\dfrac{\sinh^2 z}{z+5i}$

27. $\sqrt{z}$

28. $\operatorname{Log}(z-1)$

29. Use series to prove Euler's identity 3.6.

In Exercises 30–33 find the Maclaurin series for the function. Determine values of z for which the series converges to the function.

30. $\sin^3 z$

31. $e^z \sin z$

32. $\sqrt{z+i}$

33. $\operatorname{Tan}^{-1} z$

34. Prove that when f has a zero of order m at z_0 and g is analytic at z_0, but $g(z_0) \neq 0$, then fg has a zero of order m at z_0.

35. Find all power series $\sum_{n=0}^{\infty} a_n z^n$ that converge at $z = 4+3i$ and diverge at $z = 2-i$.

36. Expand $\operatorname{Log} z$ in a Taylor series about $-1+i$. Is the radius of convergence of the series 1 or $\sqrt{2}$? Explain.

37. When a function $f(z)$ is expanded in a Taylor series about a point, show that the radius of the circle inside of which the series converges to $f(z)$ is the distance from the point of expansion to the nearest singularity of the function. Notice that we did not call this radius the radius of convergence of the series. Why?

38. We are given a function f and a point z_0 at which f is analytic. We find the Taylor series $\sum_{n=0}^{\infty} a_n (z-z_0)^n$ of f about z_0 and determine its radius of convergence R with either of formulas 5.11. Can we always write

$$f(z) = \sum_{n=0}^{\infty} a_n (z-z_0)^n, \qquad |z-z_0| < R?$$

39. A function f has Taylor series $\sum_{n=0}^{\infty} a_n (z-z_0)^n$ about z_0 with positive radius of convergence R. Must there exist a point on the circle $|z-z_0| = R$ at which the series diverges?

40. Show that every polynomial of degree n with complex coefficients has exactly n zeros (counting multiplicities). (See Exercise 1 in Section 4.7.)

41. In this exercise we derive the binomial expansion

$$(1+z)^m = 1 + mz + \frac{m(m-1)}{2!}z^2 + \frac{m(m-1)(m-2)}{3!}z^3 + \cdots, \qquad |z| < 1.$$

(a) Verify that the radius of convergence of the power series

$$1 + mz + \frac{m(m-1)}{2!}z^2 + \frac{m(m-1)(m-2)}{3!}z^3 + \cdots$$

is $R = 1$ for any fixed real value m.

(b) Use Theorem 5.15 to prove that the sum of the series, f, satisfies the complex differential equation

$$f'(z) + zf'(z) = mf(z), \qquad f(0) = 1.$$

(c) Show that the solution of the differential equation is $f(z) = (1+z)^m$.

42. Obtain the series in Exercise 32 from that in Exercise 41.

43. Use Exercise 41 to find the Maclaurin series for $\operatorname{Sin}^{-1} z$.

44. Where is the function $f(z) = \sin(1/z)$ analytic? What are its zeros? Are they isolated?

45. Prove that if f is analytic at z_0, then there exists a neighborhood of z_0 inside of which f takes on the value $f(z_0)$ only at z_0.

46. If f is analytic inside and on the circle $C : |z - z_0| \leq R$, then it has a Taylor series $\sum_{n=0}^{\infty} a_n(z - z_0)^n$ valid inside and on C. Show that

$$\oint_C |f(z)|^2 \, |dz| = 2\pi R \sum_{n=0}^{\infty} |a_n|^2 R^{2n}.$$

47. Verify the following result due to Schwarz: Suppose f is analytic inside and on C : $|z - z_0| \leq R$, and $f(z_0) = 0$. If $|f(z)| \leq M$ on C, then $|f(z)| \leq (M/R)|z - z_0|$ for $|z| < R$. Furthermore, if $|f(z)| \leq (M/R)|z - z_0|$ for some z inside C, then $f(z) = (M/R)e^{\lambda i}(z - z_0)$, where λ is some real constant.

48. Show that coefficients in the Maclaurin series for $1/(1 - z - z^2)$ are the Fibonacci numbers

$$a_0 = 1, \qquad a_1 = 1, \qquad a_{n+1} = a_n + a_{n-1}, \qquad n \geq 1.$$

5.5 Laurent Series

In applications of real-variable calculus we are most often concerned with a function near points at which the function is well behaved, in particular, where it has derivatives. When it has derivatives of all orders and Taylor remainders approach zero, the function can be represented as its Taylor series. Likewise, a complex function can be represented by its Taylor series around points at which it is analytic. But in many applications of complex functions, it is their singularities that prove useful. Laurent series of this section represent complex functions near isolated singularities.

■ **THEOREM 5.22** (Laurent Expansion Theorem) Let f be analytic in an annulus $D : r < |z - z_0| < R$. Then $f(z)$ can be expressed in the form

$$f(z) = \sum_{n=-\infty}^{\infty} a_n(z - z_0)^n \tag{5.25a}$$

which converges and represents $f(z)$ in D. Furthermore,

$$a_n = \frac{1}{2\pi i} \oint_C \frac{f(\zeta)}{(\zeta - z_0)^{n+1}} \, d\zeta, \tag{5.25b}$$

where C is any simple, closed, piecewise-smooth curve in D that contains z_0 in its interior.

Proof Let z be any point in D. If $\rho = |z - z_0|$, construct two circles C_1 : $|\zeta - z_0| = R'$ and $C_2 : |\zeta - z_0| = r'$ where $r < r' < \rho < R' < R$, and join C_1 and C_2 by nonintersecting, piecewise-smooth curves Γ_1 and Γ_2 that do not pass through z (Figure 5.10). Because f is analytic inside and on the loop L_1

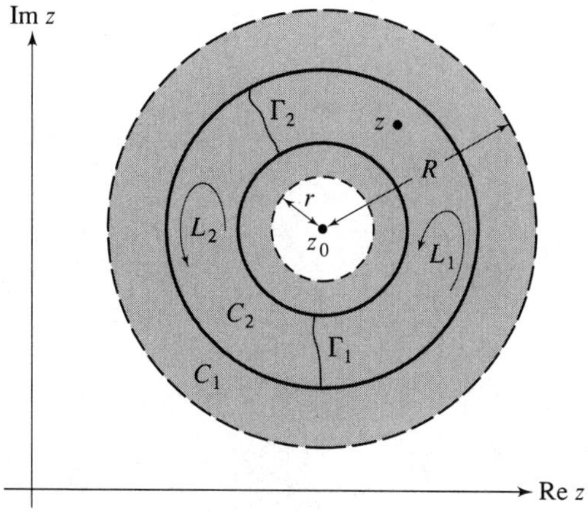

Figure 5.10

and z is interior to L_1, Cauchy's integral formula gives

$$f(z) = \frac{1}{2\pi i}\oint_{L_1} \frac{f(\zeta)}{\zeta - z}\,d\zeta.$$

Because loop L_2 contains only points of D in its interior, the Cauchy-Goursat theorem requires

$$0 = \frac{1}{2\pi i}\oint_{L_2} \frac{f(\zeta)}{\zeta - z}\,d\zeta.$$

Addition of these equations gives

$$f(z) = \frac{1}{2\pi i}\oint_{L_1} \frac{f(\zeta)}{\zeta - z}\,d\zeta + \frac{1}{2\pi i}\oint_{L_2} \frac{f(\zeta)}{\zeta - z}\,d\zeta.$$

Traversing L_1 and L_2 results in integrations along Γ_1 and Γ_2 twice each, once in one direction and once in the opposite direction. What remains are integrations around C_1 in the counterclockwise direction and C_2 in the clockwise direction,

$$f(z) = \frac{1}{2\pi i}\oint_{C_1} \frac{f(\zeta)}{\zeta - z}\,d\zeta - \frac{1}{2\pi i}\oint_{C_2} \frac{f(\zeta)}{\zeta - z}\,d\zeta.$$

As in Theorem 5.18, we may write for the integral around C_1,

$$\frac{1}{2\pi i}\oint_{C_1} \frac{f(\zeta)}{\zeta - z}\,d\zeta = \frac{1}{2\pi i}\oint_{C_1} \frac{f(\zeta)}{\zeta - z_0}\left[1 + \left(\frac{z - z_0}{\zeta - z_0}\right) + \left(\frac{z - z_0}{\zeta - z_0}\right)^2 + \cdots \right.$$

$$\left. + \left(\frac{z - z_0}{\zeta - z_0}\right)^{n-1} + \frac{\left(\dfrac{z - z_0}{\zeta - z_0}\right)^n}{1 - \left(\dfrac{z - z_0}{\zeta - z_0}\right)}\right]d\zeta$$

$$= \frac{1}{2\pi i}\oint_{C_1} \frac{f(\zeta)}{\zeta - z_0}\, d\zeta + \frac{1}{2\pi i}(z - z_0)\oint_{C_1} \frac{f(\zeta)}{(\zeta - z_0)^2}\, d\zeta + \cdots$$

$$+ \frac{(z - z_0)^{n-1}}{2\pi i}\oint_{C_1} \frac{f(\zeta)}{(\zeta - z_0)^n}\, d\zeta + R_n,$$

where

$$R_n = \frac{(z - z_0)^n}{2\pi i}\oint_{C_1} \frac{f(\zeta)}{(\zeta - z)(\zeta - z_0)^n}\, d\zeta.$$

If M is the maximum value of $|f(\zeta)|$ for ζ on C_1, then for all ζ on C_1,

$$\left|\frac{f(\zeta)}{(\zeta - z)(\zeta - z_0)^n}\right| \le \frac{M}{(R' - \rho)(R')^n}.$$

Thus,

$$|R_n| \le \frac{\rho^n}{2\pi}\frac{M}{(R' - \rho)(R')^n}(2\pi R') = \frac{R'M}{R' - \rho}\left(\frac{\rho}{R'}\right)^n.$$

Because $\rho < R'$, it follows that $R_n \to 0$ as $n \to \infty$, and we may write

$$\frac{1}{2\pi i}\oint_{C_1} \frac{f(\zeta)}{\zeta - z}\, d\zeta = a_0 + a_1(z - z_0) + a_2(z - z_0)^2 + \cdots,$$

where

$$a_n = \frac{1}{2\pi i}\oint_{C_1} \frac{f(\zeta)}{(\zeta - z_0)^{n+1}}\, d\zeta.$$

For the integral around C_2,

$$\frac{-1}{2\pi i}\oint_{C_2} \frac{f(\zeta)}{\zeta - z}\, d\zeta$$

$$= \frac{-1}{2\pi i}\oint_{C_2} \frac{f(\zeta)}{(\zeta - z_0) + (z_0 - z)}\, d\zeta$$

$$= \frac{-1}{2\pi i}\oint_{C_2} \frac{f(\zeta)}{(z_0 - z)\left(1 - \dfrac{\zeta - z_0}{z - z_0}\right)}\, d\zeta$$

$$= \frac{-1}{2\pi i}\oint_{C_2} \frac{f(\zeta)}{z_0 - z}\left[1 + \left(\frac{\zeta - z_0}{z - z_0}\right) + \left(\frac{\zeta - z_0}{z - z_0}\right)^2 + \cdots\right.$$

$$\left. + \left(\frac{\zeta - z_0}{z - z_0}\right)^{n-1} + \frac{\left(\dfrac{\zeta - z_0}{z - z_0}\right)^n}{1 - \left(\dfrac{\zeta - z_0}{z - z_0}\right)}\right] d\zeta$$

$$= \frac{1}{2\pi i(z-z_0)} \oint_{C_2} f(\zeta)\,d\zeta + \frac{1}{2\pi i(z-z_0)^2} \oint_{C_2} f(\zeta)(\zeta - z_0)\,d\zeta + \cdots$$
$$+ \frac{1}{2\pi i(z-z_0)^n} \oint_{C_2} f(\zeta)(\zeta - z_0)^{n-1}\,d\zeta + S_n,$$

where

$$S_n = \frac{1}{2\pi i(z-z_0)^n} \oint_{C_2} \frac{f(\zeta)(\zeta - z_0)^n}{z-\zeta}\,d\zeta.$$

If m is the maximum value of $|f(\zeta)|$ for ζ on C_2, then for all ζ on C_2,

$$\left|\frac{f(\zeta)(\zeta - z_0)^n}{z-\zeta}\right| \leq \frac{m(r')^n}{\rho - r'}.$$

Thus,

$$|S_n| \leq \frac{1}{2\pi\rho^n}\frac{m(r')^n}{\rho - r'}(2\pi r') = \frac{r'm}{\rho - r'}\left(\frac{r'}{\rho}\right)^n.$$

Because $r' < \rho$, it follows that $S_n \to 0$ as $n \to \infty$. Consequently,

$$\frac{-1}{2\pi i}\oint_{C_2} \frac{f(\zeta)}{\zeta - z}\,d\zeta = \frac{a_{-1}}{z-z_0} + \frac{a_{-2}}{(z-z_0)^2} + \cdots + \frac{a_{-n}}{(z-z_0)^n} + \cdots$$

where

$$a_{-n} = \frac{1}{2\pi i}\oint_{C_2} f(\zeta)(\zeta - z_0)^{n-1} d\zeta.$$

We have shown that

$$f(z) = \sum_{n=0}^{\infty} a_n (z-z_0)^n + \sum_{n=-1}^{-\infty} a_n (z-z_0)^n$$

where

$$a_n = \frac{1}{2\pi i}\oint_{C_1} \frac{f(\zeta)}{(\zeta - z_0)^{n+1}}\,d\zeta, \qquad n \geq 0,$$
$$a_n = \frac{1}{2\pi i}\oint_{C_2} \frac{f(\zeta)}{(\zeta - z_0)^{n+1}}\,d\zeta, \qquad n < 0.$$

Suppose now that C is any simple, closed, piecewise-smooth curve in D that contains z_0 in its interior. It is always possible to construct C_1 and C_2 as before such that C is contained in the interior of C_1 and C_2 is contained in the interior of C. Because $f(z)/(z-z_0)^{n+1}$ is analytic in D, it follows by

Theorem 4.8 that

$$\oint_{C_1} \frac{f(\zeta)}{(\zeta - z_0)^{n+1}}\, d\zeta = \oint_C \frac{f(\zeta)}{(\zeta - z_0)^{n+1}}\, d\zeta, \qquad n \geq 0,$$

$$\oint_{C_2} \frac{f(\zeta)}{(\zeta - z_0)^{n+1}}\, d\zeta = \oint_C \frac{f(\zeta)}{(\zeta - z_0)^{n+1}}\, d\zeta, \qquad n < 0,$$

and the proof is complete. ■

Series 5.25 is called the **Laurent expansion** of f in the annulus D. In the particular case that f is analytic for $|z - z_0| \leq r$ as well as in D, we find that for $n < 0$, $f(\zeta)/(\zeta - z_0)^{n+1}$ is analytic in $|z - z_0| < R$, and

$$\oint_C \frac{f(\zeta)}{(\zeta - z_0)^{n+1}}\, d\zeta = 0;$$

and for $n \geq 0$, the generalized Cauchy integral formula gives

$$a_n = \frac{1}{2\pi i} \oint_C \frac{f(\zeta)}{(\zeta - z_0)^{n+1}}\, d\zeta = \frac{f^{(n)}(z_0)}{n!}.$$

Thus, in this case, the Laurent series reduces to the Taylor series, as expected.

Just as the power series of a function is unique, so also is the Laurent expansion.

■ **COROLLARY** If f has a Laurent series valid in an annulus $r < |z - z_0| < R$, then it has only one such expansion. ■

Because of this corollary, we seldom use formula 5.25(b) to calculate the coefficients a_n in the Laurent expansion. Often we use properties of geometric series and/or Taylor series from Section 5.4. This is illustrated in the following examples.

EXAMPLE 5.12 ....

Find Laurent series for $f(z) = 1/(z^2 - 3z + 2)$ valid in the domains (a) $|z| < 1$, (b) $1 < |z| < 2$, and (c) $|z| > 2$.

Solution The singularities of f are $z = 1$ and $z = 2$. Figure 5.11 makes it clear that f is analytic in each of the domains $|z| < 1$, $1 < |z| < 2$, and $|z| > 2$.

(a) Because f is analytic for $|z| < 1$, the Laurent series will be the Maclaurin series. The partial fraction decomposition of $f(z)$ is

$$f(z) = \frac{1}{z-2} - \frac{1}{z-1}.$$

Because

$$\frac{1}{1-z} = 1 + z + z^2 + z^3 + \cdots, \qquad |z| < 1,$$

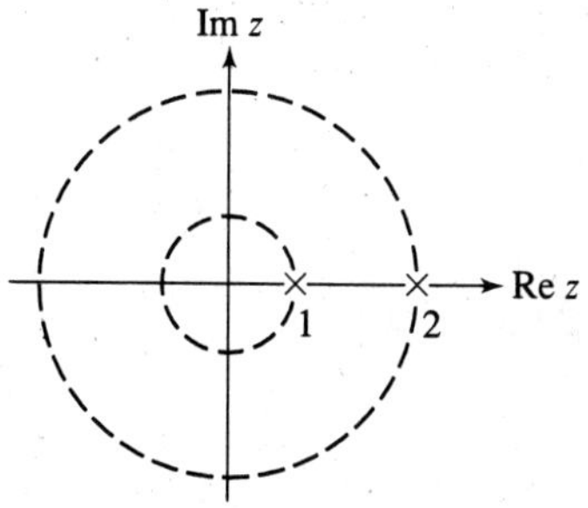

Figure 5.11

and

$$\frac{-1}{2-z} = \frac{-1}{2(1-z/2)} = -\frac{1}{2}\left(1+\frac{z}{2}+\frac{z^2}{4}+\frac{z^3}{8}+\cdots\right), \qquad |z|<2,$$

the Maclaurin series is

$$\begin{aligned} f(z) &= (1+z+z^2+z^3+\cdots) - \frac{1}{2}\left(1+\frac{z}{2}+\frac{z^2}{4}+\frac{z^3}{8}+\cdots\right), \qquad |z|<1 \\ &= \frac{1}{2}+\frac{3}{4}z+\frac{7}{8}z^2+\frac{15}{16}z^3+\cdots \\ &= \sum_{n=0}^{\infty}\frac{2^{n+1}-1}{2^{n+1}}z^n, \qquad |z|<1. \end{aligned}$$

(b) The expansion for $1/(z-2)$ obtained in (a) is valid for $|z|<2$. Because

$$\frac{-1}{z-1} = \frac{-1}{z(1-(1/z))} = -\frac{1}{z}\left(1+\frac{1}{z}+\frac{1}{z^2}+\frac{1}{z^3}+\cdots\right),$$

valid for $|1/z|<1$ or $|z|>1$, the Laurent series for $1<|z|<2$ is

$$\begin{aligned} f(z) &= -\frac{1}{z}\left(1+\frac{1}{z}+\frac{1}{z^2}+\frac{1}{z^3}+\cdots\right) - \frac{1}{2}\left(1+\frac{z}{2}+\frac{z^2}{4}+\frac{z^3}{8}+\cdots\right) \\ &= -\sum_{n=1}^{\infty}\frac{1}{z^n} - \sum_{n=0}^{\infty}\frac{1}{2^{n+1}}z^n. \end{aligned}$$

(c) The expansion for $-1/(z-1)$ obtained in (b) is valid for $|z|>1$. Because

$$\frac{1}{z-2} = \frac{1}{z(1-2/z)} = \frac{1}{z}\left(1+\frac{2}{z}+\frac{4}{z^2}+\frac{8}{z^3}+\cdots\right),$$

valid for $|2/z| < 1$ or $|z| > 2$, the Laurent series for $|z| > 2$ is

$$\begin{aligned} f(z) &= -\frac{1}{z}\left(1 + \frac{1}{z} + \frac{1}{z^2} + \frac{1}{z^3} + \cdots\right) + \frac{1}{z}\left(1 + \frac{2}{z} + \frac{4}{z^2} + \frac{8}{z^3} + \cdots\right) \\ &= \frac{1}{z^2} + \frac{3}{z^3} + \frac{7}{z^4} + \frac{15}{z^5} + \cdots \\ &= \sum_{n=2}^{\infty} \frac{2^{n-1} - 1}{z^n}. \end{aligned}$$

EXAMPLE 5.13

Find the Laurent series for $f(z) = 1/(z^2 + z)$, valid in the domain $0 < |z + 1| < 1$.

Solution Because $z = 0$ and $z = -1$ are singularities (Figure 5.12), f is analytic for $0 < |z + 1| < 1$. The Laurent series in this annulus will be in powers of $z + 1$. The second term in the partial fraction decomposition,

$$f(z) = \frac{1}{z(z+1)} = \frac{1}{z} - \frac{1}{z+1},$$

is therefore of the required form. Because

$$\frac{1}{z} = \frac{1}{(z+1) - 1} = \frac{-1}{1 - (z+1)} = -[1 + (z+1) + (z+1)^2 + \cdots], \qquad |z + 1| < 1,$$

the required Laurent series is

$$f(z) = \frac{-1}{z+1} - [1 + (z+1) + (z+1)^2 + \cdots] = \sum_{n=-1}^{\infty} -(z+1)^n, \qquad 0 < |z + 1| < 1.$$

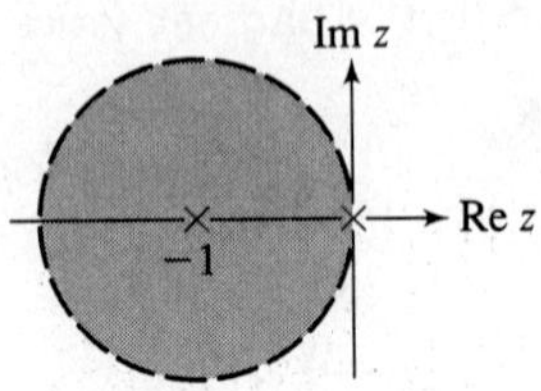

Figure 5.12

EXAMPLE 5.14

Find the Laurent series for $f(z) = 5z/(z^2 + z - 6)$ in the annulus $1 < |z - 1| < 4$.

Solution Because $z = 2$ and $z = -3$ are singularities of f (Figure 5.13), f is analytic in $1 < |z - 1| < 4$. To expand $f(z)$ we first form its partial fraction decomposition,

$$f(z) = \frac{3}{z+3} + \frac{2}{z-2}.$$

Both terms must be expressed in powers of $z - 1$. For the first we write

$$\frac{3}{z+3} = \frac{3}{(z-1)+4} = \frac{3}{4\left(1 + \dfrac{z-1}{4}\right)} = \frac{3}{4}\sum_{n=0}^{\infty}\left(-\frac{z-1}{4}\right)^n, \qquad \left|-\frac{z-1}{4}\right| < 1$$

$$= \sum_{n=0}^{\infty} \frac{3(-1)^n}{4^{n+1}}(z-1)^n, \qquad |z-1| < 4.$$

For the second term,

$$\frac{2}{z-2} = \frac{2}{(z-1)-1} = \frac{2}{(z-1)\left(1 - \dfrac{1}{z-1}\right)}$$

$$= \frac{2}{z-1}\sum_{n=0}^{\infty}\left(\frac{1}{z-1}\right)^n, \qquad \left|\frac{1}{z-1}\right| < 1$$

$$= \sum_{n=0}^{\infty} \frac{2}{(z-1)^{n+1}}, \qquad |z-1| > 1.$$

The required Laurent series is, therefore,

$$f(z) = \sum_{n=0}^{\infty} \frac{2}{(z-1)^{n+1}} + \sum_{n=0}^{\infty} \frac{3(-1)^n}{4^{n+1}}(z-1)^n, \qquad 1 < |z-1| < 4.$$

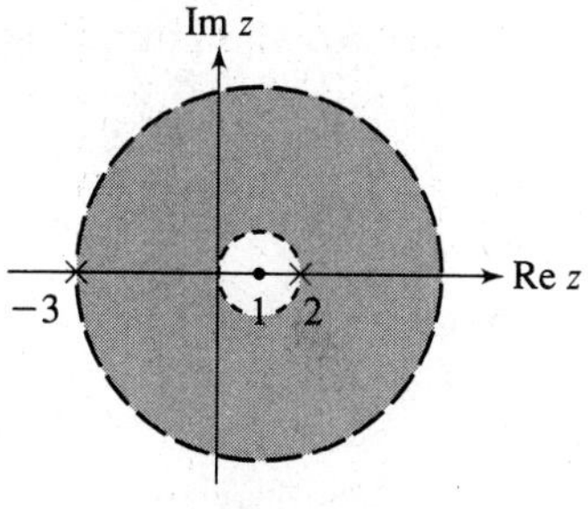

Figure 5.13

EXAMPLE 5.15

Find the Laurent series for $f(z) = z^{-3}\cos 2z$ valid in the domain $|z| > 0$.

Solution The Maclaurin series for $\cos 2z$ valid for all z is $\cos 2z = \sum_{n=0}^{\infty} \frac{(-1)^n 2^{2n}}{(2n)!} z^{2n}$. Multiplication by $1/z^3$ gives the Laurent series for $z^{-3}\cos 2z$ valid when $|z| > 0$,

$$\frac{1}{z^3}\cos 2z = \sum_{n=0}^{\infty} \frac{(-1)^n 2^{2n}}{(2n)!} z^{2n-3}.$$

Classification of Singularities

When z_0 is an isolated singularity of an analytic function f, there exists an annulus $0 < |z - z_0| < R$ centered at z_0 in which f is analytic, and therein f has a Laurent expansion. We use this expansion to classify the type of singularity at z_0.

Definition 5.4

Suppose an analytic function f has an isolated singularity at z_0, and

$$f(z) = \sum_{n=-\infty}^{\infty} a_n (z - z_0)^n$$

is the Laurent expansion of f valid in some annulus $0 < |z - z_0| < R$. Then

(i) If $a_n = 0$ for all $n < 0$, z_0 is called a **removable singularity**;
(ii) If $a_n = 0$ for $n < -m$, m a fixed positive integer, but $a_{-m} \neq 0$, z_0 is called a **pole of order m**;
(iii) If $a_n \neq 0$ for an infinity of negative integers n, z_0 is called an **essential singularity**.

Part (i) of this definition agrees with the definition of removable singularity in Section 2.3. If $a_n = 0$ for all $n < 0$, then the Laurent series of f about z_0 takes the form

$$f(z) = a_0 + a_1(z - z_0) + a_2(z - z_0)^2 + \cdots.$$

It follows that $\lim_{z\to z_0} f(z) = a_0$, and existence of the limit is the definition of removable singularity in Section 2.3. Conversely, if $\lim_{z\to z_0} f(z)$ exists, then the Laurent series of f cannot contain terms $(z - z_0)^n$, where $n < 0$. By process of elimination, the Laurent series must satisfy (i).

The classification of singularities in Definition 5.4 is illustrated in the following example.

EXAMPLE 5.16 ....

Classify singularities for the following functions:

$$\textbf{(a)}\ \frac{\sin z}{z} \qquad \textbf{(b)}\ e^{1/z} \qquad \textbf{(c)}\ \frac{1}{z^3(z+3)^2}$$

Solution **(a)** Because the Maclaurin series for $\sin z$ is

$$\sin z = z - \frac{z^3}{3!} + \frac{z^5}{5!} - \cdots, \qquad |z| < \infty,$$

it follows that

$$\frac{\sin z}{z} = 1 - \frac{z^2}{3!} + \frac{z^4}{5!} - \cdots, \qquad 0 < |z| < \infty.$$

This is the Laurent series for $z^{-1}\sin z$ in the annulus $0 < |z| < \infty$. It satisfies (i) in Definition 5.4, and therefore $z = 0$ is a removable singularity for $z^{-1}\sin z$. The Laurent series clearly demonstrates that $\lim_{z\to 0} z^{-1}\sin z = 1$.

(b) If we replace each z in the Maclaurin series

$$e^z = 1 + z + \frac{z^2}{2!} + \frac{z^3}{3!} + \cdots, \qquad |z| < \infty,$$

by $1/z$, we obtain

$$e^{1/z} = 1 + \frac{1}{z} + \frac{1}{2!\,z^2} + \frac{1}{3!\,z^3} + \cdots, \qquad 0 < |z| < \infty,$$

valid for $0 < |z| < \infty$. This is therefore the Laurent series for $e^{1/z}$ about $z = 0$, and $z = 0$ is an essential singularity.

(c) For the singularity at $z = 0$, we first expand

$$\frac{1}{z+3} = \frac{1}{3(1+z/3)} = \frac{1}{3}\left(1 - \frac{z}{3} + \frac{z^2}{9} - \cdots\right), \qquad |z| < 3.$$

Term-by-term differentiation with respect to z gives

$$-\frac{1}{(z+3)^2} = \frac{1}{3}\left(-\frac{1}{3} + \frac{2z}{9} - \frac{3z^2}{27} + \cdots\right), \qquad |z| < 3,$$

and therefore

$$\frac{1}{z^3(z+3)^2} = \frac{1}{9z^3} - \frac{2}{27z^2} + \frac{1}{27z} - \cdots, \qquad 0 < |z| < 3.$$

This is the Laurent series for $1/[z^3(z+3)^2]$ about the singularity $z = 0$. Because it satisfies part (ii) of Definition 5.4 with $m = 3$, $z = 0$ is a pole of order 3.

For the singularity at $z = -3$, we first expand

$$\frac{1}{z} = \frac{1}{(z+3)-3} = \frac{1}{-3\left(1 - \dfrac{z+3}{3}\right)} = -\frac{1}{3}\left[1 + \frac{z+3}{3} + \left(\frac{z+3}{3}\right)^2 + \cdots\right],$$

valid for $|z+3| < 3$. Term-by-term differentiation gives

$$-\frac{1}{z^2} = -\frac{1}{3}\left[\frac{1}{3} + \frac{2}{9}(z+3) + \frac{3}{27}(z+3)^2 + \cdots\right], \qquad |z+3| < 3.$$

A second differentiation yields

$$\frac{2}{z^3} = -\frac{1}{3}\left[\frac{2}{9} + \frac{6}{27}(z+3) + \cdots\right], \qquad |z+3| < 3.$$

Hence

$$\frac{1}{z^3(z+3)^2} = -\frac{1}{6}\left[\frac{2}{9(z+3)^2} + \frac{2}{9(z+3)} + \cdots\right], \qquad 0 < |z+3| < 3,$$

and $z = -3$ is a pole of order 2.

In the theory to follow, it is important to be able to determine quickly the poles (and their orders) of a function. The following result allows us to do this.

■ **THEOREM 5.23** A function f has a pole of order m at z_0 if and only if it can be written in the form

$$f(z) = \frac{g(z)}{(z-z_0)^m}, \tag{5.26}$$

where g is analytic at z_0 and $g(z_0) \neq 0$.

Proof If $f(z)$ can be expressed in form 5.26 and $g(z)$ is expanded in a Taylor series about z_0,

$$g(z) = a_0 + a_1(z-z_0) + \cdots, \qquad |z-z_0| < R \quad \text{(for some } R\text{)}$$

where $a_0 = g(z_0) \neq 0$, then

$$f(z) = \frac{a_0}{(z-z_0)^m} + \frac{a_1}{(z-z_0)^{m-1}} + \cdots, \qquad 0 < |z-z_0| < R.$$

Because the right side of this equation is the Laurent expansion of f about z_0, z_0 is a pole of order m.

Conversely, if f has a pole of order m at z_0, the Laurent expansion of f about z_0 is of the form

$$\begin{aligned} f(z) &= \frac{a_{-m}}{(z-z_0)^m} + \frac{a_{-m+1}}{(z-z_0)^{m-1}} + \cdots + \frac{a_{-1}}{z-z_0} + a_0 + a_1(z-z_0) + \cdots \\ &= \frac{1}{(z-z_0)^m}\left[a_{-m} + a_{-m+1}(z-z_0) + \cdots\right] \end{aligned}$$

valid in some annulus $0 < |z-z_0| < R$. The power series in the braces represents a function that is analytic for $|z-z_0| < R$, and the function clearly does not vanish at z_0. ■

Notice how this theorem shortens the work in part (c) of Example 5.16. The function $1/(z+3)^2$ is analytic at $z=0$ and is nonvanishing there. Consequently, $[z^3(z+3)^2]^{-1}$ has a pole of order 3 at $z=0$. Because $1/z^3$ is analytic at $z=-3$ and does not vanish there, $[z^3(z+3)^2]^{-1}$ has a pole of order 2 at $z=-3$. Further illustrations of Theorem 5.23 are contained in the following example. To appreciate the theorem, try finding the Laurent series for the functions and using Definition 5.4.

EXAMPLE 5.17 ....

Classify singularities for the following functions:

$$\text{(a)}\ \frac{\sin 3z}{(z+1)^4} \qquad \text{(b)}\ \frac{z^2+2z-1}{z^4-3z^3+3z^2-z}$$

Solution **(a)** Because $\sin 3z$ is entire and does not vanish at $z=-1$, $z=-1$ is a pole of order 4.

(b) The factored form

$$\frac{z^2+2z-1}{z^4-3z^3+3z^2-z} = \frac{z^2+2z-1}{z(z-1)^3}$$

indicates singularities at $z=0$ and $z=1$. Because $(z^2+2z-1)/(z-1)^3$ is analytic at $z=0$ and does not vanish there, $z=0$ is a pole of order 1 (also called a simple pole). Because $(z^2+2z-1)/z$ is analytic at $z=1$ and not zero there, $z=1$ is a pole of order 3. ....

Exercises 5.5

In Exercises 1–4 find the Laurent series for the function in the given domain.

1. $1/(z-3)$ for $|z|>3$
2. $1/(z^2+2z)$ for $0<|z|<2$
3. $z/(z^2-3z+2)$ for $1<|z-3|<2$
4. $\sin(1/z)$ for $|z|>0$

In Exercises 5–8 find the Laurent series about the given point. Indicate the domain of convergence of the series.

5. $1/(z^3-z^2)$ about $z=0$
6. $z/(z^2+4)$ about $z=2i$
7. $z/(z^2+4)$ about $z=2$
8. $e^z/(z+1)$ about $z=-1$
9. Find the first three nonzero terms in the Laurent series of $\csc z$ about $z=0$.
10. Find the Laurent series for $f(z)=(3z-3)/(2z^2-5z+2)$ convergent for $1/2<|z-1|<1$.
11. Expand $f(z)=z^3/(-z^2+3z-2)$ in Laurent series valid for the domains:
    (a) $|z|<1$ **(b)** $1<|z|<2$ **(c)** $|z|>2$ **(d)** $|z-1|>1$ **(e)** $0<|z-2|<1$
12. **(a)** Prove that if f is analytic at z_0 and has a zero of order m at z_0, then $1/f$ has a pole of order m at z_0.

    (b) Determine the order of the pole at $z=0$ for $1/(2\cos z-2+z^2)^2$.

In Exercises 13–25 find and classify all isolated singularities of the function.

13. $(z^2+1)/[z^2(z+1)]$

14. $z^3 e^{1/z}$

15. $(z^2-4)/(z-2)$

16. $(\sin 3z)/z^5$

17. $\operatorname{csch} z$

18. $\operatorname{Log} z$

19. $(z^3+3z)/\cos^2 z$

20. $\cot z$

21. $z \cot z$

22. $1/[(z^2+1)\cosh(z+i)]$

23. $(z-\sin z)/z^6$

24. $1/(\sin z - \sin 2z)$

25. $\sec(1/z)$

26. Does the function $f(z) = \operatorname{Re} z$ have a Laurent series valid in some domain $0 < |z| < R$? Explain.

27. Does $\operatorname{Log} z$ have a Laurent series in any annulus centered at $z = 0$? Does it have a Laurent series about any of its singularities?

28. Discuss whether $z = 2$ is an isolated singularity of the function $1/[(z-1)^{1/2}-1]$ when $(z-1)^{1/2}$ is the following branch of the double-valued function:

(a) $\sqrt{z-1}$

(b) $-\sqrt{z-1}$

(c) $(z-1)^{1/2} = |z-1|^{1/2} e^{(i/2)\arg(z-1)}, \quad \pi/2 < \arg(z-1) \le 5\pi/2$

(d) $(z-1)^{1/2} = |z-1|^{1/2} e^{(i/2)\arg(z-1)}, \quad 0 < \arg(z-1) \le 2\pi$

29. Use long division to obtain the first four nonzero terms in the Laurent series of $\csc 2z$ about $z = 0$. What do you anticipate for a domain of validity of the series?

30. **(a)** Show that the Laurent series for $e^{t(z-1/z)/2}$ (t a parameter) valid for $|z| > 0$ is

$$e^{t(z-1/z)/2} = \sum_{n=-\infty}^{\infty} J_n(t) z^n$$

where $J_n(t)$ is the Bessel function of the first kind of order n. We call $e^{t(z-1/z)/2}$ a generating function for the integer order Bessel functions.

(b) Use formula 5.25(b) to verify the following integral representation for $J_n(t)$,

$$J_n(t) = \frac{1}{2\pi} \int_0^{2\pi} \cos(n\theta - t\sin\theta)\, d\theta$$

31. Find the Laurent series for the function $f(z) = \sqrt{z-1}\sqrt{z+1}$ valid in some annulus about (a) $z = 0$ and (b) $z = 1$.

32. Show that when $b > a > 1$,

$$\operatorname{Log}\left(\frac{z-a}{z-b}\right) = \sum_{n=1}^{\infty} \frac{b^n - a^n}{n z^n}$$

for $|z| > b$. *Hint:* See Exercise 3.5–41.

5.6 Analytic Continuation

We have seen that the infinite series

$$\frac{1}{1-z} = \sum_{n=0}^{\infty} z^n, \qquad |z| < 1,$$

is a valid representation for the function $1/(1-z)$ inside the unit circle $|z| < 1$. By writing

$$\frac{1}{1-z} = \frac{1}{(1-i)-(z-i)} = \frac{1}{1-i}\left[\frac{1}{1-\left(\dfrac{z-i}{1-i}\right)}\right] = \sum_{n=0}^{\infty} \frac{1}{(1-i)^{n+1}}(z-i)^n,$$

we obtain a representation of the same function in a different, but overlapping, circle $|z-i| < \sqrt{2}$ (Figure 5.14). Inside the overlapping region, values of the two series coincide. We say that each series is an **analytic continuation** of the other. By picking further points of expansion, we could continue these representations of $1/(1-z)$ into all parts of the complex plane that do not include the point $z = 1$.

In practical problems that involve analytic continuation, we may not have a closed-form sum for a series; we may have only a series that we know represents an analytic function, call it f, inside its circle of convergence:

$$f(z) = \sum_{n=0}^{\infty} a_n (z-z_0)^n, \qquad C_0 : |z-z_0| < R.$$

If z_1 is a point interior to C_0 (Figure 5.15), Theorem 5.18 guarantees that the Taylor series of f about z_1 is analytic inside the largest circle C_1 about z_1, which is interior to C_0. It may happen, however, that this Taylor series about z_1 actually converges inside a larger circle C_2, which includes points exterior to C_0. We say then that the Taylor series about z_1 is an analytic continuation of f from C_0 into C_2. The values of f as calculated by either series are the same inside their common region of convergence. Taylor series about other points in C_0 or in C_2 may lead to further analytic continuations of f into other regions. The following theorem leads to important results about analytic continuations.

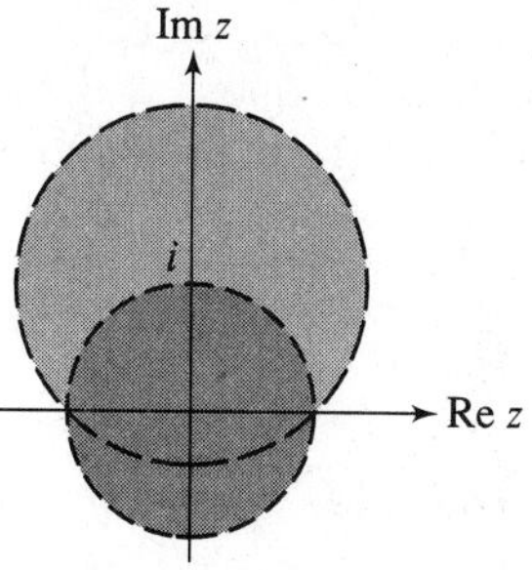

Figure 5.14

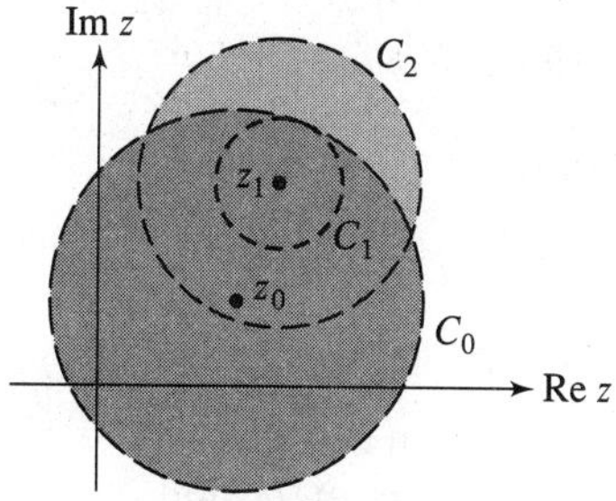

Figure 5.15

■ **THEOREM 5.24** Let f_1 and f_2 be analytic in a domain D. If $f_1(z) = f_2(z)$ at all points in a subdomain D' of D, then $f_1(z) = f_2(z)$ in D.

Proof If z_0 is any point in D' (Figure 5.16), each of $f_1(z)$ and $f_2(z)$ can be expanded in Taylor series about z_0, and both series converge in the largest circle C contained in D. Because $f_1(z) = f_2(z)$ in D', these Taylor series must be identical; that is, $f_1(z)$ must now be equal to $f_2(z)$ inside C. By expanding $f_1(z)$ and $f_2(z)$ about points inside C and outside D', the domain in which $f_1(z) = f_2(z)$ can be extended further and can eventually be made to coincide with D.

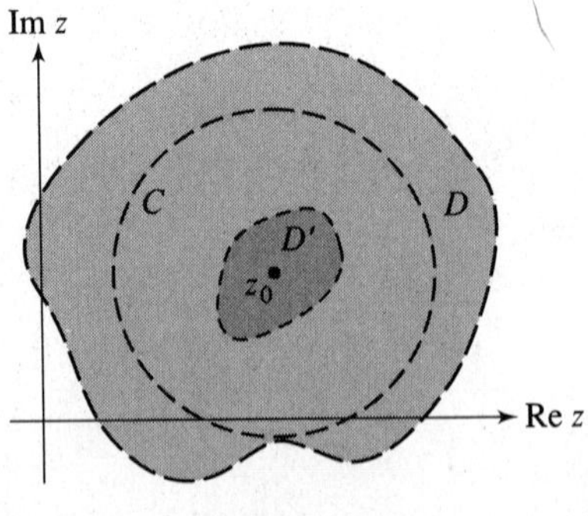

Figure 5.16

■

The requirement that $f_1(z) = f_2(z)$ in a subdomain D' of D can be relaxed to coincident values along a curve in D, or even a sequence of points having limit in D, but the proof is correspondingly more difficult.

■ **COROLLARY 1** (Local Uniqueness of Analytic Continuation) When a function f analytic in a domain D is analytically continued to a domain D' that has points in common with D, the continuation is unique. ■

■ **COROLLARY 2** If it is possible to continue analytically a function $f(x)$ defined on some interval of the x-axis, into the complex plane, it can be done in only one way. ■

In essence, this result completes the work on the trigonometric and hyperbolic functions and their inverses in Chapter 3. For example, we defined functions $\sin z = (e^{zi} - e^{-zi})/(2i)$ and $\cosh z = (e^z + e^{-z})/2$ in order that they agree with their real counterparts $\sin x$ and $\cosh x$ on the real axis. The preceding result indicates that because these complex functions are analytic, they are the only possible analytic functions that continue $\sin x$ and $\cosh x$ into the complex plane.

As a further example, recall that the complex exponential e^z was derived from the differential equation $f'(z) = f(z)$, $f(0) = 1$. Subsequently, it was shown that the Maclaurin series for e^z is $\sum_{n=0}^{\infty} z^n/n!$. Corollary 2 presents an alternative procedure — define

$$e^z = \sum_{n=0}^{\infty} \frac{1}{n!} z^n.$$

Because the function so defined is entire and coincides with e^z for $z = x$, it is the only possible analytic extension of e^x to the complex plane. Properties of e^z can then be derived from the complex power series representation.

It is important to notice that Corollary 1 is one of local uniqueness. To emphasize this, suppose a function f analytic in a domain D (Figure 5.17) is analytically continued by two separate chains of circles beginning at z_1 and z_2 in such a way that they both eventually define values for f in some common domain R.

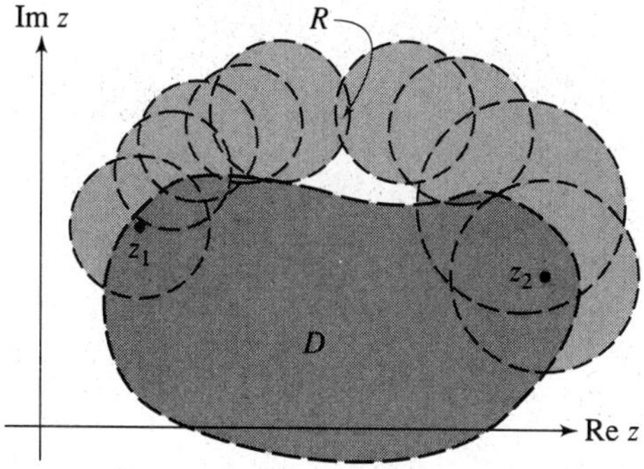

Figure 5.17

Do the continuations yield the same values for f in R? Corollary 1 does not address this global problem; it is a local result. To answer the question, we note that it is always possible to construct a chain of circles lying completely within D that joins the beginning circles at z_1 and z_2. If the simply connected domain consisting of points interior to all three chains and all points bounded by the chains is free of singularities of f, it can be shown that the values of f in R are identical for the two continuations. (This result is called the monodromy theorem.) When the domain bounded by the chains contains a singularity of f, continuations around opposite sides of the singularity may not lead to identical values for f in overlapping domains. In fact, such continuations may lead to multiple-valued functions with their associated branches. We illustrate this with the following alternative approach to the logarithm function. At the same time notice that in this section we have constantly worked with chains of circles in order to continue analytically a function from one domain into others. To indicate other possible continuation techniques, we consider the complex logarithm function from a contour integral point of view.

It is well known that the real logarithm function can be defined for all $x > 0$ by

$$\ln x = \int_1^x \frac{1}{t}\,dt.$$

We continue this function to the half-plane $D : \operatorname{Re} z > 0$ with the definition

$$\log z = \int_C \frac{1}{\zeta}\,d\zeta,$$

where C is any curve in D from $\zeta = 1$ to $\zeta = z$ (Figure 5.18). In the proof of the necessary part of Theorem 4.4 in Section 4.4, it was shown that $\int_C \zeta^{-1}\,d\zeta$ is an antiderivative of $1/z$ in D, and hence $\log z$ is an analytic continuation of $\ln x$ into D. We can continue $\log z$ into the left half-plane $\operatorname{Re} z \leq 0$ with essentially

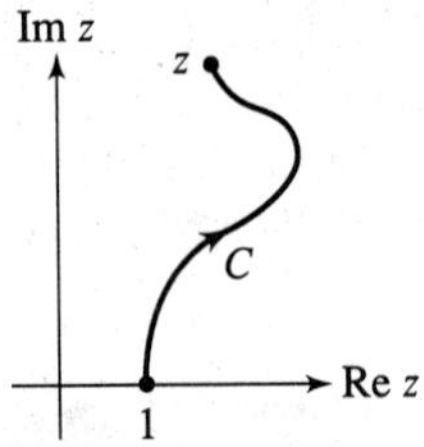

Figure 5.18

the same definition, but C is restricted so it doesn't pass through the origin. This continuation, however, does not lead to a single-valued function. For example, the upper half C' of the unit circle leads to the value

$$\log(-1) = \int_{C'} \frac{1}{\zeta}\,d\zeta = \int_0^{\pi} \frac{1}{e^{\theta i}} ie^{\theta i}\,d\theta = \pi i,$$

whereas the lower half C'' of the same circle yields

$$\log(-1) = \int_{C''} \frac{1}{\zeta}\,d\zeta = \int_{2\pi}^{\pi} \frac{1}{e^{\theta i}} ie^{\theta i}\,d\theta = -\pi i.$$

Furthermore, according to Example 4.22, by taking curves that encircle the origin, it is possible to produce values $\log(-1) = (2n+1)\pi$ for any integer n whatsoever. Continuation of $\log z$ back to D leads to multiple values there also.

In this way we have continued $\ln x$ to a function $\log z$, which although infinitely valued, is analytic at every point except $z = 0$—analytic in the sense that about any point z, there exists a neighborhood inside of which $\log z$ can be assigned values that constitute a single-valued analytic function. Indeed, in every simply connected domain that does not contain $z = 0$, it is possible to select such values. When this domain is enlarged as much as possible, we are led to branches and branch cuts of $\log z$.

Exercises 5.6

1. In Figure 5.15, what is the largest possible radius for circle C_1?
2. Prove that if a function f is analytic and not constant throughout a domain D, then it cannot be constant over any subdomain of D.
3. In this exercise we show that it is not always possible to continue a function analytically beyond its original domain of analyticity.

   (a) Show that the function $f(z) = \sum_{n=1}^{\infty} z^{n!}$ is analytic in the circle $|z| < 1$.

   (b) If $f(z)$ were continuable beyond the circle $|z| < 1$, there would necessarily be an arc of points on the circle at which $f(z)$ would be analytic. We show that this is impossible. Every arc of $|z| = 1$ contains an infinity of points representable in the form $z^* = e^{2\pi(p/q)i}$, where p and q are positive integers. Consider any one of them, and let $z = az^*$, $0 < a < 1$ be a point on the line joining 0 and z^*. Show that for this z, $f(z)$ can be expressed in the form

$$f(z) = \sum_{n=1}^{q-1} z^{n!} + \sum_{n=q}^{\infty} a^{n!}.$$

Show that $|f(z)|$ becomes unbounded as $z \to z^*$ along the radial line toward z^*.

(c) Complete the proof that $f(z)$ cannot be continued analytically beyond $|z| < 1$.

4. In this exercise we prove and illustrate the *Schwarz reflection principle.* Let f be analytic in some domain D that is symmetric about the real axis and contains a segment of this axis. If the function is real-valued when z is replaced by x (i.e., if $f(x)$ is real), then

$$\overline{f(\overline{z})} = f(z) \Leftrightarrow \overline{f(z)} = f(\overline{z}).$$

(a) Prove this result by showing that $F(z) = \overline{f(\overline{z})}$ is analytic in D and equal to $f(z)$ on that part of the x-axis in D.

(b) Prove the converse result: If $\overline{f(z)} = f(\overline{z})$, then $f(x)$ is real.

(c) Illustrate the principle with $\sin x$ and $\cosh x$.

CHAPTER 6

Residue Theory

In Chapter 4 we developed three methods for evaluating contour integrals around closed curves — the Cauchy-Goursat theorem, the Cauchy integral formulas, and Theorem 4.8. These results were not completely satisfactory. The Cauchy-Goursat theorem requires the integrand to be analytic inside and on the curve, and Cauchy's integral formulas are applicable only to integrands that have a very special form. Theorem 4.8 really does not evaluate contour integrals; it replaces a contour integral along a curve enclosing a number of singularities with contour integrals along curves each of which encloses only one singularity. With Laurent series we can develop a technique, Cauchy's residue theorem, that encompasses each of these. This theorem is far more powerful than just a synthesis of these three methods, however. It evaluates contour integrals around closed curves that enclose any number of isolated singularities, and it is independent of the form of the integrand. With this theorem we can also evaluate many real integrals that are otherwise intractable (Sections 6.2 and 6.3), and find a direct way to evaluate inverse Laplace transforms (Chapter 8).

6.1 Cauchy's Residue Theorem

Cauchy's residue theorem is a simple consequence of equation 5.25(b) for coefficients in Laurent series (5.25(a)). To illustrate, suppose $f(z) = \sum_{n=-\infty}^{\infty} a_n(z - z_0)^n$ is the Laurent series of a function f valid in some annulus $0 < |z - z_0| < R$ about an isolated singularity z_0. When C is a simple, closed, piecewise-smooth curve in this annulus that contains z_0 in its interior (Figure 6.1), coefficient a_{-1} is defined by

$$a_{-1} = \frac{1}{2\pi i} \oint_C f(z)\, dz.$$

Consequently,

$$\oint_C f(z)\, dz = 2\pi i a_{-1};$$

that is, the contour integral of f around the singularity z_0 of f is $2\pi i$ times the coefficient a_{-1} in its Laurent series about z_0.

Cauchy's residue theorem is a simple extension of this result to the case when C encloses an arbitrary number of isolated singularities. Clearly, coefficient a_{-1} in the Laurent series of a function about an isolated singularity plays a fundamental role in this theory. We give it a name in the following definition.

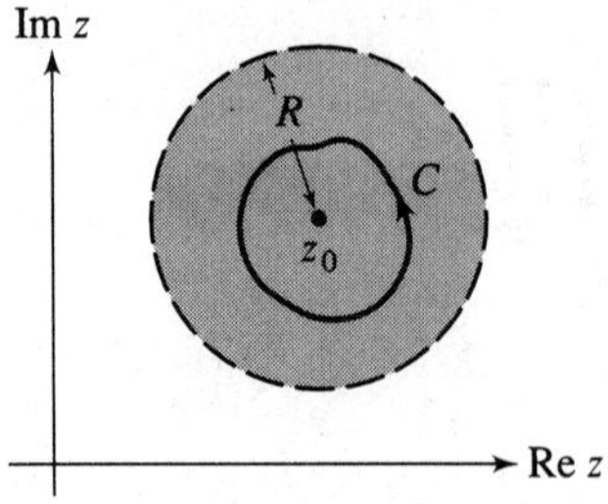

Figure 6.1

Definition 6.1
The **residue** of a function f at an isolated singularity z_0 is denoted by $\text{Res}[f, z_0]$. It is the coefficient a_{-1} in the Laurent expansion of f about z_0.

From Example 5.16, we may state that

$$\text{Res}[(\sin z)/z, 0] = 0, \qquad \text{Res}[e^{1/z}, 0] = 1,$$

$$\text{Res}[1/[z^3(z+3)^2], 0] = \tfrac{1}{27}, \qquad \text{Res}[1/[z^3(z+3)^2], -3] = -\tfrac{1}{27}.$$

How residues are used to evaluate contour integrals is contained in the next theorem.

■ **THEOREM 6.1** (Cauchy's Residue Theorem) Suppose a function f is analytic inside and on a simple, closed, piecewise-smooth curve C, except at singularities $z_1, \ldots, z_n$ in its interior. Then

$$\oint_C f(z)\, dz = 2\pi i \sum_{j=1}^{n} \text{Res}[f, z_j]. \tag{6.1}$$

Proof Construct around the z_j disjoint circles $C_j : |z - z_j| = r_j$ such that each C_j is interior to C (Figure 6.2). By Theorem 4.8, the contour integral around C can be replaced by contour integrals around the C_j,

$$\oint_C f(z)\, dz = \sum_{j=1}^{n} \oint_{C_j} f(z)\, dz.$$

Now if f is expanded in a Laurent series about z_j,

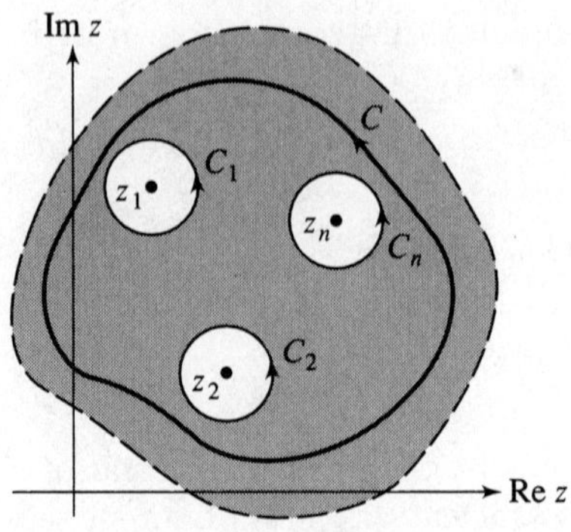

Figure 6.2

$$f(z) = \sum_{n=-\infty}^{\infty} a_n (z - z_j)^n \quad \text{where} \quad a_{-1} = \frac{1}{2\pi i} \oint_{C_j} f(z)\, dz.$$

But then $\oint_{C_j} f(z)\,dz = 2\pi i\, a_{-1} = 2\pi i \operatorname{Res}[f, z_j]$, and

$$\oint_C f(z)\,dz = 2\pi i \sum_{j=1}^{n} \operatorname{Res}[f, z_j]. \qquad \blacksquare$$

We suggested at the beginning of the chapter that Cauchy's residue theorem encompasses the Cauchy-Goursat theorem, the Cauchy integral formulas, and Theorem 4.8. If f is analytic inside a closed, piecewise-smooth curve C, then it has no residues interior to C. Hence the contour integral around C must vanish (Cauchy-Goursat theorem). Certainly residue theory encompasses Theorem 4.8; the theorem was instrumental in verifying Theorem 6.1. To illustrate that the generalized Cauchy integral formula is a special case of the Cauchy residue theorem, suppose that f is analytic inside and on a simple, closed, piecewise-smooth curve C (Figure 6.3). If z_0 is a point interior to C, then the function $f(z)/(z-z_0)^{n+1}$ is analytic inside C except at z_0. Thus,

$$\oint_C \frac{f(z)}{(z-z_0)^{n+1}}\,dz = 2\pi i \operatorname{Res}\left[\frac{f(z)}{(z-z_0)^{n+1}}, z_0\right].$$

If f is expanded in a Taylor series about z_0,

$$f(z) = f(z_0) + f'(z_0)(z-z_0) + \cdots + \frac{f^{(n)}(z_0)}{n!}(z-z_0)^n + \cdots.$$

Thus,

$$\frac{f(z)}{(z-z_0)^{n+1}} = \frac{f(z_0)}{(z-z_0)^{n+1}} + \frac{f'(z_0)}{(z-z_0)^n} + \cdots + \frac{f^{(n)}(z_0)}{n!(z-z_0)} + \cdots,$$

and

$$\operatorname{Res}\left[\frac{f(z)}{(z-z_0)^{n+1}}, z_0\right] = \frac{f^{(n)}(z_0)}{n!}.$$

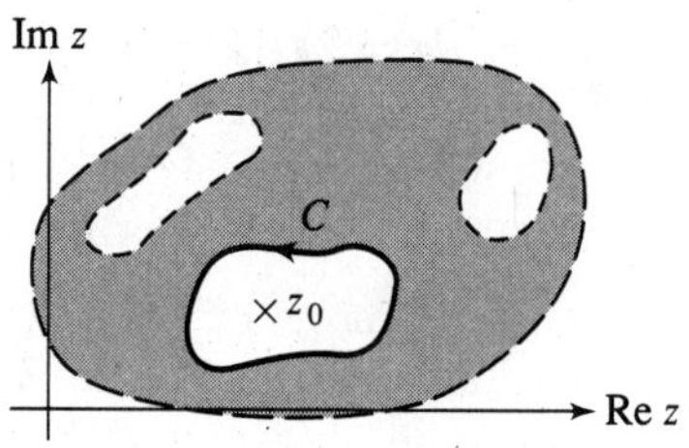

Figure 6.3

Consequently,

$$\oint_C \frac{f(z)}{(z-z_0)^{n+1}}\,dz = 2\pi i \frac{f^{(n)}(z_0)}{n!},$$

the Cauchy generalized integral formula.

EXAMPLE 6.1

Evaluate $\oint_C \frac{\sin z}{z^3}\,dz$ where $C : |z| = 1$.

Solution Because $\sin z$ is entire and $z = 0$ is interior to C, Cauchy's generalized integral formula gives

$$\oint_C \frac{\sin z}{z^3}\,dz = \frac{2\pi i}{2!}\frac{d^2}{dz^2}(\sin z)_{|z=0} = \pi i(-\sin z)_{|z=0} = 0.$$

Alternatively, $z = 0$ is the only singularity of $(\sin z)/z^3$ interior to C. Because

$$\frac{\sin z}{z^3} = \frac{1}{z^3}\left(z - \frac{z^3}{3!} + \frac{z^5}{5!} - \cdots\right) = \frac{1}{z^2} - \frac{1}{3!} + \frac{z^2}{5!} - \cdots,$$

we find that $\text{Res}[(\sin z)/z^3, 0] = 0$, and hence

$$\oint_C \frac{\sin z}{z^3}\,dz = 2\pi i(0) = 0.$$

EXAMPLE 6.2

Evaluate $\oint_C e^{1/z}dz$ where $C : |z| = 1$.

Solution Because the Laurent series for $e^{1/z}$ around $z = 0$ is

$$e^{1/z} = 1 + \frac{1}{z} + \frac{1}{2!z^2} + \frac{1}{3!z^3} + \cdots,$$

(see Example 5.16), the residue of $e^{1/z}$ at $z = 0$ is $\text{Res}[e^{1/z}, 0] = 1$. Thus,

$$\oint_C e^{1/z}dz = 2\pi i(1) = 2\pi i.$$

In order to use Cauchy's residue theorem effectively, it is essential to be able to calculate residues quickly. Using Laurent series (as in Examples 6.1 and 6.2) is not always efficient. The following theorem gives a formula for calculating residues at poles.

■ **THEOREM 6.2** If f has a pole of order m at z_0, then

$$\operatorname{Res}[f, z_0] = \lim_{z \to z_0} \frac{1}{(m-1)!} \frac{d^{m-1}}{dz^{m-1}} [(z - z_0)^m f(z)]. \tag{6.2}$$

Proof When f has a pole of order m at z_0, its Laurent expansion about z_0 takes the form

$$f(z) = \frac{a_{-m}}{(z-z_0)^m} + \frac{a_{-m+1}}{(z-z_0)^{m-1}} + \cdots + \frac{a_{-1}}{z-z_0} + a_0 + \cdots,$$

valid in some annulus $0 < |z - z_0| < R$. Thus,

$$(z-z_0)^m f(z) = a_{-m} + a_{-m+1}(z-z_0) + \cdots + a_{-1}(z-z_0)^{m-1} + \cdots$$

and

$$\frac{d^{m-1}}{dz^{m-1}} [(z-z_0)^m f(z)] = (m-1)!\, a_{-1} + m!\, a_0 (z-z_0) + \cdots.$$

Finally,

$$\lim_{z \to z_0} \frac{1}{(m-1)!} \frac{d^{m-1}}{dz^{m-1}} [(z-z_0)^m f(z)] = a_{-1}.$$ ■

EXAMPLE 6.3 ....

Evaluate $\oint_C \frac{1}{(z+5)(z^2-1)}\, dz$, where $C: |z| = 2$.

Solution Because the integrand has singularities at $z = -5$ and $z = \pm 1$, only the last two of which are interior to C,

$$\oint_C \frac{1}{(z+5)(z^2-1)}\, dz$$
$$= 2\pi i \left\{ \operatorname{Res}\left[\frac{1}{(z+5)(z^2-1)}, -1\right] + \operatorname{Res}\left[\frac{1}{(z+5)(z^2-1)}, 1\right] \right\}.$$

Now $z = 1$ is a pole of order 1, and therefore

$$\operatorname{Res}\left[\frac{1}{(z+5)(z^2-1)}, 1\right] = \lim_{z \to 1} \frac{z-1}{(z+5)(z^2-1)} = \lim_{z \to 1} \frac{1}{(z+5)(z+1)} = \frac{1}{12}.$$

Similarly, $z = -1$ is a simple pole, and

$$\operatorname{Res}\left[\frac{1}{(z+5)(z^2-1)}, -1\right] = \lim_{z \to -1} \frac{z+1}{(z+5)(z^2-1)} = \lim_{z \to -1} \frac{1}{(z+5)(z-1)} = -\frac{1}{8}.$$

Thus,

$$\oint_C \frac{1}{(z+5)(z^2-1)}\, dz = 2\pi i \left(\frac{1}{12} - \frac{1}{8}\right) = -\frac{\pi i}{12}.$$

One of the most important consequences of the residue theorem is the principle of the argument.

■ **THEOREM 6.3** (Principle of the Argument) Let f be analytic in a domain D except at a finite number of poles. If C is a simple, closed, piecewise-smooth curve in D that does not pass through any poles or zeros of f, then

$$\oint_C \frac{f'(z)}{f(z)}\, dz = 2\pi i(N - P) \tag{6.3}$$

where N and P are the sums of the orders of the zeros and poles of f inside C.

Proof According to the residue theorem,

$$\oint_C \frac{f'(z)}{f(z)}\, dz = 2\pi i \begin{Bmatrix} \text{sum of residues of } f'/f \\ \text{at its singularities interior to } C \end{Bmatrix}.$$

The possible singularities of f'/f occur at the poles and zeros of f. If z_0 is a zero of order m, then according to equation 5.23, f can be written in the form

$$f(z) = (z - z_0)^m g(z)$$

where g is analytic at z_0 and $g(z_0) \neq 0$. Consequently, $f'(z) = m(z - z_0)^{m-1} g(z) + (z - z_0)^m g'(z)$ and

$$\frac{f'(z)}{f(z)} = \frac{m}{z - z_0} + \frac{g'(z)}{g(z)}.$$

Because $g(z_0) \neq 0$, it follows that $g(z) \neq 0$ in some neighborhood of z_0. As a result, g'/g is analytic therein, and f'/f has a simple pole at z_0 with residue m.

If z_1 is a pole of order m of f, then according to equation 5.26, f can be written in the form

$$f(z) = \frac{g(z)}{(z - z_1)^m}$$

where g is analytic at z_1 and $g(z_1) \neq 0$. Consequently,

$$f'(z) = \frac{-mg(z)}{(z - z_1)^{m+1}} + \frac{g'(z)}{(z - z_1)^m},$$

and

$$\frac{f'(z)}{f(z)} = \frac{-m}{z - z_1} + \frac{g'(z)}{g(z)}.$$

Thus, f'/f has a simple pole at z_1 with residue $-m$. Clearly then, the sum of the residues of f'/f inside C is the sum of the orders of the zeros of f less the sum of the orders of the poles of f. ■

To see why this theorem is called the principle of the argument, we transform contour integral 6.3 from the z-plane to the w-plane with $w = f(z)$. It follows that

$$N - P = \frac{1}{2\pi i}\oint_C \frac{f'(z)}{f(z)}\,dz = \frac{1}{2\pi i}\oint_{f(C)} \frac{1}{w}\,dw,$$

where $f(C)$ is the image of C in the w-plane. But according to Example 4.22, the contour integral on the right is the number of times that $f(C)$ encircles $w = 0$. In other words, $N - P$ measures how many times the curve $f(C)$ encircles $w = 0$ as z traverses C once in the counterclockwise direction; that is, $N - P$ measures how many times the argument of $f(z)$ increases by 2π as z traverses C.

EXAMPLE 6.4 ....

Use the principle of the argument to evaluate $\oint_C \tan z\,dz$, where C is the circle $|z| = 5$.

Solution If we set $f(z) = \cos z$, then

$$\oint_C \tan z\,dz = -\oint_C \frac{-\sin z}{\cos z}\,dz = -\oint_C \frac{f'(z)}{f(z)}\,dz = -2\pi i(N - P),$$

where N and P are, respectively, the sums of the orders of the zeros and poles of $f(z)$ inside C. Clearly, $f(z) = \cos z$ has no poles, and its zeros are $z = (2n+1)\pi/2$ (n an integer), all of order one. Because $\pm\pi/2$ and $\pm 3\pi/2$ are the only zeros interior to C,

$$\oint_C \tan z\,dz = -2\pi i(4 - 0) = -8\pi i.$$

....

The principle of the argument leads to Rouche's theorem.

■ **THEOREM 6.4** (Rouche's Theorem) Let f and g be analytic in a domain D. If C is a simple, closed, piecewise-smooth curve in D that contains in its interior only points of D and if $|g(z)| < |f(z)|$ on C, then sums of the orders of the zeros of $f + g$ and f inside C are the same.

Proof The fact that $|g(z)| < |f(z)|$ on C implies that $|f(z)| \neq 0$ and $|f(z) + g(z)| \geq |f(z)| - |g(z)| > 0$ on C. The function $F = [f + g]/f$ does not therefore have any zeros or poles on C. If we apply the principle of the argument to F and C,

$$\frac{1}{2\pi i}\oint_C \frac{F'(z)}{F(z)}\,dz = N - P,$$

and each side may be interpreted as the number of times $w = F(z)$ encircles $w = 0$ as z traverses C. But on C,

$$|F(z)-1| = \left|\frac{f(z)+g(z)}{f(z)} - 1\right| = \left|\frac{g(z)}{f(z)}\right| < 1.$$

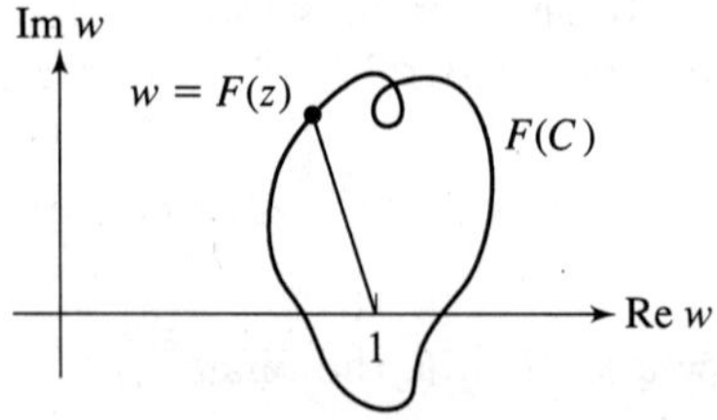

Figure 6.4

This implies that the curve $F(C)$ cannot encircle $w = 0$ (Figure 6.4), and therefore $N = P$. Because the zeros of F occur at the zeros of $f + g$ and the poles of F occur at the zeros of f, it follows that the sum of the orders of the zeros of f and $f + g$ inside C must be the same.

Rouche's theorem does not say that f and $f + g$ have the same zeros inside C, but they do have the same number of zeros inside C. ■

EXAMPLE 6.5

Show that the equation $e^z = az^n$ has exactly n solutions inside the circle $|z| = 1$ when $a > e$ is real.

Solution On $C : |z| = 1$, the modulus of $-az^n$ is $|-az^n| = a$. On the other hand,

$$|e^z| = |e^{x+yi}| = e^x \leq e,$$

because $-1 \leq x \leq 1$. Hence, $e^z - az^n$ and $-az^n$ have the same number of zeros inside C. Because $-az^n$ has a zero of order n at $z = 0$, it follows that $e^z - az^n$ also has n zeros inside C.

Exercises 6.1

In Exercises 1–22 use Cauchy's residue theorem to evaluate the contour integral.

1. Exercise 2 in Section 4.5
2. Exercise 3 in Section 4.5
3. Exercise 1 in Section 4.6
4. Exercise 2 in Section 4.6
5. Exercise 3 in Section 4.6
6. Exercise 4 in Section 4.6
7. Exercise 5 in Section 4.6
8. Exercise 6 in Section 4.6
9. Exercise 7 in Section 4.6
10. Exercise 8(c) in Section 4.6
11. Exercise 9 in Section 4.6
12. Exercise 10 in Section 4.6
13. Exercise 11 in Section 4.6
14. Exercise 12 in Section 4.6
15. $\displaystyle\oint_C \frac{\sin z}{z^2 - 4}\,dz$, where $C : |z| = 5$
16. $\displaystyle\oint_C \frac{e^z}{z(z-2)^3}\,dz$, where $C : |z| = 3$

17. $\oint_C \frac{z^2}{(z-2)^2(z^2-1)}\,dz$, where $C : |z-2| = 2$

18. $\oint_C \left(\frac{z-1}{z+1}\right)^3 dz$, where $C : |z-1| = 1$

19. $\oint_C ze^{3/z}\,dz$, where $C : |z| = 5$

20. $\oint_C \frac{1}{z^2+z+1}\,dz$, where $C : |z| = 8$

21. $\oint_C \frac{1}{z^2 \sin z}\,dz$, where $C : |z| = 1$

22. $\oint_C e^{1/z}\sin(1/z)\,dz$, where C is the square with vertices ± 1 and $\pm i$

Use the principle of the argument to evaluate the contour integral in Exercises 23–26.

23. $\oint_C \frac{2z\sin z + z^2\cos z}{z^2 \sin z}\,dz$, where $C : |z| = 4$

24. $\oint_C \frac{z^3}{z^4+1}\,dz$, where $C : |z| = 2$

25. $\oint_C \frac{z\sinh z + 3\cosh z}{z\cosh z}\,dz$, where $C : |z-i| = 7$

26. $\oint_C \frac{2z\sec^2 z^2}{\tan z^2}\,dz$, where C is the square with vertices ± 2 and $\pm 2i$

27. Use Rouche's theorem to show that $z^4 + 4z + 1$ has exactly one zero inside $|z| = 1$. *Hint:* Let $f(z) = 4z$ and $g(z) = z^4 + 1$.
28. Use Rouche's theorem to show that $z^5 + 3z^2 + 6z + 1$ has exactly one zero inside $|z| = 1$. *Hint:* Let $f(z) = 6z + 1$ and $g(z) = z^5 + 3z^2$.
29. Use Rouche's theorem to show that all zeros of $z^5 + 3z^2 + 6z + 1$ are inside $|z| = 2$. *Hint:* Let $f(z) = z^5$ and $g(z) = 3z^2 + 6z + 1$.
30. Show that all roots of the equation $z^7 - 2z^3 + 8 = 0$ satisfy $1 < |z| < 2$.
31. Show that if g is analytic inside and on $C : |z| = 1$ and if $|g(z)| < 1$ on C, then there is exactly one point z inside C for which $g(z) = z$.

6.2 Evaluation of Definite Integrals

We have used definite integrals to evaluate contour integrals. It may come as a surprise to learn that contour integrals and residues can be used to evaluate certain classes of definite integrals.

Definite Integrals Involving Trigonometric Functions

Contour integrals and residues can be useful in the evaluation of definite integrals of the form

$$\int_0^{2\pi} \frac{P(\cos\theta, \sin\theta)}{Q(\cos\theta, \sin\theta)}\, d\theta \tag{6.4}$$

where $P(\cos\theta, \sin\theta)$ and $Q(\cos\theta, \sin\theta)$ are polynomials in $\cos\theta$ and $\sin\theta$, provided Q is never equal to zero.

EXAMPLE 6.6 ....

Evaluate $\displaystyle\int_0^{2\pi} \frac{1}{2-\cos\theta}\, d\theta$.

Solution We transform the definite integral to the complex plane by setting $z = e^{\theta i}$ and $dz = ie^{\theta i} d\theta = iz\, d\theta$. As θ traces out the values 0 through 2π, z traces out the circle $C : |z| = 1$ once counterclockwise (Figure 6.5).

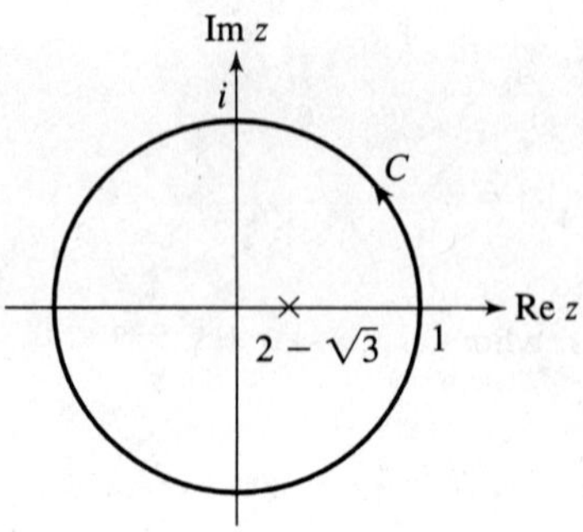

Figure 6.5

We replace $\cos\theta$ with $\cos\theta = (e^{\theta i} + e^{-\theta i})/2 = (z + z^{-1})/2$:

$$\int_0^{2\pi} \frac{1}{2-\cos\theta}\, d\theta = \oint_C \frac{1}{2 - \left(\dfrac{z+z^{-1}}{2}\right)} \frac{dz}{iz} = 2i \oint_C \frac{1}{z^2 - 4z + 1}\, dz.$$

The (real) definite integral has been replaced by a (complex) contour integral. Because $z^2 - 4z + 1 = 0$ when $z = (4 \pm \sqrt{16-4})/2 = 2 \pm \sqrt{3}$, the integrand

$$\frac{1}{z^2-4z+1} = \frac{1}{(z-2-\sqrt{3})(z-2+\sqrt{3})}$$

has simple poles at $z = 2 \pm \sqrt{3}$, only one of which is interior to C. Because

$$\operatorname{Res}\left[\frac{1}{z^2-4z+1}, 2-\sqrt{3}\right] = \lim_{z \to 2-\sqrt{3}} \frac{z-2+\sqrt{3}}{(z-2-\sqrt{3})(z-2+\sqrt{3})} = -\frac{1}{2\sqrt{3}},$$

17. $\oint_C \frac{z^2}{(z-2)^2(z^2-1)}\,dz$, where $C : |z-2| = 2$

18. $\oint_C \left(\frac{z-1}{z+1}\right)^3 dz$, where $C : |z-1| = 1$

19. $\oint_C ze^{3/z}\,dz$, where $C : |z| = 5$

20. $\oint_C \frac{1}{z^2+z+1}\,dz$, where $C : |z| = 8$

21. $\oint_C \frac{1}{z^2 \sin z} dz$, where $C : |z| = 1$

22. $\oint_C e^{1/z} \sin(1/z)\,dz$, where C is the square with vertices ± 1 and $\pm i$

Use the principle of the argument to evaluate the contour integral in Exercises 23–26.

23. $\oint_C \frac{2z \sin z + z^2 \cos z}{z^2 \sin z}\,dz$, where $C : |z| = 4$

24. $\oint_C \frac{z^3}{z^4+1}\,dz$, where $C : |z| = 2$

25. $\oint_C \frac{z \sinh z + 3 \cosh z}{z \cosh z}\,dz$, where $C : |z-i| = 7$

26. $\oint_C \frac{2z \sec^2 z^2}{\tan z^2}\,dz$, where C is the square with vertices ± 2 and $\pm 2i$

27. Use Rouche's theorem to show that $z^4 + 4z + 1$ has exactly one zero inside $|z| = 1$. *Hint:* Let $f(z) = 4z$ and $g(z) = z^4 + 1$.

28. Use Rouche's theorem to show that $z^5 + 3z^2 + 6z + 1$ has exactly one zero inside $|z| = 1$. *Hint:* Let $f(z) = 6z + 1$ and $g(z) = z^5 + 3z^2$.

29. Use Rouche's theorem to show that all zeros of $z^5 + 3z^2 + 6z + 1$ are inside $|z| = 2$. *Hint:* Let $f(z) = z^5$ and $g(z) = 3z^2 + 6z + 1$.

30. Show that all roots of the equation $z^7 - 2z^3 + 8 = 0$ satisfy $1 < |z| < 2$.

31. Show that if g is analytic inside and on $C : |z| = 1$ and if $|g(z)| < 1$ on C, then there is exactly one point z inside C for which $g(z) = z$.

6.2 Evaluation of Definite Integrals

We have used definite integrals to evaluate contour integrals. It may come as a surprise to learn that contour integrals and residues can be used to evaluate certain classes of definite integrals.

Definite Integrals Involving Trigonometric Functions

Contour integrals and residues can be useful in the evaluation of definite integrals of the form

$$\int_0^{2\pi} \frac{P(\cos\theta, \sin\theta)}{Q(\cos\theta, \sin\theta)}\, d\theta \qquad \textbf{(6.4)}$$

where $P(\cos\theta, \sin\theta)$ and $Q(\cos\theta, \sin\theta)$ are polynomials in $\cos\theta$ and $\sin\theta$, provided Q is never equal to zero.

EXAMPLE 6.6 ....

Evaluate $\displaystyle\int_0^{2\pi} \frac{1}{2-\cos\theta}\, d\theta.$

Solution We transform the definite integral to the complex plane by setting $z = e^{\theta i}$ and $dz = ie^{\theta i} d\theta = iz\, d\theta$. As θ traces out the values 0 through 2π, z traces out the circle $C : |z| = 1$ once counterclockwise (Figure 6.5).

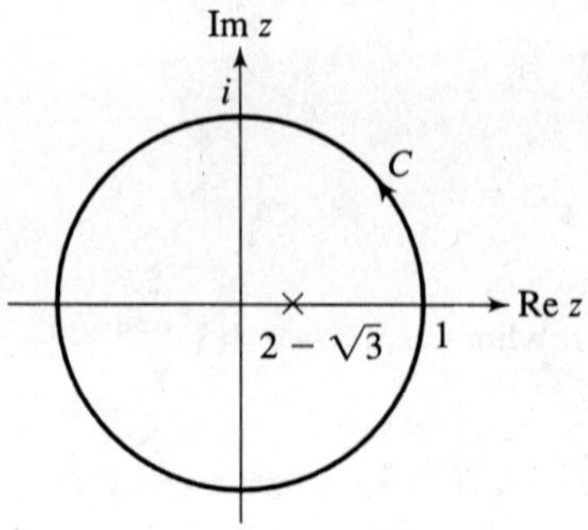

Figure 6.5

We replace $\cos\theta$ with $\cos\theta = (e^{\theta i} + e^{-\theta i})/2 = (z + z^{-1})/2$:

$$\int_0^{2\pi} \frac{1}{2-\cos\theta}\, d\theta = \oint_C \frac{1}{2 - \left(\dfrac{z+z^{-1}}{2}\right)} \frac{dz}{iz} = 2i \oint_C \frac{1}{z^2 - 4z + 1}\, dz.$$

The (real) definite integral has been replaced by a (complex) contour integral. Because $z^2 - 4z + 1 = 0$ when $z = (4 \pm \sqrt{16-4})/2 = 2 \pm \sqrt{3}$, the integrand

$$\frac{1}{z^2 - 4z + 1} = \frac{1}{(z - 2 - \sqrt{3})(z - 2 + \sqrt{3})}$$

has simple poles at $z = 2 \pm \sqrt{3}$, only one of which is interior to C. Because

$$\operatorname{Res}\left[\frac{1}{z^2 - 4z + 1}, 2 - \sqrt{3}\right] = \lim_{z \to 2-\sqrt{3}} \frac{z - 2 + \sqrt{3}}{(z - 2 - \sqrt{3})(z - 2 + \sqrt{3})} = -\frac{1}{2\sqrt{3}},$$

Cauchy's residue theorem gives

$$\int_0^{2\pi} \frac{1}{2-\cos\theta}\, d\theta = 2i\left[2\pi i\left(-\frac{1}{2\sqrt{3}}\right)\right] = \frac{2\pi}{\sqrt{3}}.$$

This method can be applied to evaluate integrals over intervals other than $[0, 2\pi]$, but residue theory may not be applicable.

EXAMPLE 6.7

Evaluate $\displaystyle\int_0^{\pi} \frac{1}{2-\cos\theta}\, d\theta.$

Solution The transformation $z = e^{\theta i}$ transforms the real integral into the contour integral of Example 6.6 over the semicircle in Figure 6.6. This time we use the partial fraction decomposition of the integrand to write

$$\int_0^{\pi} \frac{1}{2-\cos\theta}\, d\theta = 2i\int_C \frac{1}{z^2-4z+1}\, dz = 2i\int_C \left(\frac{1/(2\sqrt{3})}{z-2-\sqrt{3}} + \frac{-1/(2\sqrt{3})}{z-2+\sqrt{3}}\right) dz.$$

In the domain of Figure 6.6, the integrand has an antiderivative, and hence

$$\int_0^{\pi} \frac{1}{2-\cos\theta}\, d\theta = \frac{2i}{2\sqrt{3}}\{\log_\phi (z-2-\sqrt{3}) - \log_\phi (z-2+\sqrt{3})\}_1^{-1},$$

where we choose branches of the logarithm functions with branch cuts $\phi = -\pi/2$. Then

$$\begin{aligned}
\int_0^{\pi} \frac{1}{2-\cos\theta}\, d\theta &= \frac{i}{\sqrt{3}}\left[\log_\phi (-3-\sqrt{3}) - \log_\phi (-3+\sqrt{3})\right.\\
&\quad \left. - \log_\phi (-1-\sqrt{3}) + \log_\phi (-1+\sqrt{3})\right]\\
&= \frac{i}{\sqrt{3}}\Big\{[\ln (3+\sqrt{3}) + \pi i] - [\ln (3-\sqrt{3}) + \pi i]\\
&\quad - [\ln (1+\sqrt{3}) + \pi i] + [\ln (\sqrt{3}-1)]\Big\}\\
&= \frac{i}{\sqrt{3}}\left[-\pi i + \ln\left(\frac{3+\sqrt{3}}{3-\sqrt{3}}\frac{\sqrt{3}-1}{\sqrt{3}+1}\right)\right]\\
&= \frac{\pi}{\sqrt{3}} + \frac{i}{\sqrt{3}}\ln\left(\frac{2\sqrt{3}}{2\sqrt{3}}\right)\\
&= \frac{\pi}{\sqrt{3}}.
\end{aligned}$$

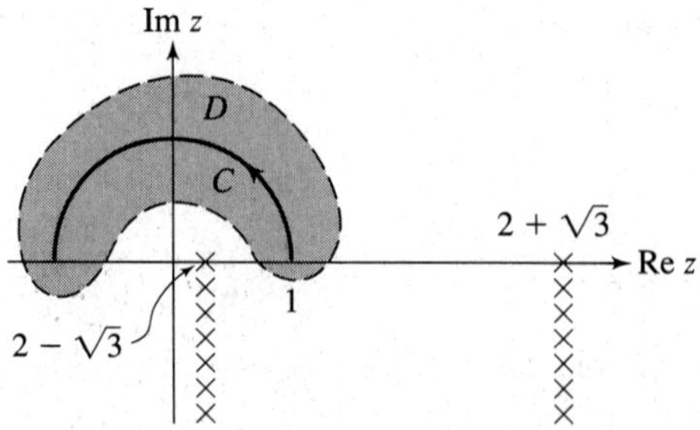

Figure 6.6

Real Improper Integrals

Residues can also be effective in evaluation of improper integrals that have infinite upper or lower limits.

EXAMPLE 6.8

Evaluate $\displaystyle\int_{-\infty}^{\infty} \frac{1}{1+x^4}\,dx.$

Solution It is fairly clear that were we to evaluate the contour integral

$$\oint_C \frac{1}{1+z^4}\,dz,$$

where C is shown in Figure 6.7(a), and were we to let $R \to \infty$, then that part of the contour integral along the real axis would give rise to the required improper integral. Let us consider this contour integral then.

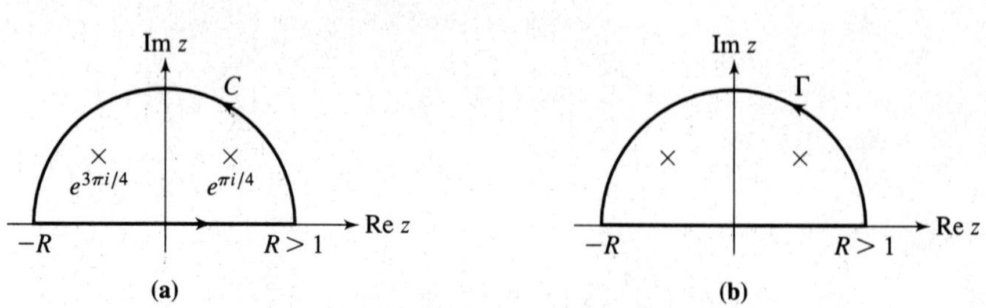

Figure 6.7

The integrand $(1+z^4)^{-1}$ has simple poles at the four fourth roots of -1,

$$e^{\pi i/4}, \qquad e^{3\pi i/4}, \qquad e^{5\pi i/4}, \qquad e^{7\pi i/4},$$

only the first two of which are interior to C. L'Hôpital's rule (Theorem 5.21) gives

$$\operatorname{Res}\left[\frac{1}{1+z^4}, e^{\pi i/4}\right] = \lim_{z\to e^{\pi i/4}} \frac{z-e^{\pi i/4}}{1+z^4} = \lim_{z\to e^{\pi i/4}} \frac{1}{4z^3} = \frac{1}{4e^{3\pi i/4}} = -\frac{\sqrt{2}}{8}(1+i).$$

Similarly, $\operatorname{Res}\left[\frac{1}{1+z^4}, e^{3\pi i/4}\right] = \frac{\sqrt{2}}{8}(1-i)$. By Cauchy's residue theorem,

$$\oint_C \frac{1}{1+z^4}\,dz = 2\pi i\left[-\frac{\sqrt{2}}{8}(1+i) + \frac{\sqrt{2}}{8}(1-i)\right] = \frac{\pi}{\sqrt{2}}.$$

Suppose we now divide C into a semicircular part Γ and a straight-line part (Figure 6.7(b)). Then

$$\frac{\pi}{\sqrt{2}} = \int_{-R}^{R} \frac{1}{1+x^4}\,dx + \int_\Gamma \frac{1}{1+z^4}\,dz.$$

If we set $z = Re^{\theta i}$, $0 \le \theta \le \pi$, on Γ, then inequality 1.33 on the semicircle gives

$$\left|\frac{1}{1+z^4}\right| \le \frac{1}{|z^4|-1} = \frac{1}{R^4-1}.$$

Hence by property 4.22,

$$\left|\int_\Gamma \frac{1}{1+z^4}\,dz\right| \le \frac{1}{R^4-1}(\pi R).$$

It is clear that the limit of this expression is zero as $R \to \infty$, and therefore

$$\frac{\pi}{\sqrt{2}} = \lim_{R\to\infty}\left(\int_{-R}^{R} \frac{1}{1+x^4}\,dx + \int_\Gamma \frac{1}{1+z^4}\,dz\right) = \int_{-\infty}^{\infty} \frac{1}{1+x^4}\,dx.$$

This example has illustrated that the contour integral of $1/(1+z^4)$ around the curve C of Figure 6.7 can be used to evaluate the improper integral of $1/(1+x^4)$ from negative infinity to infinity. The real difficulty in such problems is the choice of contour and the choice of integrand. We now do two more examples to illustrate some of these difficulties.

EXAMPLE 6.9

Evaluate $\displaystyle\int_0^\infty \frac{\cos x}{x^2+1}\,dx$.

Solution We might consider the contour integral

$$\oint_C \frac{\cos z}{z^2+1}\,dz$$

around the contour in Figure 6.7. Certainly along the straight-line portion of C the integrand reduces to $\cos x/(x^2+1)$, but if we set $z = x + yi$ along Γ, then on the semicircle,

$$\left|\frac{\cos z}{z^2+1}\right| = \left|\frac{e^{(x+yi)i} + e^{-(x+yi)i}}{2(z^2+1)}\right| = \left|\frac{e^{-y+xi} + e^{y-xi}}{2(z^2+1)}\right|,$$

which becomes infinite as $|z| \to \infty$. We shall not, therefore, be able to show that the contour integral along Γ approaches zero as $|z| \to \infty$, as in Example 6.8. This means that the choice of $\cos z/(z^2+1)$ was perhaps not a convenient one. Consider instead

$$\oint_C \frac{e^{zi}}{z^2+1}\, dz,$$

where C is again the contour in Figure 6.7. Because the integrand has simple poles at $z = \pm i$, only $z = i$ being interior to C,

$$\oint_C \frac{e^{zi}}{z^2+1}\, dz = 2\pi i \operatorname{Res}\left[\frac{e^{zi}}{z^2+1}, i\right] = 2\pi i \lim_{z \to i}\left[\frac{(z-i)e^{zi}}{(z-i)(z+i)}\right] = 2\pi i \frac{e^{i(i)}}{2i} = \frac{\pi}{e}.$$

Thus,

$$\begin{aligned}
\frac{\pi}{e} &= \int_{-R}^{R} \frac{e^{xi}}{x^2+1}\, dx + \int_\Gamma \frac{e^{zi}}{z^2+1}\, dz \\
&= \int_{-R}^{R} \frac{\cos x}{x^2+1}\, dx + i\int_{-R}^{R} \frac{\sin x}{x^2+1}\, dx + \int_\Gamma \frac{e^{zi}}{z^2+1}\, dz \\
&= \int_{-R}^{R} \frac{\cos x}{x^2+1}\, dx + \int_\Gamma \frac{e^{zi}}{z^2+1}\, dz \qquad \text{(because } \sin x/(x^2+1) \text{ is an odd function).}
\end{aligned}$$

If we set $z = x + yi = Re^{\theta i}$ on Γ, then on this semicircle,

$$\begin{aligned}
\left|\frac{e^{zi}}{z^2+1}\right| &\le \frac{|e^{(x+yi)i}|}{|z^2|-1} \qquad \text{(by inequality 1.33)} \\
&= \frac{e^{-y}}{R^2-1} \\
&\le \frac{1}{R^2-1} \qquad \text{(because } y \ge 0\text{).}
\end{aligned}$$

Hence, by inequality 4.22,

$$\left|\int_\Gamma \frac{e^{zi}}{z^2+1}\, dz\right| \le \frac{1}{R^2-1}(\pi R).$$

Because the limit of this expression is zero as $R \to \infty$, it follows that

$$\frac{\pi}{e} = \lim_{R\to\infty}\left(\int_{-R}^{R} \frac{\cos x}{x^2+1}\, dx + \int_\Gamma \frac{e^{zi}}{z^2+1}\, dz\right) = \int_{-\infty}^{\infty} \frac{\cos x}{x^2+1}\, dx = 2\int_0^{\infty} \frac{\cos x}{x^2+1}\, dx.$$

Finally,

$$\int_0^\infty \frac{\cos x}{x^2+1}\,dx = \frac{\pi}{2e}.$$

This example illustrated that we do not always replace x's by z's to obtain the appropriate contour integral. The following example indicates that the choice of contour may not always be obvious.

EXAMPLE 6.10

Evaluate $\displaystyle\int_0^\infty \frac{1}{1+x^3}\,dx$.

Solution Based on Example 6.8 we should perhaps consider

$$\oint_C \frac{1}{1+z^3}\,dz,$$

where C is some appropriate contour. Clearly a part of C should be the positive real axis and possibly a circular arc of radius $R > 1$. But we cannot take a semicircle as in Example 6.8, because $1/(1+z^3)$ has a singularity at $z = -1$ on the negative real axis. What we should like to do then is choose some other line eminating from the origin, such as $z = re^{\phi i}$, $0 < \phi < \pi$, which leads to a simple solution (Figure 6.8).

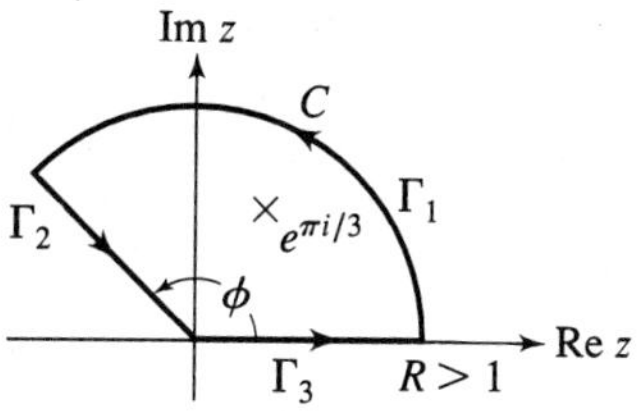

Figure 6.8

The integrand has simple poles at $z = e^{\pi i/3}$, -1, $e^{5\pi i/3}$. Suppose for the moment we stipulate that ϕ be in the interval $\pi/3 < \phi < \pi$ but leave it otherwise arbitrary. Cauchy's residue theorem and L'Hôpital's rule give

$$\oint_C \frac{1}{1+z^3}\,dz = 2\pi i \operatorname{Res}\left[\frac{1}{1+z^3}, e^{\pi i/3}\right] = 2\pi i \lim_{z\to e^{\pi i/3}} \frac{z-e^{\pi i/3}}{1+z^3}$$

$$= 2\pi i \lim_{z\to e^{\pi i/3}} \frac{1}{3z^2} = \frac{2\pi i}{3e^{2\pi i/3}} = \frac{(\sqrt{3}-i)\pi}{3}.$$

Thus,

$$\frac{(\sqrt{3}-i)\pi}{3} = \int_{\Gamma_1} \frac{1}{1+z^3}\,dz + \int_{\Gamma_2} \frac{1}{1+z^3}\,dz + \int_{\Gamma_3} \frac{1}{1+z^3}\,dz.$$

If we set $z = Re^{\theta i}$ on Γ_1, then on this arc

$$\left|\frac{1}{1+z^3}\right| \leq \frac{1}{|z|^3 - 1} = \frac{1}{R^3 - 1},$$

and hence

$$\left|\int_{\Gamma_1} \frac{1}{1+z^3}\,dz\right| \leq \frac{1}{R^3-1}(R\phi).$$

The limit of this expression is zero as $R \to \infty$. Thus,

$$\begin{aligned}\frac{(\sqrt{3}-i)\pi}{3} &= \lim_{R\to\infty}\left(\int_{\Gamma_1} \frac{1}{1+z^3}\,dz + \int_{\Gamma_2} \frac{1}{1+z^3}\,dz + \int_{\Gamma_3} \frac{1}{1+z^3}\,dz\right)\\ &= \int_{\Gamma_2} \frac{1}{1+z^3}\,dz + \int_0^\infty \frac{1}{1+x^3}\,dx,\end{aligned}$$

where $\Gamma_2 : z = re^{\phi i}$, $\infty > r \geq 0$ (and $\pi/3 < \phi < \pi$). Our problem then is to choose ϕ in order that the contour integral along Γ_2 can be evaluated. Because $z = re^{\phi i}$ on Γ_2,

$$\int_{\Gamma_2} \frac{1}{1+z^3}\,dz = \int_\infty^0 \frac{1}{1+r^3e^{3\phi i}}\,e^{\phi i}\,dr = -e^{\phi i}\int_0^\infty \frac{1}{1+r^3e^{3\phi i}}\,dr.$$

If we choose $\phi = 2\pi/3$, then

$$\int_{\Gamma_2} \frac{1}{1+z^3}\,dz = -e^{2\pi i/3}\int_0^\infty \frac{1}{1+r^3}\,dr,$$

and

$$\begin{aligned}\frac{(\sqrt{3}-i)\pi}{3} &= -e^{2\pi i/3}\int_0^\infty \frac{1}{1+r^3}\,dr + \int_0^\infty \frac{1}{1+x^3}\,dx\\ &= \left[-\left(-\frac{1}{2}+\frac{\sqrt{3}i}{2}\right)+1\right]\int_0^\infty \frac{1}{1+x^3}\,dx.\end{aligned}$$

Thus,

$$\int_0^\infty \frac{1}{1+x^3}\,dx = \frac{(\sqrt{3}-i)\pi}{3}\,\frac{2}{3-\sqrt{3}i} = \frac{2\pi}{3\sqrt{3}}.$$

....

You are asked to prove general results concerning improper integrals of the types in Examples 6.8–6.10 in Exercises 23 and 26.

Exercises 6.2

In Exercises 1–13 use a contour integral to evaluate the definite integral.

1. $\displaystyle\int_0^{2\pi} \frac{1}{3-\sin\theta}\,d\theta$

2. $\displaystyle\int_0^{2\pi} \frac{1}{3+2\cos\theta}\,d\theta$

3. $\displaystyle\int_0^{2\pi} \frac{1}{6+5\sin\theta}\,d\theta$

4. $\displaystyle\int_0^{2\pi} \frac{1}{\sin\theta+2\cos\theta+4}\,d\theta$

5. $\displaystyle\int_0^{2\pi} \frac{1}{4\cos^2\theta+3}\,d\theta$

6. $\displaystyle\int_0^{\pi} \frac{1}{3+2\cos\theta}\,d\theta$

7. $\displaystyle\int_0^{\pi/2} \frac{1}{6+5\sin\theta}\,d\theta$

8. $\displaystyle\int_0^{\pi/2} \frac{1}{3-\sin\theta}\,d\theta$

9. $\displaystyle\int_{-\pi/2}^{0} \frac{1}{5+2\cos\theta}\,d\theta$

10. $\displaystyle\int_0^{2} \frac{1}{3+2\sin\theta}\,d\theta$

11. $\displaystyle\int_0^{1} \frac{1}{4+3\cos\theta}\,d\theta$

12. $\displaystyle\int_{-\pi}^{0} \frac{1}{4+\cos\theta}\,d\theta$

13. $\displaystyle\int_{-\pi}^{\pi/2} \frac{1}{5+2\sin\theta}\,d\theta$

In Exercises 14–19 use a contour integral to evaluate the improper integral.

14. $\displaystyle\int_0^{\infty} \frac{1}{2+x^2}\,dx$

15. $\displaystyle\int_{-\infty}^{\infty} \frac{x^2}{(x^2+1)(x^2+4)}\,dx$

16. $\displaystyle\int_{-\infty}^{\infty} \frac{1}{3+x^6}\,dx$

17. $\displaystyle\int_0^{\infty} \frac{x}{1+x^3}\,dx$

18. $\displaystyle\int_{-\infty}^{\infty} \frac{\sin x}{x^2+2x+2}\,dx$

19. $\displaystyle\int_0^{\infty} \frac{x^2\cos x}{(x^2+9)^2}\,dx$

20. Show that $\displaystyle\int_0^{2\pi} \cos^{2n}\theta\,d\theta = \frac{(2n)!\,\pi}{2^{2n-1}(n!)^2}$.

21. **(a)** Use the substitution $u = 1/x$ to show that

$$I = \int_0^1 \frac{1+x^2}{1+x^4}\,dx = \int_1^{\infty} \frac{1+u^2}{1+u^4}\,du$$

and hence

$$I = \frac{1}{4}\int_{-\infty}^{\infty} \frac{1+x^2}{1+x^4}\,dx.$$

(b) Now use contour integration to calculate I.

22. Use contour integrals to prove the result of Exercise 3 in Section 4.8.

23. Example 6.8 and Exercises 14–17 are examples of improper integrals of the form

$$\int_{-\infty}^{\infty} \frac{P(x)}{Q(x)}\, dx,$$

where $P(x)$ and $Q(x)$ are polynomials (of degrees m and n) and $Q(x) \neq 0$ for all real x. Show that when $n \geq m+2$,

$$\int_{-\infty}^{\infty} \frac{P(x)}{Q(x)}\, dx = 2\pi i \begin{Bmatrix} \text{sum of the residues of } P(z)/Q(z) \text{ at} \\ \text{its poles in the half-plane } \operatorname{Im} z > 0 \end{Bmatrix}.$$

With this result, it is no longer necessary to introduce the contour of Figure 6.7. The fact that $n \geq m+2$ guarantees that the contour integral along Γ vanishes as $R \to \infty$.

Use Exercise 23 to evaluate each improper integral in Exercises 24 and 25.

24. $\displaystyle\int_{-\infty}^{\infty} \frac{1}{(x^2+4x+5)^2}\, dx$

25. $\displaystyle\int_{-\infty}^{\infty} \frac{x^2+3}{(x^2+1)(x^2-x+1)}\, dx$

26. Example 6.9 and Exercises 18 and 19 are examples of improper integrals of the form

$$\int_{-\infty}^{\infty} \frac{P(x)\cos ax}{Q(x)}\, dx \quad \text{or} \quad \int_{-\infty}^{\infty} \frac{P(x)\sin ax}{Q(x)}\, dx,$$

where $P(x)$ and $Q(x)$ are polynomials (of degrees m and n), $a > 0$ is real, and $Q(x) \neq 0$ for any real x.

(a) Use the following figure to verify that

$$\sin\theta \geq \frac{2\theta}{\pi}, \qquad 0 \leq \theta \leq \pi/2,$$

called Jordan's inequality. Prove that for $a > 0$,

$$\int_0^{\pi} e^{-aR\sin\theta}\, d\theta \leq \frac{\pi}{aR}(1 - e^{-aR}).$$

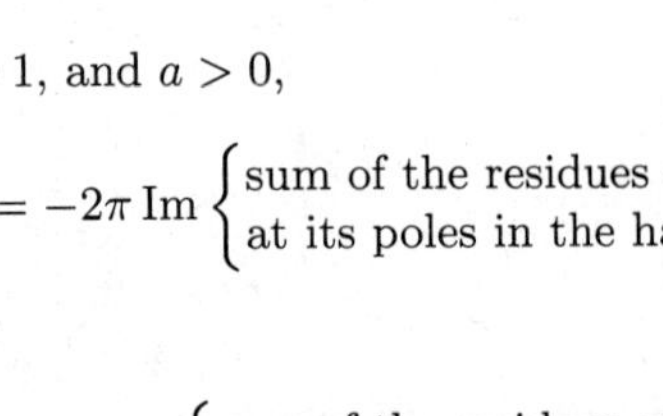

(b) Show that when $n \geq m+1$, and $a > 0$,

$$\int_{-\infty}^{\infty} \frac{P(x)\cos ax}{Q(x)}\, dx = -2\pi \operatorname{Im} \begin{Bmatrix} \text{sum of the residues of } P(z)e^{azi}/Q(z) \\ \text{at its poles in the half-plane } \operatorname{Im} z > 0 \end{Bmatrix}$$

and

$$\int_{-\infty}^{\infty} \frac{P(x)\sin ax}{Q(x)}\, dx = 2\pi \operatorname{Re} \begin{Bmatrix} \text{sum of the residues of } P(z)e^{azi}/Q(z) \\ \text{at its poles in the half-plane } \operatorname{Im} z > 0 \end{Bmatrix}.$$

Hint: Use the contour in Figure 6.7, and show that R can be chosen sufficiently large that on Γ,

$$\left|\frac{P(z)e^{azi}}{Q(z)}\right| \leq \frac{(|a_m|R^m + \cdots + |a_0|)e^{-aR\sin\theta}}{|b_n|R^n - \cdots - |b_0|},$$

where $P(z) = a_m z^m + \cdots + a_0$ and $Q(z) = b_n z^n + \cdots + b_0$. Now use the result in (a).

In Exercises 27–30 use the result in Exercise 26 to evaluate the improper integral.

27. $\displaystyle\int_0^\infty \frac{x\sin 2x}{x^2+5}\,dx$

28. $\displaystyle\int_{-\infty}^\infty \frac{x^2\cos 3x}{x^4+4}\,dx$

29. $\displaystyle\int_0^\infty \frac{\sin^2 x}{x^2+1}\,dx$

30. $\displaystyle\int_{-\infty}^\infty \frac{\cos x}{(x+a)^2+b^2}\,dx$

31. Evaluate $\displaystyle\int_0^{\pi/2} \frac{1}{(3-\sin\theta)^2}\,d\theta.$

32. Show that

$$\int_0^{2\pi} \frac{1}{d + a\cos\theta + b\sin\theta}\,d\theta = \frac{2\pi \operatorname{sgn} d}{\sqrt{d^2-a^2-b^2}}, \qquad \text{where } \operatorname{sgn} d = \begin{cases} 1, & d>0 \\ -1, & d<0 \end{cases}$$

when a, b, and d are real with $d^2 > a^2 + b^2$.

6.3 More Improper Integrals by Residues

In this section we use residues to evaluate more difficult improper integrals. Before doing so, however, we discuss the concept of Cauchy's principal value for an improper integral. The improper integral of a continuous function $f(x)$ over the real line is defined by

$$\int_{-\infty}^\infty f(x)\,dx = \lim_{a\to-\infty}\int_a^0 f(x)\,dx + \lim_{b\to\infty}\int_0^b f(x)\,dx, \tag{6.5}$$

provided both limits exist. The **Cauchy principal value** of the integral is defined as

$$\int_{-\infty}^\infty f(x)\,dx = \lim_{b\to\infty}\int_{-b}^b f(x)\,dx, \tag{6.6}$$

provided this limit exists. When an improper integral has a value in the ordinary sense of 6.5, it must have an identical Cauchy principal value. However, an improper integral may have a Cauchy principal value but no value in the ordinary sense. For example, the first of the integrals

$$\int_{-\infty}^\infty \frac{x}{x^4+1}\,dx, \qquad \int_{-\infty}^\infty \frac{x}{x^2+1}\,dx$$

has value zero in both senses; the second has a Cauchy principal value of zero but does not converge in the sense of 6.5.

Cauchy principal values are also defined for improper integrals due to discontinuities in integrands. When the only discontinuity of a function $f(x)$ in the interval $a < x < b$ is $x = d$, its improper integral over the interval $a \le x \le b$ is defined by

$$\int_a^b f(x)\,dx = \lim_{\epsilon \to 0+} \int_a^{d-\epsilon} f(x)\,dx + \lim_{\epsilon \to 0+} \int_{d+\epsilon}^b f(x)\,dx, \tag{6.7}$$

provided both limits exist. The integral is said to have a **Cauchy principal value** when the limit

$$\int\!\!\!\!\!\!\!\times_a^b f(x)\,dx = \lim_{\epsilon \to 0+} \left[\int_a^{d-\epsilon} f(x)\,dx + \int_{d+\epsilon}^b f(x)\,dx \right] \tag{6.8}$$

exists. Once again existence of 6.7 implies that of 6.8, but not conversely. The integral of $1/x$ from $x = -1$ to $x = 1$ has a Cauchy principal value of zero but no value in the sense of 6.7.

Examination of proofs in Exercises 23 and 26 in Section 6.2 shows that convergence of the improper integrals was verified only in the Cauchy principal value sense. However, with the degrees of the polynomials indicated, convergence in the ordinary sense is guaranteed also. In this section we will be careful to distinguish those improper integrals that converge in the sense of Cauchy's principal value but do not converge in the ordinary sense. The first type of integral we consider is

$$\int_{-\infty}^{\infty} \frac{P(x)}{Q(x)}\,dx, \tag{6.9}$$

where $P(x)$ and $Q(x)$ are polynomials and $Q(x)$ has real, simple zeros. We illustrate with the following example.

EXAMPLE 6.11 ....

Evaluate $\int\!\!\!\!\!\!\times_{-\infty}^{\infty} \frac{x}{x^3 - 1}\,dx$.

Solution This integral does not have a value in the ordinary sense, as can be seen by writing

$$\int_{-\infty}^{\infty} \frac{x}{x^3-1}\,dx = \int_{-\infty}^{0} \frac{x}{x^3-1}\,dx + \int_0^1 \frac{x}{x^3-1}\,dx + \int_1^2 \frac{x}{x^3-1}\,dx + \int_2^{\infty} \frac{x}{x^3-1}\,dx.$$

When the partial fraction decomposition of the integrand is substituted into the second of these,

$$\int_0^1 \frac{x}{x^3-1}\,dx = \frac{1}{3}\int_0^1 \left(\frac{1}{x-1} + \frac{1-x}{x^2+x+1} \right) dx,$$

and the integral of $1/(x-1)$ diverges at $x = 1$. So also will the third integral diverge at $x = 1$, but the first and last integrals converge. In other words, the Cauchy principal value in this example is only in the sense of the discontinuity at

$x = 1$, not as regards the infinite limits. To evaluate the integral in this sense we consider the contour integral

$$\oint_C \frac{z}{z^3 - 1}\, dz$$

where C is the contour in Figure 6.9.

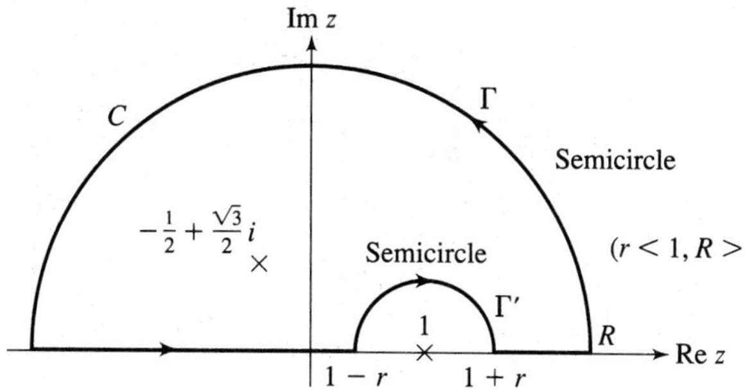

Figure 6.9

The integrand has a simple pole at $z = 1$ (on the real axis) and two other simple poles at $z = (-1/2) \pm (\sqrt{3}/2)i$, but only $-1/2 + \sqrt{3}i/2$ is inside C. By Cauchy's residue theorem

$$\oint_C \frac{z}{z^3 - 1}\, dz = 2\pi i \operatorname{Res}\left[\frac{z}{z^3 - 1}, -\frac{1}{2} + \frac{\sqrt{3}i}{2}\right]$$

$$= 2\pi i \lim_{z \to -1/2 + \sqrt{3}i/2} \frac{(z + 1/2 - \sqrt{3}i/2)z}{z^3 - 1}.$$

When we use L'Hôpital's rule on this limit,

$$\oint_C \frac{z}{z^3 - 1}\, dz = 2\pi i \lim_{z \to -1/2 + \sqrt{3}i/2} \frac{(z + \frac{1}{2} - \sqrt{3}i/2) + z}{3z^2} = 2\pi i \lim_{z \to -1/2 + \sqrt{3}i/2} \frac{1}{3z}$$

$$= \frac{2\pi i}{3(-\frac{1}{2} + \sqrt{3}i/2)} = \frac{\pi}{3}(\sqrt{3} - i).$$

On the semicircle Γ of radius $R > 2$, $\left|\dfrac{z}{z^3 - 1}\right| \le \dfrac{R}{R^3 - 1}$, and it follows that

$$\left|\int_\Gamma \frac{z}{z^3 - 1}\, dz\right| \le \frac{R}{R^3 - 1}(\pi R).$$

The limit of this expression as $R \to \infty$ is zero. The partial fraction decomposition of $z/(z^3 - 1)$ gives

$$\int_{\Gamma'} \frac{z}{z^3 - 1}\, dz = -\frac{1}{3}\int_{-\Gamma'} \frac{1}{z - 1}\, dz + \frac{1}{3}\int_{\Gamma'} \frac{1 - z}{z^2 + z + 1}\, dz.$$

If we set $z = 1 + re^{\theta i}$, $0 \le \theta \le \pi$, on $-\Gamma'$,

$$-\frac{1}{3}\int_{-\Gamma'} \frac{1}{z-1}\, dz = -\frac{1}{3}\int_0^{\pi} \frac{1}{re^{\theta i}} rie^{\theta i}\, d\theta = -\frac{\pi i}{3}.$$

Because $(1-z)/(z^2+z+1)$ is analytic in the domain $|z-1| < 1$, it must be bounded therein; that is, there exists a constant M such that for z in $|z-1| < 1$, $|(1-z)/(z^2+z+1)| \le M$. Because Γ' lies within this domain, it follows that

$$\left|\int_{\Gamma'} \frac{1-z}{z^2+z+1}\, dz\right| \le M\pi r,$$

and this expression approaches zero as $r \to 0$. Consequently,

$$\lim_{r\to 0}\int_{\Gamma'} \frac{z}{z^3-1}\, dz = -\frac{\pi i}{3}.$$

It now follows that

$$\begin{aligned}
&\frac{\pi}{3}(\sqrt{3} - i) \\
&\quad = \lim_{\substack{r\to 0 \\ R\to\infty}} \oint_C \frac{z}{z^3-1}\, dz \\
&\quad = \lim_{\substack{r\to 0 \\ R\to\infty}} \left(\int_{-R}^{1-r} \frac{x}{x^3-1}\, dx + \int_{1+r}^{R} \frac{x}{x^3-1}\, dx + \int_{\Gamma} \frac{z}{z^3-1}\, dz + \int_{\Gamma'} \frac{z}{z^3-1}\, dz\right) \\
&\quad = \int_{-\infty}^{\infty} \frac{x}{x^3-1}\, dx + 0 - \frac{\pi i}{3},
\end{aligned}$$

and, therefore, $\displaystyle\int_{-\infty}^{\infty} \frac{x}{x^3-1}\, dx = \frac{\pi}{\sqrt{3}}$. ....

General results for integrals of the type of 6.9 are discussed in Exercise 1 below. Integrals of the form

$$\int_{-\infty}^{\infty} \frac{P(x)}{Q(x)} \cos ax\, dx \quad \text{and} \quad \int_{-\infty}^{\infty} \frac{P(x)}{Q(x)} \sin ax\, dx, \tag{6.10}$$

where $P(x)$ and $Q(x)$ are polynomials and $Q(x)$ has simple, real zeros, can be handled in a similar way.

EXAMPLE 6.12 ....

Evaluate $\displaystyle\int_0^{\infty} \frac{\cos 3x}{16 - x^4}\, dx.$

Solution Because the integrand is an even function, we may write

$$\int_0^{\infty} \frac{\cos 3x}{16-x^4}\, dx = \frac{1}{2}\int_{-\infty}^{\infty} \frac{\cos 3x}{16-x^4}\, dx.$$

Improper integrals for the discontinuities of the latter integral at $x = \pm 2$ do not converge in the ordinary sense, so this integral must be regarded in the sense of Cauchy's principal value. (The integral does exist in the ordinary sense as regards the infinite limits.) We consider the contour integral

$$\oint_C \frac{e^{3zi}}{16 - z^4}\, dz,$$

where C is the curve in Figure 6.10.

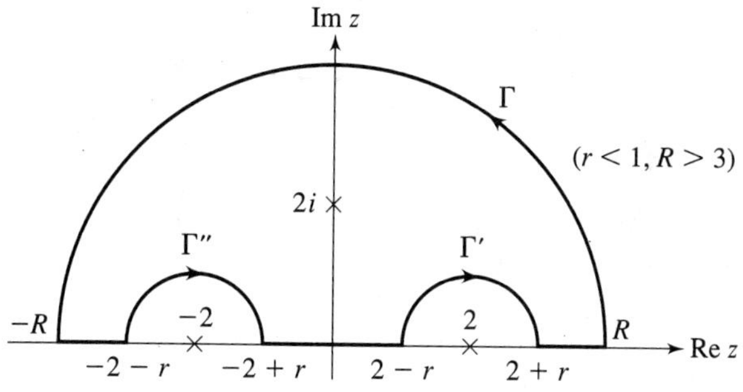

Figure 6.10

The integrand has simple poles at $z = \pm 2$ and $z = \pm 2i$, but only $2i$ is interior to C. By Cauchy's residue theorem,

$$\oint_C \frac{e^{3zi}}{16 - z^4}\, dz = 2\pi i \operatorname{Res}\left[\frac{e^{3zi}}{16 - z^4}, 2i\right] = 2\pi i \lim_{z \to 2i} \frac{e^{3zi}(z - 2i)}{16 - z^4}$$

$$= 2\pi i \lim_{z \to 2i} \frac{e^{3zi} + 3i(z - 2i)e^{3zi}}{-4z^3} = 2\pi i \lim_{z \to 2i} \frac{e^{3zi}}{-4z^3}$$

$$= 2\pi i \frac{e^{-6}}{-4(2i)^3} = \frac{\pi e^{-6}}{16}.$$

On the semicircle Γ of radius $R > 3$,

$$\left|\frac{e^{3zi}}{16 - z^4}\right| = \left|\frac{e^{3(x+yi)i}}{z^4 - 16}\right| \le \frac{e^{-3y}}{R^4 - 16} \le \frac{1}{R^4 - 16},$$

and it follows that

$$\left|\int_\Gamma \frac{e^{3zi}}{16 - z^4}\, dz\right| \le \frac{\pi R}{R^4 - 16}.$$

The limit of this expression as $R \to \infty$ is zero. Because

$$\operatorname{Res}\left[\frac{e^{3zi}}{16 - z^4}, 2\right] = \lim_{z \to 2} \frac{(z - 2)e^{3zi}}{(2 - z)(2 + z)(4 + z^2)} = -\frac{e^{6i}}{32},$$

$e^{3zi}/(16-z^4)$ has a Laurent expansion about $z=2$ of the form

$$\frac{e^{3zi}}{16-z^4} = \frac{-e^{6i}/32}{z-2} + g(z),$$

where g is analytic in the domain $|z-2|<1$. (This expression takes the place of the partial fraction decomposition in Example 6.11.) The contour integral along Γ' can therefore be expressed in the form

$$\int_{\Gamma'} \frac{e^{3zi}}{16-z^4}\,dz = \int_{\Gamma'} \frac{-e^{6i}/32}{z-2}\,dz + \int_{\Gamma'} g(z)\,dz = \frac{e^{6i}}{32}\int_{-\Gamma'} \frac{1}{z-2}\,dz + \int_{\Gamma'} g(z)\,dz.$$

If we set $z = 2 + re^{\theta i}$, $0 \le \theta \le \pi$, on $-\Gamma'$, then

$$\frac{e^{6i}}{32}\int_{-\Gamma'} \frac{1}{z-2}\,dz = \frac{e^{6i}}{32}\int_0^{\pi} \frac{1}{re^{\theta i}} ire^{\theta i}\,d\theta = \frac{\pi i e^{6i}}{32}.$$

Because g is analytic in $|z-2|<1$, it is bounded therein; that is, there exists a constant M such that for z in $|z-2|<1$, $|g(z)|<M$. Thus,

$$\left|\int_{\Gamma'} g(z)\,dz\right| \le M\pi r,$$

and this expression approaches zero as $r \to 0$. Consequently,

$$\lim_{r\to 0}\int_{\Gamma'} \frac{e^{3zi}}{16-z^4}\,dz = \frac{\pi i e^{6i}}{32}.$$

It can be shown similarly that

$$\lim_{r\to 0}\int_{\Gamma''} \frac{e^{3zi}}{16-z^4}\,dz = \frac{-\pi i e^{-6i}}{32}.$$

It now follows that

$$\begin{aligned}
\frac{\pi e^{-6}}{16} &= \lim_{\substack{r\to 0\\ R\to\infty}} \oint_C \frac{e^{3zi}}{16-z^4}\,dz \\
&= \lim_{\substack{r\to 0\\ R\to\infty}} \left(\int_{-R}^{-2-r} \frac{e^{3xi}}{16-x^4}\,dx + \int_{-2+r}^{2-r} \frac{e^{3xi}}{16-x^4}\,dx + \int_{2+r}^{R} \frac{e^{3xi}}{16-x^4}\,dx \right. \\
&\qquad \left. + \int_{\Gamma} \frac{e^{3zi}}{16-z^4}\,dz + \int_{\Gamma'} \frac{e^{3zi}}{16-z^4}\,dz + \int_{\Gamma''} \frac{e^{3zi}}{16-z^4}\,dz \right) \\
&= \int_{-\infty}^{\infty} \frac{e^{3xi}}{16-x^4}\,dx + 0 + \frac{\pi i e^{6i}}{32} - \frac{\pi i e^{-6i}}{32},
\end{aligned}$$

and when we take real parts,

$$\frac{\pi e^{-6}}{16} = \int_{-\infty}^{\infty} \frac{\cos 3x}{16-x^4}\,dx - \frac{\pi}{16}\sin 6.$$

Thus,

$$\int_0^\infty \frac{\cos 3x}{16 - x^4}\,dx = \frac{1}{2}\int_{-\infty}^\infty \frac{\cos 3x}{16 - x^4}\,dx = \frac{\pi}{32}(e^{-6} + \sin 6).$$

General results for integrals of the type of 6.10 are discussed in Exercise 6 below. We now consider integrals that lead to branch points of multiple-valued functions.

EXAMPLE 6.13

Evaluate $\displaystyle\int_0^\infty \frac{\ln x}{x^2 + a^2}\,dx$ when $a > 0$.

Solution This integral exists in the ordinary sense, because both integrals on the right side of the following equation converge:

$$\int_0^\infty \frac{\ln x}{x^2 + a^2}\,dx = \int_0^1 \frac{\ln x}{x^2 + a^2}\,dx + \int_1^\infty \frac{\ln x}{x^2 + a^2}\,dx.$$

Because

$$\int_0^\infty \frac{\ln x}{x^2 + a^2}\,dx = \int_{-\infty}^0 \frac{\ln(-x)}{x^2 + a^2}\,dx,$$

we are led to consider

$$\oint_C \frac{\log_{-\pi/2} z}{z^2 + a^2}\,dz$$

where C is the contour in Figure 6.11.

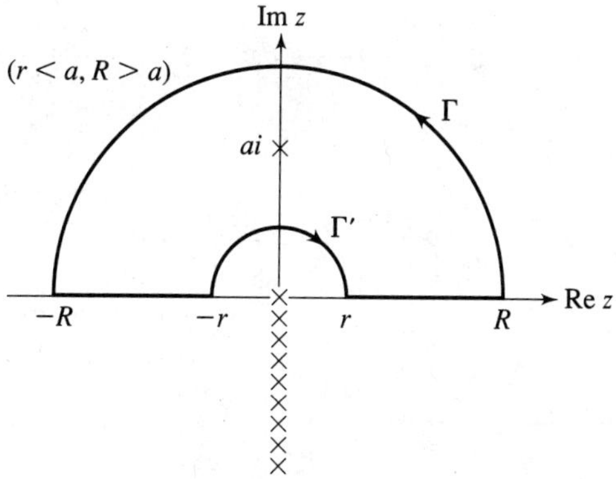

Figure 6.11

The integrand has simple poles at $z = \pm ai$, so that

$$\oint_C \frac{\log_{-\pi/2} z}{z^2+a^2}\,dz = 2\pi i \operatorname{Res}\left[\frac{\log_{-\pi/2} z}{z^2+a^2}, ai\right] = 2\pi i \lim_{z\to ai} \frac{(z-ai)\log_{-\pi/2} z}{(z-ai)(z+ai)}$$

$$= \frac{\pi}{a}\log_{-\pi/2}(ai) = \frac{\pi}{a}[\ln a + \pi i/2].$$

On the semicircle Γ, we set $z = Re^{\theta i}$, $0 \le \theta \le \pi$, in which case

$$\left|\frac{\log_{-\pi/2} z}{z^2+a^2}\right| \le \frac{|\ln R + \theta i|}{R^2-a^2} \le \frac{\ln R + \pi}{R^2-a^2}$$

(provided $R > a$ and $R > 1$). Consequently,

$$\left|\int_\Gamma \frac{\log_{-\pi/2} z}{z^2+a^2}\,dz\right| \le \frac{\ln R+\pi}{R^2-a^2}(\pi R),$$

and this expression has limit zero as $R \to \infty$. On the semicircle $-\Gamma'$, we set $z = re^{\theta i}$, $0 \le \theta \le \pi$, in which case

$$\left|\frac{\log_{-\pi/2} z}{z^2+a^2}\right| \le \frac{|\ln r + \theta i|}{a^2-r^2} \le \frac{|\ln r| + \pi}{a^2-r^2}.$$

Thus,

$$\left|\int_{\Gamma'} \frac{\log_{-\pi/2} z}{z^2+a^2}\,dz\right| = \left|\int_{-\Gamma'} \frac{\log_{-\pi/2} z}{z^2+a^2}\,dz\right| \le \frac{|\ln r|+\pi}{a^2-r^2}(\pi r),$$

and this value has limit zero as $r \to 0$. It follows, therefore, that

$$\frac{\pi}{a}[\ln a + \pi i/2] = \lim_{\substack{r\to 0\\ R\to\infty}} \oint_C \frac{\log_{-\pi/2} z}{z^2+a^2}\,dz = \int_{-\infty}^{\infty} \frac{\log_{-\pi/2} x}{x^2+a^2}\,dx$$

$$= \int_{-\infty}^{0} \frac{\ln(-x)+\pi i}{x^2+a^2}\,dx + \int_0^\infty \frac{\ln x}{x^2+a^2}\,dx$$

$$= 2\int_0^\infty \frac{\ln x}{x^2+a^2}\,dx + \int_{-\infty}^0 \frac{\pi i}{x^2+a^2}\,dx,$$

and real parts give

$$\int_0^\infty \frac{\ln x}{x^2+a^2}\,dx = \frac{\pi}{2a}\ln a.$$

EXAMPLE 6.14

Evaluate $\displaystyle\int_0^\infty \frac{1}{x^a(x+1)}\,dx$, where $0 < a < 1$.

Solution To evaluate this improper integral by contour integration it would seem reasonable to replace the real integrand with the complex one, $[z^a(z+1)]^{-1}$. Because z^a, for fractional a, is multiple-valued, we must choose some branch of the function. This choice will be dictated by the choice of contour. To obtain the required real improper integral, part of the contour should consist of a portion of the positive real axis. To avoid the branch point, we include a small circle (of radius r) around $z = 0$. In addition, we should have a circle of large radius R centered at $z = 0$. So far, then, we have the curves in Figure 6.12(a). To complete a contour C we add that part of the real axis joining the circles but in the negative direction (Figure 6.12(b)). It might be argued at this point that the two integrals along the real axis cancel, and therefore this contour cannot possibly yield the required improper integral. In addition, the contour intersects the branch cut of z^a for any branch whatsoever. To overcome these difficulties, we divide C into two parts, C_1 and C_2, as shown in Figure 6.12(c), where straight line γ is in the third quadrant.

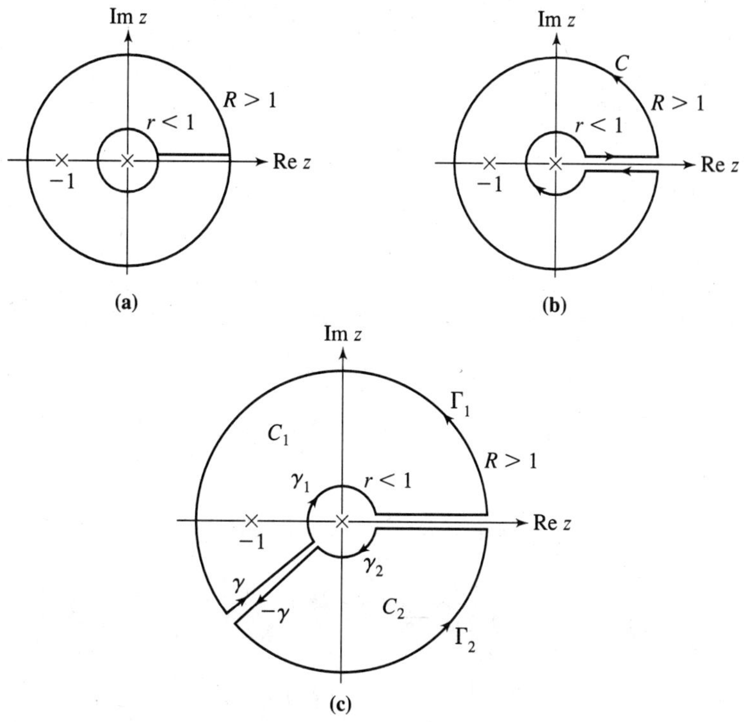

Figure 6.12

For the integral around C_2, we choose a branch of z^a with branch cut along the positive imaginary axis $z^a = e^{a \log_{\pi/2} z}$. With this branch, $[z^a(z+1)]^{-1}$ is analytic inside and on C_2, and therefore

$$\oint_{C_2} \frac{1}{z^a(z+1)}\, dz = 0.$$

For the integral around C_1, we choose a branch of z^a with branch cut along the negative imaginary axis, $z^a = e^{a \log_{-\pi/2} z}$. With this branch, $[z^a(z+1)]^{-1}$ has only the simple pole $z=-1$ interior to C_1, so that

$$\oint_{C_1} \frac{1}{z^a(z+1)}\,dz = 2\pi i \operatorname{Res}\left[\frac{1}{z^a(z+1)}, -1\right] = \frac{2\pi i}{(-1)^a} = \frac{2\pi i}{e^{a\log_{-\pi/2}(-1)}} = 2\pi i e^{-a\pi i}.$$

We now add these two contour integrals together:

$$2\pi i e^{-a\pi i} = \oint_{C_1} \frac{1}{(z+1)e^{a\log_{-\pi/2} z}}\,dz + \oint_{C_2} \frac{1}{(z+1)e^{a\log_{\pi/2} z}}\,dz.$$

On Γ_1, $\left|\dfrac{1}{(z+1)e^{a\log_{-\pi/2} z}}\right| \le \dfrac{1}{R^a(R-1)}$, so that

$$\left|\int_{\Gamma_1} \frac{1}{(z+1)e^{a\log_{-\pi/2} z}}\,dz\right| \le \frac{1}{R^a(R-1)}(R\phi),$$

where ϕ is the angle between the positive real axis and the line along γ. But clearly this expression approaches zero as $R \to \infty$. Similarly, the contour integral along Γ_2 vanishes in the limit as $R \to \infty$. Furthermore,

$$\left|\int_{\gamma_1} \frac{1}{(z+1)e^{a\log_{-\pi/2} z}}\,dz\right| \le \frac{1}{r^a(1-r)}(r\phi),$$

and this approaches zero as $r \to 0$. Likewise, the integral along γ_2 approaches zero as $r \to 0$. Next we consider contributions from the straight lines γ and $-\gamma$. If we set $z = \rho e^{\phi i}$, $r \le \rho \le R$ on $-\gamma$, then

$$\int_{-\gamma} \frac{1}{(z+1)e^{a\log_{\pi/2} z}}\,dz = \int_r^R \frac{1}{(\rho e^{\phi i}+1)\rho^{a-1}}\, i e^{\phi(1-a)i}\,d\rho.$$

The contour integral along γ is precisely the negative of this definite integral, so that integrals along γ and $-\gamma$ cancel. As a result, we may write

$$\begin{aligned}
2\pi i e^{-a\pi i} &= \lim_{\substack{r\to 0\\ R\to\infty}} \oint_{C_1} \frac{1}{(z+1)e^{a\log_{-\pi/2} z}}\,dz + \lim_{\substack{r\to 0\\ R\to\infty}} \oint_{C_2} \frac{1}{(z+1)e^{a\log_{\pi/2} z}}\,dz \\
&= \int_0^\infty \frac{1}{(x+1)e^{a\log_{-\pi/2} x}}\,dx + \int_\infty^0 \frac{1}{(x+1)e^{a\log_{\pi/2} x}}\,dx \\
&= \int_0^\infty \frac{1}{(x+1)e^{a\ln x}}\,dx - \int_0^\infty \frac{1}{(x+1)e^{a(\ln x + 2\pi i)}}\,dx \\
&= \int_0^\infty \frac{1}{x^a(x+1)}\,dx - \frac{1}{e^{2a\pi i}}\int_0^\infty \frac{1}{x^a(x+1)}\,dx \\
&= (1-e^{-2a\pi i})\int_0^\infty \frac{1}{x^a(x+1)}\,dx.
\end{aligned}$$

Thus,

$$\int_0^\infty \frac{1}{x^a(x+1)}\,dx = \frac{2\pi i e^{-a\pi i}}{1-e^{-2a\pi i}} = \frac{\pi}{\sin a\pi}.$$

Exercises 6.3

1. In this exercise we prove the following result: When $P(x)$ and $Q(x)$ are polynomials of degrees m and n where $n \geq m+2$ and the only zeros of $Q(x)$ on the real axis are simple zeros,

$$\int_{-\infty}^{\infty} \frac{P(x)}{Q(x)}\,dx = 2\pi i \begin{Bmatrix} \text{sum of the residues of } P(z)/Q(z) \text{ at} \\ \text{its poles in the half-plane } \operatorname{Im} z > 0 \end{Bmatrix}$$

$$+ \pi i \begin{Bmatrix} \text{sum of the residues of } P(z)/Q(z) \text{ at} \\ \text{its simple poles on the real axis} \end{Bmatrix}.$$

(a) Let $x_1, \ldots, x_n$ be the simple zeros of $Q(x)$ on the x-axis and $z_1, \ldots, z_m$ be the zeros of $Q(z)$ with positive imaginary parts. (See the accompanying figure.) Use the Cauchy residue theorem to evaluate the contour integral of $P(z)/Q(z)$ around the contour C. Radius R of the outer semicircle is chosen so large and radius r of the semicircles Γ_k around x_k is chosen so small that the semicircles are disjoint, and all z_k are interior to C.

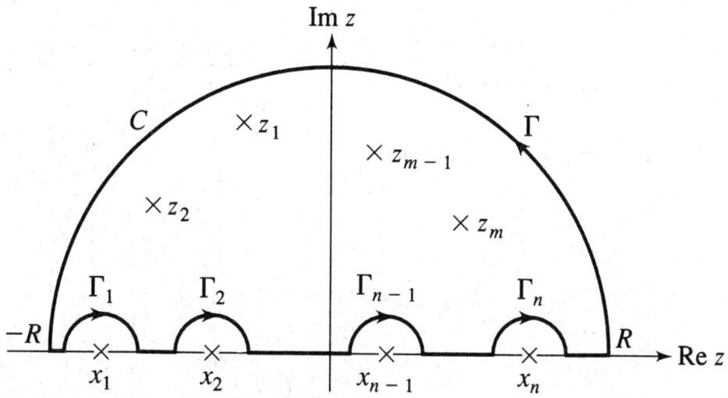

(b) Use Laurent series to show that for $k = 1, \ldots, n$,

$$\int_{\Gamma_k} \frac{P(z)}{Q(z)}\,dz = \int_{\Gamma_k} \frac{\operatorname{Res}[P(z)/Q(z), x_k]}{z - x_k}\,dz + \int_{\Gamma_k} g_k(z)\,dz$$

where $g_k(z)$ is analytic in a domain containing Γ_k. Hence deduce that

$$\lim_{r\to 0} \int_{\Gamma_k} \frac{P(z)}{Q(z)}\,dz = -\pi i \operatorname{Res}\left[\frac{P(z)}{Q(z)}, x_k\right].$$

(c) Verify that the limit of the contour integral along Γ is zero as $R \to \infty$.

(d) Combine (a), (b), and (c) to prove the required result.

(e) Can the Cauchy principal value in (a) be dropped?

In Exercises 2–5 use Exercise 1, or follow the procedure of Example 6.11, to evaluate the improper integral.

2. $\displaystyle\int_{-\infty}^{\infty} \frac{x^2}{x^4-1}\,dx$

3. $\displaystyle\int_{-\infty}^{\infty} \frac{x+1}{x(x^2+4)^2}\,dx$

4. $\displaystyle\int_{-\infty}^{\infty} \frac{1}{x^5+x^4+x+1}\,dx$

5. $\displaystyle\int_{-\infty}^{\infty} \frac{x^2+2x}{(x-1)(x^3+1)}\,dx$

6. In this exercise we prove the following result: When $P(x)$ and $Q(x)$ are polynomials of degrees m and n where $n \geq m+1$ and the only zeros of $Q(x)$ on the real axis are simple zeros, then for $a > 0$,

$$\int_{-\infty}^{\infty} \frac{P(x)\cos ax}{Q(x)}\,dx = -2\pi\,\mathrm{Im}\left\{\begin{array}{l}\text{sum of the residues of } P(z)e^{azi}/Q(z)\\ \text{at its poles in the half-plane } \mathrm{Im}\,z > 0\end{array}\right\}$$
$$-\pi\,\mathrm{Im}\left\{\begin{array}{l}\text{sum of the residues of } P(z)e^{azi}/Q(z)\\ \text{at its simple poles on the real axis}\end{array}\right\},$$

and

$$\int_{-\infty}^{\infty} \frac{P(x)\sin ax}{Q(x)}\,dx = 2\pi\,\mathrm{Re}\left\{\begin{array}{l}\text{sum of the residues of } P(z)e^{azi}/Q(z)\\ \text{at its poles in the half-plane } \mathrm{Im}\,z > 0\end{array}\right\}$$
$$+\pi\,\mathrm{Re}\left\{\begin{array}{l}\text{sum of the residues of } P(z)e^{azi}/Q(z)\\ \text{at its simple poles on the real axis}\end{array}\right\}.$$

(a) Let $x_1, \ldots, x_n$ be the simple zeros of $Q(x)$ on the real axis and $z_1, \ldots, z_m$ be the zeros of $Q(z)$ with positive imaginary parts. Use the Cauchy residue theorem to evaluate the contour integral of $P(z)e^{azi}/Q(z)$ around the contour in Exercise 1.

(b) Use Laurent series to show that for $k = 1, \ldots, n$,

$$\int_{\Gamma_k} \frac{P(z)e^{azi}}{Q(z)}\,dz = \int_{\Gamma_k} \frac{\mathrm{Res}[P(z)e^{azi}/Q(z), x_k]}{z-x_k}\,dz + \int_{\Gamma_k} g_k(z)\,dz$$

where $g_k(z)$ is analytic in a domain containing Γ_k. Hence deduce that

$$\lim_{r\to 0}\int_{\Gamma_k} \frac{P(z)e^{azi}}{Q(z)}\,dz = -\pi i\,\mathrm{Res}\left[\frac{P(z)e^{azi}}{Q(z)}, x_k\right].$$

(c) Use the result of Exercise 26 in Section 6.2 to conclude that the limit of the contour integral along Γ is zero as $R \to \infty$.

(d) Combine (a), (b), and (c) to prove the required result.

(e) Can the Cauchy principal value in (a) be dropped?

In Exercises 7–10 use Exercise 6, or follow the procedure of Example 6.12, to evaluate the improper integral.

7. $\displaystyle\int_{-\infty}^{\infty} \frac{\sin x}{x}\, dx$

8. $\displaystyle⨍_{-\infty}^{\infty} \frac{\cos 2x}{1-3x^2}\, dx$

9. $\displaystyle⨍_{-\infty}^{\infty} \frac{x^2 \cos x}{x^4-1}\, dx$

10. $\displaystyle⨍_{-\infty}^{\infty} \frac{\sin x \cos x}{x^3-8}\, dx$

In Exercises 11–14 evaluate the improper integral.

11. $\displaystyle\int_0^{\infty} \frac{\ln x}{(x^2+a^2)^2}\, dx, \quad a>0$

12. $\displaystyle\int_0^{\infty} \frac{\ln x}{x^4+a^4}\, dx, \quad a>0$

13. $\displaystyle\int_0^{\infty} \frac{1}{\sqrt{x}(x^2+a^2)}\, dx, \quad a>0$

14. $\displaystyle\int_0^{\infty} \frac{\sin^2 x}{x^2}\, dx$ *Hint:* Consider the contour integral of $(1-e^{2zi})/z^2$ around the curve in Figure 6.11.

15. The free-space Green's function for the one-dimensional Helmholtz operator is sometimes quoted as

$$U = \frac{1}{2\pi} ⨍_{-\infty}^{\infty} \frac{\cos \omega|x-X|}{k^2-\omega^2}\, d\omega,$$

where $k>0$ and X are constants. Show that a simpler expression is

$$U = \frac{1}{2k} \sin k|x-X|.$$

16. **(a)** Use Cauchy's residue theorem to evaluate

$$\oint_C \frac{e^{az}}{\cosh z}\, dz$$

where C is the contour in the following diagram and $-1<a<1$.

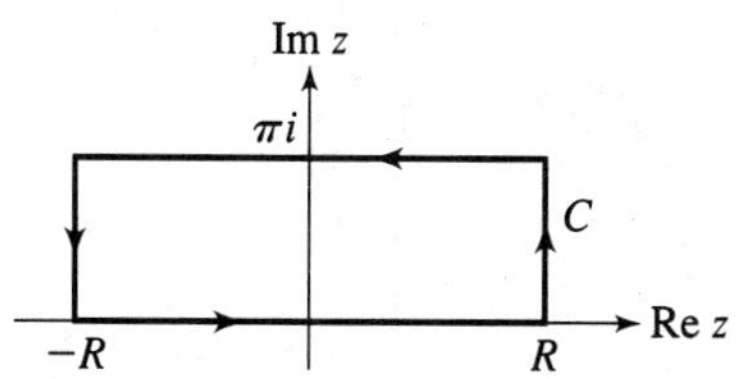

(b) Show that contributions to the contour integral along the two vertical lines vanish in the limit as $R \to \infty$.

(c) Take limits as $R \to \infty$ on the contour integral in (a) to obtain

$$\int_{-\infty}^{\infty} \frac{e^{ax}}{\cosh x}\, dx = \frac{\pi}{\cos(\pi a/2)}.$$

(d) Now evaluate $\displaystyle\int_0^{\infty} \frac{\cosh ax}{\cosh x}\, dx$, $-1 < a < 1$.

17. Show that

$$\int_0^{\infty} e^{-x^2} \cos 2bx\, dx = \frac{\sqrt{\pi}}{2} e^{-b^2}$$

by integrating $f(z) = e^{-z^2}$ around the edge of the rectangle $|x| \le R$, $0 \le y \le b$, and using the fact that

$$\int_{-\infty}^{\infty} e^{-x^2}\, dx = \sqrt{\pi}.$$

18. The Fresnel integrals

$$\int_0^{\infty} \sin x^2\, dx \qquad \text{and} \qquad \int_0^{\infty} \cos x^2\, dx$$

are important in optics. Evaluate them using the contour integral of $e^{z^2 i}$ around the boundary of the sector $0 \le r \le R$, $0 \le \theta \le \pi/4$.

CHAPTER 7

Conformal Mapping

In Chapters 4–6 we concentrated on algebraic properties of analytic functions. In this chapter we concentrate on functions $w = f(z)$ as mappings from the z-plane to the w-plane and uncover geometric consequences of analyticity. We confirm what was suggested in Section 2.4. The modulus of $f'(z)$ is a length-magnification factor of the mapping, and the argument of $f'(z)$ is the rotational effect of the mapping. In Sections 2.6–2.8 we indicated how complex variables, through harmonic functions, fit into discussions on two-dimensional heat and fluid flow and electrostatics. The power of complex variables in these applications is brought out in this chapter.

7.1 One-to-One Mappings

We have already seen that a function f can be regarded as a mapping from the z-plane to the w-plane, where $w = f(z)$. Of particular importance to our discussion are mappings that are one-to-one.

Definition 7.1
A mapping $w = f(z)$ is said to be **one-to-one** (1-1) in a domain D if for any two points $z_1 \neq z_2$ in D, $f(z_1) \neq f(z_2)$.

What this implies is that given any point w in the image $f(D)$ of D, there is exactly one point z in D that is mapped by $f(z)$ onto w. We may therefore define an inverse mapping f^{-1}, which maps points w in $f(D)$ onto points in D:

$$z = f^{-1}(w) \quad \text{if} \quad w = f(z). \tag{7.1}$$

The following four examples indicate that a mapping may be 1-1 in some domains but not in others.

EXAMPLE 7.1

Discuss the mapping $w = f(z) = z^2$. Is it one-to-one?

Solution If we set $z = re^{\theta i}$ in the mapping, then $w = r^2 e^{2\theta i}$. Consequently, the modulus of w is the square of the modulus of z, and the argument of w is double that of z. It follows therefore that the angle in a sector-shaped domain (Figure 7.1) is doubled under $w = z^2$.

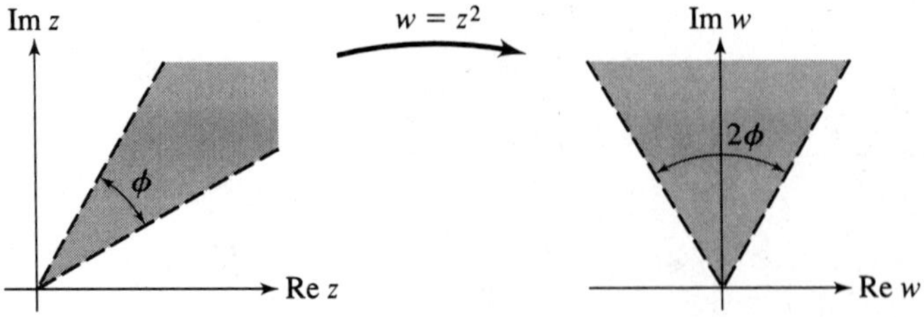

Figure 7.1

In particular, any half-space formed by a line through $z = 0$ (Figure 7.2) maps onto the entire w-plane. Another way of saying the same thing is to note that because $(-z)^2 = z^2$, the mapping is 2-1 on some domains. It is, in fact, 2-1 on any open circle $|z| < R$ centered at $z = 0$. On the other hand, for any open circle $|z - z_0| < R$ that does not contain $z = 0$ (Figure 7.3), the mapping is 1-1. Thus, $z = 0$ is the only point about which it is impossible to construct a neighborhood in which $w = z^2$ is 1-1.

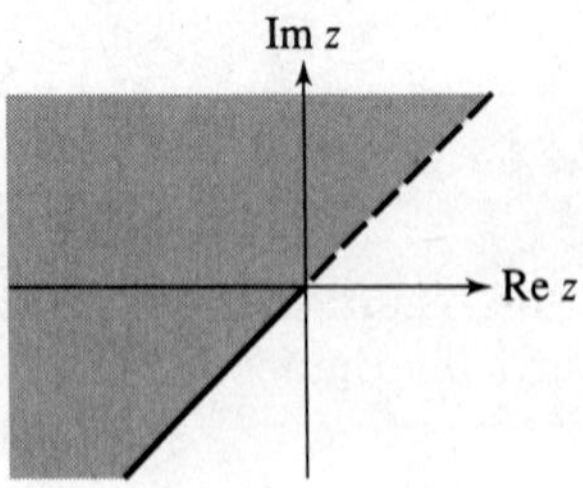

Figure 7.2

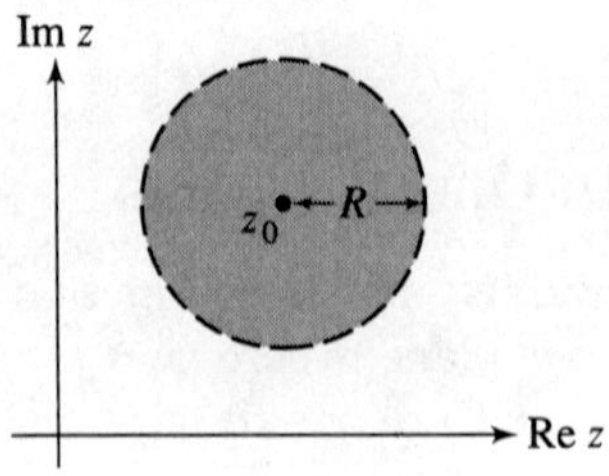

Figure 7.3

EXAMPLE 7.2

Discuss the mapping $w = f(z) = e^z$. Is it one-to-one?

Solution We have already seen that the function $f(z) = e^z$ is $2\pi i$-periodic, so that each point in the w-plane is the image of an infinite number of points in the z-plane. However, the discussion in Section 3.2 indicates that $w = e^z$ is 1-1 on any strip $y_0 < \operatorname{Im} z \leq y_0 + 2\pi$ (Figure 7.4). It follows then that $w = e^z$ is 1-1 on any

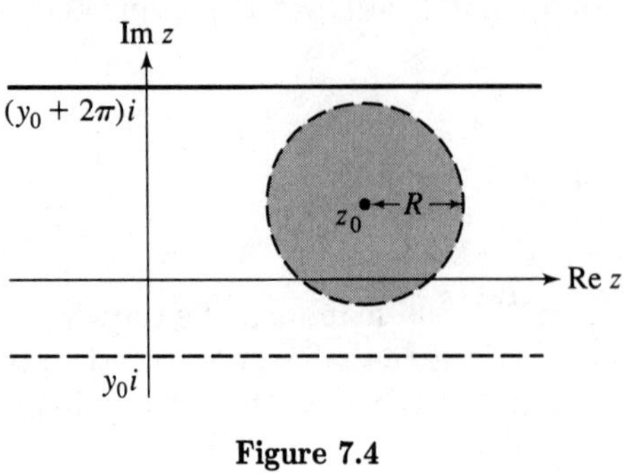

Figure 7.4

open circle $|z - z_0| < R$ with radius $R \leq \pi$. In other words, about any point z_0, we can construct a neighborhood in which $w = e^z$ is 1-1.

EXAMPLE 7.3

Discuss the mapping $w = f(z) = \log_\phi z$. Is it one-to-one?

Solution The branch

$$\log_\phi z = \ln|z| + \arg z\, i, \qquad z \neq 0, \qquad \phi < \arg z \leq \phi + 2\pi,$$

of the logarithm function maps the z-plane onto the horizontal strip in the w-plane shown in Figure 7.5. It maps circles $|z| = R$ onto vertical line segments and radial lines $\arg z =$ constant onto horizontal lines in the w-plane. When $z_1 \neq z_2$, either $|z_1| \neq |z_2|$, $\arg z_1 \neq \arg z_2$, or both. In any case, $\log_\phi z_1 \neq \log_\phi z_2$, and $w = \log_\phi z$ is 1-1.

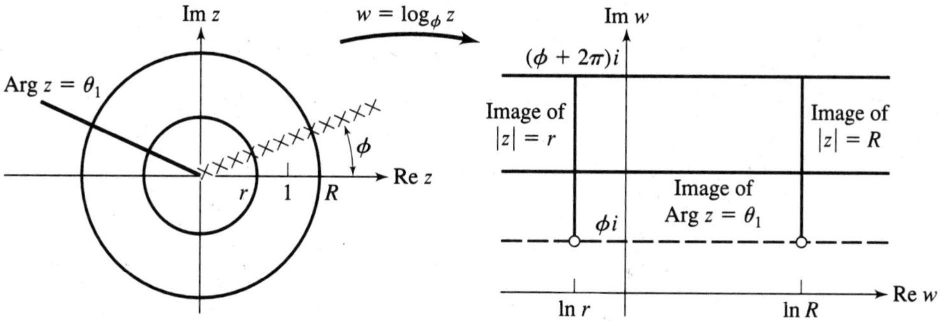

Figure 7.5

EXAMPLE 7.4

Discuss the mapping $w = f(z) = |z|$. Is it one-to-one?

Solution The function $w = |z|$ maps the entire z-plane onto the nonnegative real axis. All points on the circle $|z| = R$ are mapped by $w = |z|$ onto the single point $w = R$ (Figure 7.6). Furthermore, it is impossible to construct about any point z a circle inside of which the mapping is 1-1.

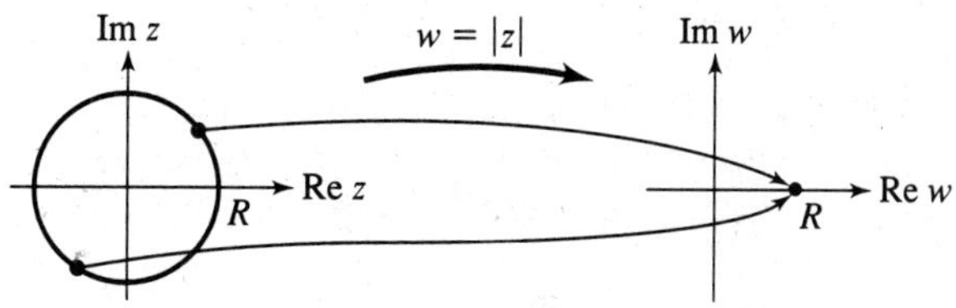

Figure 7.6

We have just seen that about any point $z \neq 0$ we can find circles inside of which z^2 is one-to-one; about any point we can find circles inside of which e^z and $\log_\phi z$ are one-to-one. On the other hand, about no point can we find circles inside of which $|z|$ is one-to-one. If we note that $d(z^2)/dz = 0$ at $z = 0$, $d(e^z)/dz$ never vanishes, and $d(\log_\phi z)/dz$ never vanishes, we might suspect the following result.

■ **THEOREM 7.1** If f is analytic at z_0 and $f'(z_0) \neq 0$, then there is a neighborhood of z_0 in which f is one-to-one. ■

We now establish that analytic functions map domains to domains.

■ **THEOREM 7.2** If f is analytic and not constant in a domain D, then the image $f(D)$ of D in the $w = f(z)$ plane is a domain.

Proof We first establish that $f(D)$ is an open set by showing that every point w_0 in $f(D)$ has a neighborhood that is contained in $f(D)$. Let z_0 be any point in D for which $w_0 = f(z_0)$. The function $F(z) = w_0 - f(z)$ has an isolated zero at z_0, $F(z)$ being analytic and nonconstant (see the corollary to Theorem 5.20). There therefore exists a circle $C : |z - z_0| = \epsilon$ in D such that $F(z)$ does not vanish on C and the only point inside C at which it vanishes is z_0 (Figure 7.7).

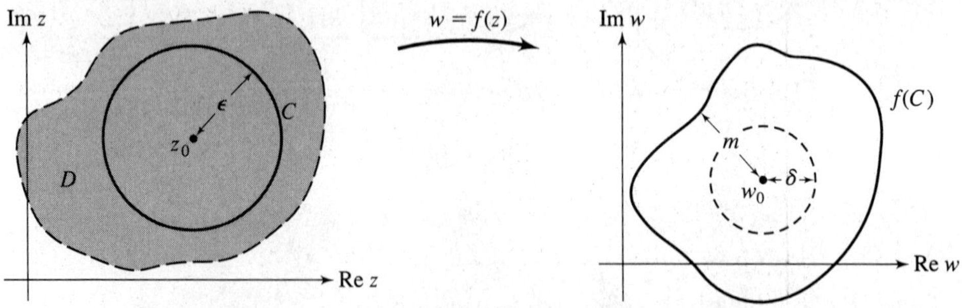

Figure 7.7

It follows that $f(C)$ cannot pass through w_0, and we let m be the distance from w_0 to the point on $f(C)$ closest to w_0,

$$m = \underset{w \text{ on } f(C)}{\text{minimum}} |w - w_0| = \underset{z \text{ on } C}{\text{minimum}} |f(z) - f(z_0)|.$$

Let $\delta < m$ and consider the open set $|w - w_0| < \delta$. We show that the open set $|w - w_0| < \delta < m$ is contained in the image of $|z - z_0| < \epsilon$. To do this we show that for any point w^* in $|w - w_0| < \delta$, there exists a point z in $|z - z_0| < \epsilon$ for which $w^* = f(z)$.

Consider the function

$$f(z) - w^* = [f(z) - w_0] + [w_0 - w^*].$$

For any point on C, we have

$$|w_0 - w^*| < \delta < |f(z) - w_0|.$$

By Rouche's theorem (Theorem 6.4), the functions $f(z) - w^*$ and $f(z) - w_0$ have the same number of zeros inside C. Because $f(z) - w_0$ has at least one zero inside C (counting multiplicities, it could be more than one), it follows that $f(z) - w^*$ has at least one zero inside C; that is, there exists at least one point inside C at which $w^* = f(z)$. This completes the proof that $f(D)$ is an open set.

We now prove that $f(D)$ is connected. Let $w_1 = f(z_1)$ and $w_2 = f(z_2)$ be any two points in $f(D)$. Because D is connected, there is a finite number of line segments C joining z_1 and z_2 that lie entirely within D (Figure 7.8).

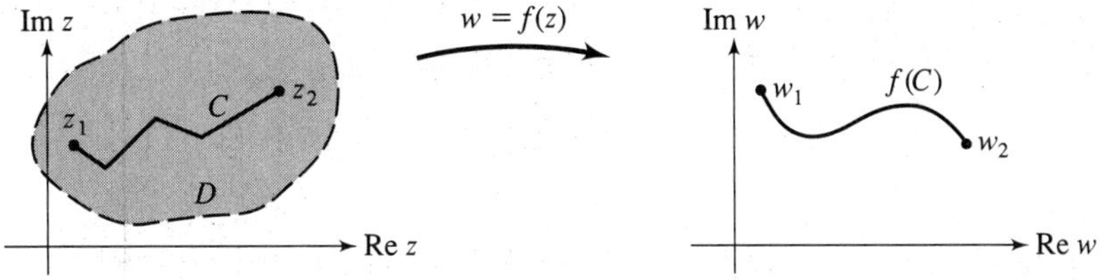

Figure 7.8

Because a differentiable mapping always maps a curve to a curve, it follows that the image $f(C)$ of C is a curve in $f(D)$ joining w_1 and w_2. This curve can be approximated by line segments in $f(D)$. ∎

To summarize Theorems 7.1 and 7.2, suppose $w = f(z)$, where f is analytic in some domain D, and z_0 is a point in D at which $f'(z_0) \neq 0$. Then about z_0 we can find a neighborhood $N : |z - z_0| < \delta$ in which $w = f(z)$ is one-to-one. The image $f(N)$ of N is a domain in the w-plane (Figure 7.9), and the inverse mapping f^{-1} that maps points w in $f(N)$ onto their preimages z in N exists:

$$z = f^{-1}(w) \quad \text{if} \quad w = f(z).$$

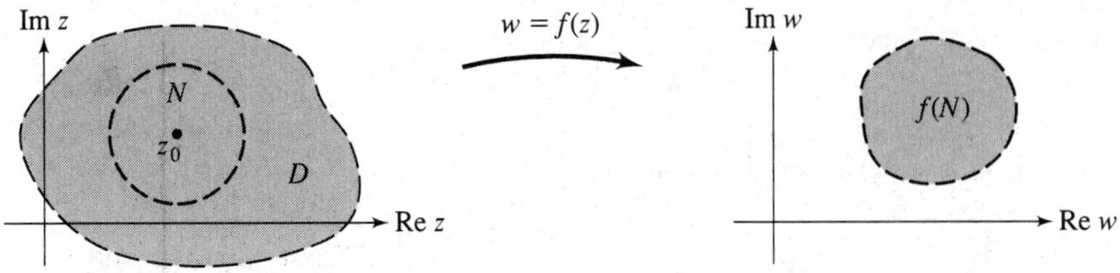

Figure 7.9

These are local properties; they indicate what can be done in a neighborhood around a point at which $f'(z_0) \neq 0$.

When mappings are used in the solution of boundary-value problems associated with the two-dimensional Laplacian (and we shall see how in Section 7.4), the following global problem always arises: Given a domain D_z in the z-plane, is it possible to find a one-to-one mapping $w = f(z)$ that maps D_z onto a prescribed domain D_w in the w-plane? Fortunately, there exist theorems that reply affirmatively. One such is Theorem 7.3.

■ **THEOREM 7.3** (Riemann's Mapping Theorem) If D is a simply connected domain in the z-plane (but not the z-plane itself), there exist 1-1 mappings $w = f(z)$ of D onto the unit circle $|w| < 1$. ■

With the unit circle as a reference domain, this theorem indicates that there are one-to-one mappings that map any simply connected domain onto any other simply connected domain.

Exercises 7.1

1. Find images under the mapping $w = z^2 = (x + yi)^2$ of the following domains:
   (a) $x > 0,\ y > 0$
   (b) $0 < \operatorname{Arg} z < 3\pi/4,\ |z| < 2$
   (c) $0 < \operatorname{Arg} z < \pi/2,\ xy < 1$
   (d) $x^2 - y^2 < 4,\ x > 0$
   (e) $x^2 - y^2 > 4,\ x > 0$
2. Find images under the mapping $w = e^z = e^{x+yi}$ of the following domains:
   (a) $1 < x < 2,\ -\pi/2 < y < \pi/2$
   (b) $|x| < 1,\ 0 < y < 2$
   (c) $0 < y < \pi$
   (d) $x > 0,\ 0 < y < 2\pi$
   (e) $x > 0,\ 0 < y < \pi$
3. Find images under the mapping $w = \operatorname{Log} z = \operatorname{Log}(x + yi)$ of the following domains:
   (a) $x > 0,\ y > 0$
   (b) $y > x,\ y > -x,\ x^2 + y^2 < 4$
   (c) $1 < x^2 + y^2 < 3,\ y \neq 0$ when $x < 0$
4. Illustrate that the mapping $w = \log_\phi z$ maps an open sector centered at $z = 0$ onto two rectangles in the w-plane, only one of which is open when the sector contains the branch cut.
5. Give a discussion of the mapping $w = z^n$, where $n \geq 3$ is an integer, that parallels Example 7.1.
6. Find images under the mapping $w = z^{\pi/\alpha}$ $(0 < \alpha < 2\pi)$ of the following domains:
   (a) $0 < \operatorname{Arg} z < \alpha,\ |z| < 1$
   (b) $0 < \operatorname{Arg} z < \alpha$

7. Find images under the mapping $w = \sqrt{z} = \sqrt{x + yi}$ of the following domains:

(a) $x^2 + y^2 < R^2$, $y \neq 0$ when $x \leq 0$

(b) $y > 0$

(c) $x > 0$, $x^2 + y^2 > 4$

(d) $x > 1$

(e) $y > 1$

8. Illustrate that the mapping $w = \sin z$ maps

(a) A vertical line onto one-half a hyperbola.

(b) A horizontal line onto an ellipse.

(c) The rectangular domain $-\pi < x < \pi$, $1 < y < 2$ onto an elliptic domain.

(d) The half-strip $-\pi/2 < x < \pi/2$, $y > 0$ onto the half-plane $v > 0$.

9. Illustrate that the mapping $w = \cosh z$ maps

(a) A vertical line onto an ellipse.

(b) A horizontal line onto one-half a hyperbola.

(c) The rectangular domain $0 < x < 2$, $0 < y < 2\pi$ onto an elliptic domain with a slit.

10. (a) What is the branch cut for $\text{Log}(z - 2i)$?

(b) What is the image of the domain $y > 2$ when $w = \text{Log}(z - 2i)$?

11. (a) Show that the function $w = \cos^2 z$ maps the semi-infinite strip $0 < x < \pi/2$, $y > 0$ onto the half-plane $v < 0$.

(b) What does $w = \sin^2 z$ do to the same strip?

12. Show that the Joukowski transformation $w = z + a^2/z$, where $a > 0$ is a constant, maps

(a) Circles $|z| = R$ $(R \neq a)$ to ellipses with foci at $(\pm 2a, 0)$.

(b) Line segments $\theta = \alpha$ to hyperbolas with foci at $(\pm 2a, 0)$.

(c) The circle $|z| = a$ to that part of the $\text{Re}\, w$ axis for which $-2a \leq u \leq 2a$.

(d) The domain $|z| > a$ to the w-plane slit along the line segment $-2a < u < 2a$ (that is, points on the slit are not images of points outside the circle).

(e) The circle $|z| < a$ to the w-plane slit along the line segment $-2a < u < 2a$.

7.2 Bilinear Transformations

We introduced bilinear mappings in Section 2.2, transformations of the form

$$w = f(z) = \frac{az + b}{cz + d}, \tag{7.2}$$

where a, b, c, and d are complex constants for which $ad - bc \neq 0$. The condition $ad - bc \neq 0$ guarantees that $f(z)$ is not a constant. Because

$$f'(z) = \frac{(cz + d)(a) - (az + b)(c)}{(cz + d)^2} = \frac{ad - bc}{(cz + d)^2}$$

never vanishes, Theorem 7.1 indicates that every point except the pole $z = -d/c$ has a neighborhood in which the function is one-to-one. The function is globally one-to-one also. We can solve 7.2 for z in terms of w:

$$z = \frac{-dw + b}{cw - a},$$

so that $f^{-1}(w)$ is defined for all $w \neq a/c$. In other words, bilinear transformations are globally invertible.

Every bilinear transformation is a composition of translations,

$$w = z + b, \tag{7.3}$$

magnifications,

$$w = az, \qquad a \text{ real}, \tag{7.4}$$

rotations,

$$w = e^{\phi i} z, \qquad \phi \text{ real}, \tag{7.5}$$

and reciprocations,

$$w = \frac{1}{z}. \tag{7.6}$$

If $c = 0$ in 7.2, we may absorb d into a and b, and the transformation becomes

$$w = f(z) = az + b, \tag{7.7}$$

a combination of a magnification, rotation, and translation. When $c \neq 0$, 7.2 can be expressed in the form

$$w = f(z) = \frac{az+b}{cz+d} = \frac{\frac{a}{c}(cz+d) + b - \frac{ad}{c}}{cz+d} = \frac{a}{c} + \frac{b - \frac{ad}{c}}{cz+d}, \tag{7.8}$$

and $f(z)$ may be regarded successively as a rotation and magnification and a translation

$$w_1 = cz + d,$$

a reciprocation,

$$w_2 = \frac{1}{w_1},$$

a rotation and magnification,

$$w_3 = \left(b - \frac{ad}{c}\right) w_2,$$

and finally a translation,

$$w = w_3 + \frac{a}{c}.$$

There are many properties of bilinear transformations that have considerable importance in physical problems. We discuss some of these in the following examples.

EXAMPLE 7.5 ....

Show that circles and straight lines in the z-plane are mapped by bilinear transformations onto circles and straight lines in the w-plane.

Solution Every bilinear transformation can be expressed as a finite sequence of translations, rotations, magnifications, and/or reciprocations. Clearly, translations, rotations, and magnifications map circles to circles and lines to lines. The only question concerns that of a reciprocation $w = 1/z$. If we set $w = u + vi$ and $z = x + yi$,

$$z = x + yi = \frac{1}{w} = \frac{1}{u + vi} = \frac{u - vi}{u^2 + v^2}.$$

Thus,

$$x = \frac{u}{u^2 + v^2}, \qquad y = \frac{-v}{u^2 + v^2}.$$

All circles in the z-plane have real and imaginary parts satisfying

$$x^2 + y^2 + Ax + By + C = 0,$$

where $A^2 + B^2 > 4C$. Consequently, the real and imaginary parts of the image of such a circle under $w = 1/z$ must satisfy

$$\frac{u^2}{(u^2 + v^2)^2} + \frac{v^2}{(u^2 + v^2)^2} + \frac{Au}{u^2 + v^2} - \frac{Bv}{u^2 + v^2} + C = 0,$$

or

$$C(u^2 + v^2) + Au - Bv + 1 = 0.$$

If $C \neq 0$ (so that the circle in the z-plane does not pass through the pole of the reciprocation), then

$$u^2 + v^2 + \frac{A}{C}u - \frac{B}{C}v + \frac{1}{C} = 0,$$

a circle in the w-plane. If $C = 0$ (so that the circle in the z-plane does pass through the pole),

$$Au - Bv + 1 = 0,$$

a straight line in the w-plane. Circles in the z-plane are therefore mapped by the reciprocation $w = 1/z$ onto either circles or straight lines in the w-plane.

Straight lines in the z-plane have real and imaginary parts satisfying

$$Ax + By + C = 0,$$

where $A^2 + B^2 > 0$. Consequently, the real and imaginary parts of the image of such a straight line under $w = 1/z$ must satisfy

$$A\left(\frac{u}{u^2+v^2}\right) + B\left(\frac{-v}{u^2+v^2}\right) + C = 0,$$

or

$$Au - Bv + C(u^2+v^2) = 0.$$

If $C \neq 0$ (in which case the line does not pass through the pole),

$$u^2 + v^2 + \frac{A}{C}u - \frac{B}{C}v = 0,$$

a circle in the w-plane. If $C = 0$ (so that the line passes through the pole),

$$Au - Bv = 0,$$

a straight line in the w-plane. Straight lines in the z-plane are therefore mapped by $w = 1/z$ onto either circles or straight lines in the w-plane.

EXAMPLE 7.6

If z, z_1, z_2, and z_3 are points in the z-plane, their cross ratio is defined as the quotient

$$\frac{(z-z_1)(z_2-z_3)}{(z-z_3)(z_2-z_1)}.$$

Show that the cross ratio is invariant under a bilinear mapping.

Solution If $w = f(z) = (az+b)/(cz+d)$, then the cross ratio of w, $w_1 = f(z_1)$, $w_2 = f(z_2)$, and $w_3 = f(z_3)$ is

$$\begin{aligned}
&\frac{(w-w_1)(w_2-w_3)}{(w-w_3)(w_2-w_1)} \\
&\quad = \frac{\left(\dfrac{az+b}{cz+d} - \dfrac{az_1+b}{cz_1+d}\right)\left(\dfrac{az_2+b}{cz_2+d} - \dfrac{az_3+b}{cz_3+d}\right)}{\left(\dfrac{az+b}{cz+d} - \dfrac{az_3+b}{cz_3+d}\right)\left(\dfrac{az_2+b}{cz_2+d} - \dfrac{az_1+b}{cz_1+d}\right)} \\
&\quad = \frac{[(az+b)(cz_1+d) - (az_1+b)(cz+d)][(az_2+b)(cz_3+d) - (cz_2+d)(az_3+b)]}{[(az+b)(cz_3+d) - (cz+d)(az_3+b)][(az_2+b)(cz_1+d) - (az_1+b)(cz_2+d)]} \\
&\quad = \frac{[(ad-bc)(z-z_1)][(ad-bc)(z_2-z_3)]}{[(ad-bc)(z-z_3)][(ad-bc)(z_2-z_1)]} \\
&\quad = \frac{(z-z_1)(z_2-z_3)}{(z-z_3)(z_2-z_1)}.
\end{aligned}$$

EXAMPLE 7.7 ....

Show that any three distinct points z_1, z_2, and z_3 can be mapped by a bilinear transformation onto any three distinct points w_1, w_2, and w_3.

Solution Under any bilinear transformation, the cross ratios of z, z_1, z_2, and z_3 and their images w, w_1, w_2, and w_3 are identical; that is,

$$\frac{(w-w_1)(w_2-w_3)}{(w-w_3)(w_2-w_1)} = \frac{(z-z_1)(z_2-z_3)}{(z-z_3)(z_2-z_1)},$$

or

$$(w-w_1)(w_2-w_3)(z-z_3)(z_2-z_1) = (w-w_3)(w_2-w_1)(z-z_1)(z_2-z_3).$$

If we define

$$W_{23} = w_2 - w_3, \qquad W_{21} = w_2 - w_1, \qquad Z_{23} = z_2 - z_3, \qquad Z_{21} = z_2 - z_1,$$

then

$$W_{23}Z_{21}(w-w_1)(z-z_3) = W_{21}Z_{23}(w-w_3)(z-z_1),$$

from which

$$[W_{23}Z_{21}(z-z_3) - W_{21}Z_{23}(z-z_1)]w = W_{23}Z_{21}w_1(z-z_3) - W_{21}Z_{23}w_3(z-z_1).$$

Thus,

$$w = \frac{(W_{23}Z_{21}w_1 - W_{21}Z_{23}w_3)z + (W_{21}Z_{23}w_3z_1 - W_{23}Z_{21}w_1z_3)}{(W_{23}Z_{21} - W_{21}Z_{23})z + (W_{21}Z_{23}z_1 - W_{23}Z_{21}z_3)}.$$

This equation defines a transformation $w = f(z)$ that is bilinear (provided we can establish that it is nonconstant). From the equality of cross ratios (from which the transformation was derived), it is clear that if $z = z_1$, then

$$0 = \frac{(w-w_1)(w_2-w_3)}{(w-w_3)(w_2-w_1)},$$

and this implies that $w = w_1$. If $z = z_2$, then

$$1 = \frac{(w-w_1)(w_2-w_3)}{(w-w_3)(w_2-w_1)},$$

and hence

$$ww_2 - ww_3 - w_1w_2 + w_1w_3 = ww_2 - ww_1 - w_3w_2 + w_3w_1,$$

or

$$0 = (w-w_2)(w_1-w_3).$$

But this implies that $w = w_2$. Finally if $z = z_3$, the only way equality of cross ratios can hold is if $w = w_3$. Consequently the transformation defined by the equality of the cross ratios is bilinear (because we have now established that $f(z) \neq$ constant) and maps z_1, z_2, and z_3 onto w_1, w_2, and w_3.

A most important consequence of Examples 7.5 and 7.7 is that any circle in the z-plane can be mapped onto any circle in the w-plane by a bilinear transformation. Indeed, let C be any circle in the z-plane (Figure 7.10). The circle is characterized by any three distinct points z_1, z_2, and z_3 on it. Suppose we want to map C onto some circle C' in the w-plane that is also characterized by any three points w_1, w_2, and w_3. By Example 7.7, there is a bilinear transformation that carries z_1, z_2, and z_3 onto w_1, w_2, and w_3, respectively. But according to Example 7.5, bilinear transformations always map circles to either circles or straight lines. Consequently, C must be mapped onto a circle (or line) in the w-plane. But there is only one circle in the w-plane containing w_1, w_2, and w_3 (and certainly no line) and this circle is C'. Thus, C is mapped onto C' by the bilinear transformation that takes z_1, z_2, and z_3 onto w_1, w_2, and w_3.

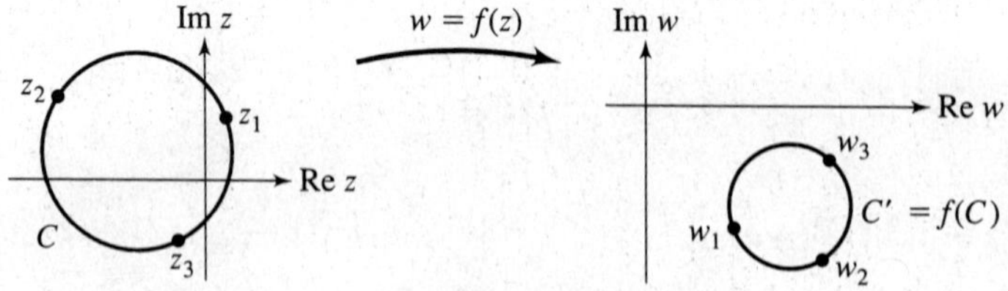

Figure 7.10

It can be shown that if w_1, w_2, and w_3 have the same orientation around C' as z_1, z_2, and z_3 do around C, then the interior of C is mapped onto the interior of C', whereas for opposite orientations, the interior of C is mapped onto the exterior of C'.

It is also true that particular straight lines in the z-plane can be mapped onto particular circles in the w-plane, particular circles onto particular lines, and particular lines onto particular lines. We illustrate the second of these in the next example.

EXAMPLE 7.8

Find all bilinear transformations that map the circle $|z| < 1$ onto the left half-plane $\operatorname{Re} w < 0$.

Solution According to Example 7.5, circles are mapped by bilinear transformations onto either circles or straight lines, and a circle is mapped onto a straight line when the circle passes through the pole of the transformation. Suppose we take $z_0 = e^{\theta_0 i}$, any point on $|z| = 1$, as the pole of the required bilinear transformation $w = f(z)$ (Figure 7.11). Then

$$w = \frac{az + b}{z - z_0}.$$

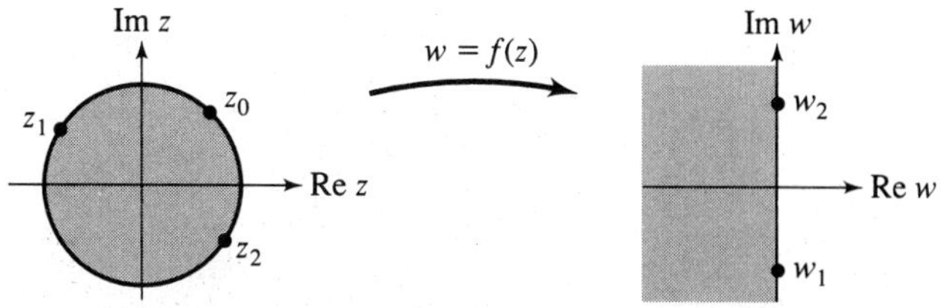

Figure 7.11

Let $z_1 = e^{\theta_1 i}$ and $z_2 = e^{\theta_2 i}$, $0 < \theta_1 < \theta_2 \neq \theta_0$ be any two other points on $|z| = 1$ and $w_1 = v_1 i$ and $w_2 = v_2 i$ be any two points on the v-axis for which $v_2 > v_1$. For $w_1 = f(z_1)$ and $w_2 = f(z_2)$,

$$w_1 = \frac{az_1 + b}{z_1 - z_0}, \qquad w_2 = \frac{az_2 + b}{z_2 - z_0}.$$

If these equations are solved for a and b, the result is

$$a = \frac{w_2(z_2 - z_0) - w_1(z_1 - z_0)}{z_2 - z_1}, \qquad b = \frac{w_1 z_2(z_1 - z_0) - w_2 z_1(z_2 - z_0)}{z_2 - z_1}.$$

Thus, any bilinear transformation of the form

$$w = \frac{[w_2(z_2 - z_0) - w_1(z_1 - z_0)]z + [w_1 z_2(z_1 - z_0) - w_2 z_1(z_2 - z_0)]}{(z_2 - z_1)(z - z_0)}$$

maps $|z| < 1$ onto $\operatorname{Re} w < 0$. For example, if we choose $z_0 = 1$, $z_1 = i = -z_2$, $w_2 = i = -w_1$,

$$w = \frac{[i(-i - 1) - (-i)(i - 1)]z + [(-i)(-i)(i - 1) - i(i)(-i - 1)]}{(-i - i)(z - 1)} = \frac{z + 1}{z - 1}.$$

Exercises 7.2

In Exercises 1–4 find the bilinear transformation that maps the circle passing through the points z_j into the circle passing through the corresponding points w_j.

1. $z_1 = 1$, $z_2 = i$, $z_3 = -1$; $w_1 = 2i$, $w_2 = -2$, $w_3 = -2i$
2. $z_1 = 2$, $z_2 = 1 + i$, $z_3 = 0$; $w_1 = 1$, $w_2 = i$, $w_3 = -i$
3. $z_1 = -1 + 2i$, $z_2 = -i$, $z_3 = 3 + 4i$; $w_1 = 0$, $w_2 = 1 + i$, $w_3 = -1 + i$
4. $z_1 = -1 - i$, $z_2 = 4 + 5i$, $z_3 = 1 - i$; $w_1 = 3 - 2i$, $w_2 = -1$, $w_3 = -1 + 3i$

In Exercises 5–8 find the bilinear transformation that maps the straight line passing through the points z_1, z_2, and z_3 onto the line passing through w_1 and w_2 in such a way that z_3 is the pole of the transformation.

5. $z_1 = 0$, $z_2 = 1+i$, $z_3 = 2+2i$; $w_1 = 3$, $w_2 = 3+4i$
6. $z_1 = 2-i$, $z_2 = -1+2i$, $z_3 = 1$; $w_1 = -1$, $w_2 = 1$
7. $z_1 = -3+i$, $z_2 = 3+i$, $z_3 = i$; $w_1 = i$, $w_2 = -i$
8. $z_1 = 2+5i$, $z_2 = -1+i$, $z_3 = -4-3i$; $w_1 = 1-3i$, $w_2 = -3+6i$
9. What bilinear transformation results in Exercise 3 if w_2 and w_3 are interchanged? How does it differ from the result in Exercise 3?

In Exercises 10–13 find a bilinear transformation that performs the indicated mapping.

10. Maps the interior of the circle in the z-plane defined by $1+i$, $-1+i$, and $-\sqrt{2}i$ onto the half-plane $\operatorname{Im} w > 0$
11. Maps the exterior of the circle $|z+1| = 2$ onto the interior of the circle $|w| = 1$
12. Maps the left half-plane $\operatorname{Re} z < 0$ onto the circle $|w| < 1$
13. Maps the exterior of the circle $|z+1-2i| = 3$ onto the half-plane $\operatorname{Re} w > 0$
14. Show that the bilinear mapping $w = (1+z)/(1-z)$ maps the half-plane $\operatorname{Re} z < 0$ onto the circle $|w| < 1$.
15. A bilinear transformation maps a circle onto a straight line when the pole of the transformation is on the circle. If two circles (one inside the other) share a common tangent line and the point of tangency is chosen as pole for a bilinear transformation, then the crescent-shaped domain between the circles is mapped onto an infinite strip. Illustrate this by finding a bilinear transformation that maps the circle $|z-1| = 1$ onto the u-axis in the w-plane and has $z = 0$ as pole. Now show that any other circle $|z-R| = R$ is mapped by your transformation onto a line parallel to the u-axis.
16. If the bilinear transformation $w = R(1-z)/(1+z)$ maps the half-plane $\operatorname{Re} z > 0$ onto the circle $|w| < R$, what bilinear transformation maps $|z| < R$ onto $\operatorname{Re} w > 0$?
17. Show that the composition of two bilinear transformations is a bilinear transformation.
18. When a bilinear transformation maps a circle to a circle, is the center of the circle mapped onto the center of the circle?
19. **(a)** Show that bilinear transformations that map the circle $|z| < R$ onto the half-plane $\operatorname{Im} w > 0$ so that z_0 on $|z| = R$ is mapped to $w = 0$ and $-z_0$ is the pole of the transformation are of the form $w = a(z-z_0)/(z+z_0)$. Is a arbitrary?
    (b) What is a if a point z_1 on $|z| = R$ is to map to w_1? Are there any restrictions on z_1 and w_1?
    (c) What is the transformation when $z_0 = R$, $z_1 = Ri$, and $w_1 = 1$?
20. **(a)** Points z_1 and z_2 are inverses with respect to the circle $|z-z_0| = R$ if they lie on the same half-line through the center of the circle ($\operatorname{Arg}(z_1 - z_0) = \operatorname{Arg}(z_2 - z_0)$) and the product of their distances from the center of the circle is equal to the square of the radius ($|z_1 - z_0|\,|z_2 - z_0| = R^2$) (see the accompanying figure). We say that the center of the circle and the point at infinity are inverses. Show that

$$z_2 = z_0 + \frac{R^2(z_1 - z_0)}{|z_1 - z_0|^2}.$$

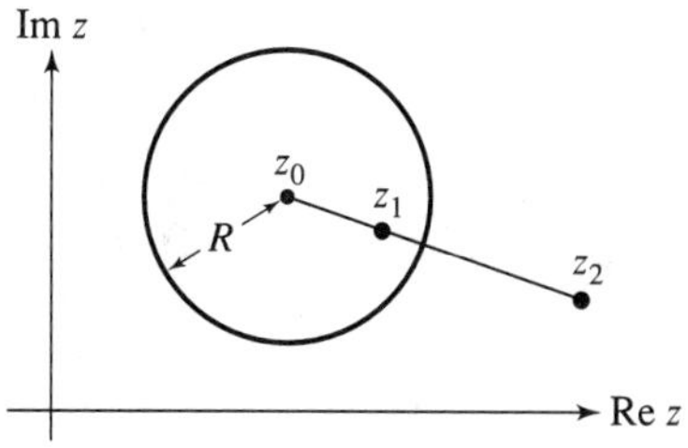

(b) Points z_1 and z_2 are inverses with respect to a straight line if they are mirror images of one another in the line. Determine whether the following pairs of points are inverses with respect to the given line:

(i) $3i$ and $2+i$ with respect to the line through -1 and i

(ii) $3i$ and $2+i$ with respect to the line through -1 and $-i$

(iii) 2 and $-2+2i$ with respect to the line through 1 and $2i$

21. Bilinear mappings preserve inverse points of Exercise 20; that is, they map inverse points to inverse points. Prove this for translations, rotations, and magnifications.

22. (a) Show that all bilinear transformations that map the half-plane $\operatorname{Im} z > 0$ onto the circle $|w| < R$ so that some given point z_0 maps onto $w = 0$ can be expressed in the form $w = Re^{\lambda i}(z - z_0)/(z - \overline{z_0})$, where λ is a real number. *Hint:* Use the fact that bilinear transformations preserve inverse points (Exercise 21).

(b) Show that all bilinear transformations that map the circle $|z| < R$ onto the half-plane $\operatorname{Im} w > 0$ so that the center $z = 0$ of the circle maps onto $w = w_0$ can be expressed in the form $w = (\overline{w_0} z - w_0 R e^{\lambda i})/(z - Re^{\lambda i})$.

(c) What form does the transformation in (b) take if w_0 is purely imaginary?

23. Show that all bilinear transformations that map the unit circle $|z| = 1$ onto the unit circle $|w| = 1$ so that some given point z_0 maps onto $w = 0$ can be expressed in the form $w = e^{\lambda i}(z - z_0)/(1 - \overline{z_0} z)$, where λ is a real number. *Hint:* Use the fact that bilinear transformations preserve inverse points (Exercise 21).

24. (a) If a bilinear transformation 7.2 maps the unit circle $|z| < 1$ onto some simply connected domain D, what transformation maps $|z| < R$ onto D?

(b) What transformation maps the circle $|z - z_0| < R$ onto D?

7.3 Mappings to the Half-Plane $\operatorname{Im} w > 0$

Frequently in applications it is necessary to map a domain D in the z-plane onto the half-plane $\operatorname{Im} w > 0$. Figures 7.12–7.17 illustrate mappings that do this.

Composition of these transformations with others leads to further mappings of domains in the z-plane onto the half-plane $\operatorname{Im} w > 0$. We illustrate in the following examples.

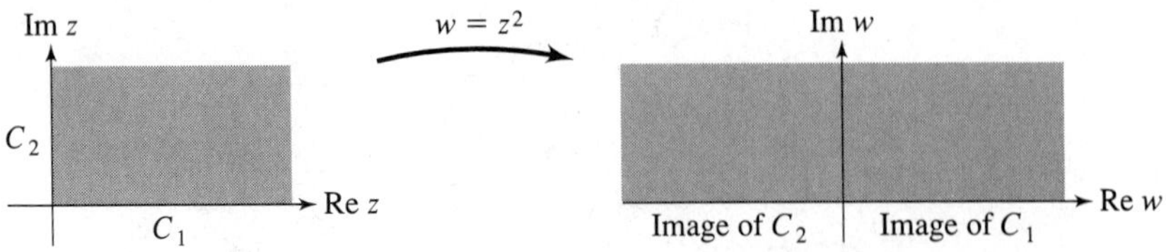

Figure 7.12 (see Exercise 7.1–1(a))

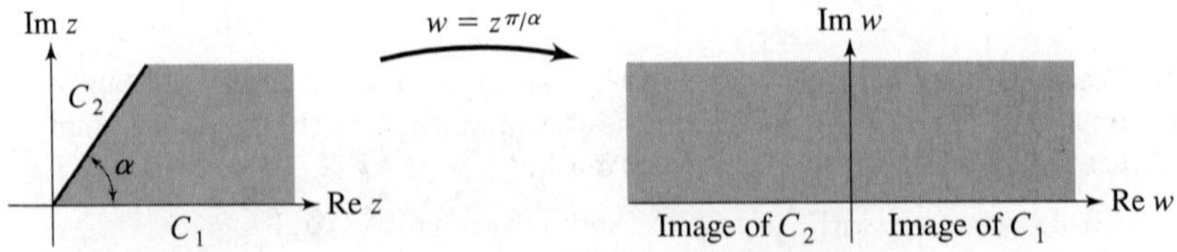

Figure 7.13 (see Exercise 7.1–6(b))

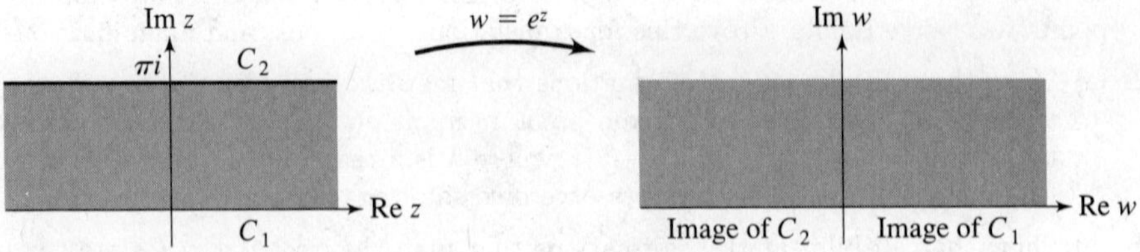

Figure 7.14 (see Figure 3.5)

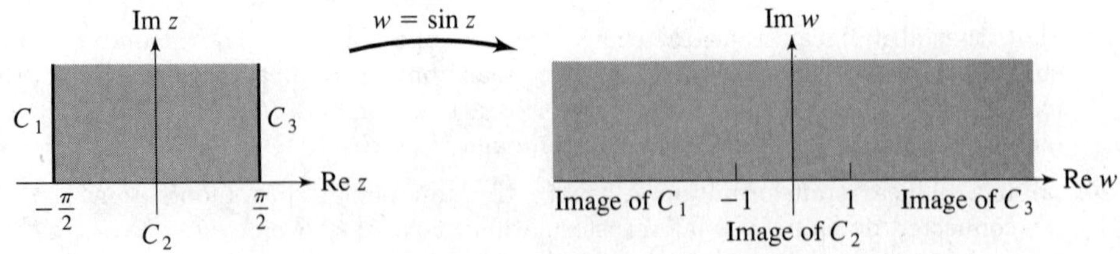

Figure 7.15 (see Exercise 7.1–8(d))

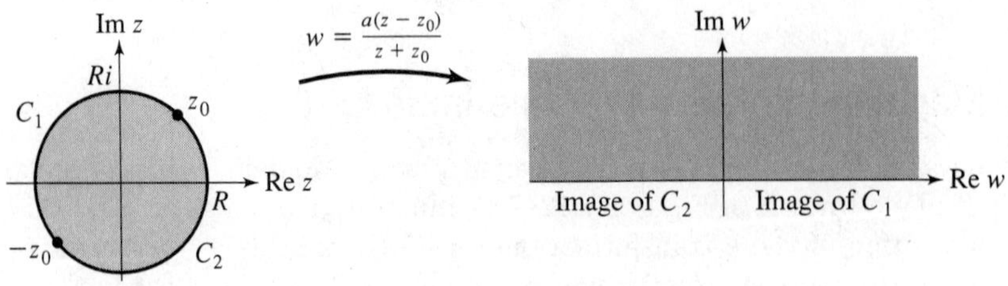

Figure 7.16 (see Exercise 7.2–19)

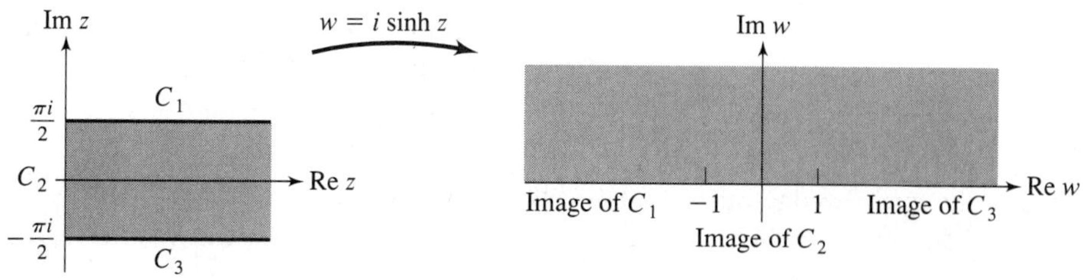

Figure 7.17

EXAMPLE 7.9 ....

Find a transformation that maps the semi-infinite strip in Figure 7.18 onto the half-plane Im $w > 0$.

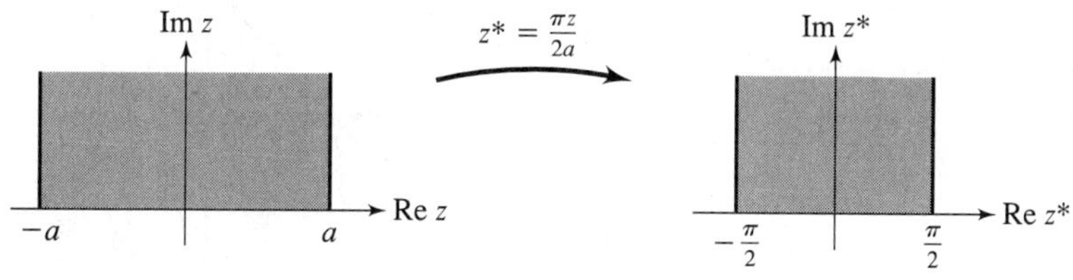

Figure 7.18

Solution The magnification $z^* = \pi z/(2a)$ maps the given semi-infinite strip onto the semi-infinite strip $-\pi/2 < \operatorname{Re} z^* < \pi/2$, $\operatorname{Im} z^* > 0$. When this is composed with the mapping in Figure 7.15, we obtain the mapping

$$w = \sin z^* = \sin\left(\frac{\pi z}{2a}\right).$$

It maps the given semi-infinite strip onto the upper half-plane.

....

EXAMPLE 7.10 ....

Find a transformation that maps the wedge in Figure 7.19 onto the half-plane Im $w > 0$.

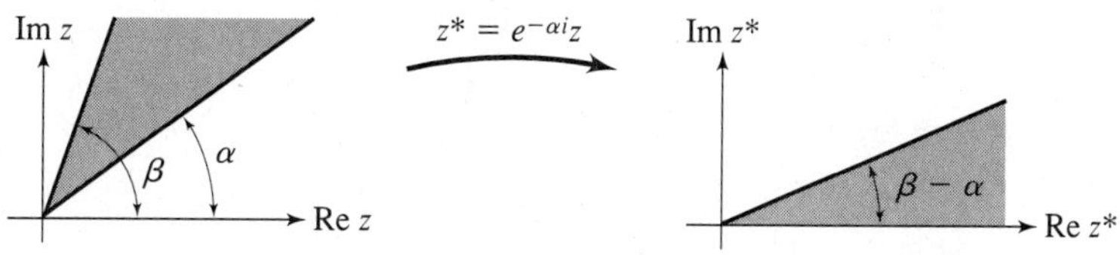

Figure 7.19

Solution The rotation $z^* = e^{-\alpha i}z$ rotates the wedge clockwise through angle α. When this is composed with the mapping in Figure 7.13, the result is

$$w = (z^*)^{\pi/(\beta-\alpha)} = (e^{-\alpha i}z)^{\pi/(\beta-\alpha)} = e^{-\alpha\pi i/(\beta-\alpha)}z^{\pi/(\beta-\alpha)}.$$

It maps the given wedge onto the half-plane $\operatorname{Im} w > 0$.

EXAMPLE 7.11

Find a transformation that maps the channel $0 < \operatorname{Im} z < a$ onto the half-plane $\operatorname{Im} w > 0$.

Solution The magnification $z^* = \pi z/a$ maps the channel $0 < \operatorname{Im} z < a$ onto the channel $0 < \operatorname{Im} z^* < \pi$ (Figure 7.20).

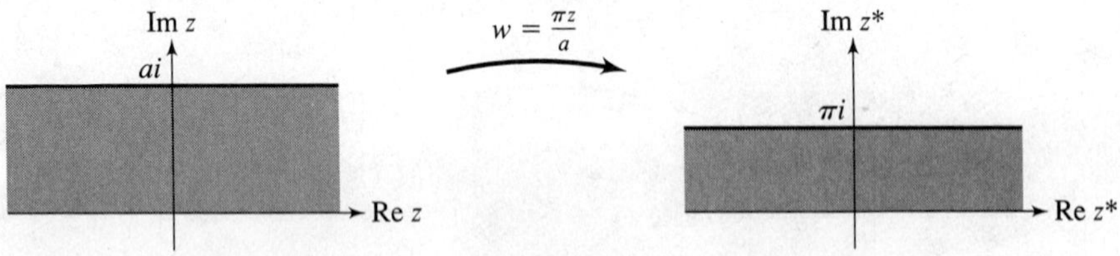

Figure 7.20

When this is composed with the function $w = e^z$ (Figure 7.14), the result is

$$w = e^{z^*} = e^{\pi z/a}.$$

It maps the channel $0 < \operatorname{Im} z < a$ onto $\operatorname{Im} w > 0$.

Exercises 7.3

In Exercises 1–12 show that the transformation maps the domain D onto the half-plane $\operatorname{Im} w > 0$.

1. $w = e^{iz}, \quad D: 0 < \operatorname{Re} z < \pi$
2. $w = e^{\pi z}, \quad D: 0 < \operatorname{Im} z < 1$
3. $w = \sin(\pi z/6), \quad D: -3 < \operatorname{Re} z < 3, \operatorname{Im} z > 0$
4. $w = \sin^2 z, \quad D: 0 < \operatorname{Re} z < \pi/2, \operatorname{Im} z > 0$
5. $w = i\sinh[\pi z/(2a)], \quad D: -a < \operatorname{Im} z < a, \operatorname{Re} z > 0$
6. $w = \sinh^2[\pi z/(2a)], \quad D: 0 < \operatorname{Im} z < a, \operatorname{Re} z > 0$
7. $w = z^4, \quad D: 0 < \operatorname{Arg} z < \pi/4$
8. $w = z^{2/3}, \quad D: 0 < \operatorname{Arg} z < 3\pi/2$
9. $w = i(R-z)/(R+z), \quad D: |z| < R$
10. $w = i(z-R)/(z+R), \quad D: |z| > R$

11. $w = (R+z)^2/(R-z)^2$, $D: |z| < R$, $\operatorname{Im} z > 0$
12. $w = z + R^2/z$, $D: |z| > R$, $\operatorname{Im} z > 0$

In Exercises 13–19 find a transformation that maps the domain D onto the half-plane $\operatorname{Im} w > 0$.

13. $D: -1 < \operatorname{Re} z < 1$, $\operatorname{Im} z > 0$
14. $D: 0 < \operatorname{Re} z < a$, $\operatorname{Im} z > 0$
15. $D: -1 < \operatorname{Im} z < 1$, $\operatorname{Re} z > 0$
16. $D: -1 < \operatorname{Im} z < 1$, $\operatorname{Re} z < 0$
17. $D: 0 < \operatorname{Im} z < 4$
18. $D: 0 < \operatorname{Re} z < 2$
19. $D: 1 < \operatorname{Im} z < 2$

20. **(a)** Show that the transformation $w = \sqrt{z - \dfrac{1}{4a}} - \dfrac{i}{2\sqrt{a}}$ maps the domain below to the quarter-plane $\operatorname{Re} w > 0$, $\operatorname{Im} w > 0$.
    (b) Find a transformation that maps D to the half-plane $\operatorname{Im} w > 0$.

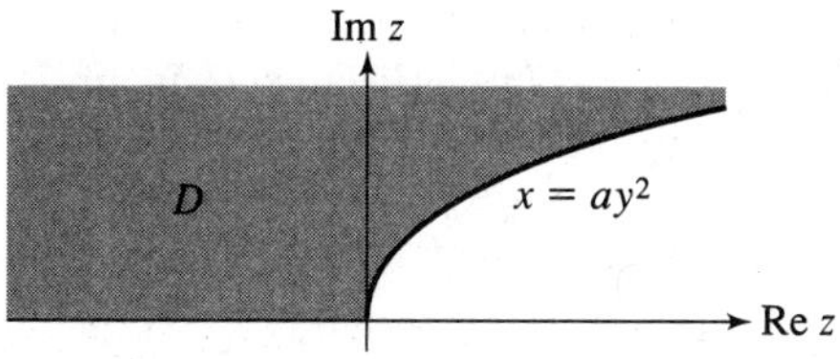

21. Show that $w = \coth(\pi a/z)$ maps the half-plane $\operatorname{Im} z > 0$ outside the circle $|z - ai| = a$ to the half-plane $\operatorname{Im} w > 0$.
22. Find the transformation that maps that part of the first quadrant outside the circle $x^2 + y^2 = a^2$ to the half-plane $\operatorname{Im} w > 0$. *Hint:* See Exercise 12.

7.4 Conformal Mappings

Suppose that f is analytic in a domain D and C is a smooth curve in D passing through a point z_0 at which $f'(z_0) \neq 0$ (Figure 7.21). If C is defined parametrically by

$$C: z(t) = x(t) + y(t)i, \qquad t_A \leq t \leq t_B, \tag{7.9}$$

then real and imaginary parts of $z'(t_0) = x'(t_0) + y'(t_0)i$ define the components of a tangent vector to C at z_0. We assume that $z'(t_0)$ does not vanish.

Under the mapping $w = f(z)$, C is mapped onto a curve $f(C)$ in the w-plane defined by $w(t) = f[z(t)]$. The components of a tangent vector to $f(C)$ at $f(z_0)$ are the real and imaginary parts of

$$w'(t_0) = \left(\frac{df}{dz}\frac{dz}{dt}\right)_{|z_0} = f'(z_0)z'(t_0), \tag{7.10}$$

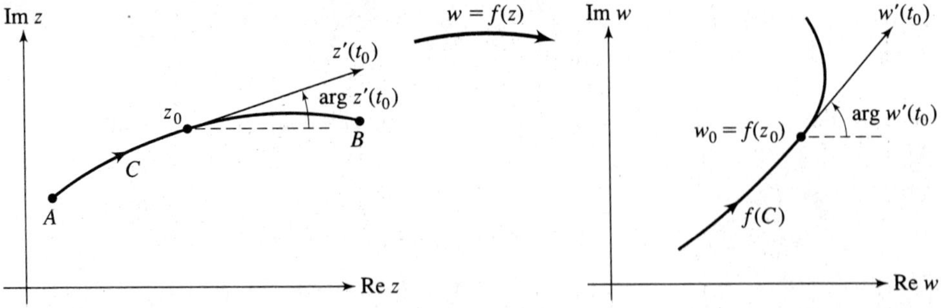

Figure 7.21

which cannot vanish since both terms on the right are nonzero. When we take arguments of both sides of this equation

$$\arg[w'(t_0)] = \arg[f'(z_0)] + \arg[z'(t_0)], \tag{7.11}$$

which relates the angle of inclination of the tangent vector to C at z_0 to that of the tangent vector to $f(C)$ at $w_0 = f(z_0)$. Because C is any curve whatsoever through z_0, we have shown that every curve through z_0 is rotated by an amount $\arg[f'(z_0)]$ under the mapping $w = f(z)$. Consequently, if C_1 and C_2 are two curves through z_0, the angle between their images $f(C_1)$ and $f(C_2)$ is exactly the same as the angle between C_1 and C_2 (Figure 7.22).

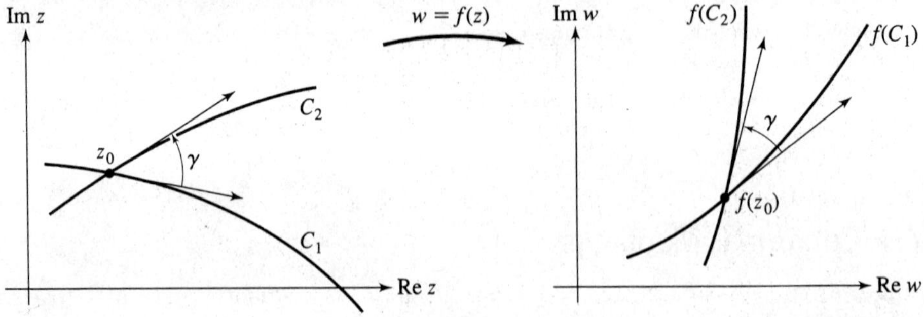

Figure 7.22

Mappings that preserve angles between curves and their orientation are said to be **conformal**. We have shown the following result.

■ **THEOREM 7.4** An analytic function f is conformal at every point z_0 at which $f'(z_0) \neq 0$. ■

We have just illustrated that at a point z_0 where an analytic function f has a nonvanishing derivative, curves are all rotated through an angle $\arg[f'(z_0)]$. We now give a geometric interpretation of $|f'(z_0)|$. Length along the curve C defined by 7.9 as measured from z_0 is given by

$$s = \int_{z_0}^{z} \sqrt{(dx)^2 + (dy)^2}.$$

But $dz = dx + dy\, i$, so that

$$s = \int_{z_0}^{z} |dz|.$$

Similarly, if $f(C)$ is the image of C under $w = f(z) = u + vi$, length along $f(C)$ measured from $w_0 = f(z_0)$ is

$$s' = \int_{w_0}^{w} |dw| = \int_{w_0}^{w} \sqrt{(du)^2 + (dv)^2}.$$

But $dw/dz = f'(z)$, and hence $|dw| = |f'(z)|\,|dz|$. In particular, at z_0,

$$|dw| = |f'(z_0)|\,|dz|; \tag{7.12}$$

that is, at w_0 the element of length $|dw|$ along $f(C)$ is $|f'(z_0)|$ times that at z_0 along C. In other words, the modulus $|f'(z_0)|$ is the length magnification factor at z_0 for the mapping $w = f(z)$.

We now have a geometric interpretation for $f'(z)$ in terms of the mapping $w = f(z)$ — its modulus is a length-magnification factor, and its argument represents the rotational effect of the mapping.

The mapping $w = z^2$ of Example 7.1 is conformal at every point except $z = 0$. At $z = 0$, $w = z^2$ preserves the sense of angles but doubles them. The functions e^z and $\log_\phi z$ of Examples 7.2 and 7.3 are conformal at every point at which they are differentiable. In particular, notice that in Example 7.2, vertical and horizontal lines (which intersect at right angles) are mapped onto circles and radial lines (that also intersect orthogonally). In Example 7.3, circles and radial lines are mapped by $\log_\phi z$ (the inverse of e^z) onto vertical and horizontal lines.

EXAMPLE 7.12 ....

Illustrate the conformality of the mapping $w = z^2$ at every point (except $z = 0$) by showing that horizontal and vertical lines, which intersect at right angles, are mapped onto curves that also intersect at right angles.

Solution If we set $w = u + vi$, then $u + vi = (x + yi)^2$, implying that

$$u = x^2 - y^2, \qquad v = 2xy.$$

A horizontal line $y = y_0$ in the z-plane is mapped by $w = z^2$ onto a curve defined parametrically by

$$u = x^2 - y_0^2, \qquad v = 2xy_0.$$

If x is eliminated,

$$u = \frac{v^2}{4y_0^2} - y_0^2 \qquad (y_0 \neq 0),$$

a parabola (Figure 7.23).

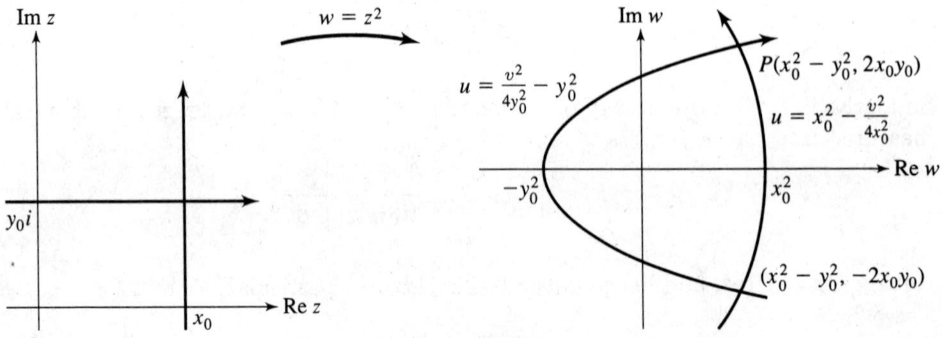

Figure 7.23

A vertical line $x = x_0$ is mapped by $w = z^2$ onto the curve

$$u = x_0^2 - y^2, \qquad v = 2x_0 y,$$

or

$$u = x_0^2 - \frac{v^2}{4x_0^2}, \qquad (x_0 \neq 0).$$

This is also a parabola. The two parabolas intersect in the points $(x_0^2 - y_0^2, \pm 2x_0y_0)$, and the point $P(x_0^2 - y_0^2, 2x_0y_0)$ is the image of (x_0, y_0) under $w = z^2$. The slope dv/du of the former parabola at any point is defined by

$$1 = \frac{2v}{4y_0^2}\frac{dv}{du},$$

and hence the slope of this parabola at P is

$$\frac{dv}{du}_{|P} = \frac{2y_0^2}{v}_{|P} = \frac{2y_0^2}{2x_0y_0} = \frac{y_0}{x_0}.$$

Similarly, the slope of $u = x_0^2 - v^2/(4x_0^2)$ at P is

$$\frac{dv}{du}_{|P} = -\frac{2x_0^2}{v}_{|P} = -\frac{2x_0^2}{2x_0y_0} = -\frac{x_0}{y_0}.$$

Clearly then, these parabolas intersect at right angles at P.

We have shown that horizontal and vertical lines in the z-plane are mapped by $w = z^2$ onto parabolas in the w-plane that also intersect orthogonally. Notice, however, that the argument is not valid at $z = 0$ (where $x_0 = 0$ and $y_0 = 0$). The real z-axis $y = 0$ is mapped onto the positive part of the real w-axis ($y = 0$ implies $u = x^2$, $v = 0$), and the imaginary z-axis $x = 0$ is mapped onto the negative part of the real w-axis. Thus, the right angle between the axes at $z = 0$ is doubled to an angle of π radians at $w = 0$.

Exercises 7.4

1. Prove that if the argument of $f'(z)$ is constant along a straight-line segment in the z-plane, then the image of the line segment in the $w = f(z)$ plane is also a straight-line segment.

In Exercises 2–5 illustrate conformality of the mapping $w = f(z)$ by showing that horizontal and vertical lines, which intersect at right angles, are mapped onto curves that also intersect at right angles.

2. $f(z) = e^{2z}$
3. $f(z) = 2z^2 + 1$
4. $f(z) = \cos z$
5. $f(z) = \sinh z$
6. Illustrate conformality of the mapping $w = \text{Log}(z - 2 - i)$ by showing that circles centered at $2 + i$ and radial lines from this point are mapped onto vertical and horizontal lines.
7. Verify that $w = \cos z$ is not a conformal mapping in any neighborhood of the points $z = n\pi$ where n is an integer.

In Exercises 8–11 find the length-magnification factor of the mapping at the point.

8. $f(z) = 3z^2 + 2z$ at $z = 1 - i$
9. $f(z) = (z + 1)/(2z - i)$ at $z = 3i$
10. $f(z) = \cosh(3z - 1)$ at $z = 1 - i$
11. $f(z) = z \sin z$ at $z = -2i$
12. When a certain curve C is mapped by a conformal mapping $w = f(z)$ onto a curve C', its length is doubled. Does this imply that every curve with the same length as C has its length doubled under the mapping? Explain.
13. If $|f'(z)|$ is a length-magnification factor for a conformal mapping, then $|f'(z)|^2$ is an area-magnification factor. To illustrate this consider the mapping $w = e^{2z}$.
    (a) What is the length-magnification factor of $f(z)$ at $z = 0$?
    (b) The area of the rectangle $-a \le x \le a$, $-b \le y \le b$ is $4ab$. Find the area of the image of this rectangle in the w-plane when $a < 1$ and $b < \pi/2$.
    (c) Show that the limit of the ratio of the areas in (b) as $a \to 0$ and $b \to 0$ is 4.
    (d) Does this mean that regions in the z-plane have their areas quadrupled when mapped to the w-plane by $w = e^{2z}$?
14. A mapping $w = f(z)$ that is analytic is not conformal at a point z_0 at which $f'(z_0) = 0$. Show that when z_0 is a zero of order m, then curves with angle β between their tangents at z_0 are mapped to curves with angle $m\beta$ between their tangents at $w_0 = f(z_0)$.
15. Show that if $w = f(z)$ is 1-1 and conformal in a domain D, then $z = f^{-1}(w)$ is conformal in $f(D)$.

7.5 Conformal Mappings and Harmonic Functions

Conformal mappings are important in the solution of boundary-value problems for the two-dimensional Laplace equation. To see how, suppose that $\phi(x, y)$ is harmonic in some domain D of the xy-plane. Let $w = f(z)$ be a 1-1 conformal mapping from D to $f(D)$. We set $f(z) = u + vi$, where, as usual,

$$u = u(x, y), \qquad v = v(x, y). \tag{7.13}$$

Because the mapping is 1-1, the inverse function $f^{-1}(w)$ is defined throughout $f(D)$, and it follows that equations 7.13 can be solved for x and y in terms of u and v:

$$x = x(u, v), \qquad y = y(u, v). \tag{7.14}$$

We use $\phi(x, y)$ to define a function $\psi(u, v)$ in $f(D)$ according to

$$\psi(u, v) = \phi[x(u, v), y(u, v)]. \tag{7.15}$$

Alternatively, $\psi(u, v)$ is that function satisfying

$$\psi[u(x, y), v(x, y)] = \phi(x, y). \tag{7.16}$$

Now, $\phi(x, y)$ is harmonic in D, so that in D

$$\frac{\partial^2 \phi}{\partial x^2} + \frac{\partial^2 \phi}{\partial y^2} = 0. \tag{7.17}$$

By chain rules for partial derivatives, we may write

$$\frac{\partial \phi}{\partial x} = \frac{\partial \psi}{\partial u}\frac{\partial u}{\partial x} + \frac{\partial \psi}{\partial v}\frac{\partial v}{\partial x}$$

and

$$\frac{\partial^2 \phi}{\partial x^2} = \left(\frac{\partial^2 \psi}{\partial u^2}\frac{\partial u}{\partial x} + \frac{\partial^2 \psi}{\partial v \partial u}\frac{\partial v}{\partial x}\right)\frac{\partial u}{\partial x} + \frac{\partial \psi}{\partial u}\frac{\partial^2 u}{\partial x^2} + \left(\frac{\partial^2 \psi}{\partial u \partial v}\frac{\partial u}{\partial x} + \frac{\partial^2 \psi}{\partial v^2}\frac{\partial v}{\partial x}\right)\frac{\partial v}{\partial x} + \frac{\partial \psi}{\partial v}\frac{\partial^2 v}{\partial x^2}.$$

Similarly,

$$\frac{\partial^2 \phi}{\partial y^2} = \left(\frac{\partial^2 \psi}{\partial u^2}\frac{\partial u}{\partial y} + \frac{\partial^2 \psi}{\partial u \partial v}\frac{\partial v}{\partial y}\right)\frac{\partial u}{\partial y} + \frac{\partial \psi}{\partial u}\frac{\partial^2 u}{\partial y^2} + \left(\frac{\partial^2 \psi}{\partial u \partial v}\frac{\partial u}{\partial y} + \frac{\partial^2 \psi}{\partial v^2}\frac{\partial v}{\partial y}\right)\frac{\partial v}{\partial y} + \frac{\partial \psi}{\partial v}\frac{\partial^2 v}{\partial y^2}.$$

Consequently,

$$\begin{aligned}
0 &= \frac{\partial^2 \phi}{\partial x^2} + \frac{\partial^2 \phi}{\partial y^2} \\
&= \frac{\partial^2 \psi}{\partial u^2}\left[\left(\frac{\partial u}{\partial x}\right)^2 + \left(\frac{\partial u}{\partial y}\right)^2\right] + 2\frac{\partial^2 \psi}{\partial u \partial v}\left(\frac{\partial u}{\partial x}\frac{\partial v}{\partial x} + \frac{\partial u}{\partial y}\frac{\partial v}{\partial y}\right) \\
&\quad + \frac{\partial^2 \psi}{\partial v^2}\left[\left(\frac{\partial v}{\partial x}\right)^2 + \left(\frac{\partial v}{\partial y}\right)^2\right] + \frac{\partial \psi}{\partial u}\left(\frac{\partial^2 u}{\partial x^2} + \frac{\partial^2 u}{\partial y^2}\right) + \frac{\partial \psi}{\partial v}\left(\frac{\partial^2 v}{\partial x^2} + \frac{\partial^2 v}{\partial y^2}\right).
\end{aligned}$$

Because $f(z) = u + vi$ is analytic, u and v are harmonic and satisfy the Cauchy-Riemann equations in D; that is, in D,

$$0 = \frac{\partial^2 u}{\partial x^2} + \frac{\partial^2 u}{\partial y^2}; \qquad 0 = \frac{\partial^2 v}{\partial x^2} + \frac{\partial^2 v}{\partial y^2}; \qquad \frac{\partial u}{\partial x} = \frac{\partial v}{\partial y}; \qquad \frac{\partial u}{\partial y} = -\frac{\partial v}{\partial x}.$$

Hence, we may write

$$\begin{aligned}
0 &= \frac{\partial^2 \psi}{\partial u^2}\left[\left(\frac{\partial u}{\partial x}\right)^2 + \left(\frac{\partial v}{\partial x}\right)^2\right] + 2\frac{\partial^2 \psi}{\partial u \partial v}\left(\frac{\partial u}{\partial x}\frac{\partial v}{\partial x} - \frac{\partial v}{\partial x}\frac{\partial u}{\partial x}\right) \\
&\quad + \frac{\partial^2 \psi}{\partial v^2}\left[\left(\frac{\partial v}{\partial x}\right)^2 + \left(\frac{\partial u}{\partial x}\right)^2\right] \\
&= \left(\frac{\partial^2 \psi}{\partial u^2} + \frac{\partial^2 \psi}{\partial v^2}\right)\left[\left(\frac{\partial u}{\partial x}\right)^2 + \left(\frac{\partial v}{\partial x}\right)^2\right].
\end{aligned}$$

Because $f'(z) = \frac{\partial u}{\partial x} + \frac{\partial v}{\partial x} i \neq 0$ in D,

$$0 \neq |f'(z)|^2 = \left(\frac{\partial u}{\partial x}\right)^2 + \left(\frac{\partial v}{\partial x}\right)^2,$$

and therefore in $f(D)$,

$$0 = \frac{\partial^2 \psi}{\partial u^2} + \frac{\partial^2 \psi}{\partial v^2}. \tag{7.18}$$

Thus, $\psi(u, v)$ is harmonic in $f(D)$. We have shown, therefore, that conformal mappings carry harmonic functions to harmonic functions.

But why is this so important? To answer this, consider the problem of finding a function $\phi(x, y)$ satisfying Laplace's equation 7.17 in the domain D of the xy-plane in Figure 7.24. Suppose we interpret this plane as the complex z-plane, and also suppose that we can find a 1-1 conformal mapping $w = f(z) = u + vi$ that takes D into a simpler domain, say perhaps the half-plane $\operatorname{Im} w > 0$. Then the function

$$\psi(u, v) = \phi[x(u, v), y(u, v)],$$

where $x = x(u, v)$ and $y = y(u, v)$ are the inverses of $u = u(x, y)$ and $v = v(x, y)$,

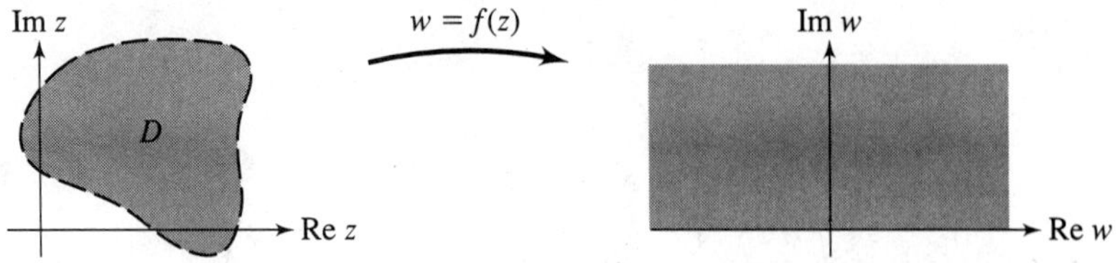

Figure 7.24

satisfies Laplace's equation in $\operatorname{Im} w > 0$. Further, any conditions that $\phi(x, y)$ satisfies on the boundary of D impose conditions that $\psi(u, v)$ must satisfy on $\operatorname{Im} w = 0$. Consequently, the problem of finding $\phi(x, y)$ harmonic in D and satisfying conditions on the rather unpleasant boundary of D can be replaced by the problem of finding $\psi(u, v)$ harmonic in $\operatorname{Im} w > 0$ and satisfying conditions on the line $\operatorname{Im} w = 0$. If we solve the latter problem, then the solution of the original problem is

$$\phi(x, y) = \psi[u(x, y), v(x, y)].$$

Clearly then, the ability to map a domain D in the z-plane onto a simpler domain in the w-plane under a conformal mapping $w = f(z)$ could be very useful in solving boundary-value problems involving Laplace's equation. We illustrate with the following example.

EXAMPLE 7.13 ....

Find the solution of Laplace's equation in the circle $x^2 + y^2 < 1$ that has value 0 on the semicircle $x^2 + y^2 = 1$, $y > 0$, and value 1 on the semicircle $x^2 + y^2 = 1$, $y < 0$.

Solution According to Poisson's integral formula 4.35, the solution can be expressed in polar coordinates as

$$\phi(r, \theta) = \frac{1 - r^2}{2\pi} \int_{-\pi}^{0} \frac{1}{1 + r^2 - 2r\cos(\theta - \zeta)} \, d\zeta.$$

It is difficult, but not impossible, to evaluate this integral in closed form.

Instead, we map the unit circle $|z| \leq 1$ onto the half-plane $v = \operatorname{Im} w \geq 0$ with a bilinear transformation in such a way that the upper semicircle is mapped onto the positive $u = \operatorname{Re} w$ axis and the lower semicircle onto the negative u-axis. If we choose to map $z = 1$ onto $w = 0$, then $z = -1$ maps to points infinitely far out the real u-axis (Figure 7.25). Because $z = -1$ is the pole of the mapping, the transformation takes the form

$$w = \frac{a(z - 1)}{z + 1}.$$

(See also Figure 7.16 with $z_0 = 1$.) If we arbitrarily demand that $z = i$ map to $w = 1$, then $1 = a(i - 1)/(i + 1)$, and this requires that $a = -i$. The bilinear transformation is, therefore,

$$w = \frac{i(1 - z)}{1 + z}.$$

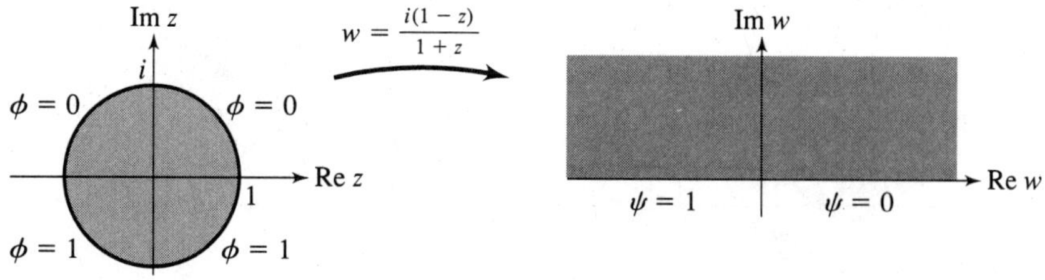

Figure 7.25

According to Poisson's integral formula 4.38 for a half-space, the solution for the function $\psi(u, v)$ harmonic in $v > 0$ is

$$\psi(u, v) = \frac{v}{\pi}\int_{-\infty}^{0} \frac{1}{(U-u)^2 + v^2}\, dU = \frac{v}{\pi}\left\{\frac{1}{v}\operatorname{Tan}^{-1}\left(\frac{U-u}{v}\right)\right\}_{-\infty}^{0}$$

$$= \frac{1}{\pi}\left[\operatorname{Tan}^{-1}\left(-\frac{u}{v}\right) + \frac{\pi}{2}\right] = \frac{1}{2} - \frac{1}{\pi}\operatorname{Tan}^{-1}\left(\frac{u}{v}\right).$$

Because

$$w = u + vi = i\left(\frac{1-z}{1+z}\right) = \frac{2y + i(1 - x^2 - y^2)}{(1+x)^2 + y^2},$$

it follows that $u = 2y/[(1+x)^2 + y^2]$ and $v = (1 - x^2 - y^2)/[(1+x)^2 + y^2]$, and therefore

$$\phi(x, y) = \psi[u(x, y), v(x, y)] = \frac{1}{2} - \frac{1}{\pi}\operatorname{Tan}^{-1}\left(\frac{2y}{1 - x^2 - y^2}\right).$$

Boundary conditions always accompany Laplace's equation in a domain D of the xy-plane, the most common being Dirichlet and Neumann conditions. A Dirichlet boundary condition specifies the value of the harmonic function $\phi(x, y)$ on the boundary $\beta(D)$ of D,

$$\phi(x, y) = F(x, y), \qquad (x, y) \text{ on } \beta(D). \tag{7.19a}$$

Transformation 7.14, induced by a conformal mapping $w = f(z)$, which maps D onto $f(D)$, replaces condition 7.19(a) on $\phi(x, y)$ with the following condition on the harmonic function $\psi(u, v) = \phi[x(u, v), y(u, v)]$:

$$\psi(u, v) = F[x(u, v), y(u, v)], \qquad (u, v) \text{ on } \beta[f(D)]. \tag{7.19b}$$

In other words, conformal mappings transform Dirichlet boundary conditions to Dirichlet boundary conditions.

A Neumann boundary condition specifies the directional derivative of $\phi(x, y)$ in the direction outwardly normal to $\beta(D)$ (Figure 7.26):

$$\nabla\phi \cdot \hat{\mathbf{n}} = \frac{\partial \phi}{\partial n} = F(x, y), \qquad (x, y) \text{ on } \beta(D). \tag{7.20a}$$

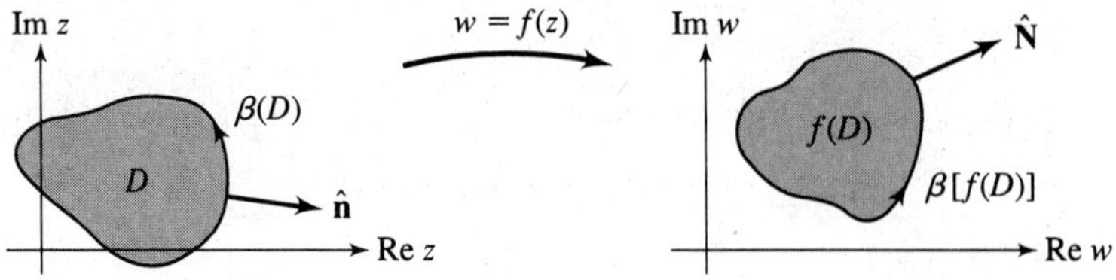

Figure 7.26

We now investigate the effect of a conformal mapping $w = f(z)$ on such a boundary condition. When s measures length counterclockwise around $\beta(D)$, the unit normal vector is defined by $\hat{\mathbf{n}} = (dy/ds)\hat{\mathbf{i}} - (dx/ds)\hat{\mathbf{j}}$, and therefore

$$\nabla\phi \cdot \hat{\mathbf{n}} = \frac{\partial\phi}{\partial x}\frac{dy}{ds} - \frac{\partial\phi}{\partial y}\frac{dx}{ds}.$$

Chain rules based on 7.16 give

$$\frac{\partial\phi}{\partial x} = \frac{\partial\psi}{\partial u}\frac{\partial u}{\partial x} + \frac{\partial\psi}{\partial v}\frac{\partial v}{\partial x}, \qquad \frac{\partial\phi}{\partial y} = \frac{\partial\psi}{\partial u}\frac{\partial u}{\partial y} + \frac{\partial\psi}{\partial v}\frac{\partial v}{\partial y}.$$

Furthermore, based on 7.14 and the fact that x and y can be expressed in terms of length s along $\beta(D)$, we may write $x = x[u(s), v(s)]$ and $y = y[u(s), v(s)]$. Chain rules for these give

$$\frac{dx}{ds} = \frac{\partial x}{\partial u}\frac{du}{ds} + \frac{\partial x}{\partial v}\frac{dv}{ds}, \qquad \frac{dy}{ds} = \frac{\partial y}{\partial u}\frac{du}{ds} + \frac{\partial y}{\partial v}\frac{dv}{ds}.$$

When these are all substituted into the expression for $\nabla\phi \cdot \hat{\mathbf{n}}$,

$$\begin{aligned}
\nabla\phi \cdot \hat{\mathbf{n}} &= \frac{\partial\phi}{\partial x}\frac{dy}{ds} - \frac{\partial\phi}{\partial y}\frac{dx}{ds} \\
&= \left(\frac{\partial\psi}{\partial u}\frac{\partial u}{\partial x} + \frac{\partial\psi}{\partial v}\frac{\partial v}{\partial x}\right)\left(\frac{\partial y}{\partial u}\frac{du}{ds} + \frac{\partial y}{\partial v}\frac{dv}{ds}\right) \\
&\quad - \left(\frac{\partial\psi}{\partial u}\frac{\partial u}{\partial y} + \frac{\partial\psi}{\partial v}\frac{\partial v}{\partial y}\right)\left(\frac{\partial x}{\partial u}\frac{du}{ds} + \frac{\partial x}{\partial v}\frac{dv}{ds}\right) \\
&= \frac{\partial\psi}{\partial u}\frac{du}{ds}\left(\frac{\partial u}{\partial x}\frac{\partial y}{\partial u} - \frac{\partial u}{\partial y}\frac{\partial x}{\partial u}\right) + \frac{\partial\psi}{\partial u}\frac{dv}{ds}\left(\frac{\partial u}{\partial x}\frac{\partial y}{\partial v} - \frac{\partial u}{\partial y}\frac{\partial x}{\partial v}\right) \\
&\quad + \frac{\partial\psi}{\partial v}\frac{du}{ds}\left(\frac{\partial v}{\partial x}\frac{\partial y}{\partial u} - \frac{\partial v}{\partial y}\frac{\partial x}{\partial u}\right) + \frac{\partial\psi}{\partial v}\frac{dv}{ds}\left(\frac{\partial v}{\partial x}\frac{\partial y}{\partial v} - \frac{\partial v}{\partial y}\frac{\partial x}{\partial v}\right).
\end{aligned}$$

Differentiation of $u = u[x(u, v), y(u, v)]$ with respect to u gives $1 = \dfrac{\partial u}{\partial x}\dfrac{\partial x}{\partial u} + \dfrac{\partial u}{\partial y}\dfrac{\partial y}{\partial u}$, and if we use the Cauchy-Riemann equations,

$$1 = \left(\frac{\partial v}{\partial y}\right)\frac{\partial x}{\partial u} + \left(-\frac{\partial v}{\partial x}\right)\frac{\partial y}{\partial u}.$$

Similar results for the other three terms in parentheses lead to

$$\nabla\phi\cdot\hat{\mathbf{n}} = \frac{\partial\psi}{\partial u}\frac{dv}{ds} - \frac{\partial\psi}{\partial v}\frac{du}{ds} = \left(\frac{\partial\psi}{\partial u}\frac{dv}{dS} - \frac{\partial\psi}{\partial v}\frac{du}{dS}\right)\frac{dS}{ds} = \nabla\psi\cdot\hat{\mathbf{N}}\left(\frac{dS}{ds}\right),$$

or

$$\nabla\psi\cdot\hat{\mathbf{N}} = |f'(z)|^{-1}\nabla\phi\cdot\hat{\mathbf{n}},$$

where $\hat{\mathbf{N}}$ is the unit outward-pointing normal to $f[\beta(D)]$, and S is length thereon. Neumann condition 7.20(a) along $\beta(D)$ is therefore transformed by $w = f(z)$ into a Neumann condition along $\beta[f(D)]$,

$$\nabla\psi\cdot\hat{\mathbf{N}} = \frac{\partial\psi}{\partial N} = \frac{F[x(u,v),y(u,v)]}{|f'[z(w)]|}, \qquad (u,v) \text{ on } \beta[f(D)]. \tag{7.20b}$$

Very often in applications it is necessary to solve Laplace's equation in a domain D together with a Dirichlet boundary condition that is piecewise constant on $\beta(D)$. Suppose, for example, that $\phi(x,y)$, harmonic in the simply connected domain of Figure 7.27, is to be equal to constant values a_j $(j = 1,\ldots,n)$ on the n parts C_j of the boundary $\beta(D)$.

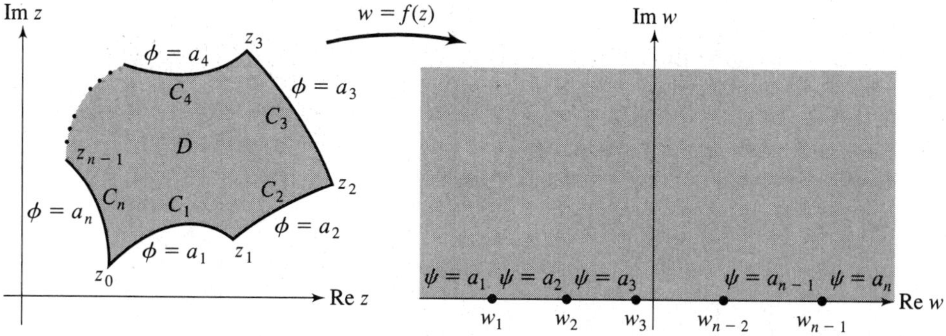

Figure 7.27

We can solve this problem very easily if a one-to-one conformal mapping $w = f(z)$ can be found that maps D onto the half-plane $\operatorname{Im} w > 0$. Indeed, suppose z_0 is the pole of the mapping (so that z_0 is mapped infinitely far out the $\operatorname{Re} w$ axis), and the z_j $(j = 1,\ldots,n-1)$ are mapped onto points $w_1,\ldots,w_{n-1}$. Once we find the function $\psi(u,v)$ satisfying Laplace's equation in $\operatorname{Im} w > 0$ and the boundary conditions along $\operatorname{Im} w = 0$, the solution of the original problem in D is $\phi(x,y) = \psi[u(x,y),v(x,y)]$. In other words, function $\psi(u,v)$ together with the conformal mapping $w = f(z)$ that maps D onto $\operatorname{Im} w > 0$ provide $\phi(x,y)$. The function $\psi(u,v)$ is derived in the following theorem.

■ **THEOREM 7.5** The solution of Laplace's equation $\nabla^2\psi = 0$ in the domain $\operatorname{Im} w > 0$ subject to the piecewise-constant Dirichlet boundary condition in

Figure 7.27 is

$$\psi(u,v) = \frac{a_1 + a_n}{2} + \frac{1}{\pi}\sum_{j=1}^{n-1}(a_j - a_{j+1})\,\mathrm{Tan}^{-1}\left(\frac{w_j - u}{v}\right). \tag{7.21}$$

Proof According to Poisson's integral formula 4.38,

$$\begin{aligned}\psi(u,v) &= \frac{v}{\pi}\int_{-\infty}^{w_1}\frac{a_1}{(U-u)^2+v^2}\,dU + \frac{v}{\pi}\sum_{j=2}^{n-1}\int_{w_{j-1}}^{w_j}\frac{a_j}{(U-u)^2+v^2}\,dU \\ &\quad + \frac{v}{\pi}\int_{w_{n-1}}^{\infty}\frac{a_n}{(U-u)^2+v^2}\,dU \\ &= \frac{a_1 v}{\pi}\left\{\frac{1}{v}\mathrm{Tan}^{-1}\left(\frac{U-u}{v}\right)\right\}_{-\infty}^{w_1} + \frac{a_j v}{\pi}\sum_{j=2}^{n-1}\left\{\frac{1}{v}\mathrm{Tan}^{-1}\left(\frac{U-u}{v}\right)\right\}_{w_{j-1}}^{w_j} \\ &\quad + \frac{a_n v}{\pi}\left\{\frac{1}{v}\mathrm{Tan}^{-1}\left(\frac{U-u}{v}\right)\right\}_{w_{n-1}}^{\infty} \\ &= \frac{a_1}{\pi}\left[\mathrm{Tan}^{-1}\left(\frac{w_1-u}{v}\right) + \frac{\pi}{2}\right] \\ &\quad + \frac{a_j}{\pi}\sum_{j=2}^{n-1}\left[\mathrm{Tan}^{-1}\left(\frac{w_j-u}{v}\right) - \mathrm{Tan}^{-1}\left(\frac{w_{j-1}-u}{v}\right)\right] \\ &\quad + \frac{a_n}{\pi}\left[\frac{\pi}{2} - \mathrm{Tan}^{-1}\left(\frac{w_{n-1}-u}{v}\right)\right]. \end{aligned} \tag{7.22}$$

Regrouping terms according to inverse tangents gives 7.21. ■

The angles represented by the inverse tangents in 7.22 are easily interpreted in Figure 7.28. We may write that

$$\psi(u,v) = \frac{1}{\pi}\sum_{j=1}^{n} a_j\delta_j. \tag{7.23}$$

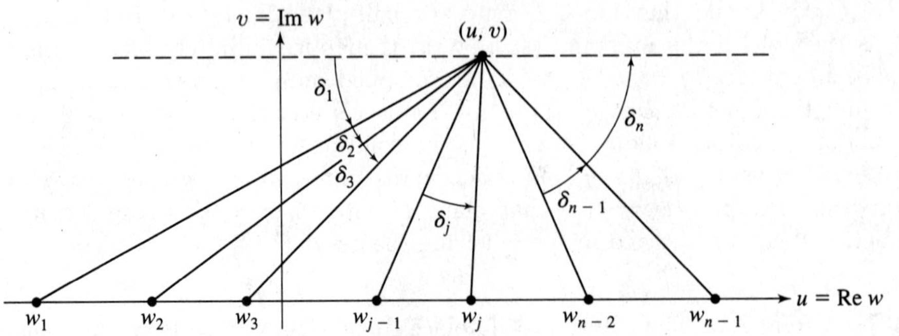

Figure 7.28

We use formula 7.21 along with a conformal mapping to solve the boundary-value problem in the following example. Many more examples are given in Sections 7.7–7.10.

EXAMPLE 7.14 ....

Find the solution of Laplace's equation in the semi-infinite strip $-\pi/2 < x < \pi/2$, $y > 0$ (Figure 7.29) given that its values along the edges are

$$\phi(-\pi/2, y) = \phi(\pi/2, y) = -1, \qquad y > 0,$$
$$\phi(x, 0) = 1, \qquad -\pi/2 < x < \pi/2.$$

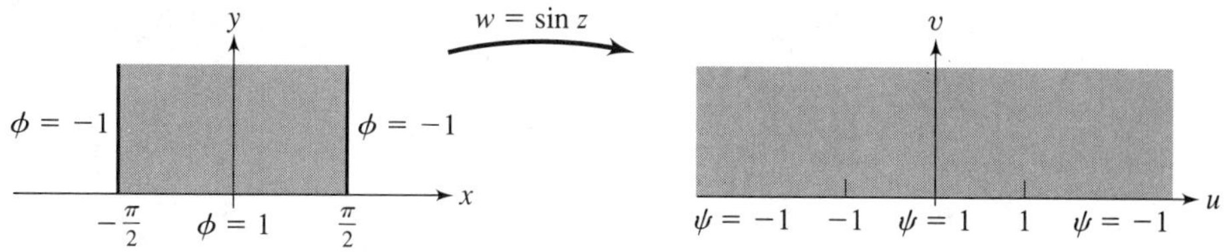

Figure 7.29

Solution The transformation $w = \sin z$ maps the semi-infinite strip onto the half-plane $\operatorname{Im} w > 0$ (see Figure 7.15). According to 7.21, the solution of Laplace's equation $\psi(u, v)$ with boundary conditions $\psi(u, 0) = \begin{cases} -1, & |u| > 1 \\ 1, & |u| < 1 \end{cases}$ is

$$\begin{aligned}\psi(u, v) &= \frac{1}{2}(-1 - 1) + \frac{1}{\pi}\left[-2\operatorname{Tan}^{-1}\left(\frac{-1-u}{v}\right) + 2\operatorname{Tan}^{-1}\left(\frac{1-u}{v}\right)\right] \\ &= -1 + \frac{2}{\pi}\left[\operatorname{Tan}^{-1}\left(\frac{1+u}{v}\right) + \operatorname{Tan}^{-1}\left(\frac{1-u}{v}\right)\right].\end{aligned}$$

Because $w = u + vi = \sin z = \sin x \cosh y + \cos x \sinh y\, i$, it follows that $u = \sin x \cosh y$ and $v = \cos x \sinh y$. The solution of Laplace's equation in the strip is, therefore,

$$\phi(x, y) = -1 + \frac{2}{\pi}\left[\operatorname{Tan}^{-1}\left(\frac{1 + \sin x \cosh y}{\cos x \sinh y}\right) + \operatorname{Tan}^{-1}\left(\frac{1 - \sin x \cosh y}{\cos x \sinh y}\right)\right].$$

We can simplify this solution by bringing the inverse tangents together. We return to the expression for $\psi(u, v)$ and rewrite it in the form

$$\frac{\pi}{2}(\psi + 1) = \operatorname{Tan}^{-1}\left(\frac{1+u}{v}\right) + \operatorname{Tan}^{-1}\left(\frac{1-u}{v}\right).$$

When we take tangents on both sides,

$$\tan\left[\frac{\pi}{2}(\psi+1)\right] = \frac{\dfrac{1+u}{v}+\dfrac{1-u}{v}}{1-\left(\dfrac{1+u}{v}\right)\left(\dfrac{1-u}{v}\right)} = \frac{2v}{u^2+v^2-1}.$$

Because $-1 \le \psi \le 1$, it follows that $0 \le \pi(\psi+1)/2 \le \pi$. If we use the notation $\tan^{-1}$ to denote values of the inverse tangent function between 0 and π (whereas Tan^{-1} denotes values between $\pm\pi/2$), then

$$\frac{\pi}{2}(\psi+1) = \tan^{-1}\left(\frac{2v}{u^2+v^2-1}\right),$$

and

$$\psi(u,v) = -1+\frac{2}{\pi}\tan^{-1}\left(\frac{2v}{u^2+v^2-1}\right).$$

Hence,

$$\begin{aligned}\phi(x,y) &= -1+\frac{2}{\pi}\tan^{-1}\left(\frac{2\cos x\,\sinh y}{\sin^2 x\,\cosh^2 y+\cos^2 x\,\sinh^2 y-1}\right)\\ &= -1+\frac{2}{\pi}\tan^{-1}\left(\frac{2\cos x\,\sinh y}{\sinh^2 y-\cos^2 x}\right)\\ &= -1+\frac{2}{\pi}\tan^{-1}\left(\frac{\dfrac{2\cos x}{\sinh y}}{1-\dfrac{\cos^2 x}{\sinh^2 y}}\right).\end{aligned}$$

If we set $\alpha = \tan^{-1}\left(\dfrac{\cos x}{\sinh y}\right)$, then

$$\begin{aligned}\phi(x,y) &= -1+\frac{2}{\pi}\tan^{-1}\left(\frac{2\tan\alpha}{1-\tan^2\alpha}\right) = -1+\frac{2}{\pi}\tan^{-1}(\tan 2\alpha)\\ &= -1+\frac{4\alpha}{\pi} = -1+\frac{4}{\pi}\tan^{-1}\left(\frac{\cos x}{\sinh y}\right)\\ &= -1+\frac{4}{\pi}\mathrm{Tan}^{-1}\left(\frac{\cos x}{\sinh y}\right),\end{aligned}$$

because $\cos x/\sinh y > 0$. ....

Exercises 7.5

In Exercises 1–6 use formula 7.21 to find the solution $\psi(u,v)$ of Laplace's equation in the half-plane $v > 0$ subject to the given boundary conditions.

1. $\psi(u,0) = \begin{cases} 0, & u < 0 \\ a, & u > 0, \end{cases}$ a a constant

2. $\psi(u,0) = \begin{cases} -a, & u < 0 \\ a, & u > 0, \end{cases}$ a a constant

3. $\psi(u,0) = \begin{cases} a, & |u| < c \\ 0, & |u| > c, \end{cases}$ a and c constants

4. $\psi(u,0) = \begin{cases} a, & |u| < c \\ -a, & |u| > c, \end{cases}$ a and c constants

5. $\psi(u,0) = \begin{cases} a|u|/u, & 0 < |u| < c \\ 0, & |u| > c, \end{cases}$ a and c constants

6. $\psi(u,0) = \begin{cases} a|u|/u, & 0 < |u| < c \\ b|u|/u, & |u| > c, \end{cases}$ a, b, and c constants

In Exercises 7–10 use the transformation $w = f(z)$ to map the domain D onto the half-plane $\operatorname{Im} w > 0$. Then use 7.21 to find the solution $\phi(x,y)$ of Laplace's equation in D subject to the given boundary conditions.

7.

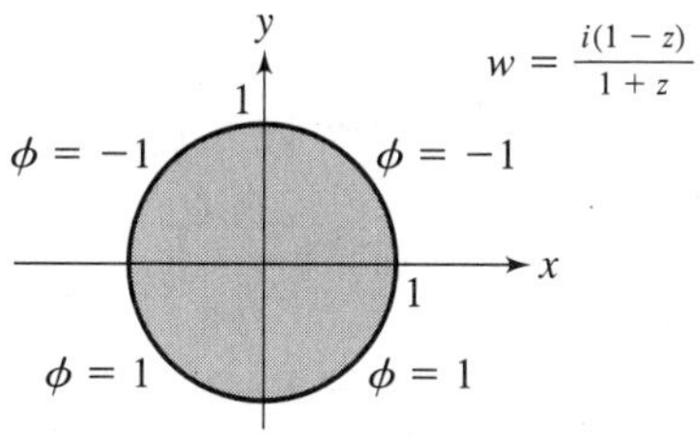

8.

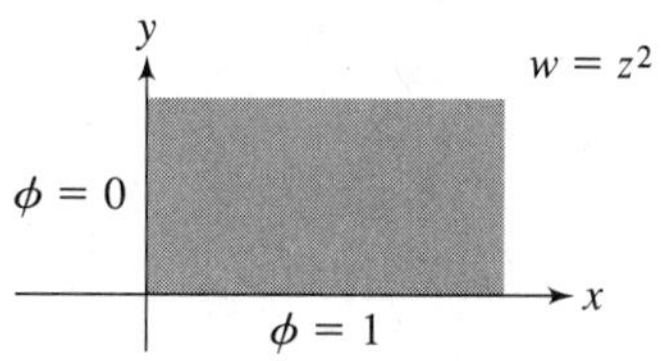

9.

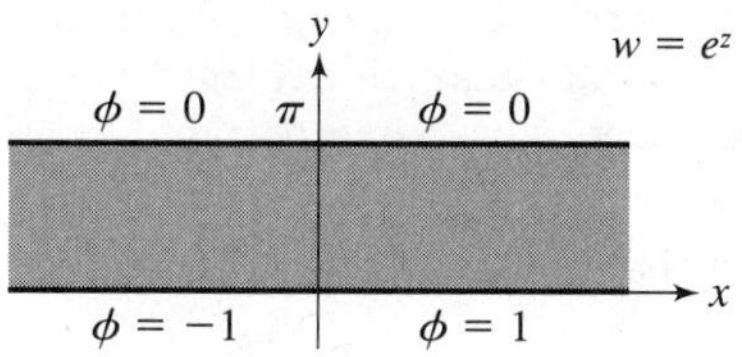

10.

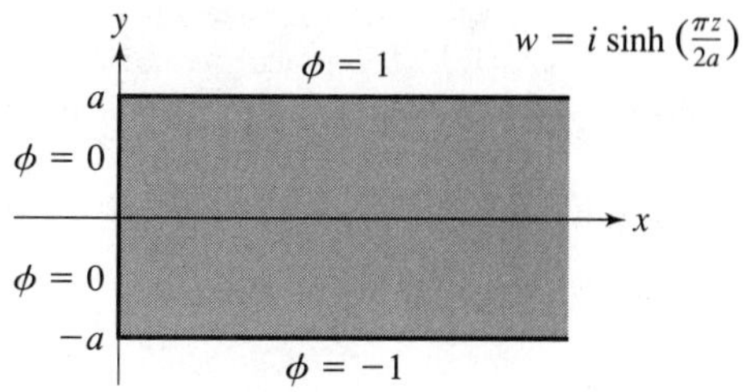

In Exercises 11–15 find a transformation that maps the domain shown onto the half-plane $\operatorname{Im} w > 0$. Then use 7.21 to find the solution $\phi(x,y)$ of Laplace's equation subject to the given boundary conditions.

11.

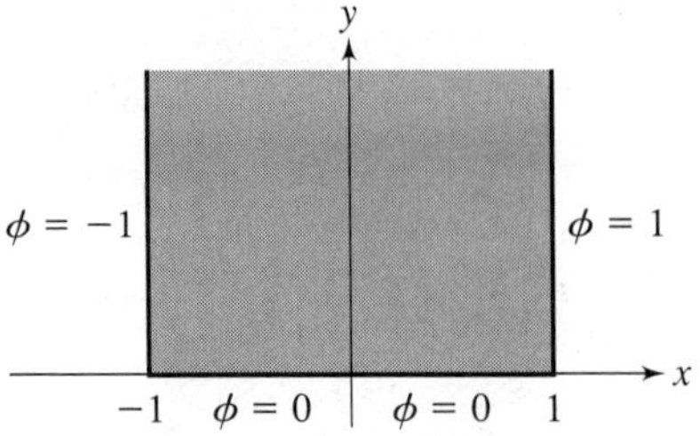

12.

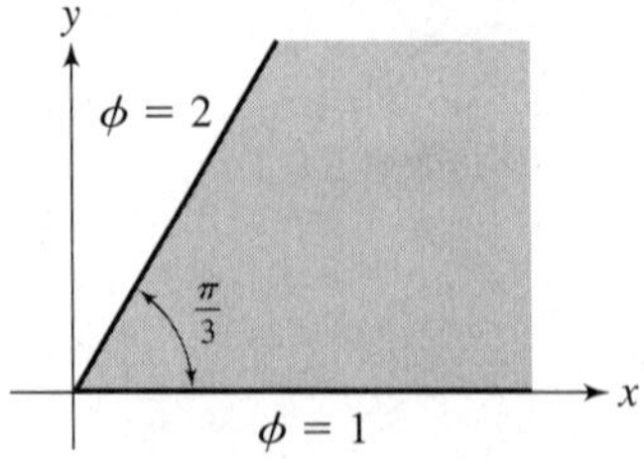

13.

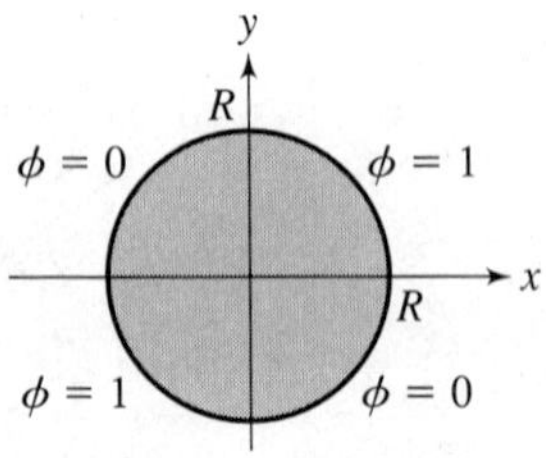

14.

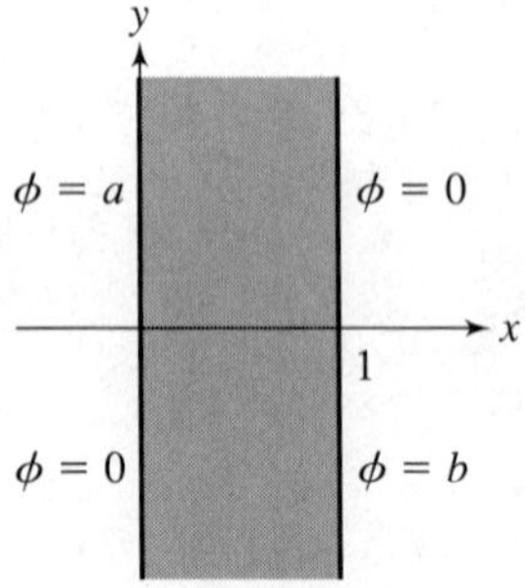

15.

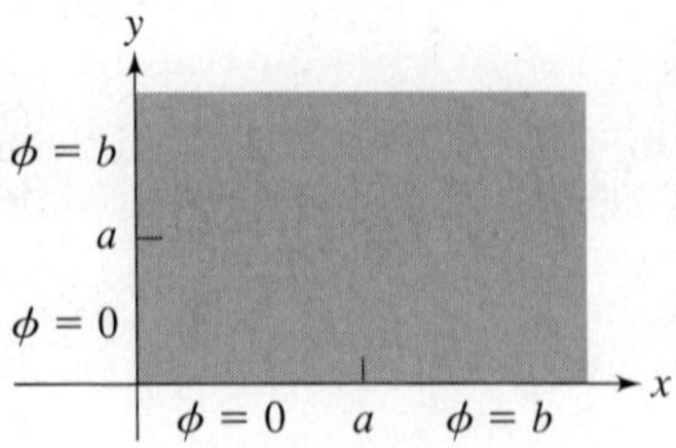

16. Show that the solution $\phi(x, y)$ of Laplace's equation in the quarter plane $x > 0$, $y > 0$ with boundary values

$$\phi(0, y) = c, \qquad y > 0,$$

$$\phi(x, 0) = b, \qquad x > 0,$$

is $\phi(x, y) = b + \dfrac{2}{\pi}(c - b)\,\mathrm{Tan}^{-1}\left(\dfrac{y}{x}\right)$.

17. **(a)** Use the mapping $w = (R + z)^2/(R - z)^2$ of Exercise 11 in Section 7.3 to find the solution of Laplace's equation in the semicircular domain $|z| < R$, $\operatorname{Im} z > 0$, given that on the semicircle the solution has a constant value b and on the diameter it has a constant value c.

(b) Use the mapping $w = i(R - z)/(R + z)$ and the result of Exercise 16 to solve the problem in (a).

7.6 The Schwarz-Christoffel Transformation

In this section we consider the problem of mapping the interior of a polygon in the z-plane onto the half-plane $\operatorname{Im} w > 0$ (Figure 7.30). Riemann's mapping theorem (Theorem 7.3) essentially guarantees existence of such mappings. Unfortunately,

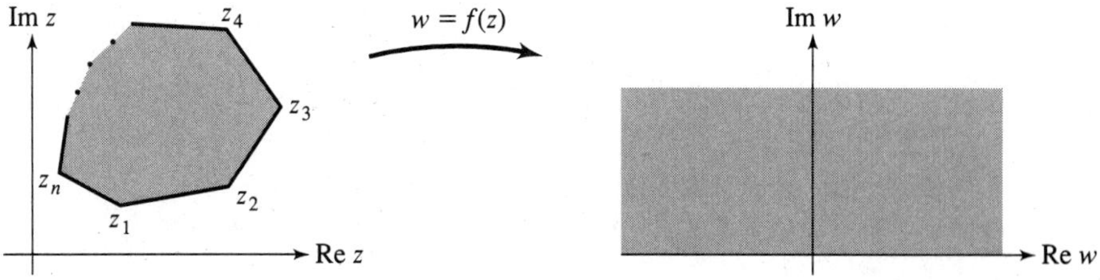

Figure 7.30

as we shall see, the mappings cannot be found in terms of standard elementary functions even for the simplest of polygons — triangles and rectangles. In spite of this, the information that we uncover in attempting to find such mappings proves very useful in closely related problems.

It is easier to begin discussions by considering the inverse mapping, but in a less specific sense. Given n points $u_1, \ldots, u_n$ on the real axis in the w-plane, what mappings $z = f^{-1}(w)$ map the upper half-plane $\operatorname{Im} w > 0$ into a polygon with the images of the u_j as vertices? For the moment, we do not specify positions of the vertices, we simply require that the mapping produce a polygon with the u_j mapped onto its vertices. To find such mappings, we first discuss the transformation

$$z = f^{-1}(w) = A(w - u_1)^{\alpha/\pi} + B, \tag{7.24}$$

where A and B are complex constants and u_1 and $\alpha > 0$ are real. (The branch cut may be anywhere in $\operatorname{Im} w < 0$.) Points on Γ_2, that part of the real w-axis to the right of u_1 (Figure 7.31), are mapped onto points on the half-line Γ_2', and those to the left of u_1 are mapped onto

$$\begin{aligned} \Gamma_1' : z &= A(w - u_1)^{\alpha/\pi} + B \\ &= A[(u_1 - w)e^{\pi i}]^{\alpha/\pi} + B = A(u_1 - w)^{\alpha/\pi} e^{\alpha i} + B. \end{aligned}$$

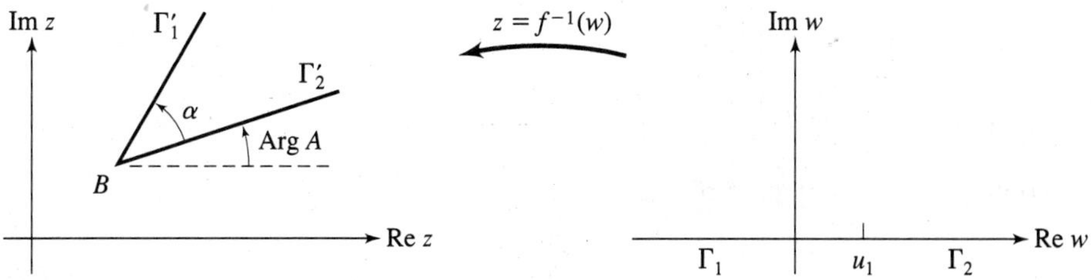

Figure 7.31

In other words, the half-space $\operatorname{Im} w > 0$ is mapped onto a sector with angle α at $z = B$. The effect of A is to rotate the sector by $\arg A$ and to introduce an expansion (or contraction) equal to $|A|$. Functions of form 7.24 result from integrations of differential equations of the form

$$\frac{dz}{dw} = A(w - u_1)^{\alpha/\pi - 1}. \tag{7.25}$$

With this in mind, we now consider a mapping that results from integration of a differential equation containing the product of two such factors,

$$\frac{dz}{dw} = A(w - u_1)^{\alpha_1/\pi - 1}(w - u_2)^{\alpha_2/\pi - 1}, \tag{7.26}$$

where $u_1 < u_2$ and $\alpha_1 > 0$ and $\alpha_2 > 0$ are all real constants. We can express the solution as an antiderivative,

$$z = A\int (w - u_1)^{\alpha_1/\pi - 1}(w - u_2)^{\alpha_2/\pi - 1}dw + B \tag{7.27a}$$

(but in general, the integration is impossible), or as a contour integral,

$$z = A\int_{w_0}^{w} (\zeta - u_1)^{\alpha_1/\pi - 1}(\zeta - u_2)^{\alpha_2/\pi - 1}d\zeta + B, \tag{7.27b}$$

where w_0 is constant. Suppose that such a mapping carries u_1 and u_2 onto points z_1 and z_2 in the z-plane (Figure 7.32).

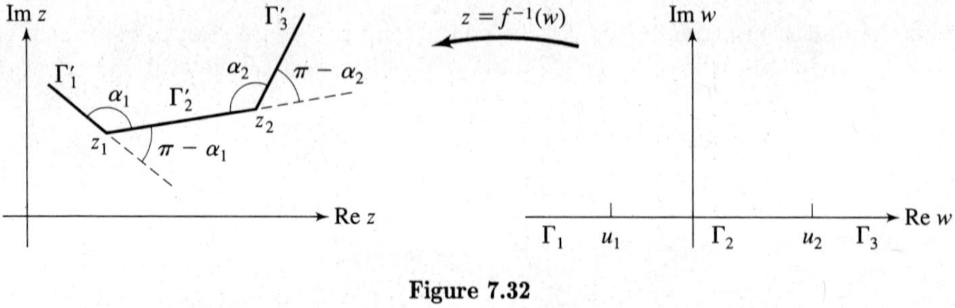

Figure 7.32

If we take arguments of 7.26,

$$\arg\frac{dz}{dw} = \arg A + \left(\frac{\alpha_1}{\pi} - 1\right)\arg(w - u_1) + \left(\frac{\alpha_2}{\pi} - 1\right)\arg(w - u_2). \tag{7.28}$$

For points on Γ_1, that part of the real w-axis to the left of u_1, we find

$$\arg\frac{dz}{dw} = \arg A + \left(\frac{\alpha_1}{\pi} - 1\right)(\pi) + \left(\frac{\alpha_2}{\pi} - 1\right)(\pi) = \arg A + \alpha_1 + \alpha_2 - 2\pi. \tag{7.29a}$$

The fact that this is constant implies that the image Γ_1' of Γ_1 is a straight line (see Exercise 1 in Section 7.4). For points on Γ_2,

$$\arg\frac{dz}{dw} = \arg A + \left(\frac{\alpha_1}{\pi} - 1\right)(0) + \left(\frac{\alpha_2}{\pi} - 1\right)(\pi) = \arg A + \alpha_2 - \pi. \tag{7.29b}$$

This is also constant and differs from 7.29(a) by $\pi - \alpha_1$. In other words, Γ_2' is a straight line at angle $\pi - \alpha_1$ relative to Γ_1'. Similarly, for points on Γ_3,

$$\arg\frac{dz}{dw} = \arg A + \left(\frac{\alpha_1}{\pi} - 1\right)(0) + \left(\frac{\alpha_2}{\pi} - 1\right)(0) = \arg A, \tag{7.29c}$$

indicating that Γ_3' is a straight line at angle $\pi - \alpha_2$ relative to Γ_2'.

In general, any mapping $z = f^{-1}(w)$ defined by the differential equation

$$\frac{dz}{dw} = A(w - u_1)^{\alpha_1/\pi - 1} \cdots (w - u_n)^{\alpha_n/\pi - 1} \tag{7.30}$$

maps the real axis in the w-plane onto a succession of straight-line segments joining the images z_j of u_j in such a way that the angle between the segments at z_j is α_j. We can express the solution of 7.30 in antiderivative form,

$$z = A\int (w - u_1)^{\alpha_1/\pi - 1} \cdots (w - u_n)^{\alpha_n/\pi - 1} dw + B, \tag{7.31a}$$

or as a contour integral,

$$z = A\int_{w_0}^{w} (\zeta - u_1)^{\alpha_1/\pi - 1} \cdots (\zeta - u_n)^{\alpha_n/\pi - 1} d\zeta + B. \tag{7.31b}$$

For such a mapping to take the real axis in the w-plane onto a polygon in the z-plane, the straight lines must close. Because the sum of the interior angles of a polygon with n sides is $(n-2)\pi$ radians, we require that

$$\sum_{j=1}^{n} \alpha_j = (n-2)\pi. \tag{7.32}$$

The shape of the polygon is dictated by 7.31 with $A = 1$ and $B = 0$. Constant A rotates and expands (or contracts) this polygon, and B translates it.

Suppose now that we are given a polygon in the z-plane with vertices $z_1, \ldots, z_n$, and we want to find a mapping $z = f^{-1}(w)$ that maps the half-plane $\operatorname{Im} w > 0$ onto the interior of the polygon. Such a mapping is called a **Schwarz–Christoffel** transformation. In view of the preceding remarks, this would entail solving 7.30 so that $\operatorname{Im} w = 0$ maps onto the sides of a polygon similar to the given polygon, similar in the sense that the interior angles of the polygons are identical, as are ratios of corresponding sides. Naturally, then, we would choose the α_j in 7.30 to be the interior angles of the given polygon. May we choose the u_j arbitrarily? If the polygon is a triangle, then the fact that equality of angles implies equality of ratios of sides means that u_1, u_2, and u_3 may be chosen arbitrarily. But for a quadrilateral, say a rectangle, equality of interior angles is not sufficient to guarantee equality of ratios of corresponding sides. We must specify the ratio of lengths of two adjacent sides. This means that only three of the four constants u_1, u_2, u_3, and u_4 can be assigned arbitrarily. The fourth must be chosen to guarantee similarity to the given quadrilateral. In general, for an n-sided polygon, only three of the points u_j may be chosen arbitrarily; the remaining $n - 3$ must be determined in order to guarantee the required similarity. Considerable ingenuity is often required in arriving at the remaining u_j. We illustrate this in the following example.

EXAMPLE 7.15 ....

Find a Schwarz-Christoffel transformation $z = f^{-1}(w)$ that maps the half-plane $\operatorname{Im} w > 0$ onto the interior of the rectangle with vertices $z = \pm 1, \pm 1 + i$.

Solution According to the preceding discussion only three of the points u_1, u_2, u_3, and u_4 that map onto the vertices of the rectangle can be assigned arbitrarily; the fourth must be chosen so that the ratio of sides is 2 to 1. Because of the symmetry of the rectangle, it seems likely that we may choose the u_j as shown in Figure 7.33 for some value $0 < a < 1$.

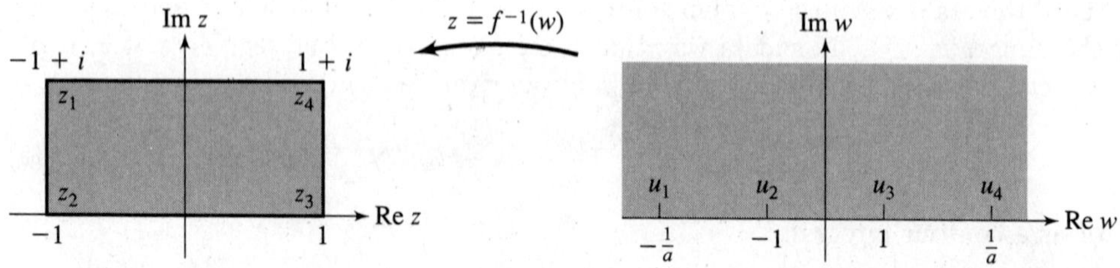

Figure 7.33

With interior angles of $\pi/2$ radians, equation 7.30 for the transformation gives

$$\frac{dz}{dw} = A(w + 1/a)^{-1/2}(w+1)^{-1/2}(w-1)^{-1/2}(w-1/a)^{-1/2}$$

$$= \frac{A}{(w+1)^{1/2}(w-1)^{1/2}(w+1/a)^{1/2}(w-1/a)^{1/2}}.$$

We cannot integrate this equation analytically, but we can express the solution as a contour integral,

$$z = A\int_0^w \frac{1}{(\zeta+1)^{1/2}(\zeta-1)^{1/2}(\zeta+1/a)^{1/2}(\zeta-1/a)^{1/2}}d\zeta + B$$

$$= aA\int_0^w \frac{1}{(\zeta+1)^{1/2}(\zeta-1)^{1/2}(a\zeta+1)^{1/2}(a\zeta-1)^{1/2}}d\zeta + B.$$

Because integrations are in the half-plane $\operatorname{Im} w > 0$, branch cuts for the root functions are chosen in the lower half-plane, such as in Figure 7.34.

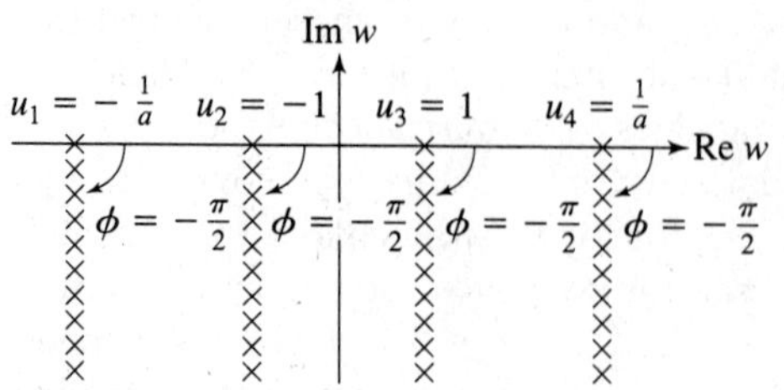

Figure 7.34

To determine a, A, and B, we specify the conditions for the u_j to map to the vertices of the rectangle:

$$-1+i = aA\int_0^{-1/a} \frac{1}{(\zeta+1)^{1/2}(\zeta-1)^{1/2}(a\zeta+1)^{1/2}(a\zeta-1)^{1/2}}\,d\zeta + B, \qquad \text{(a)}$$

$$-1 = aA\int_0^{-1} \frac{1}{(\zeta+1)^{1/2}(\zeta-1)^{1/2}(a\zeta+1)^{1/2}(a\zeta-1)^{1/2}}\,d\zeta + B, \qquad \text{(b)}$$

$$1 = aA\int_0^{1} \frac{1}{(\zeta+1)^{1/2}(\zeta-1)^{1/2}(a\zeta+1)^{1/2}(a\zeta-1)^{1/2}}\,d\zeta + B, \qquad \text{(c)}$$

$$1+i = aA\int_0^{1/a} \frac{1}{(\zeta+1)^{1/2}(\zeta-1)^{1/2}(a\zeta+1)^{1/2}(a\zeta-1)^{1/2}}\,d\zeta + B. \qquad \text{(d)}$$

A change of variable of integration in (b) together with (c) shows that $B = 0$. Condition (c) can then be expressed in the form

$$\begin{aligned} 1 &= aA\int_0^1 \frac{1}{\sqrt{\zeta+1}\left(\sqrt{1-\zeta}\,i\right)\sqrt{a\zeta+1}\left(\sqrt{1-a\zeta}\,i\right)}\,d\zeta \\ &= aA\int_0^1 \frac{-1}{\sqrt{(1-\zeta^2)(1-a^2\zeta^2)}}\,d\zeta, \end{aligned}$$

where all square roots are applied to positive real numbers. For real ζ,

$$F(a) = \int_0^1 \frac{1}{\sqrt{(1-\zeta^2)(1-a^2\zeta^2)}}\,d\zeta$$

is the well-known and tabulated complete elliptic integral of the first kind. In terms of this function, a and A must satisfy

$$-1 = aA\,F(a). \qquad \text{(e)}$$

Condition (d) can be expressed as

$$\begin{aligned} 1+i = aA\Bigg[&\int_0^1 \frac{1}{(\zeta+1)^{1/2}(\zeta-1)^{1/2}(a\zeta+1)^{1/2}(a\zeta-1)^{1/2}}\,d\zeta \\ &+\int_1^{1/a} \frac{1}{(\zeta+1)^{1/2}(\zeta-1)^{1/2}(a\zeta+1)^{1/2}(a\zeta-1)^{1/2}}\,d\zeta\Bigg]. \end{aligned}$$

When we substitute from (c),

$$1+i = 1 + aA\int_1^{1/a} \frac{-i}{\sqrt{(\zeta^2-1)(1-a^2\zeta^2)}}\,d\zeta.$$

Hence,

$$-1 = aA \int_1^{1/a} \frac{1}{\sqrt{(\zeta^2 - 1)(1 - a^2\zeta^2)}}\, d\zeta,$$

or

$$-1 = aA\, G(a), \tag{f}$$

where

$$G(a) = \int_1^{1/a} \frac{1}{\sqrt{(\zeta^2 - 1)(1 - a^2\zeta^2)}}\, d\zeta.$$

Equations (e) and (f) require $G(a) = F(a)$, which can be solved numerically for a. Either (e) or (f) now yields A, and the Schwarz-Christoffel transformation is essentially defined.

EXAMPLE 7.16

Illustrate, but do not find, the Schwarz-Christoffel transformation $z = f^{-1}(w)$ that maps the half-plane $\operatorname{Im} w > 0$ onto the interior of the equilateral triangle with vertices $z = \pm 1$ and $z = \sqrt{3}\, i$ in such a way that $w = -1$, 0, and 1 are mapped onto $z = -1$, 1, and $\sqrt{3}\, i$, respectively (Figure 7.35). As suggested earlier, we may (and have) assigned the three points on the $\operatorname{Re} w$–axis that map to the vertices of the triangle.

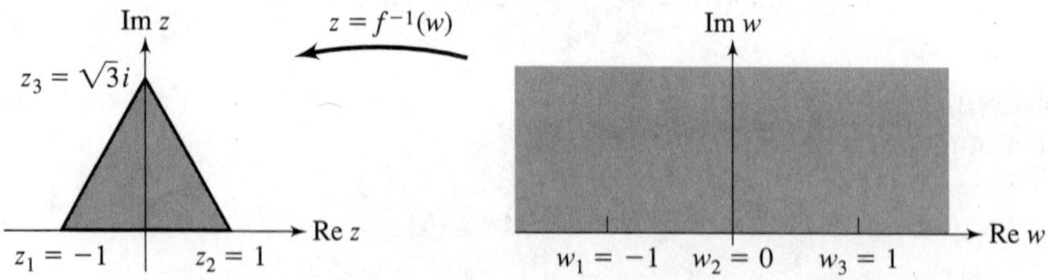

Figure 7.35

Solution Because the interior angles of the triangle are $\pi/3$ rad, the mapping can be found by solving

$$\frac{dz}{dw} = A(w+1)^{1/3-1}w^{1/3-1}(w-1)^{1/3-1} = \frac{A}{w^{2/3}(w+1)^{2/3}(w-1)^{2/3}}.$$

We cannot integrate this equation analytically, but we can express the solution as a contour integral

$$z = A \int_i^{w} \frac{1}{\zeta^{2/3}(\zeta+1)^{2/3}(\zeta-1)^{2/3}}\, d\zeta + B.$$

To determine A and B we use the fact that $w = -1$, 0, and 1 are mapped onto $z = -1$, 1, and $\sqrt{3}\,i$,

$$-1 = A \int_i^{-1} \frac{1}{\zeta^{2/3}(\zeta+1)^{2/3}(\zeta-1)^{2/3}}\, d\zeta + B,$$

$$1 = A \int_i^{0} \frac{1}{\zeta^{2/3}(\zeta+1)^{2/3}(\zeta-1)^{2/3}}\, d\zeta + B,$$

$$\sqrt{3}\,i = A \int_i^{1} \frac{1}{\zeta^{2/3}(\zeta+1)^{2/3}(\zeta-1)^{2/3}}\, d\zeta + B.$$

We shall not attempt to solve these for A and B. Using the "point at infinity" in the w-plane in place of $w = 0$ leads to a much simpler transformation (Example 7.17).

It is often more convenient to use the point at infinity on the real axis in the w-plane to map onto one of the vertices of a polygon. This leaves $n-1$ finite points u_j to map onto the remaining vertices. We claim that mappings defined by

$$\frac{dz}{dw} = A(w-u_1)^{\alpha_1/\pi-1}\cdots(w-u_{n-1})^{\alpha_{n-1}/\pi-1} \tag{7.33}$$

map the real axis in the w-plane onto a polygon in the z-plane with the point at infinity as one of the vertices of the polygon. To verify this we suppose that a transformation defined by this equation maps the u_j $(j = 1,\ldots,n-1)$ onto points z_j and maps $w = \infty$ onto z_n. If we take arguments of 7.33,

$$\arg\frac{dz}{dw} = \arg A + \sum_{j=1}^{n-1}\left(\frac{\alpha_j}{\pi}-1\right)\arg(w-u_j).$$

The difference between limits of this expression as $w \to \pm\infty$ for real w represents the angle of rotation, β, of the sides of the polygon at z_n (Figure 7.36):

$$\begin{aligned}\beta &= \lim_{w\to-\infty}\arg\frac{dz}{dw} - \lim_{w\to\infty}\arg\frac{dz}{dw}\\ &= \sum_{j=1}^{n-1}\left(\frac{\alpha_j}{\pi}-1\right)(\pi) - \sum_{j=1}^{n-1}\left(\frac{\alpha_j}{\pi}-1\right)(0)\\ &= \sum_{j=1}^{n-1}\alpha_j - (n-1)\pi.\end{aligned}$$

If α_n is the angle that satisfies 7.32, then $\sum_{j=1}^{n-1}\alpha_j = (n-2)\pi - \alpha_n$, and $\beta = (n-2)\pi - \alpha_n - (n-1)\pi = -(\alpha_n+\pi)$. Rotationally, this angle is equivalent to

$2\pi - (\alpha_n + \pi) = \pi - \alpha_n$, and therefore the polygon has a vertex at z_n with interior angle α_n.

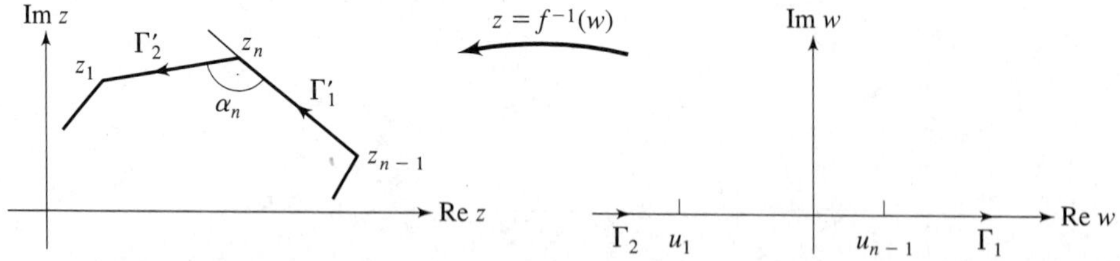

Figure 7.36

When the point at infinity maps to a vertex of the polygon, there is one less term in product 7.33 as compared to 7.30. The resulting integration may be simplified considerably. We illustrate by repeating Example 7.16, using the point at infinity as one of the source points in the w-plane.

EXAMPLE 7.17

Repeat Example 7.16 using the points $w = -1$, 1, and ∞ to map onto $z = -1$, 1, and $\sqrt{3}\,i$ (Figure 7.37).

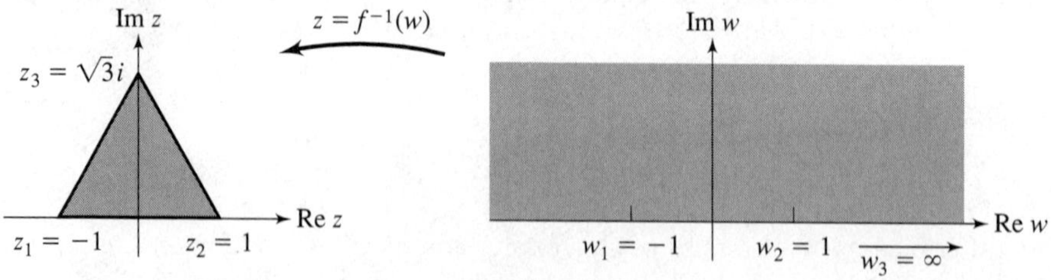

Figure 7.37

Solution Because interior angles are $\pi/3$ rad, we consider

$$\frac{dz}{dw} = A(w+1)^{1/3-1}(w-1)^{1/3-1} = \frac{A}{(w+1)^{2/3}(w-1)^{2/3}}.$$

The solution can be expressed in the form

$$z = A\int_0^w \frac{1}{(\zeta+1)^{2/3}(\zeta-1)^{2/3}}\,d\zeta + B.$$

To determine A and B, we use that fact that $w = \pm 1$ and ∞ are mapped onto $z = \pm 1$ and $\sqrt{3}\,i$:

$$1 = A\int_0^1 \frac{1}{(\zeta+1)^{2/3}(\zeta-1)^{2/3}}\,d\zeta + B,$$

$$-1 = A\int_0^{-1} \frac{1}{(\zeta+1)^{2/3}(\zeta-1)^{2/3}}\,d\zeta + B,$$

$$\sqrt{3}\,i = A\int_0^{\infty} \frac{1}{(\zeta+1)^{2/3}(\zeta-1)^{2/3}}\,d\zeta + B.$$

A change of variable of integration on the second of these, combined with the first, implies that $B = 0$. From the first condition,

$$1 = A\int_0^1 \frac{1}{(\zeta+1)^{2/3}[(1-\zeta)e^{\pi i}]^{2/3}}\,d\zeta = A\int_0^1 \frac{1}{(1-\zeta^2)^{2/3}e^{2\pi i/3}}\,d\zeta$$

$$= Ae^{-2\pi i/3}\int_0^1 \frac{1}{(1-\zeta^2)^{2/3}}\,d\zeta.$$

If we set $\zeta = \sqrt{t}$ in this definite integral,

$$\int_0^1 \frac{1}{(1-\zeta^2)^{2/3}}\,d\zeta = \frac{1}{2}\int_0^1 t^{-1/2}(1-t)^{-2/3}\,dt = \frac{1}{2}\,\mathrm{B}\left(\frac{1}{2},\frac{1}{3}\right),$$

where $\mathrm{B}(p,q)$ is the beta function

$$\mathrm{B}(p,q) = \int_0^1 t^{p-1}(1-t)^{q-1}\,dt.$$

In other words, $A = 2e^{2\pi i/3}/\,\mathrm{B}(\frac{1}{2},\frac{1}{3})$. ....

Degenerate Polygons

Quite often in applications we must map a *degenerate* polygon in the z-plane onto the half-plane $\operatorname{Im} w > 0$. A polygon is degenerate if one or more of its vertices is at infinity. Examples of this are the infinite strip and the semi-infinite strip in Figure 7.38.

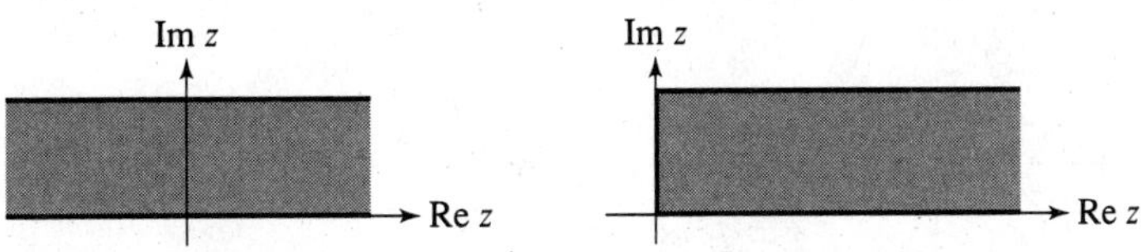

Figure 7.38

What we have just learned about Schwarz-Christoffel transformations can be used to advantage in such problems. We illustrate with the following examples.

EXAMPLE 7.18

Find a Schwarz-Christoffel transformation that maps the semi-infinite strip in Figure 7.39 onto the half-plane $\operatorname{Im} w > 0$.

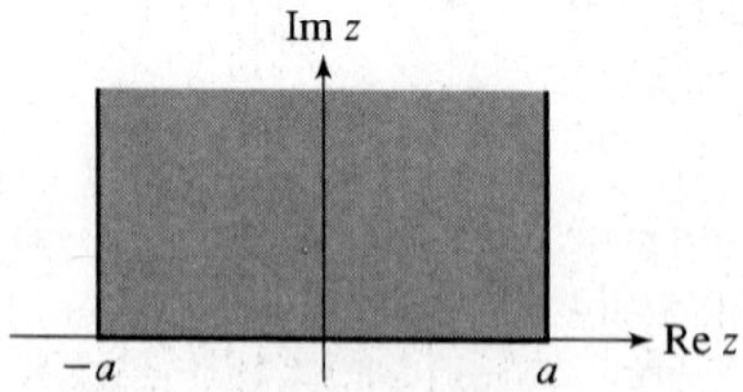

Figure 7.39

Solution The semi-infinite strip can be regarded as a degenerate isosceles triangle with vertex infinitely far up the imaginary axis (Figure 7.40).

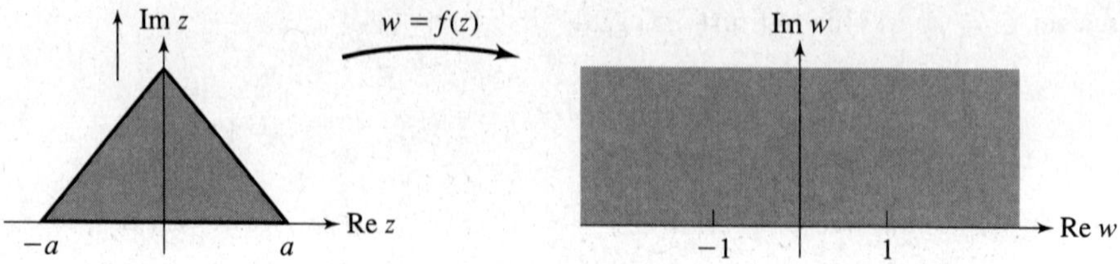

Figure 7.40

Let us choose to map vertices $z = \pm a$ onto $w = \pm 1$ and regard the point at infinity in the w-plane as mapped to the vertex up the imaginary z-axis. To obtain changes in direction of $\pi/2$ rad at $z = \pm a$, equation 7.33 suggests that we solve

$$\frac{dz}{dw} = A(w-1)^{-1/2}(w+1)^{-1/2},$$

and because A is as yet undetermined, we write

$$\frac{dz}{dw} = \frac{A}{(1-w^2)^{1/2}}.$$

Integration gives $z = A \operatorname{Sin}^{-1} w + B$, or

$$w = \sin\left[\frac{1}{A}(z-B)\right].$$

For $z = \pm a$ to map onto $w = \pm 1$, we require

$$1 = \sin\left(\frac{a-B}{A}\right), \qquad -1 = \sin\left(\frac{-a-B}{A}\right).$$

These can be solved for $A = 2a/\pi$ and $B = 0$, and therefore the required Schwarz-Christoffel transformation is

$$w = \sin\left(\frac{\pi z}{2a}\right).$$

On the basis of Example 7.9, we should have anticipated this mapping. Notice that we did not concern ourselves with the fact that branch cuts of the functions $1/[(w-1)^{1/2}(w+1)^{1/2}]$ and $1/(1-w^2)^{1/2}$ may not agree. Ultimately we invert the function $z = f^{-1}(w)$ for $w = f(z)$, and $w = f(z)$ has no branch cuts. What we normally do is verify that the final transformation $w = f(z)$ satisfies the original mapping requirements. In this case, it means checking that $w = \sin[\pi z/(2a)]$ maps the semi-infinite strip in Figure 7.39 onto $\operatorname{Im} w > 0$. ....

EXAMPLE 7.19 ....

Find a Schwarz-Christoffel transformation that maps the half-plane $\operatorname{Im} z > 0$, slit along the imaginary axis from $z = 0$ to $z = ai$ $(a > 0)$, onto the half-plane $\operatorname{Im} w > 0$ (Figure 7.41).

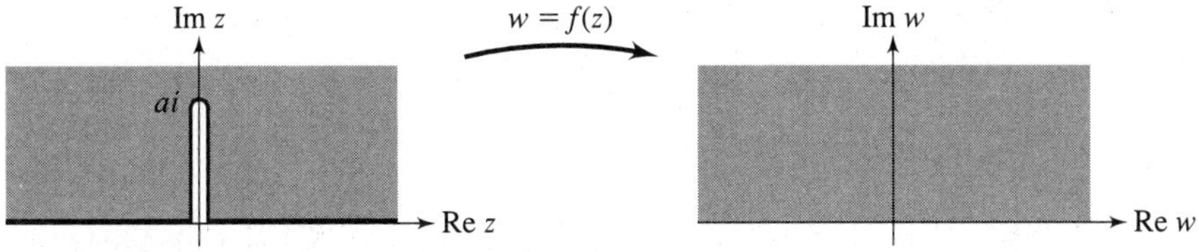

Figure 7.41

Solution Consider the transformation that maps the half-plane $\operatorname{Im} w > 0$ onto the degenerate quadrilateral in Figure 7.42, where $w = \pm 1$ and $w = 0$ are mapped onto $z = \pm b$ and $z = ai$. Angles at $z = \pm b$ and $z = ai$ require

$$\frac{dz}{dw} = A(w+1)^{\alpha/\pi - 1} w^{\beta/\pi - 1} (w-1)^{\alpha/\pi - 1}.$$

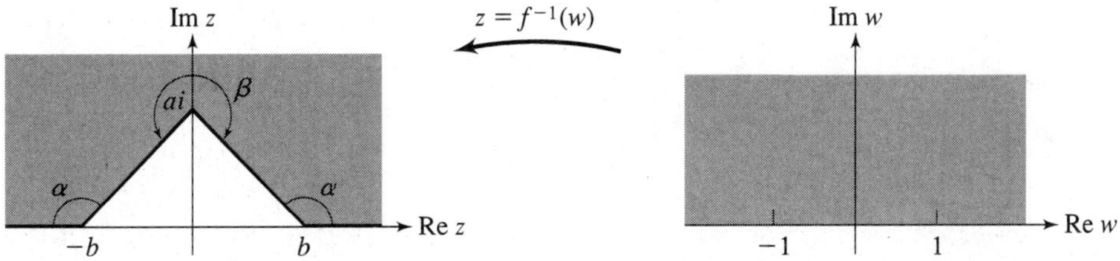

Figure 7.42

To obtain the slit in Figure 7.41, we let $\alpha \to \pi/2$ and $\beta \to 2\pi$. The limit of the derivative is

$$\frac{dz}{dw} = A(w+1)^{-1/2} w (w-1)^{-1/2}, \quad \text{or} \quad \frac{dz}{dw} = \frac{Aw}{(w^2-1)^{1/2}}.$$

Integration gives $z = A(w^2 - 1)^{1/2} + B$. The fact that $w = 0$ and $w = \pm 1$ are mapped to $z = ai$ and $z = 0$ requires

$$ai = A(-1)^{1/2} + B, \qquad 0 = B.$$

We must choose a branch of the square root function for which $(-1)^{1/2} = i$ (because upper half-plane is mapped to upper half-plane), in which case these equations give $A = a$ and $B = 0$—that is, $z = a(w^2 - 1)^{1/2}$. When this equation is solved for w, the result is $w = (1 + z^2/a^2)^{1/2}$. We must choose a branch of this multivalued function. In order to restrict the branch cut to the imaginary axis below $z = ai$, we separate the function in the form

$$w = \frac{1}{a}(z + ai)^{1/2}(z - ai)^{1/2}$$

and choose branches of the root functions corresponding to $\log_{-\pi/2} z$.

We have not verified that it is acceptable to take limits on angles in equation 7.30 in order to produce degenerate polygons. Once again, if we verify that the mapping obtained does indeed produce the desired effect, then justification of limits is unnecessary. This is done for Example 7.19 in Exercise 1 below.

Exercises 7.6

1. Verify that the transformation $w = a^{-1}(z + ai)^{1/2}(z - ai)^{1/2}$ in Example 7.19 maps the half-plane $\operatorname{Im} z > 0$, slit along the imaginary axis from $z = 0$ to $z = ai$ $(a > 0)$, onto the half-plane $\operatorname{Im} w > 0$. Do this by showing that
   (a) Points $z = x < 0$ are mapped to $w = u < -1$.
   (b) Points $z = x > 0$ are mapped to $w = u > 1$.
   (c) In the limit as $x \to 0^+$, the point z_2 to the right of the imaginary axis in the diagram to the right is mapped to a point $w = u$ where $0 < u < 1$.
   (d) In the limit as $x \to 0^-$, the point z_1 to the left of the imaginary axis in the diagram to the right is mapped to a point $w = u$ where $-1 < u < 0$.

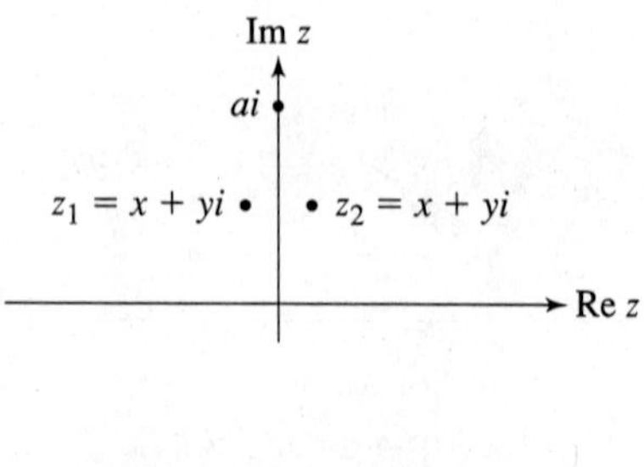

2. In this exercise we use a Schwarz-Christoffel transformation to derive the mapping $w = e^{\pi z/a}$ for the infinite channel $0 < \operatorname{Im} z < a$ to the half-plane $\operatorname{Im} w > 0$.
   (a) Regard the infinite channel as a degeneration of the rhombus shown in the accompanying figure as $b \to \infty$. Choose images of z_1, z_2, z_3, and z_4 as $u_1 < 0$ (but as yet undetermined), $u_2 = 0$, $u_3 = 1$, and u_4 at infinity. Show that using 7.33 and letting

$\alpha \to \pi$ and $\beta \to 0$, the mapping $z = f^{-1}(w)$ must satisfy

$$\frac{dz}{dw} = Aw.$$

(b) Integrate the differential equation in (a), determine A and u_1, and hence show that $z = (a/\pi) \log_\phi w$ for appropriately chosen ϕ. This can be solved for $w = e^{\pi z/a}$.

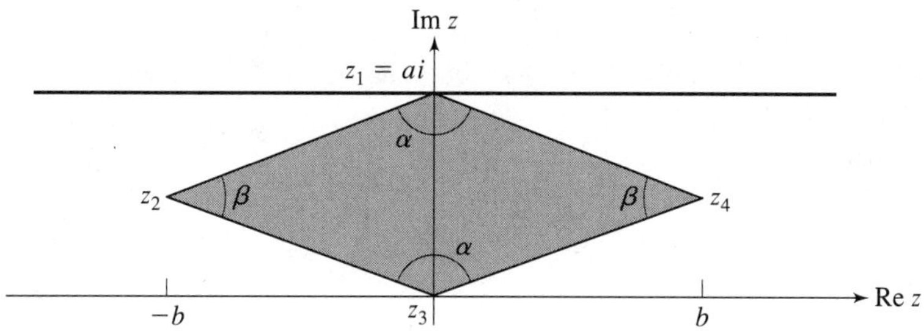

3. Repeat Example 7.15 for the following rectangle.

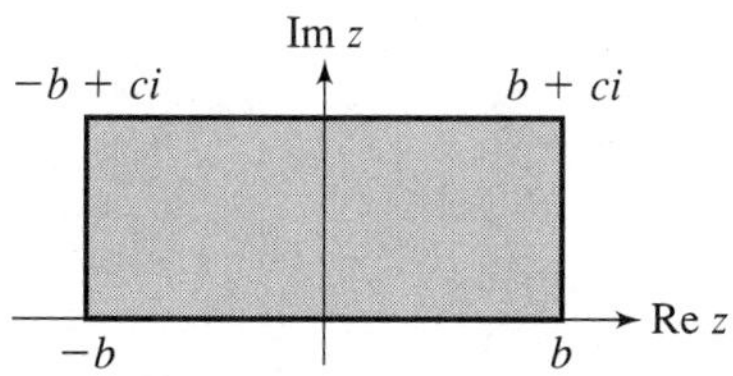

4. In this exercise we use the Schwarz-Christoffel transformation to derive the mapping $w = z^2$ from a quarter-plane to a half-plane.

(a) Show that differential equation 7.33 takes the form

$$\frac{dz}{dw} = Aw^{-1/2}(w+1)^{\alpha_1/\pi - 1}(w-1)^{\alpha_2/\pi - 1}$$

when the quadrilateral with vertices 0, 1, i, and $x + yi$, where $x > 0$ and $y > 0$, is mapped to the half-plane $\operatorname{Im} w > 0$ if these vertices map to the points $0, 1, -1, \infty$, respectively.

(b) To map the quarter-plane $x > 0$, $y > 0$ to $\operatorname{Im} w > 0$, take limits of the differential equation as $\alpha_1 \to \pi$ and $\alpha_2 \to \pi$.

(c) Solve the differential equation in (b) and show that $w = z^2$.

5. Derive the transformation $z = a(w + \log_{-\pi/2} w)$ that maps the half-plane $\operatorname{Im} w > 0$ to the half-plane $\operatorname{Im} z > 0$ slit along the line $y = a > 0$, $x \leq -a$ (left diagram, next page). *Hint:* Regard the slit z-plane as the limit of the polygon $ABCD$ as $C \to -\infty$ along the real axis (right diagram). Points A and D are also at infinity. Map B to $w = -1$ and C to $w = 0$ with angles α and β as shown. Let $\alpha \to 0$ and $\beta \to 2\pi$.

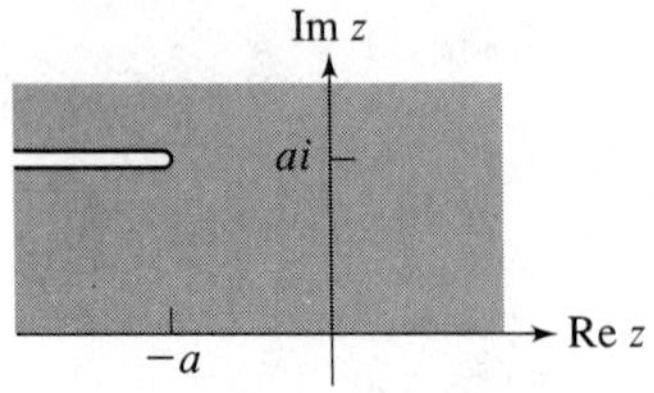

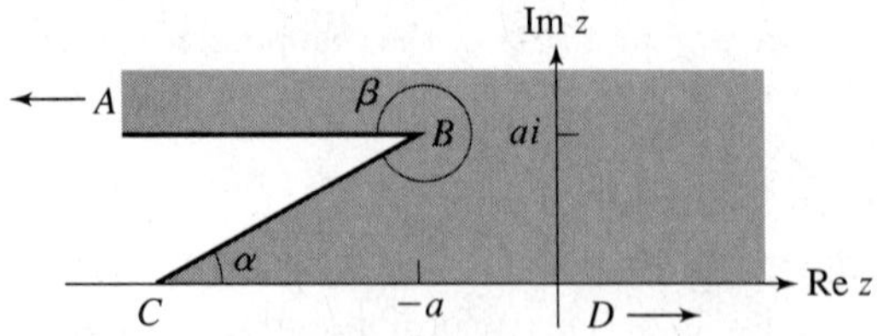

6. Use the technique of Exercise 5 to find the Schwarz-Christoffel transformation that maps the half-plane $\operatorname{Im} w > 0$ to the half-plane $\operatorname{Im} z > 0$ slit along the line $y = a > 0$, $x \leq 0$.

7. Derive the transformation

$$z = \frac{a}{\pi}\{(w+1)^{1/2}(w-1)^{1/2} + \log_{-\pi/2}[w + (w+1)^{1/2}(w-1)^{1/2}]\} - ai$$

that maps the half-plane $\operatorname{Im} w > 0$ to the half-plane with a step shown in the left figure. *Hint:* Regard the step in the z-plane as the limit of the polygon $ABCD$ as A and D move to infinity in the right diagram. Map B to $w = -1$ and C to $w = 1$.

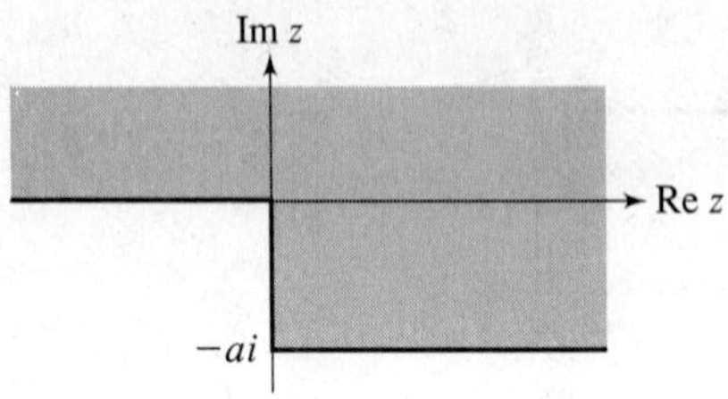

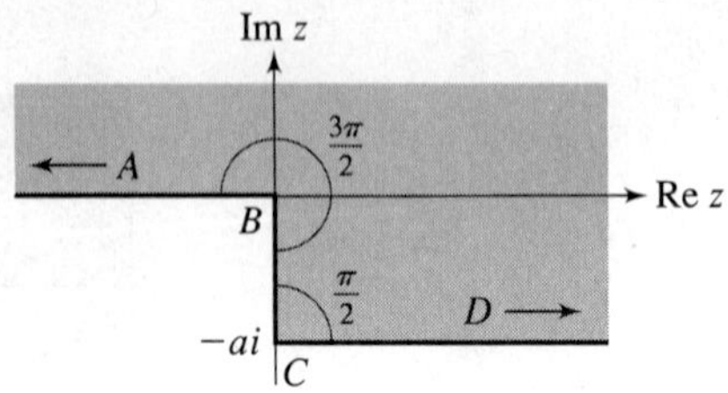

8. Derive the transformation $z = \dfrac{-ia}{2} w^{1/2}(w-3)$ that maps the half-plane $\operatorname{Im} w > 0$ to the domain in the left figure — quadrants 1, 2, and 4 with the y-axis slit to height a. *Hint:* Map C, D, and E to $w = 0$, $w = 1$, and $w = \infty$, respectively. Let $\alpha \to 2\pi$ and $\beta \to \pi/2$. To evaluate the constant A and B in 7.31, use the fact that D is mapped to $w = 1$, and let $z = 0$ map to $w = d > 1$. Show that a convenient choice for d is $\sqrt{3}$.

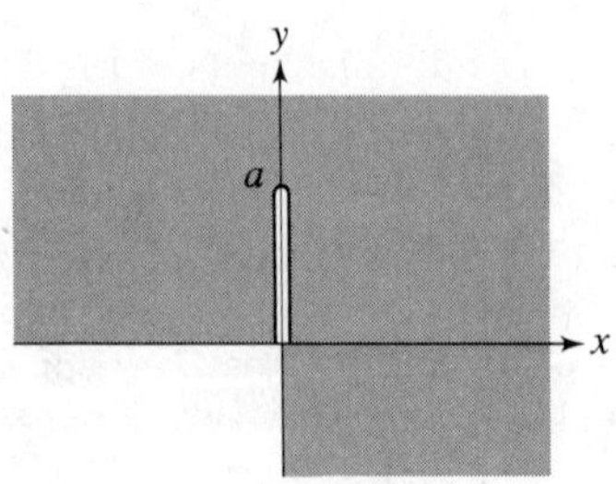

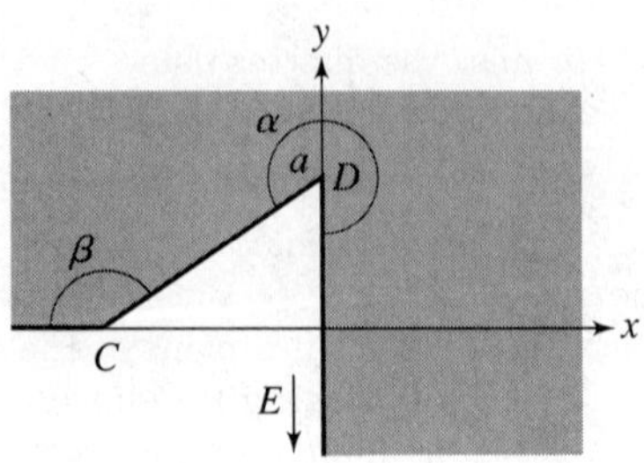

7.7 Steady-State Two-Dimensional Heat Flow

In Section 2.6 we indicated that when a thin plate occupying a domain D in the xy-plane is insulated top and bottom, any steady-state temperature distribution in the plate must satisfy Laplace's equation,

$$\frac{\partial^2 T}{\partial x^2} + \frac{\partial^2 T}{\partial y^2} = 0. \tag{7.34a}$$

Accompanying 7.34(a) will be Dirichlet and/or Neumann boundary conditions specifying values of temperature or its normal derivative along the boundary $\beta(D)$ of D. Curves $T(x, y) =$ constant are isothermal curves, curves joining all points with the same temperature (Figure 7.43). The harmonic conjugate $\tilde{T}(x, y)$ defines a one-parameter family of curves $\tilde{T}(x, y) =$ constant that are orthogonal trajectories of the isothermal curves. They represent curves along which heat flows.

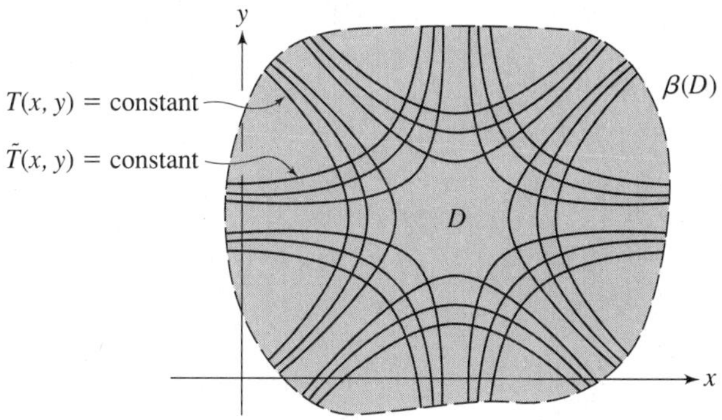

Figure 7.43

Suppose for the moment that D is a bounded domain (as shown in Figure 7.43) and we must solve 7.34(a) subject to prescribed values for temperature on the boundary $\beta(D)$ of D,

$$T(x, y) = F(x, y), \qquad (x, y) \text{ on } \beta(D), \tag{7.34b}$$

where $F(x, y)$ is a given function. According to Corollary 3 to Theorem 4.24, there is one and only one steady-state temperature distribution $T(x, y)$ satisfying this Dirichlet problem. We have mathematical confirmation of what we would expect physically. Theorem 4.24 states that there can be no relative maxima or minima for temperature inside the plate; hottest and coldest points of the plate must be on the edge of the plate. This means that no isothermal could be a closed curve as shown in Figure 7.44; all isothermals must begin and end on the edge of the plate, as in Figure 7.43.

In the remainder of this section we solve various heat conduction problems using equation 7.21 and conformal mappings.

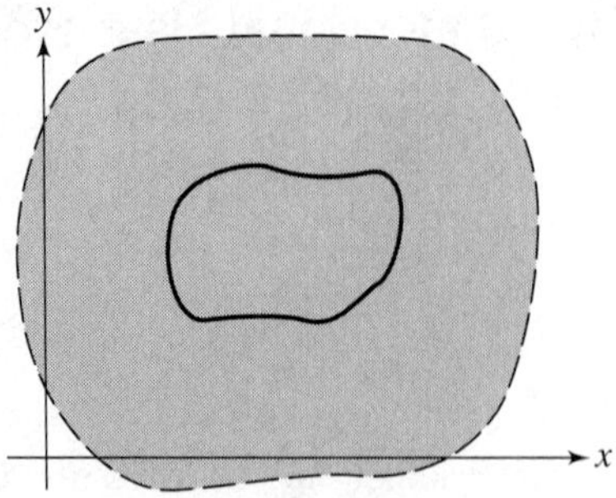

Figure 7.44

EXAMPLE 7.20

Find steady-state temperature $T(x, y)$ in a thin semi-infinite plate $y > 0$ with insulated surfaces, given that the temperature along its edge $y = 0$ is

$$T(x, 0) = \begin{cases} T_0 > 0, & |x| < 1 \\ 0, & |x| > 1. \end{cases}$$

Solution Equation 7.21 gives

$$\begin{aligned} T(x, y) &= \frac{1}{\pi}\left[-T_0 \operatorname{Tan}^{-1}\left(\frac{-1-x}{y}\right) + T_0 \operatorname{Tan}^{-1}\left(\frac{1-x}{y}\right)\right] \\ &= \frac{T_0}{\pi}\left[\operatorname{Tan}^{-1}\left(\frac{1+x}{y}\right) + \operatorname{Tan}^{-1}\left(\frac{1-x}{y}\right)\right]. \end{aligned}$$

Using the technique in Example 7.14, the solution can also be expressed in the form

$$T(x, y) = \frac{T_0}{\pi} \tan^{-1}\left[\frac{2y}{x^2 + y^2 - 1}\right],$$

where values of $\tan^{-1}$ are between 0 and π (see Exercise 19).

Isothermal curves $T(x, y) = C$ simplify to $x^2 + \left[y - \cot\left(\frac{\pi C}{T_0}\right)\right]^2 = \csc^2\left(\frac{\pi C}{T_0}\right)$, circles through the points $(\pm 1, 0)$ with centers on the y-axis. Isothermals for $T = T_0/8$, $3T_0/16$, and $T_0/4$ are shown in Figure 7.45.

To determine curves along which heat flows, we find the harmonic conjugate $\tilde{T}(x, y)$ of $T(x, y)$. The Cauchy-Riemann equations require

$$\frac{\partial \tilde{T}}{\partial y} = \frac{\partial T}{\partial x} = \frac{T_0}{\pi}\left[\frac{y}{y^2 + (1+x)^2} - \frac{y}{y^2 + (1-x)^2}\right],$$

$$\frac{\partial \tilde{T}}{\partial x} = -\frac{\partial T}{\partial y} = \frac{T_0}{\pi}\left[\frac{1+x}{y^2 + (1+x)^2} + \frac{1-x}{y^2 + (1-x)^2}\right].$$

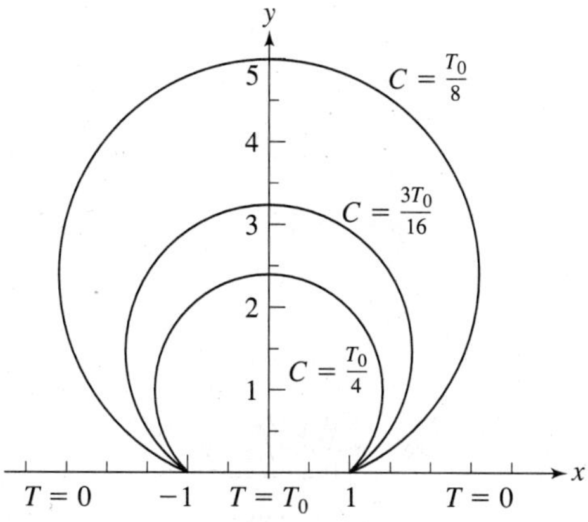

Figure 7.45

These expressions give

$$\tilde{T}(x, y) = \frac{T_0}{2\pi} \ln \left[\frac{y^2 + (1+x)^2}{y^2 + (1-x)^2} \right].$$

Heat-flow lines are defined by

$$C = \frac{T_0}{2\pi} \ln \left[\frac{y^2 + (1+x)^2}{y^2 + (1-x)^2} \right],$$

and this equation can be written in the form

$$\left[x - \coth\left(\frac{\pi C}{T_0} \right) \right]^2 + y^2 = \operatorname{csch}^2\left(\frac{\pi C}{T_0} \right).$$

Described are circles with centers on the x-axis (see Exercise 18). Heat-flow lines for $C = \pm T_0/8$, $\pm 3T_0/16$, and $\pm T_0/4$ are shown in Figure 7.46.

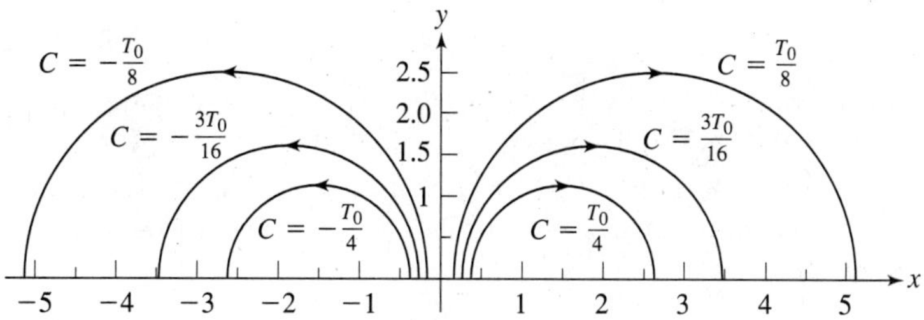

Figure 7.46

Superposition of Figures 7.45 and 7.46 in Figure 7.47 illustrates orthogonality of isothermals and heat-flow lines.

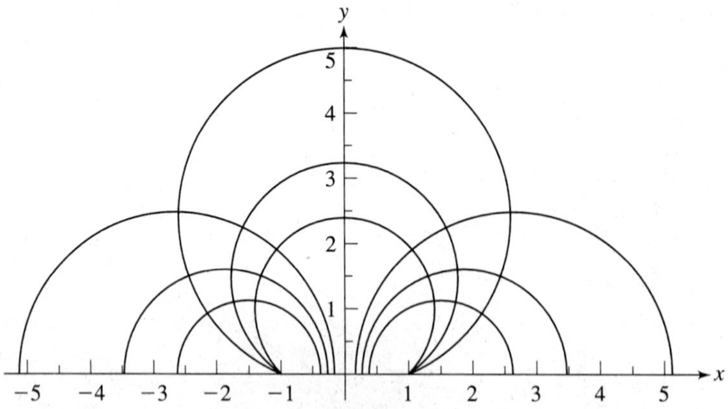

Figure 7.47

EXAMPLE 7.21

Find steady-state temperature $T(x, y)$ in a thin plate in the form of a semi-infinite strip $-a < x < a$, $y > 0$ (Figure 7.48) that satisfies the Dirichlet boundary conditions

$$T(\pm a, y) = T_0, \quad y > 0; \qquad T(x, 0) = 0, \quad -a < x < a.$$

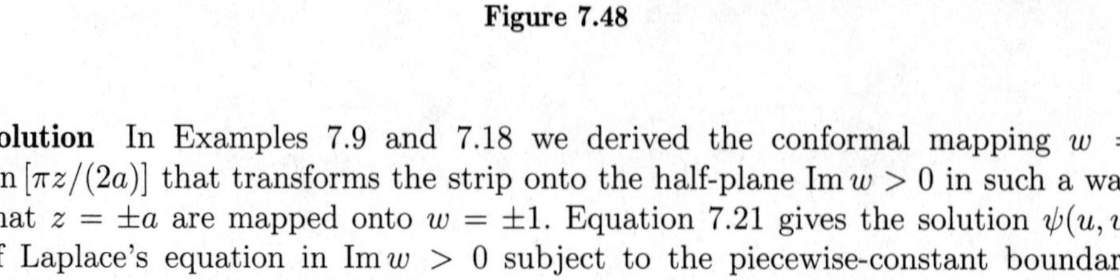

Figure 7.48

Solution In Examples 7.9 and 7.18 we derived the conformal mapping $w = \sin[\pi z/(2a)]$ that transforms the strip onto the half-plane $\operatorname{Im} w > 0$ in such a way that $z = \pm a$ are mapped onto $w = \pm 1$. Equation 7.21 gives the solution $\psi(u, v)$ of Laplace's equation in $\operatorname{Im} w > 0$ subject to the piecewise-constant boundary condition along $\operatorname{Im} w = 0$,

$$\begin{aligned}\psi(u, v) &= \frac{1}{2}(T_0 + T_0) + \frac{1}{\pi}\left[T_0 \operatorname{Tan}^{-1}\left(\frac{-1-u}{v}\right) - T_0 \operatorname{Tan}^{-1}\left(\frac{1-u}{v}\right)\right] \\ &= T_0 - \frac{T_0}{\pi}\left[\operatorname{Tan}^{-1}\left(\frac{1+u}{v}\right) + \operatorname{Tan}^{-1}\left(\frac{1-u}{v}\right)\right].\end{aligned}$$

Because $w = u + vi = \sin \dfrac{\pi z}{2a} = \sin \dfrac{\pi x}{2a} \cosh \dfrac{\pi y}{2a} + \cos \dfrac{\pi x}{2a} \sinh \dfrac{\pi y}{2a}\, i$, it follows that

$$T(x, y) = \psi[u(x,y), v(x,y)] = T_0 - \frac{T_0}{\pi}\left[\operatorname{Tan}^{-1}\left(\frac{1 + \sin \frac{\pi x}{2a} \cosh \frac{\pi y}{2a}}{\cos \frac{\pi x}{2a} \sinh \frac{\pi y}{2a}}\right) + \operatorname{Tan}^{-1}\left(\frac{1 - \sin \frac{\pi x}{2a} \cosh \frac{\pi y}{2a}}{\cos \frac{\pi x}{2a} \sinh \frac{\pi y}{2a}}\right)\right].$$

For a simpler form of the solution, see Exercise 21. Isothermals are shown in Figure 7.49.

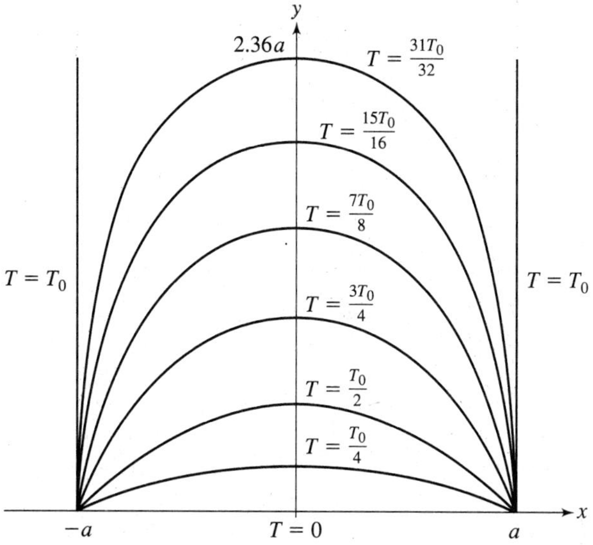

Figure 7.49

EXAMPLE 7.22

Find steady-state temperature $T(x, y)$ in a thin plate in the form of a quarter-plane $x > 0$, $y > 0$ given that edge temperatures are

$$T(0, y) = T_1, \qquad T(x, 0) = T_2.$$

Solution We give two solutions to this problem. First the conformal mapping $w = \operatorname{Log} z$ maps the first quadrant in the z-plane onto the infinite strip $0 < v < \pi/2$ in the w-plane (Figure 7.50).

The solution of Laplace's equation in this strip subject to constant conditions on the horizontal boundaries is

$$\psi(u, v) = T_2 + \frac{2(T_1 - T_2)}{\pi} v.$$

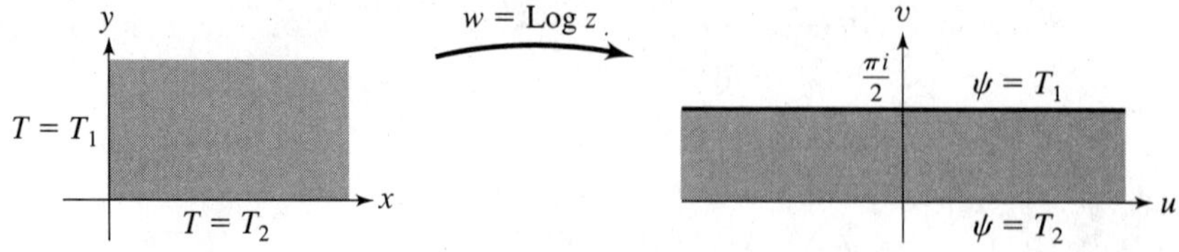

Figure 7.50

Because $v = \text{Tan}^{-1}(y/x)$, temperature in the plate is given by

$$T(x,y) = T_2 + \frac{2}{\pi}(T_1 - T_2)\,\text{Tan}^{-1}\left(\frac{y}{x}\right).$$

Alternatively, the conformal mapping $w = z^2$ maps the first quadrant in the z-plane onto the half-plane $v > 0$ (Figure 7.51).

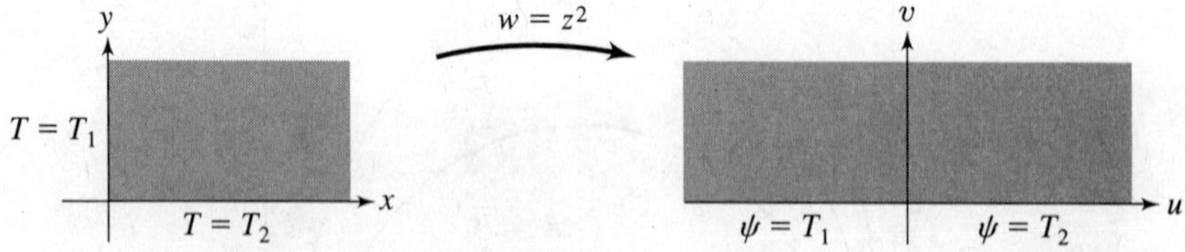

Figure 7.51

According to 7.21, the solution of Laplace's equation in the uv-plane subject to the boundary conditions $\psi(u,0) = T_1$ for $u < 0$ and $\psi(u,0) = T_2$ for $u > 0$ is

$$\psi(u,v) = \frac{1}{2}(T_1 + T_2) + \frac{1}{\pi}(T_2 - T_1)\,\text{Tan}^{-1}\left(\frac{u}{v}\right).$$

Because $u = x^2 - y^2$ and $v = 2xy$, it follows that

$$T(x,y) = \frac{1}{2}(T_1 + T_2) + \frac{1}{\pi}(T_2 - T_1)\,\text{Tan}^{-1}\left(\frac{x^2 - y^2}{2xy}\right).$$

Clearly the former solution is simpler. (They are shown to be the same in Exercise 16.) Isothermal curves are radial lines and heat-flow lines are quarter circles centered at $(0,0)$.

EXAMPLE 7.23

Find steady-state temperature $T(x,y)$ in a thin plate $y > 0$ with insulated faces, given that the temperature along the edge $y = 0$ is T_2 for $x > 1$; that it is $T_1 < T_2$ for $x < -1$; and that part of the x-axis between $x = -1$ and $x = 1$ is insulated (Figure 7.52).

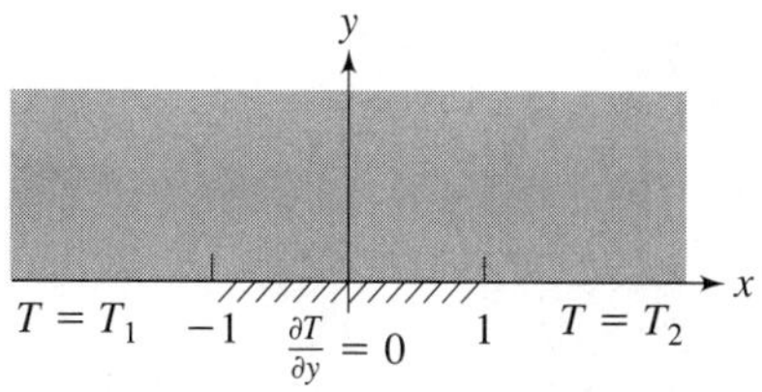

Figure 7.52

Solution It is advantageous to map the plate to a region such that the Neumann condition falls on a boundary perpendicular to the boundaries with Dirichlet conditions. Recall from Example 7.9 in Section 7.3 that the function $w = \sin[\pi z/(2a)]$ maps the semi-infinite strip in Figure 7.53 onto the upper half-plane with $z = \pm a$ mapped onto $w = \pm 1$. This result suggests that for the present problem we consider the mapping $w = \operatorname{Sin}^{-1} z$; it maps the half-plane $y > 0$ onto the semi-infinite strip in Figure 7.54.

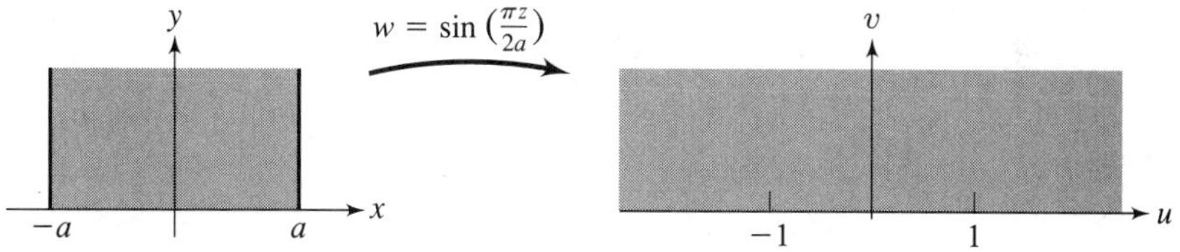

Figure 7.53

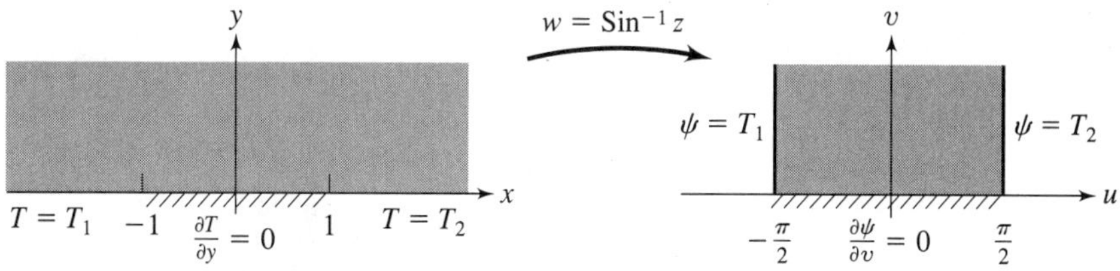

Figure 7.54

The solution $\psi(u, v)$ of Laplace's equation in the strip is

$$\psi(u, v) = T_1 + \frac{1}{\pi}(T_2 - T_1)\left(u + \frac{\pi}{2}\right) = \frac{T_1 + T_2}{2} + \frac{1}{\pi}(T_2 - T_1)u.$$

To express u in terms of x and y, we set $z = x + yi$ and $w = u + vi$ in $z = \sin w$:

$$x + yi = \sin(u + vi) = \sin u \cosh v + \cos u \sinh v\, i.$$

It follows that

$$x = \sin u \cosh v, \qquad y = \cos u \sinh v.$$

For any value of u, these equations are parametric equations for the upper half of the hyperbola

$$\frac{x^2}{\sin^2 u} - \frac{y^2}{\cos^2 u} = 1.$$

Foci are $(\pm 1, 0)$ and the transverse axis has length $2|\sin u|$ (Figure 7.55). Any point (x, y) on the hyperbola satisfies the equation

$$2\sin u = \sqrt{(x+1)^2 + y^2} - \sqrt{(x-1)^2 + y^2},$$

and hence

$$u = \operatorname{Sin}^{-1}\left[\frac{\sqrt{(x+1)^2 + y^2} - \sqrt{(x-1)^2 + y^2}}{2}\right].$$

The steady-state temperature function is, therefore,

$$T(x, y) = \frac{T_1 + T_2}{2} + \frac{1}{\pi}(T_2 - T_1)\operatorname{Sin}^{-1}\left[\frac{\sqrt{(x+1)^2 + y^2} - \sqrt{(x-1)^2 + y^2}}{2}\right].$$

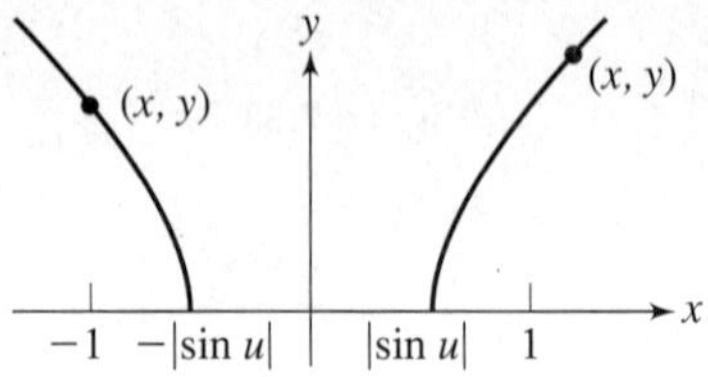

Figure 7.55

Exercises 7.7

In Exercises 1–12 find steady-state temperature $T(x, y)$ in the thin plate with insulated faces and edge temperatures shown. Find isothermal curves and lines of heat flow in exercises marked with an asterisk. Appropriate mappings are suggested at the end of the exercises. Try to discover them yourself; if necessary, consult the list.

*1.

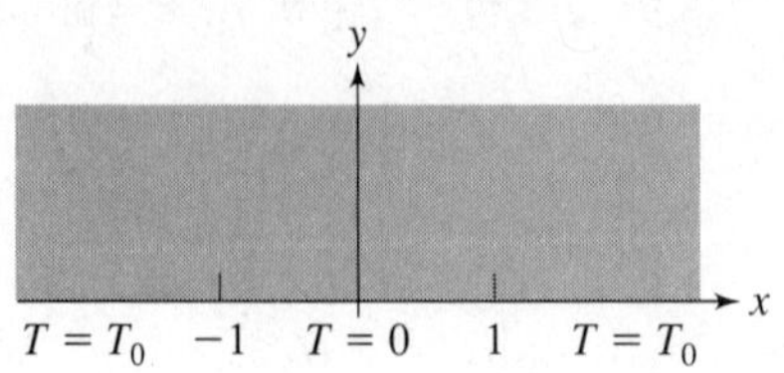

2.

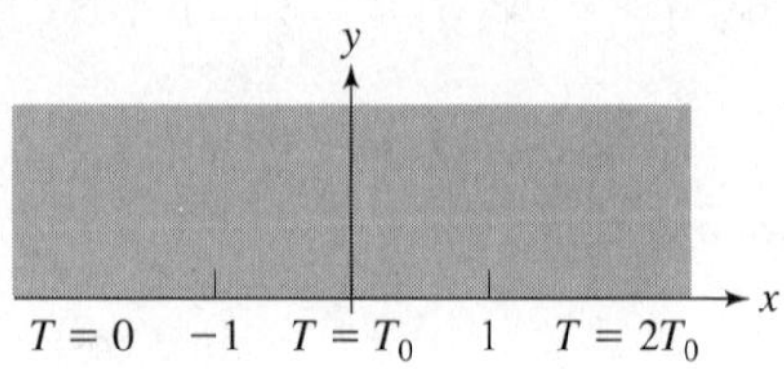

3.

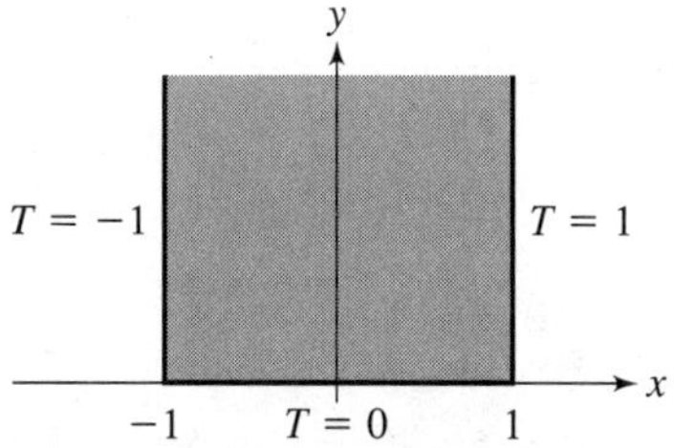

4.

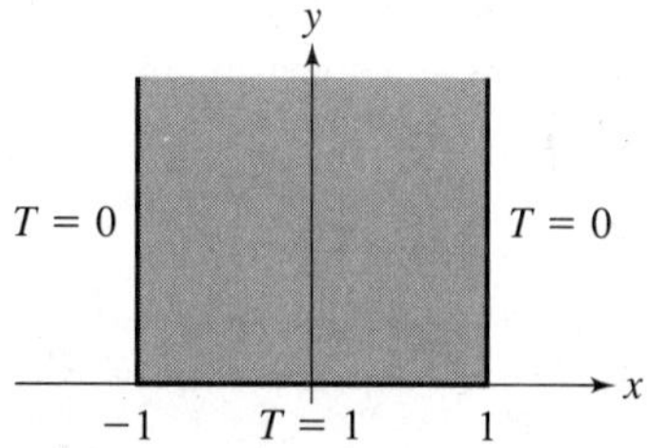

5.

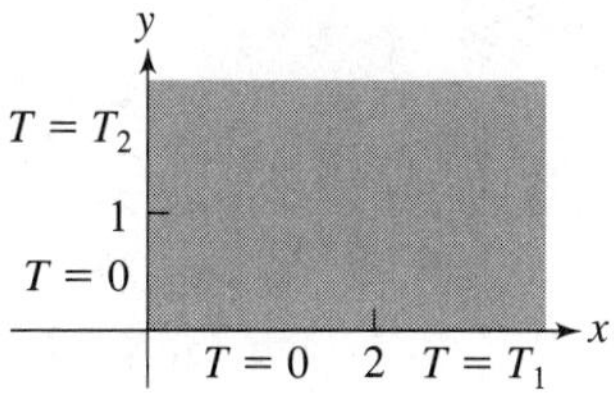

6.

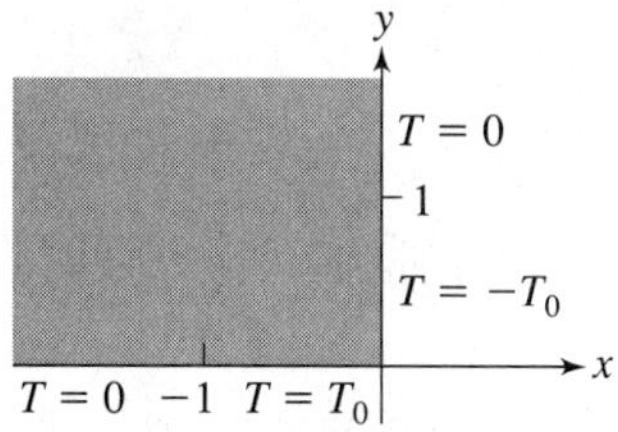

***7.**

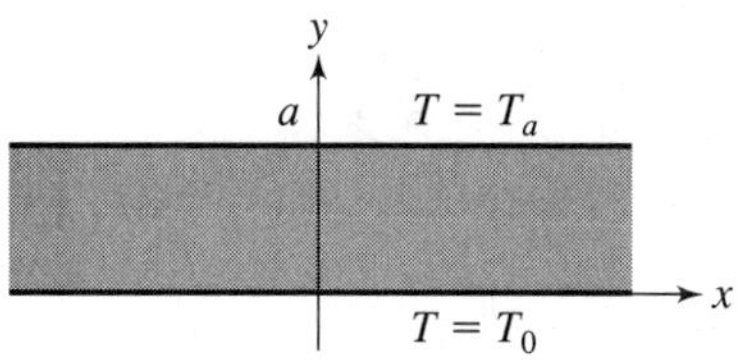

8.

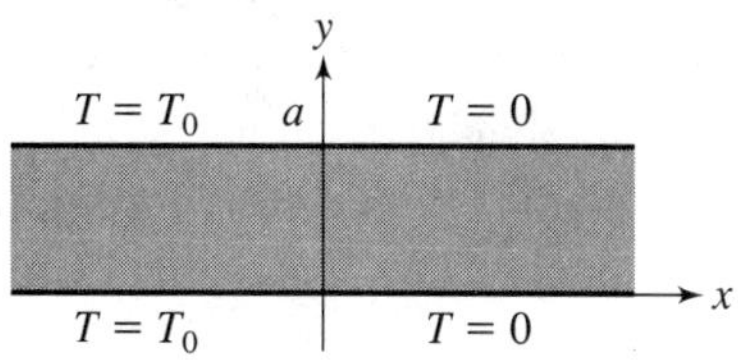

***9.**

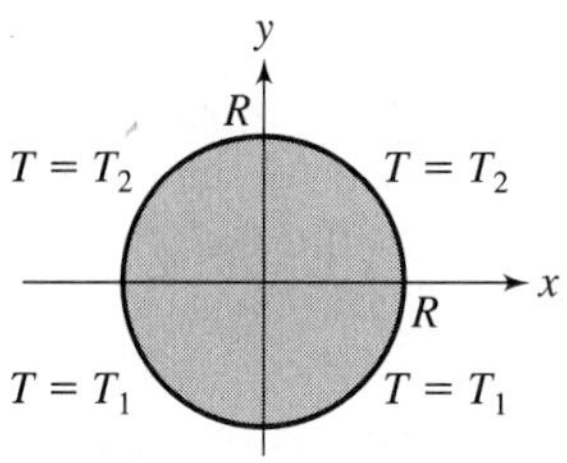

***10.**

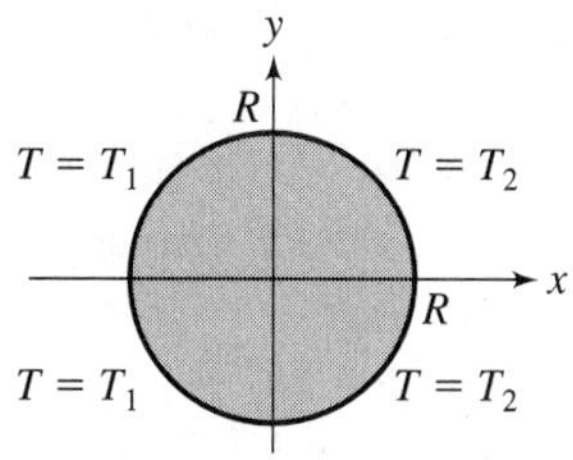

***11.**

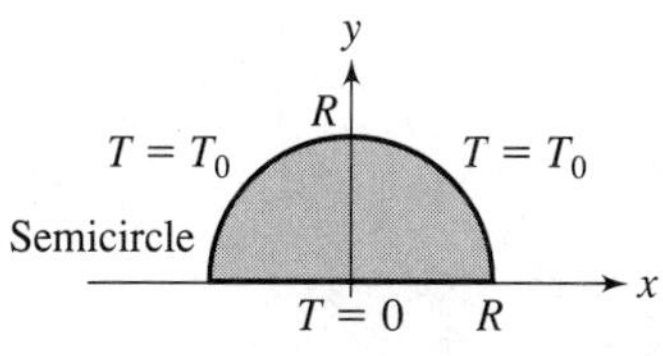

12.

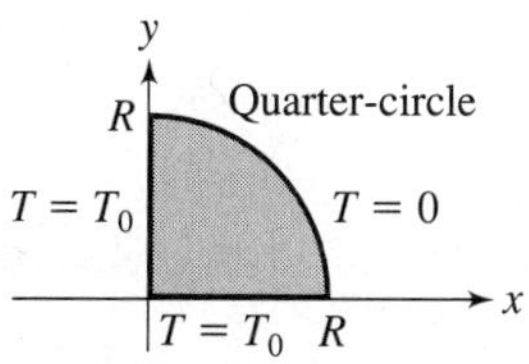

13. **(a)** Use equation 7.21 to find steady-state temperature $T(x, y)$ in a thin semi-infinite plate $y > 0$ with insulated surfaces, given that temperature along its edge $y = 0$ is T_2 for $x > 0$ and $T_1 < T_2$ for $x < 0$.

(b) Show that $T(x, y)$ can be expressed in the form

$$T(x, y) = T_2 + \frac{1}{\pi}(T_1 - T_2)\,\mathrm{Arg}(x + yi).$$

(c) What are isothermals and heat-flow lines?

14. Find the steady-state temperature in the quarter-plane $x > 0$, $y > 0$ when boundary conditions are as shown in the accompanying figure. *Hint:* Use $w = \operatorname{Sin}^{-1} z$ and see Example 7.23.

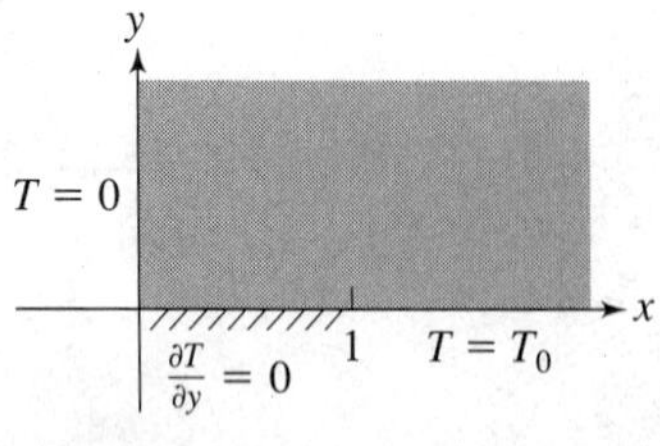

15. **(a)** Sketch isothermal curves in Example 7.20 for values $C = 0, T_0/4, T_0/2, 3T_0/4, T_0$.

(b) Sketch heat-flow lines in Example 7.20 for values $C = 0, \pm T_0/4, \pm T_0/2, \pm 3T_0/4$.

16. Verify that the solutions in Example 7.22 are the same.

17. A steady-state temperature situation exists in a circular plate $x^2 + y^2 \leq R^2$. Is it possible for heat to flow along circles $x^2 + y^2 = r^2$?

18. Verify the simplification for lines of heat flow in Example 7.20.

19. Show that the temperature function in Example 7.20 can be expressed in the form

$$T(x, y) = \frac{T_0}{\pi} \tan^{-1}\left(\frac{2y}{x^2 + y^2 - 1}\right),$$

where values of $\tan^{-1}$ are between 0 and π.

20. **(a)** Show that $w = f(z) = z + 1/z$ maps the half-plane $\operatorname{Im} z > 0$ exterior to $|z| = 1$ to the half-plane $\operatorname{Im} w > 0$.

(b) Find the steady-state temperature in the half-plane $\operatorname{Im} z > 0$ exterior to $|z| = 1$ if the temperature on the linear boundary is 100°C and that on the semicircle is 0°C.

21. Show that the temperature function in Example 7.21 can be expressed in the form

$$T(x, y) = T_0 - \frac{T_0}{\pi} \tan^{-1}\left(\frac{2\cos\dfrac{\pi x}{2a}\sinh\dfrac{\pi y}{2a}}{\sin^2\dfrac{\pi x}{2a} + \sinh^2\dfrac{\pi y}{2a} - 1}\right)$$

where values of $\tan^{-1}$ are between 0 and π.

Transformations: 3. $w = \sin(\pi z/2)$ **4.** $w = \sin(\pi z/2)$ **5.** $w = z^2$ **6.** $w = -z^2$ **8.** $w = e^{\pi z/a}$ **9.** $w = i(R - z)/(R + z)$ **10.** $w = i(R - iz)/(R + iz)$ **11.** $w = i(R - z)/(R + z)$ **(See Exercise 9 in Section 7.3.) 12.** $w = i(R^2 - z^2)/(R^2 + z^2)$ **(See Exercise 9 in Section 7.3.)**

7.8 Electrostatic Potential

In Section 2.7, we associated an analytic complex potential $f(z) = V(x, y) + W(x, y)i$ with two-dimensional electrostatic fields $V(x, y)$. The components of the electric field intensity $\mathbf{E}$ are the real and imaginary parts of $-\overline{f'(z)}$. Curves (cylinders) in the one-parameter family $V(x, y) = C$ are equipotentials, and curves $W(x, y) = C$ are lines of force.

In domains D that are free of charge, the potential satisfies Laplace's equation

$$\frac{\partial^2 V}{\partial x^2} + \frac{\partial^2 V}{\partial y^2} = 0. \tag{7.35a}$$

What completes the characterization of $V(x, y)$ is specification of a boundary condition on the boundary $\beta(D)$ of D, usually a Dirichlet or Neumann condition (see equations 7.19(a) and 7.20(a)).

Suppose for the moment that D is a bounded domain as shown in Figure 7.56, and we want to solve 7.35(a) subject to a Dirichlet condition on $\beta(D)$,

$$V(x, y) = F(x, y), \qquad (x, y) \text{ on } \beta(D), \tag{7.35b}$$

where $F(x, y)$ is a given function. According to Corollary 3 to Theorem 4.24, the solution of this problem is unique; there cannot be two different solutions. Furthermore, Theorem 4.24 indicates that $V(x, y)$ cannot have a relative maximum or minimum inside D; maximum and minimum potentials must occur on $\beta(D)$.

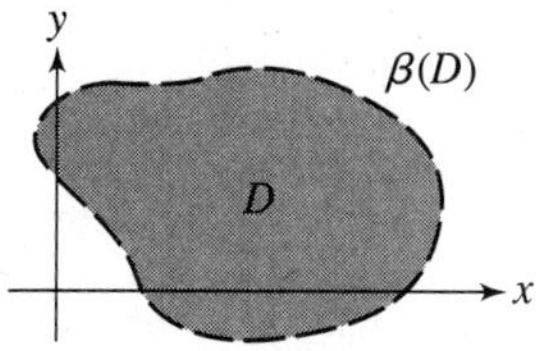

Figure 7.56

When D is a complicated domain, it may be possible to use a conformal mapping to map D onto a simpler domain, solve the simple problem, and then invert the transformation to obtain the solution to the original problem. We illustrate in the following examples.

EXAMPLE 7.24 ....

A cylindrical conductor of infinite length and radius R is centered around the z-axis in space. One half, $-\pi < \theta < 0$, is held at potential V_1, and the other half, $0 < \theta < \pi$, is held at V_2, the parts being separated by a thin piece of insulation (Figure 7.57).

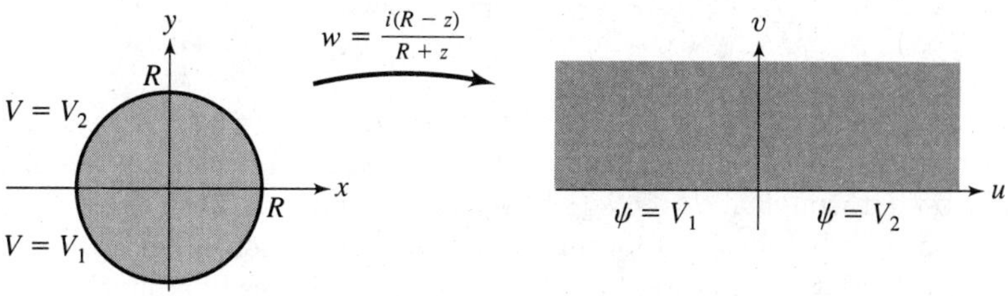

Figure 7.57

Find the potential interior to the cylinder. Describe the equipotential surfaces and lines of force interior to the cylinder.

Solution We map the circle $x^2 + y^2 < R^2$ to the half-plane $\operatorname{Im} w > 0$ with a bilinear transformation in such a way that the upper semicircle is mapped to the positive $u = \operatorname{Re} w$–axis and the lower semicircle is mapped to the negative u-axis. If we choose to map $z = R$ to $w = 0$, then $z = -R$ maps to points infinitely far out the real u-axis. Because $z = -R$ is the pole of the mapping, the transformation takes the form $w = a(z - R)/(z + R)$. If we arbitrarily demand that $z = Ri$ map to $w = 1$, then $1 = a(Ri - R)/(Ri + R)$, and this requires that $a = -i$. The bilinear transformation is therefore

$$w = \frac{i(R - z)}{R + z}.$$

According to equation 7.21, the solution $\psi(u, v)$ of Laplace's equation in the domain $\operatorname{Im} w > 0$ subject to the boundary conditions in Figure 7.57 is

$$\begin{aligned}\psi(u, v) &= \frac{V_1 + V_2}{2} + \frac{1}{\pi}(V_1 - V_2)\operatorname{Tan}^{-1}\left(-\frac{u}{v}\right)\\ &= \frac{V_1 + V_2}{2} + \frac{1}{\pi}(V_2 - V_1)\operatorname{Tan}^{-1}\left(\frac{u}{v}\right).\end{aligned}$$

If we set $w = u + vi$ and $z = x + yi$ in the bilinear transformation,

$$u + vi = \frac{i(R - x - yi)}{R + x + yi} = \frac{i[(R - x) - yi]}{(R + x) + yi}\,\frac{(R + x) - yi}{(R + x) - yi} = \frac{2yR + (R^2 - x^2 - y^2)i}{(R + x)^2 + y^2}.$$

Thus,

$$u = \frac{2yR}{(R + x)^2 + y^2}, \qquad v = \frac{R^2 - x^2 - y^2}{(R + x)^2 + y^2},$$

and the electrostatic potential is

$$V = \frac{V_1 + V_2}{2} + \frac{1}{\pi}(V_2 - V_1)\operatorname{Tan}^{-1}\left(\frac{2Ry}{R^2 - x^2 - y^2}\right).$$

Equipotential curves (really surfaces) are defined by

$$C = \frac{V_1 + V_2}{2} + \frac{1}{\pi}(V_2 - V_1)\operatorname{Tan}^{-1}\left(\frac{2Ry}{R^2 - x^2 - y^2}\right).$$

This equation can be expressed in the form $x^2 + (y + kR)^2 = R^2(1 + k^2)$, where k is a constant involving V_1, V_2, and C (see Exercise 1). Equipotentials are therefore circular arcs passing through the points $(\pm R, 0)$ with centers on the y-axis (Figure 7.58). Lines of force can be obtained by finding the harmonic conjugate of $V(x, y)$ or by solving the differential equation for orthogonal trajectories of equipotential curves. They are circles $(x - C)^2 + y^2 = C^2 - R^2$ (see Exercise 2).

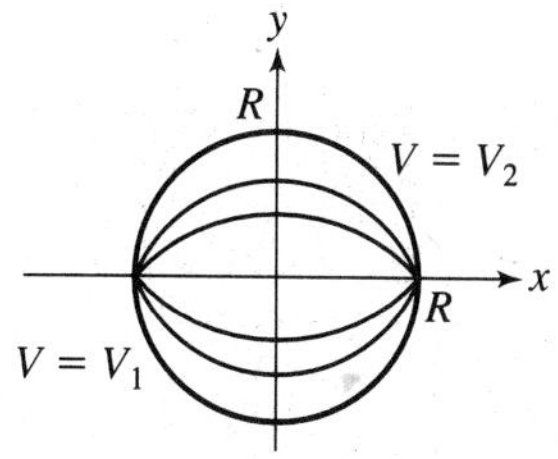

Figure 7.58

EXAMPLE 7.25

In Figure 7.59(a), the horizontal lines $y = \pm d$ for $x < 0$ represent the cross section of a semi-infinite parallel plate capacitor. If potentials on top and bottom plates are V_0 and $-V_0$, respectively, find equipotentials and lines of force.

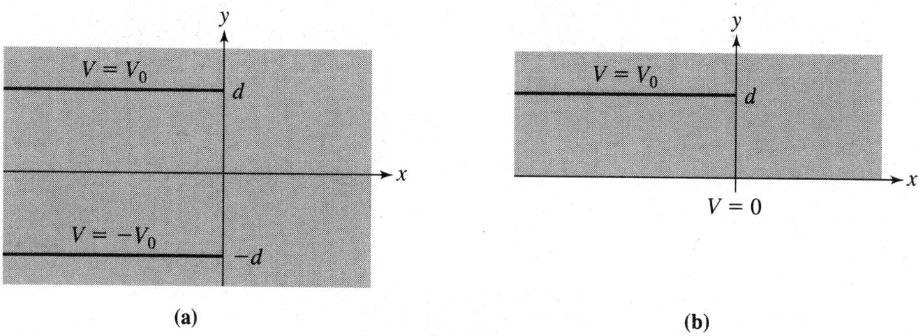

Figure 7.59

Solution Symmetry indicates that potential along the x-axis is zero for $x < 0$ and $x > 0$ and that potential below the x-axis is the negative of that above the x-axis. We therefore solve the problem in Figure 7.59(b), where potential is V_0 along the half-line $y = d$ ($x < 0$) and 0 along $y = 0$ ($-\infty < x < \infty$). We begin by mapping the region onto the upper half-plane $\operatorname{Im} \tilde{w} > 0$ with a Schwarz-Christoffel transformation. We regard the region in the z-plane as a degenerate triangle with vertices z_1, z_2, and z_3 in Figure 7.60. Angles α and β at z_2 and z_3 must be made to approach 2π and 0, respectively. Difficulty with the angle at z_1 is eliminated by choosing $\tilde{u}_1$, the preimage of z_1 as the point at infinity, in which case the angle does not enter the Schwarz-Christoffel transformation. If we choose $\tilde{u}_2 = -1$ and $\tilde{u}_3 = 0$, then taking limits of

$$\frac{dz}{d\tilde{w}} = A(\tilde{w} + 1)^{\alpha/\pi - 1}\tilde{w}^{\beta/\pi - 1}$$

as $\alpha \to 2\pi$ and $\beta \to 0$ gives

$$\frac{dz}{d\tilde{w}} = A\left(\frac{\tilde{w}+1}{\tilde{w}}\right) = A\left(1+\frac{1}{\tilde{w}}\right).$$

Integration gives

$$z = A(\tilde{w} + \log_\phi \tilde{w}) + B.$$

If we choose $\phi = -\pi/2$, then for $\tilde{u}_2 = -1$ to map onto $z_2 = di$, A and B must satisfy

$$di = A[-1 + \log_{-\pi/2}(-1)] + B = A(-1+\pi i) + B.$$

This equation is satisfied if we choose $A = B = d/\pi$, in which case

$$z = \frac{d}{\pi}(1 + \tilde{w} + \log_{-\pi/2} \tilde{w}).$$

It is straightforward to check that this transformation maps the positive $\tilde{u}$-axis onto the x-axis and maps both parts $-\infty < \tilde{u} < -1$ and $-1 < \tilde{u} < 0$ of the negative $\tilde{u}$-axis onto the half-line $y = d$, $x < 0$.

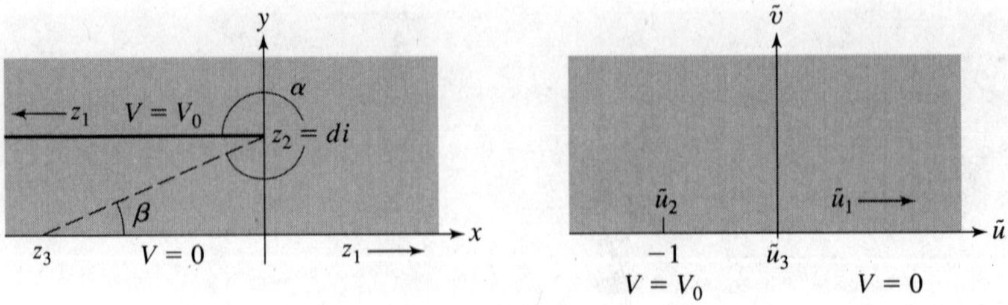

Figure 7.60

We could now use equation 7.21 to find potential in the half-plane $\operatorname{Im} \tilde{w} > 0$ with value V_0 on the negative $\tilde{u}$-axis and value 0 on the positive $\tilde{u}$-axis. Unfortunately, it would not yield a convenient representation for equipotentials and/or lines of force. Instead, we map $\operatorname{Im} \tilde{w} > 0$ to the infinite strip in Figure 7.61 in such a way that the negative $\tilde{u}$-axis is mapped onto the line $V = V_0 i$ and the positive $\tilde{u}$-axis is mapped onto the u-axis. This is accomplished by the inverse of $\tilde{w} = e^{\pi w/V_0}$ (see Example 7.11 in Section 7.3). The transformation

$$z = \frac{d}{\pi}\left(1 + e^{\pi w/V_0} + \log_{-\pi/2}(e^{\pi w/V_0})\right)$$

therefore maps the strip in Figure 7.61 onto the half-space, with u-axis mapped to x-axis and line $V_0 i$ mapped to half-line $y = d$, $x < 0$.

The solution of Laplace's equation in the strip is $\psi(u,v) = v$. To find potential $V(x,y)$, we would set

$$x + yi = z = \frac{d}{\pi}\left\{1 + e^{\pi(u+vi)/V_0} + \log_{-\pi/2}\left[e^{\pi(u+vi)/V_0}\right]\right\},$$

and solve for v in terms of x and y, an impossibility. We can, however, find equations for equipotential curves and lines of force. If we take take real and imaginary parts of the preceding equation,

$$x = \frac{d}{\pi}\left[1 + e^{\pi u/V_0}\cos\frac{\pi v}{V_0} + \frac{\pi u}{V_0}\right] \quad \text{and} \quad y = \frac{d}{\pi}\left[e^{\pi u/V_0}\sin\frac{\pi v}{V_0} + \frac{\pi v}{V_0}\right].$$

Equipotential curves for $y \geq 0$ are therefore defined parametrically by

$$x = \frac{d}{\pi}\left[1 + e^{\pi u/V_0}\cos\frac{\pi V}{V_0} + \frac{\pi u}{V_0}\right], \qquad y = \frac{d}{\pi}\left[e^{\pi u/V_0}\sin\frac{\pi V}{V_0} + \frac{\pi V}{V_0}\right]$$

for fixed $V \geq 0$ and u as parameter.

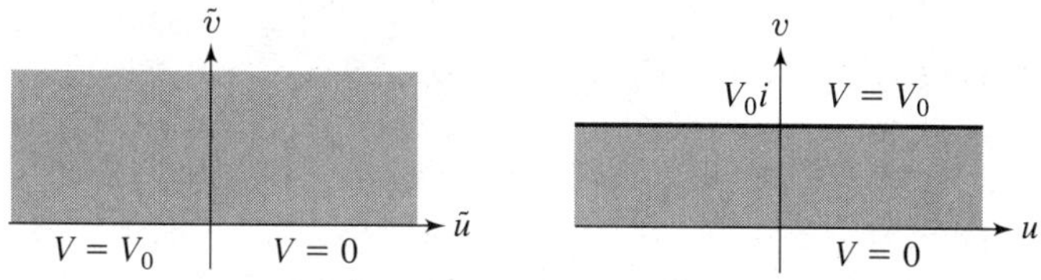

Figure 7.61

Orthogonal trajectories in the strip are $u = U =$ constant, and therefore lines of force in the xy-plane are given parametrically by

$$x = \frac{d}{\pi}\left[1 + e^{\pi U/V_0}\cos\frac{\pi v}{V_0} + \frac{\pi U}{V_0}\right], \qquad y = \frac{d}{\pi}\left[e^{\pi U/V_0}\sin\frac{\pi v}{V_0} + \frac{\pi v}{V_0}\right],$$

for fixed U and v as parameter. Both sets of curves are shown in Figure 7.62.

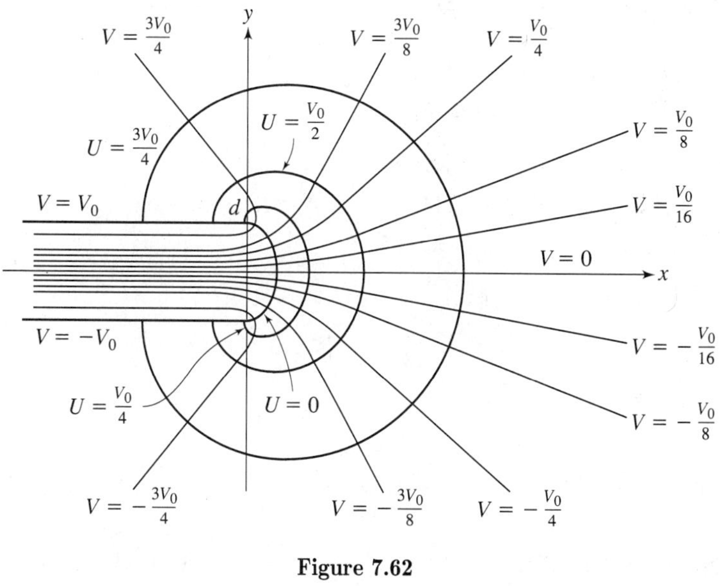

Figure 7.62

Exercises 7.8

1. (a) Show that equipotential curves in Example 7.24 are circles with centers on the y-axis.
   (b) What is the equipotential for $C = (V_1 + V_2)/2$?
   (c) Draw equipotential curves for $C = -3/4, -1/2, -1/4, 0, 1/4, 1/2, 3/4$ when $V_1 = -1$ and $V_2 = 1$.

2. Show that the lines of force in Example 7.24 must satisfy the differential equation $dy/dx = (R^2 + y^2 - x^2)/(2xy)$ and that the solution of this equation is $(x - C)^2 + y^2 = C^2 - R^2$.

3. Find the potential in the first quadrant if the x- and y-axes are held at constant potentials V_1 and V_2, respectively.

4. Generalize Exercise 3 to find the potential interior to the wedge $0 \le \theta \le \alpha < 2\pi$ if the potential along $\theta = 0$, $r > 0$, is V_0 and that along $\theta = \alpha$, $r > 0$, is V_α.

5. Find the potential in the semicircle $x^2 + y^2 < R^2$, $y > 0$, when $V = V_0$ on $y = 0$ and $V = V_1$ on $x^2 + y^2 = R^2$. *Hint:* Use the mapping $w = i(R - z)/(R + z)$.

6. Find potential in the circle $x^2 + y^2 < R^2$ when potential on that part of the circle in the first quadrant is V_0 and potential on the remainder of the circle is zero. *Hint:* Use a bilinear transformation that maps the points $z = Ri$, $z = -R$, and $z = R$ to $w = 0$, $w = 1$, and $w = \infty$. What is the potential at the center of the circle?

7. Find potential in the half-plane $y > 0$ outside the circle $x^2 + y^2 = R^2$ given that potential on the semicircle is V_0 and potential on the x-axis for $|x| > R$ is zero. *Hint:* Use the bilinear transformation $w = (z - R)/(R + z)$.

8. Find potential in the infinite strip $0 < y < a$ if potential on $y = 0$ and $y = a$ for $x > 0$ is $V = 0$ and on $y = 0$ and $y = a$ for $x < 0$, it is $V = V_0$, a constant.

9. Find potential in the domain bounded by that part of $x^2 + y^2 = a^2$ above the x-axis and that part of $x^2 + (y - a)^2 = 2a^2$ below the x-axis. Potential on $x^2 + y^2 = a^2$ is V_0, and potential on $x^2 + (y - a)^2 = 2a^2$ is V_1. *Hint:* First map the domain to the sector $\pi/2 < \arg w^* < 5\pi/4$ by $w^* = (z - a)/(z + a)$, and then map the sector to an infinite strip with $w = \log_{-\pi/2} w^* - \pi i/2$.

10. Find potential at points in the half-plane $y > 0$ that are outside the circle $x^2 + (y - a)^2 = a^2$ given that potential on the circle is a constant V_0 and potential on $y = 0$ is a constant V_1. *Hint:* Try $w = -2a/z$.

11. Find potential $V(x, y)$ in the semicircle $x^2 + y^2 < R^2$, $y > 0$, that satisfies the boundary conditions

$$V(x, 0) = 0, \qquad 0 < x < R,$$

$$V(x, 0) = V_0, \qquad -R < x < 0,$$

$$\frac{\partial V(x, y)}{\partial n} = 0, \qquad \text{on } x^2 + y^2 = R^2,\ y > 0.$$

(See equation 7.20(a).) *Hint:* $w = \log_{-\pi/2} z$ could help.

All exercises in Section 7.7 may be treated as potential problems. Simply replace the word *temperature* with *potential* in each exercise.

7.9 Fluid Flow

In Section 2.8 we associated a complex potential

$$F(z) = \phi(x, y) + \psi(x, y)i \tag{7.36}$$

in domains of incompressible, irrotational, steady-state, two-dimensional flow. Function $\phi(x, y)$ is the velocity potential, and velocity $\mathbf{q}$ of the flow at any point is the gradient of $\phi(x, y)$,

$$\mathbf{q} = \frac{\partial \phi}{\partial x}\hat{\mathbf{i}} + \frac{\partial \phi}{\partial y}\hat{\mathbf{j}}. \tag{7.37}$$

Alternatively, components of $\mathbf{q}$ are real and imaginary parts of $\overline{F'(z)}$. The harmonic conjugate $\psi(x, y)$ of $\phi(x, y)$ is the stream function. Motion of the fluid is along stream curves $\psi(x, y) = C$. In this section and the next we use conformal mappings to discuss flow patterns associated with various geometries. Fundamental to our discussions is the complex potential for uniform flow over a half-plane.

When fluid flows with constant speed U in the positive x-direction over the half-plane $y > 0$, its velocity is $\mathbf{q} = U\hat{\mathbf{i}}$ (Figure 7.63). The velocity potential is $\phi(x, y) = Ux + A$, A a real constant, and the stream function is $\psi(x, y) = Uy + B$, B a real constant. The complex potential is

$$F(z) = \phi + \psi i = Uz + A + Bi.$$

By choosing $B = 0$ we obtain the natural situation where the boundary $y = 0$ of the flow corresponds to the stream function $\psi = 0$. With $A = 0$, the velocity equipotential $\phi = 0$ corresponds to the line $x = 0$. The complex potential for uniform flow in the positive x-direction over the half-plane $y > 0$ can therefore be taken in the form

$$F(z) = Uz. \tag{7.38}$$

This is also the complex potential for uniform flow in the positive x-direction for the whole xy-plane.

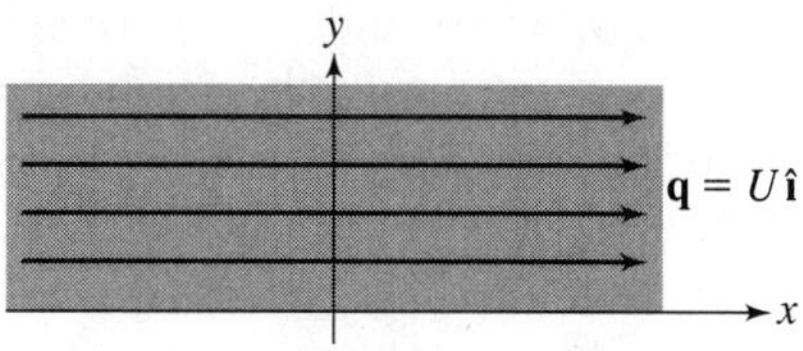

Figure 7.63

Conformal mappings are used to map flows to flows. To illustrate, suppose that $F(z)$ is the complex potential for flow in some domain D. Suppose further that $w = g(z)$ is a 1-1 analytic function conformally mapping D onto $g(D)$, the boundary $\beta(D)$ of D onto the boundary $\beta[g(D)]$ of $g(D)$, and the interior of D onto the interior of $g(D)$. Then the function $F[g^{-1}(w)]$ is the complex potential for a flow in $g(D)$. To verify this we must show that $F[g^{-1}(w)]$ is analytic in $g(D)$ and that the boundary $\beta[g(D)]$ is a stream curve. Because $g(z)$ is conformal, its derivative does not vanish in D, and therefore the derivative of $g^{-1}(w)$ exists at each point in $g(D)$. Because $F(z)$ is analytic, it follows that so also is $F[g^{-1}(w)]$. We now show that the imaginary part of $F[g^{-1}(w)]$ is constant along any parts of the $\beta[g(D)]$. Suppose $F(z) = \phi(x,y) + \psi(x,y)i$, so that the stream function $\psi(x,y)$ is constant along parts of $\beta(D)$. Let its value along a part C_j of the boundary be $\psi(x,y) = k$. We may regard $\psi(x,y) = k$ as the equation of C_j, and therefore the image $g(C_j)$ of C_j under $w = g(z)$ has equation $\psi[x(u,v), y(u,v)] = k$. But the real and imaginary parts of $F[g^{-1}(w)]$ can be expressed in terms of real and imaginary parts of w as follows:

$$F[g^{-1}(w)] = \phi[x(u,v), y(u,v)] + \psi[x(u,v), y(u,v)]i.$$

Hence, the imaginary part of $F[g^{-1}(w)]$ is constant along $g(C_j)$. We illustrate these ideas in the following example.

EXAMPLE 7.26

Use the transformation $w = \sqrt{z}$ to map uniform flow left to right in the half-plane $y > 0$ with speed U to flow in a quarter plane. Draw stream curves and find the velocity and speed of the flow.

Solution The transformation $w = \sqrt{z}$ maps the half-plane $y > 0$ to the quarter-plane $u > 0$, $v > 0$, as shown in Figure 7.64. Because the complex potential for uniform flow in the z-plane is $F(z) = Uz$, the complex potential for flow in the w-plane is $f(w) = Uw^2$. Stream curves are $C = \psi(u,v) = \text{Im}(Uw^2) = 2Uuv$, rectangular hyperbolas asymptotic to the positive u- and v-axes (Figure 7.65). Because $\overline{f'(w)} = 2U\overline{w} = 2U(u - vi)$, the velocity of the flow at any point is

$$\mathbf{q} = 2U(u\hat{\mathbf{i}} - v\hat{\mathbf{j}}).$$

The u-component is always positive, and the v-component is always negative. In other words, flow is always downward and to the right. Along the line $v = u$, $\mathbf{q} = 2Uu(\hat{\mathbf{i}} - \hat{\mathbf{j}})$, and this vector is perpendicular to the line. The speed of the flow is $|\mathbf{q}| = 2U\sqrt{u^2 + v^2}$; it is proportional to the distance from the origin. The origin is a stagnation point of the flow.

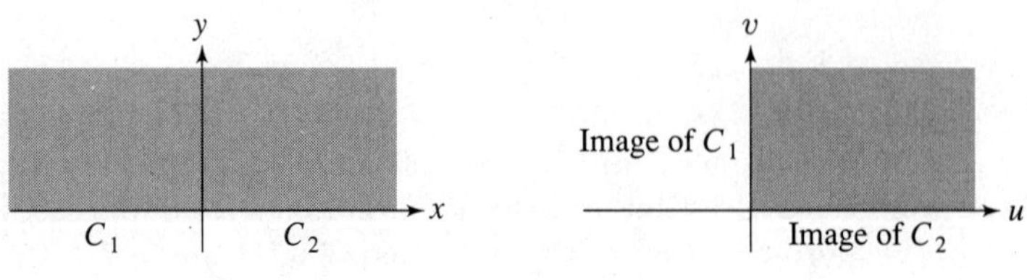

Figure 7.64

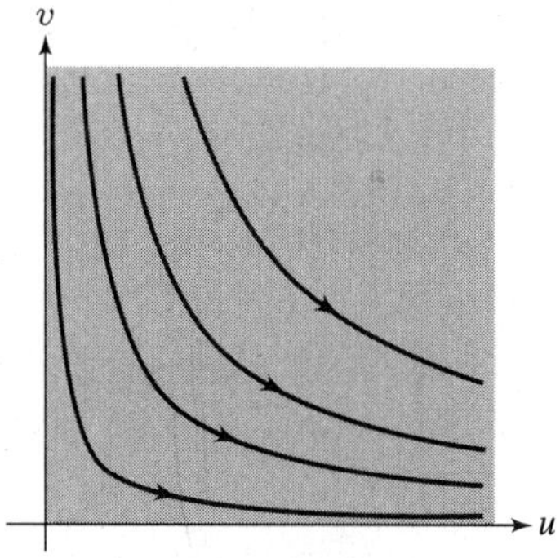

Figure 7.65

Conformal transformations map flows to flows; they map stream curves to stream curves and complex potentials to complex potentials. To use this concept in applications, we need one additional fact: For a given geometry, flow and its mathematical description by means of the complex potential are in some sense unique. Clearly flow itself is not unique for a given geometry. For example, uniform flow left to right in the xy-plane is not unique; it is always along horizontal lines, but the speed of the flow is arbitrary. Speed is constant, but arbitrary. In other words, for given geometry, the flow pattern is unique; variability lies in some overall strength factor. For uniform flow in the positive x-direction, this is manifested as far as the complex potential $F(z) = Uz$ is concerned in the multiplicative factor U. Thus, the complex potential for a given flow geometry is subject to the variability of the strength of the flow. A complex constant may also be added to the complex potential. It has no effect on the flow pattern (or its strength), because velocity of the flow is related to the real and imaginary parts of the derivative of $F(z)$. Customarily this constant is determined by specifying that the stream function vanish on some boundary of the flow and also specifying that one of the equipotentials be identified by $\phi = 0$.

Consider now the problem of finding the flow pattern in a given domain D of the xy-plane, which, based on the preceding discussion, is unique. The flow itself is subject to a strength factor and an additive constant, but the pattern is unique. Suppose that we can find a conformal transformation $w = g(z)$ mapping D onto a domain $g(D)$ in the w-plane and that the flow pattern, through the complex potential $F(w)$, is known in $g(D)$. It follows then that the flow pattern for D is described by the complex potential $f(z) = F[g(z)]$.

In many practical problems we have the situation where an otherwise uniform flow in the z-plane is interrupted by an object. To determine the flow pattern around the obstacle, we find a conformal transformation $w = g(z)$ that maps the flow domain in the z-plane to a known flow pattern. The transformation must map any boundaries of the flow in the z-plane to boundaries of the flow in the w-plane. Quite often the image is uniform flow left to right over the half-plane $\operatorname{Im} w > 0$. Because 7.38 gives $F(w) = \tilde{U}w$ as a complex potential for uniform flow in the

w-plane, a complex potential for the flow pattern in the z-plane is $\tilde{U}g(z)$. Constant $\tilde{U}$ must be related to the speed of the flow in the z-plane (usually as $|z| \to \infty$). We illustrate this in the following two examples.

EXAMPLE 7.27 ....

Uniform flow of a fluid from left to right with speed U across the complex plane is interrupted by a circle $|z| = a > 0$ (Figure 7.66). Find the resulting flow pattern.

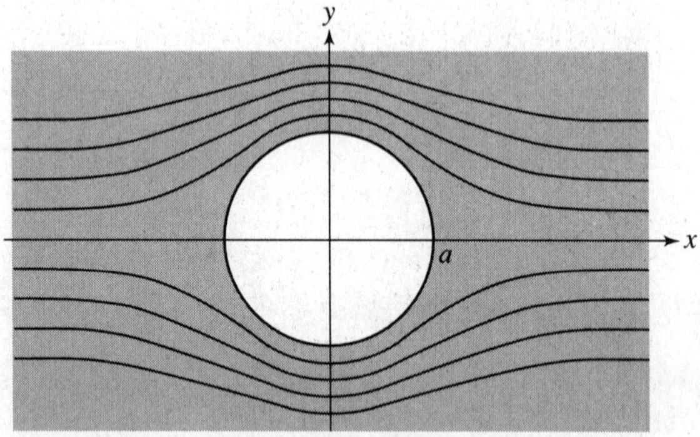

Figure 7.66

Solution By symmetry, the flow is the same above and below the x-axis. We therefore treat the x-axis exterior to the circle and the upper semicircle as a boundary and consider only the upper half of the flow (Figure 7.67). It is straightforward to show that

$$w = z + \frac{a^2}{z}$$

maps the upper flow region onto $\operatorname{Im} w > 0$, with the boundary of the flow mapped onto the real axis in the w-plane. In particular, $z = x \leq -a$ is mapped onto $w = u \leq -2a$; $z = x \geq a$ is mapped onto $w = u \geq 2a$; and $z = ae^{\theta i}$, $0 \leq \theta \leq \pi$ is mapped onto $w = u$, $-2a \leq u \leq 2a$ (Figure 7.68).

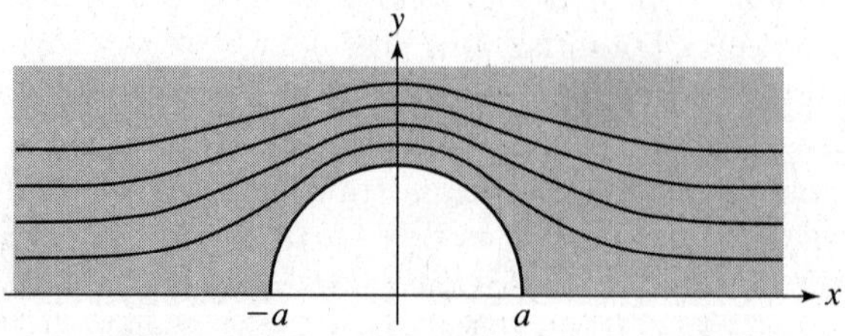

Figure 7.67

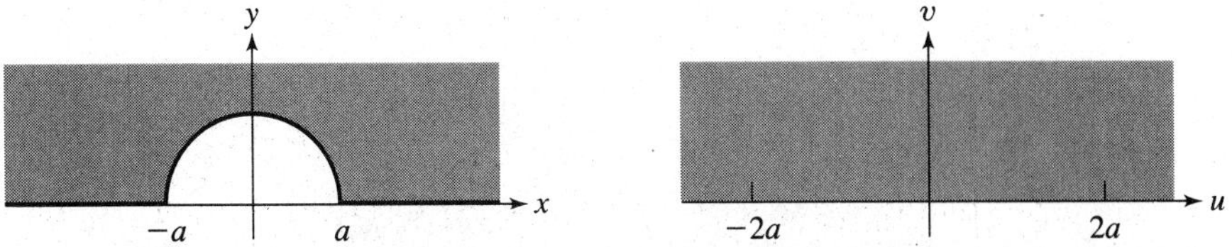

Figure 7.68

Uniform flow over the half-plane $\operatorname{Im} w > 0$ is described by $F(w) = \tilde{U}w$. Consequently, a complex potential for the flow in the z-plane is

$$f(z) = \tilde{U}\left(z + \frac{a^2}{z}\right).$$

For large $|z|$, this function is approximated by $\tilde{U}z$. Because flow in the z-plane far from $z = 0$ should essentially be uniform, with complex potential Uz, we identify $\tilde{U}$ and U, and

$$f(z) = U\left(z + \frac{a^2}{z}\right).$$

Stream curves for this complex potential were discussed in Example 2.24. If we set $z = re^{\theta i}$, then stream curves are defined by

$$\begin{aligned} C &= \operatorname{Im}\left[U\left(re^{\theta i} + \frac{a^2}{r}e^{-\theta i}\right)\right], \qquad C \text{ a constant} \\ &= U\left(r\sin\theta - \frac{a^2}{r}\sin\theta\right), \\ &= \frac{U}{r}(r^2 - a^2)\sin\theta. \end{aligned}$$

The stream curve corresponding to $C = 0$ consists of the upper half of the circle $r = a$ and those parts of the lines $\theta = 0$ and $\theta = \pi$ for which $r \geq a$. Other stream curves are shown in Figure 7.67. Stream curves below the x-axis are obtained by symmetry.

EXAMPLE 7.28

Find equations for stream curves for the flow pattern due to a uniform flow with speed U across the upper half of the xy-plane from left to right that is interrupted by a vertical barrier of height a at $x = 0$ (Figure 7.69).

Solution We derived the transformation $w = a^{-1}(z + ai)^{1/2}(z - ai)^{1/2}$ that maps the z-plane, slit along the y-axis for $0 \leq y \leq a$, onto the half-plane $\operatorname{Im} w > 0$ in Example 7.19. With $F(w) = \tilde{U}w$ representing uniform flow over the w-plane, the

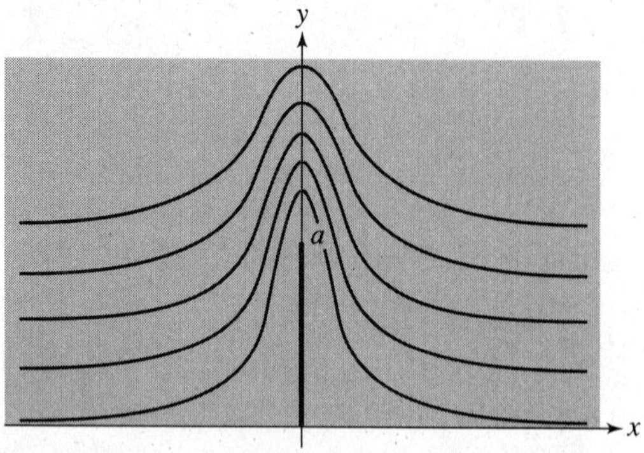

Figure 7.69

complex potential for the required flow is

$$f(z) = \frac{\tilde{U}}{a}(z + ai)^{1/2}(z - ai)^{1/2}.$$

For large $|z|$, this flow can be approximated by $\tilde{U}z/a$. We must therefore identify $\tilde{U}/a$ with U, and

$$f(z) = U(z + ai)^{1/2}(z - ai)^{1/2}.$$

For $x > 0$, stream curves are given by

$$C = \psi(x, y)$$

$$= U[x^2 + (y + a)^2]^{1/4}[x^2 + (y - a)^2]^{1/4} \sin\left[\frac{\operatorname{Tan}^{-1}\left(\frac{y+a}{x}\right) + \operatorname{Tan}^{-1}\left(\frac{y-a}{x}\right)}{2}\right]$$

(see Exercise 1). To find a simpler representation, we note that stream curves in the w-plane are horizontal lines $v = C$. We can represent each one parametrically in the form $w = u + Ci$, $-\infty < u < \infty$. The corresponding stream curve in the z-plane can be found by solving

$$u + Ci = U(z + ai)^{1/2}(z - ai)^{1/2}$$

for z in terms of u. If we set $z = x + yi$ and square both sides,

$$\frac{1}{U^2}(u + Ci)^2 = (x + yi + ai)(x + yi - ai).$$

Equating real and imaginary parts gives

$$\frac{1}{U^2}(u^2 - C^2) = x^2 - (y^2 - a^2), \qquad \frac{2uC}{U^2} = 2xy.$$

If we solve the second for u and substitute into the first,

$$\frac{1}{U^2}\left(\frac{x^2y^2U^4}{C^2} - C^2\right) = x^2 - y^2 + a^2,$$

and this can be solved for y in terms of x:

$$y = \frac{C}{U}\sqrt{\frac{x^2 + a^2 + C^2/U^2}{x^2 + C^2/U^2}}.$$

This is the equation for the stream curves in Figure 7.69. Notice that for large x, $y \approx C/U$, indicating that the stream curve corresponding to C is asymptotic to the line $y = C/U$; that is, far from the barrier, stream curves are essentially horizontal.

Exercises 7.9

1. Verify the stream function in Example 7.28.
2. (a) Fluid flows closely along the side $\theta = \alpha$ of the wedge $0 < \theta < \alpha$ toward the origin (see the following figure). It then flows away from the corner. Find the flow pattern.
   (b) What is the fluid velocity for $\theta = \alpha$?
   (c) What is the fluid velocity for $\theta = \alpha/2$? Is it perpendicular to the line $\theta = \alpha/2$?
   (d) Are components of the fluid velocity negative for $\theta > \alpha/2$?
   (e) Is the x-component of the fluid velocity positive for $\theta < \alpha/2$? Is the y-component negative?

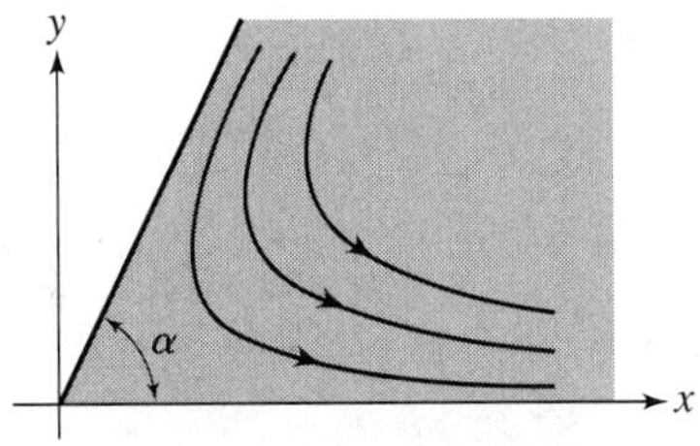

3. Show that the complex potential for the flow resulting from placing a circle of radius a at the origin in an otherwise uniform flow with speed U making angle α $(0 < \alpha < \pi)$ with the positive x-axis is

$$f(z) = U\left(e^{-\alpha i}z + \frac{a^2e^{\alpha i}}{z}\right).$$

4. (a) According to Exercise 12 in Section 7.1, the Joukowski transformation $w = z + c^2/z$ maps the exterior of the circle $|z| = R$ to the exterior of an ellipse. Use this result to find c and R so that the transformation maps the exterior of $|z| = R$ to the exterior of $u^2/a^2 + v^2/b^2 = 1$ for given a and b.

(b) Use Example 7.27 to find the complex potential when uniform flow from left to right in the xy-plane is interrupted by an ellipse $x^2/a^2 + y^2/b^2 = 1$.

5. **(a)** Use Exercises 3 and 4 to show that the complex potential for the flow due to the ellipse $x^2/a^2 + y^2/b^2 = 1$ in an otherwise uniform flow with speed U making angle α $(0 < \alpha < \pi)$ with the positive x-axis (see the following figure) is

$$f(z) = \frac{U(a+b)}{2}\left[\frac{e^{-\alpha i}(z+\sqrt{z^2+b^2-a^2})}{a+b} + \frac{e^{\alpha i}(z-\sqrt{z^2+b^2-a^2})}{a-b}\right].$$

(b) Take the limit of the potential in (a) as $b \to 0$. Would you expect it to represent angular flow past a plate of length $2a$?

(c) What is the speed of the flow at the ends of the plate?

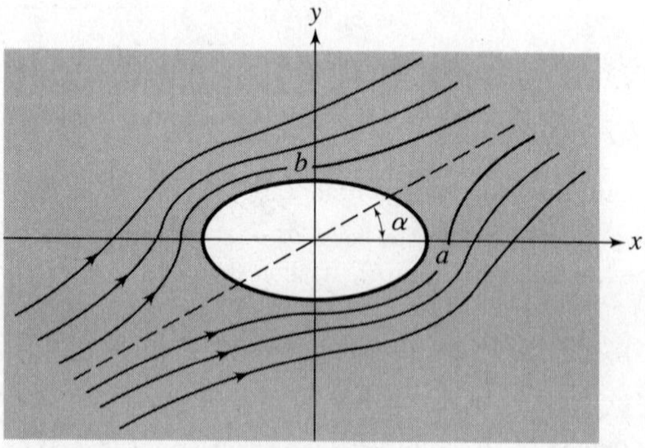

6. **(a)** Use the mapping in Exercise 20 of Section 7.3 to find the complex potential when uniform flow with speed U in the x-direction is deflected by a parabolic wall, as shown in the following figure.

(b) Show that $z = 0$ is a stagnation point of the flow.

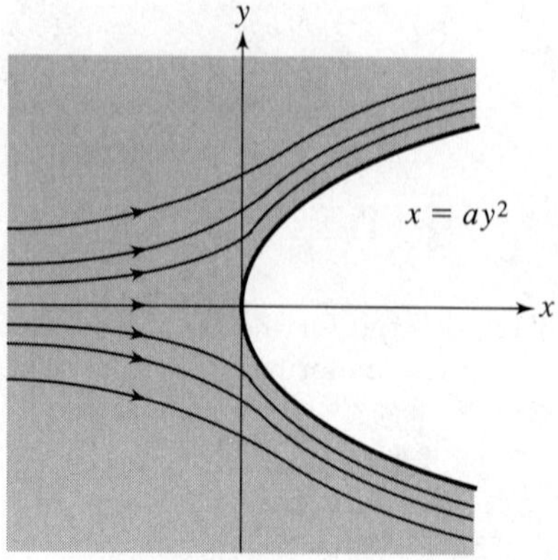

7. Use the transformation in Exercise 21 of Section 7.3 to find the complex potential for uniform flow left to right with speed U in the half-plane $y > 0$ over the circle $x^2 + (y-a)^2 = a^2$.

8. Fluid flows down the left side of the semi-infinite channel $-a < x < a$, $y > 0$, is deflected at $y = 0$, and flows up the right side. Find the stream function for the flow. What is the velocity and speed of the flow?

9. Uniform flow with speed U in the positive x-direction is deflected by a wedge with angle 2α, as shown in the following figure. Find the complex potential for the flow.

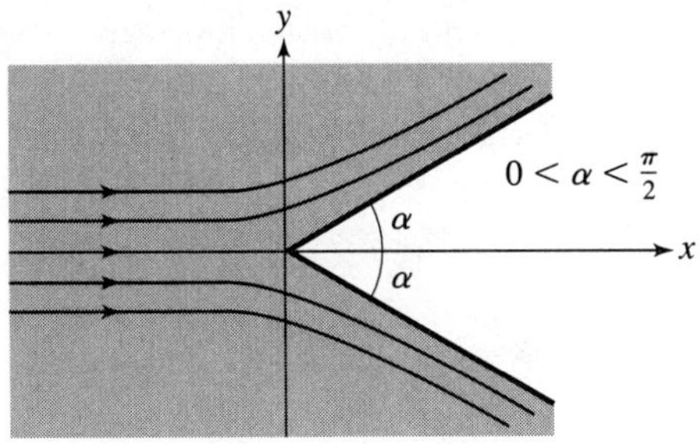

10. Find stream curves for the flow with complex potential $f(z) = Ua^3/(2z^2)$, where U and a are positive constants.

11. Find a complex potential for first-quadrant flow around the quarter circle $x^2 + y^2 = a^2$ (see the following figure). *Hint:* See Exercise 22 in Section 7.3. Are there any stagnation points?

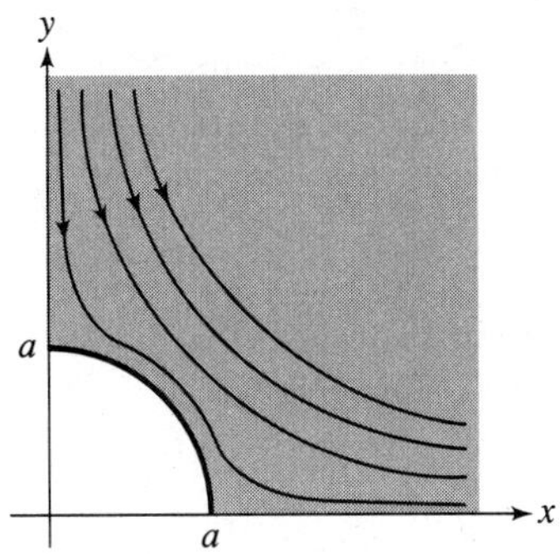

12. Fluid with uniform speed U exits from the channel to the right, as shown in the figure. Find equations for the stream curves. *Hint:* See Example 7.25.

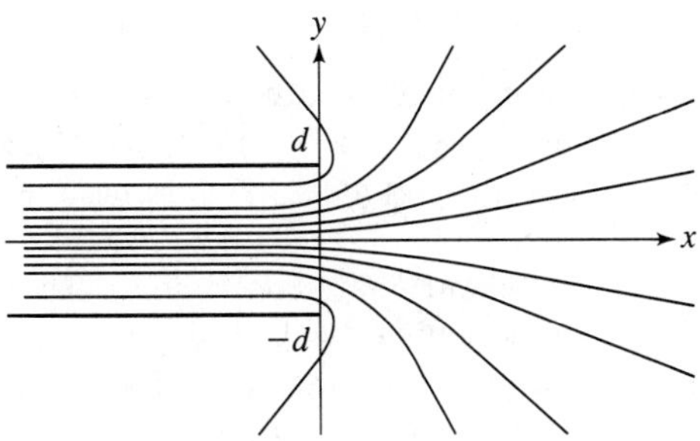

13. **(a)** Show that the function

$$z = g^{-1}(w)$$
$$= \frac{a}{\pi}\{(w+1)^{1/2}(w-1)^{1/2} + \log_{-\pi/2}[w + (w+1)^{1/2}(w-1)^{1/2}]\} - ai, \qquad a > 0,$$

maps the intervals $u < -1$, $-1 < u < 1$, and $u > 1$ on the real axis in the w-plane to the boundaries C_1, C_2, and C_3 of the flow region shown. See also Exercise 7 in Section 7.6.

(b) Find the complex potential for uniform flow with speed U over the step.

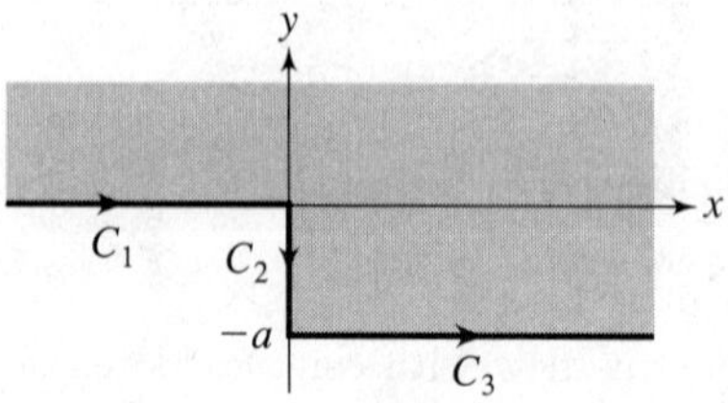

7.10 Sources and Sinks

If motion of a fluid is radially outward from a point (x_0, y_0), symmetric in all directions, (x_0, y_0) is called a **source** (Figure 7.70). A source is an abstract concept; it does not occur in nature, but it is useful in describing many flows. When motion is radially toward (x_0, y_0), we speak of a **sink** at (x_0, y_0).

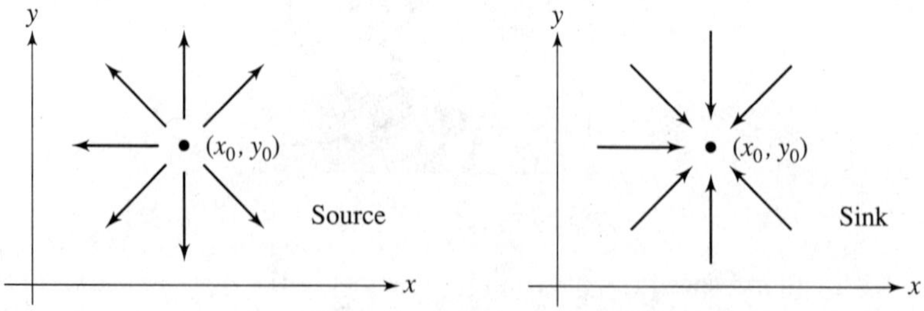

Figure 7.70

To find the complex potential associated with a source, we note that stream curves are radial lines from (x_0, y_0), and equipotentials (the orthogonal trajectories of stream curves) are circles centered at (x_0, y_0). Functions $f(z) = \phi + \psi i$ for which $\psi = C$ are radial lines through $z_0 = x_0 + y_0 i$ and $\phi = C$ are circles with centers at z_0 are $k \log(z - z_0) + D$, where k and D are constants. Because a flow is characterized by its velocity field and components of velocity are real and imaginary parts of the derivative of the complex potential, we may dispense with constant D. In addition, because branches of the logarithm function differ only by an additive constant, every

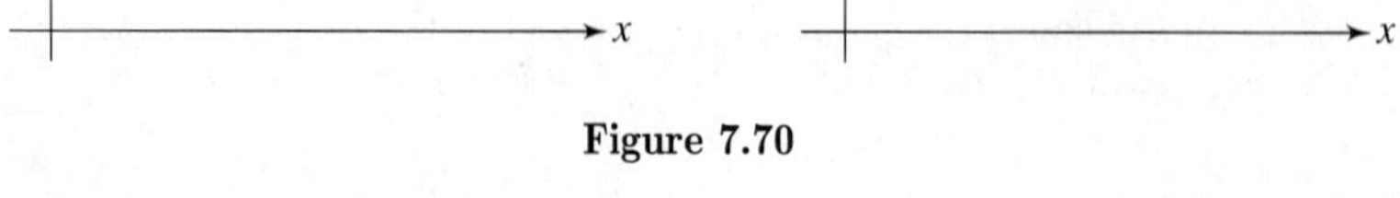

branch of $k \log(z - z_0)$ leads to the same velocity field for the flow. We therefore write

$$f(z) = k \log(z - z_0) \tag{7.39}$$

as the complex potential for a source at z_0 (when $k > 0$) or a sink (when $k < 0$), without specifying the particular branch. But we regard $\log(z - z_0)$ as a single-valued function (with unspecified branch). Constant k is related to the mass flow rate from the source. If C is a circle centered at z_0, then flux (mass flow rate) through C is given by integral 2.47. Because

$$f'(z) = \frac{k}{z - z_0} = \frac{k}{(x - x_0) + (y - y_0)i} = \frac{k[(x - x_0) - (y - y_0)i]}{(x - x_0)^2 + (y - y_0)^2},$$

velocity at each point in the flow is $\mathbf{q} = \dfrac{k}{(x - x_0)^2 + (y - y_0)^2}[(x - x_0)\hat{\mathbf{i}} + (y - y_0)\hat{\mathbf{j}}]$. Consequently,

$$\text{Flux} = \rho \int_C u\,dy - v\,dx = \rho k \int_C \frac{(x - x_0)\,dy - (y - y_0)\,dx}{(x - x_0)^2 + (y - y_0)^2}.$$

If we set $x = x_0 + R\cos t$, $y = y_0 + R\sin t$, $-\pi \le t \le \pi$ on C,

$$\text{Flux} = \rho k \int_{-\pi}^{\pi} \frac{R\cos t(R\cos t\,dt) - (R\sin t)(-R\sin t\,dt)}{R^2} = 2\pi\rho k.$$

Alternatively, using the result in Exercise 13 of Section 4.3,

$$\text{Flux} = \text{Im}\left[\rho \int_C f'(z)\,dz\right] = \rho\,\text{Im} \int_C \frac{k}{z - z_0}dz = \rho k\,\text{Im}(2\pi i) = 2\pi\rho k.$$

No matter how small the circle, flux is $2\pi\rho k$. This quantity is therefore the rate at which fluid is emitted by the source. We call k the **strength** of the source. In other words, strength of a source is its mass flow rate divided by $2\pi\rho$. When $k < 0$, we say that the sink has strength $-k$.

When a source occurs on the boundary of a flow, it may not emit fluid in all directions. For example, a source on the x-axis for half-plane flow $y > 0$ emits fluid only into the half-plane; a source at the origin in quarter-plane flow $x > 0$, $y > 0$ emits fluid only into the quarter plane. In these situations, strength is related to flux by a different formula.

A more general situation is for a source to emit fluid into a sector with angle α from its vertex (Figure 7.71). Because flow continues to be radially away from the source and equipotentials are circular arcs, the complex potential for the flow is still of form 7.39. The mass flow rate of fluid from the source is equal to the flux through any equipotential C. It is straightforward to show that

$$\text{Flux} = \rho\alpha k.$$

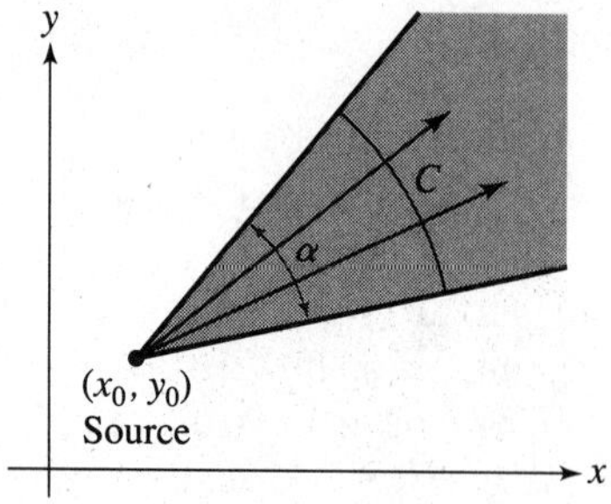

Figure 7.71

In other words, if we continue to call k in 7.39 the strength of the source, then strength is mass flow rate from the source divided by $\rho\alpha$. When $\alpha = 2\pi$, we have the case of a source emitting fluid in all directions.

EXAMPLE 7.29

A source of strength k is placed at point $(h, 0)$ on the x-axis $(h > 0)$, and a sink of equal strength is placed at $(-h, 0)$. Find stream curves for the flow.

Solution The complex potential for the flow is

$$f(z) = k \log(z - h) - k \log(z + h).$$

Stream curves are

$$C = \mathrm{Im}[f(z)] = k \tan^{-1}\left(\frac{y}{x-h}\right) - k \tan^{-1}\left(\frac{y}{x+h}\right),$$

where values of $\tan^{-1}$ are between $\pm\pi$. If we take tangents, then

$$\tan\left(\frac{C}{k}\right) = \frac{\dfrac{y}{x-h} - \dfrac{y}{x+h}}{1 + \left(\dfrac{y}{x-h}\right)\left(\dfrac{y}{x+h}\right)} = \frac{2hy}{x^2 + y^2 - h^2}.$$

This equation can be simplified to

$$x^2 + \left[y - h\cot\left(\frac{C}{k}\right)\right]^2 = h^2 \csc^2\left(\frac{C}{k}\right).$$

Stream curves are therefore circles through the source and sink with centers on the y-axis. Fluid flows from source to sink along these circular arcs (Figure 7.72).

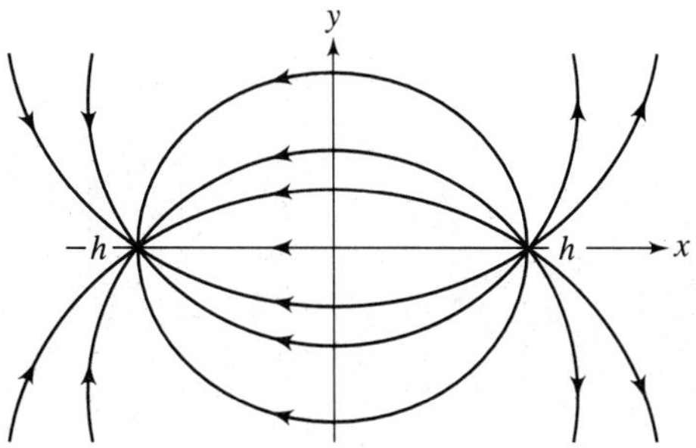

Figure 7.72

When a source is inserted into an existing flow that has no boundary, the complex potential of the resulting flow is the sum of the complex potentials of the existing flow and the source. For instance, if a source at $z = 0$ of strength k is inserted into uniform flow left to right with speed U, the resulting flow has complex potential $f(z) = Uz + k \log z$ (see Exercise 1). When the existing flow has boundaries, however, the complex potential of the resulting flow may or may not be the sum of the complex potentials of the existing flow and the source. For example, uniform horizontal flow past a circle of radius a has complex potential $U(z + a^2/z)$. If to this is added $k \log(z - z_0)$, where $|z_0| > a$, the result $U(z+a^2/z)+k \log(z - z_0)$ is the complex potential for some flow, but it is not that due to insertion of a source into the flow $U(z + a^2/z)$ (see Example 7.30). It would be only if the circle $|z| = a$ continued to be a stream curve of the flow. In other words, complex potentials can be added only when boundaries continue to be stream curves.

EXAMPLE 7.30

Uniform horizontal flow past a circle of radius a has complex potential $U(z+a^2/z)$. Show that when a source of strength k is placed at position $(-h, 0)$, where $h > a$, the complex potential for the combined flow is not $f(z) = U(z + a^2/z) + k \log(z + h)$.

Solution Function $f(z)$ is the complex potential for the flow if the boundary of the flow, the circle $|z| = a$, is a stream curve for $f(z)$. Stream curves are given by

$$C = \operatorname{Im}[U(z + a^2/z) + k \log(z + h)] = U\left(r \sin\theta - \frac{a^2}{r} \sin\theta\right) + k\phi,$$

where θ and ϕ are shown in Figure 7.73. When the equation of the circle $r = a$ is substituted into this equation, $C = k\phi$, which is not constant. Hence, $f(z)$ is not the complex potential for the flow.

To find the complex potential for the flow we note that because the flow pattern is symmetric about the x-axis, we may treat the x-axis exterior to the circle as a flow boundary and consider only flow in the half-plane $y > 0$ exterior to the

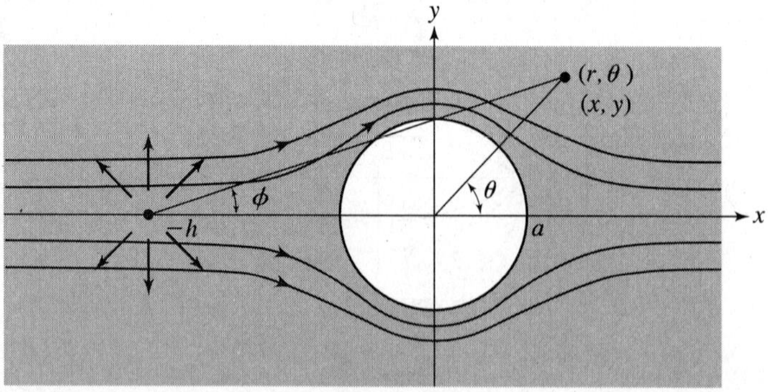

Figure 7.73

circle (Figure 7.74). The source at $z = -h$ still has strength k. The transformation $w = z + a^2/z$, which maps the flow domain to the half-plane $\operatorname{Im} w > 0$, maps the source at $z = -h$ to a source at $w_1 = -(h + a^2/h)$ with strength k.

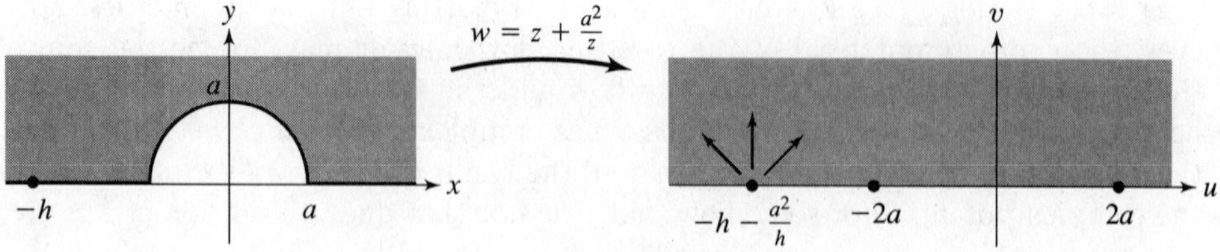

Figure 7.74

The complex potential for the flow pattern in the w-plane is

$$F(w) = \tilde{U}w + k\log(w - w_1).$$

(It is straightforward to show that $v = 0$ is a stream curve for this potential.) The complex potential for the flow pattern in the z-plane is

$$f(z) = \tilde{U}\left(z + \frac{a^2}{z}\right) + k\log\left(z + \frac{a^2}{z} + h + \frac{a^2}{h}\right).$$

Because $f'(z)$ can be approximated by $\tilde{U}$ for large $|z|$, it follows that we may identify $\tilde{U}$ and U, and

$$f(z) = U\left(z + \frac{a^2}{z}\right) + k\log\left(z + \frac{a^2}{z} + h + \frac{a^2}{h}\right).$$

We now show that $F(z) = Uz$, the complex potential for uniform horizontal flow over the z-plane, is also the complex potential for uniform flow in a horizontal channel.

EXAMPLE 7.31 ....

An infinitely long channel is bounded by the horizontal lines $y = 0$ and $y = a$ (Figure 7.75). At the "end" $x = -\infty$, flow enters the channel with constant speed U across the channel. Show that 7.38 is a complex potential for the flow.

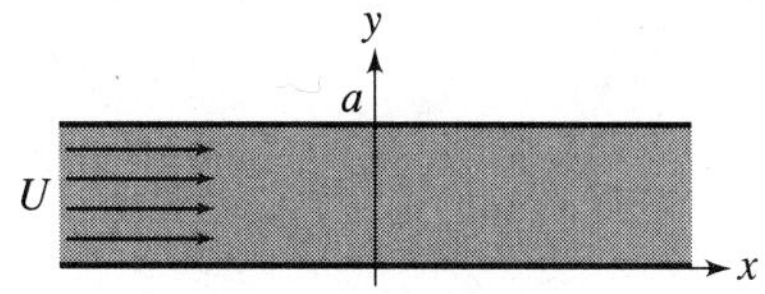

Figure 7.75

Solution The stream function $\psi(x, y)$ for the flow must satisfy Laplace's equation, $\partial^2\psi/\partial x^2 + \partial^2\psi/\partial y^2 = 0$. We also know that $\psi(x, y)$ must be constant along $y = 0$ and $y = a$ (they are boundaries of the flow). We can arbitrarily set $\psi(x, 0) = 0$, and to find the value of ψ on $y = a$, we note that flux (mass flow rate) through every vertical cross section of the channel is ρUa. But according to Exercise 10 in Section 2.8, flux along any curve from $y = 0$ to $y = a$ is equal to

$$\rho[\psi(0, a) - \psi(0, 0)] = \rho\psi(0, a).$$

Hence, $\rho Ua = \rho\psi(0, a)$, and along the stream curve $y = a$, the stream function has value $\psi(x, a) = Ua$. The solution of Laplace's equation subject to $\psi(x, 0) = 0$ and $\psi(x, a) = Ua$ is $\psi(x, y) = Uy$. The Cauchy-Riemann equations then give $\phi(x, y) = Ux + C$, and the complex potential is $F(z) = Ux + C + Uyi = Uz + C$. If we choose the y-axis as the equipotential $\phi = 0$, then $C = 0$, and $F(z) = Uz$. It is often helpful to regard the left end of the channel as a source with mass flow rate ρUa, and the right end as a sink with mass flow rate $-\rho Ua$. This is evident in the following alternative derivation of the complex potential.

The transformation $w = e^{\pi z/a}$ maps the channel to the half-plane $\operatorname{Im} w > 0$ with the left end of the channel mapping to the origin (Figure 7.76). We now have a source at $w = 0$ emitting fluid into the half-plane, the complex potential for which is $F(w) = k \log w$. Constant k is equal to the mass flow rate from the source divided by $\rho\pi$, and the mass flow rate from $w = 0$ is equal to that from the left end of the

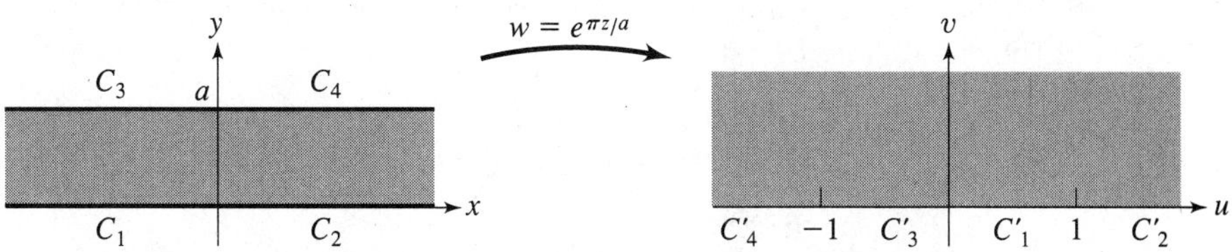

Figure 7.76

channel, namely, ρUa. Hence, $k = Ua/\pi$. It follows that the complex potential for flow in the channel is

$$f(z) = \frac{Ua}{\pi} \log (e^{\pi z/a}) = Uz.$$

We consider one more example of conformal mapping in conjunction with sources.

EXAMPLE 7.32

A source at the origin emits fluid into the channel $0 < y < a$ (Figure 7.77). Find a complex potential for the flow. Is there a stagnation point in the flow?

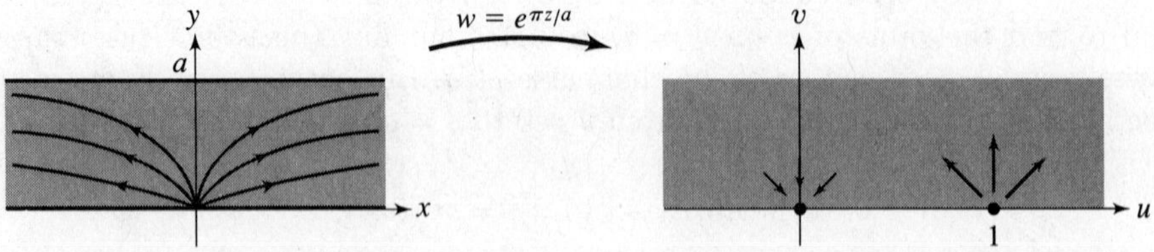

Figure 7.77

Solution The transformation $w = e^{\pi z/a}$ maps the channel to the half-plane $\operatorname{Im} w > 0$ with the source at $z = 0$ mapped to $w = 1$. In addition, it maps the sink at the left end of the channel to the origin $w = 0$. The complex potential for a source at $w = 1$ is $k \log (w - 1)$, where k is the strength of the source at $z = 0$ in the z-plane. Symmetry of the flow in the channel indicates that the strength of the sink at the left end of the channel is $k/2$. A complex potential for the sink at $w = 0$ is therefore $-(k/2) \log w$. The sum of these potentials is

$$F(w) = k \log (w - 1) - \frac{k}{2} \log w.$$

This is a complex potential for flow in the w-plane if the u-axis is a stream curve. This is indeed the case; the u-axis is the stream curve $\psi(u, v) = 0$. The complex potential in the channel is, therefore,

$$\begin{aligned} f(z) &= k \log (e^{\pi z/a} - 1) - \frac{k}{2} \log (e^{\pi z/a}) = k \log \left[\frac{e^{\pi z/a} - 1}{e^{\pi z/(2a)}} \right] \\ &= k \log \left[e^{\pi z/(2a)} - e^{-\pi z/(2a)} \right] = k \log \left(\sinh \frac{\pi z}{2a} \right), \end{aligned}$$

where we have omitted the constant $k \log 2$. Stagnation points occur where

$$0 = f'(z) = \frac{k\pi}{2a} \coth \frac{\pi z}{2a},$$

and this results yields $z = ai$.

Exercises 7.10

1. Uniform flow with speed U parallel to the x-axis fills the xy-plane. If a source of strength k is placed at the origin, find the stream function for the resulting flow. Is there a stagnation point?
2. Find a complex potential for flow due to a source at the origin into a semi-infinite channel $x > 0$, $0 < y < a$ by
   (a) Noting that in Example 7.32, the y-axis inside the channel is a stream curve.
   (b) Using the transformation $w = \sinh^2 [\pi z/(2a)]$.
3. What is the complex potential from two sources of equal strengths located at the real points $z = \pm a$? Find stream curves and equipotentials for the resulting flow. Is there a stagnation point in the flow?
4. (a) What is the complex potential due to a source of strength $Ua/(2\pi h)$ at $z = h > 0$ and a sink of equal strength at $z = -h$?
   (b) When we take the limit of the flow as $h \to 0$, the flow is said to be due to a dipole or doublet. Find its complex potential, stream curves, and equipotentials.
5. Into uniform flow in the positive x-direction is placed a source of strength k at position $(-a, 0)$, where $a > 0$, and a sink of equal strength at $(a, 0)$. Show that stagnation points in the flow are always on the x-axis.
6. Use the result in Exercise 3 to find the complex potential for the flow in the half-plane $x > 0$ due to a source with strength k at the point $z = a$ $(a > 0)$, the y-axis being a boundary for the flow. Considering only points on the wall $x = 0$, at which points is the speed of the flow smallest and greatest?
7. Show that the complex potential for flow in a wedge $0 < \operatorname{Arg} z < \alpha < \pi/2$ due to a source of strength k at z_0 (see the following figure) is

$$f(z) = k \log\left(z^{\pi/\alpha} - z_0^{\pi/\alpha}\right) + k \log(z^{\pi/\alpha} - \overline{z_0}^{\pi/\alpha}).$$

   Hint: See Exercise 6.

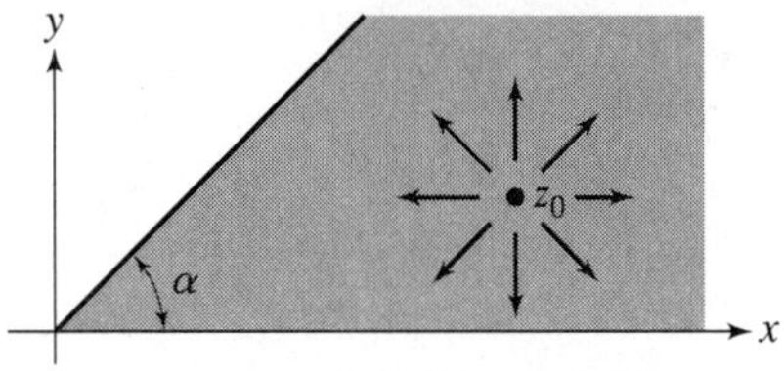

8. Find the complex potential for flow due to a source of strength k at position $z = c > a > 0$ on the x-axis in the presence of a circular obstacle $|z| \leq a$.

9. Two sources of equal strength k are placed at points $(0, \pm a)$ in uniform flow left to right with speed U. Find the stagnation points of the resulting flow. When is there only one stagnation point? When are they on the x-axis?

10. If source and sink in Exercise 5 are reversed, show that stagnation points are either on the x-axis or the y-axis.

11. The positive x- and y-axes are boundaries for first-quadrant flow due to a source of strength k at a distance a from the origin on the line $y = x$. Show that the complex potential for the flow is $k \log(z^4 + a^4)$. *Hint:* Consider flow in the xy-plane from sources at the points $(\pm a/\sqrt{2}, \pm a/\sqrt{2})$ and $(\pm a/\sqrt{2}, \mp a/\sqrt{2})$.

12. A source with strength k is placed at position $(0, 0)$ in the infinite channel $-a/2 < y < a/2$. Find a complex potential for the flow. *Hint:* Example 7.32 may be useful.

CHAPTER 8

Laplace Transforms

The Laplace transform is a mathematical operation that replaces differentiation problems with algebraic ones, an important simplification for ordinary and partial differential equations. In most elementary approaches to Laplace transforms, discussions are restricted to real variables. As a result, finding inverse Laplace transforms usually requires the practitioner to be proficient at calculating partial fraction decompositions of rational functions and to have a table of transforms and/or their inverses. We follow this approach in Sections 8.1 and 8.2, but in Section 8.3 we show how complex variables lead to a direct method for finding inverse transforms.

8.1 Laplace Transform

When f is a function of t, its Laplace transform, denoted by $F = \mathcal{L}\{f\}$, is a function with values defined by

$$F(s) = \int_0^\infty e^{-st} f(t)\, dt \tag{8.1}$$

provided the improper integral converges. In Sections 8.1 and 8.2, the variable s is real; in Section 8.3, it is complex. When f is piecewise continuous on every finite interval $0 \le t \le T$ and f is of exponential order α, its Laplace transform exists for $s > \alpha$. A function f is said to be of exponential order α, written $O(e^{\alpha t})$, if there exist constants T and M such that $|f(t)| < Me^{\alpha t}$ for all $t > T$. For example, e^{2t} is $O(e^{2t})$, $\sin t$ is $O(e^{0t})$, and t^n, n a nonnegative integer, is $O(e^{\epsilon t})$ for arbitrarily small ϵ.

When F is the Laplace transform of f, we call f the inverse Laplace transform of F, and write

$$f = \mathcal{L}^{-1}\{F\}. \tag{8.2}$$

The following Laplace transforms (Table 8.1) are fundamental to applications of the transform to ordinary and partial differential equations; more extensive tables are contained in such references as *Tables of Integral Transforms* (Volume 1) by A. Erdelyi, W. Magnus, F. Oberhettinger, and F. G. Tricomi (New York: McGraw-Hill, 1954). All but the last entry are straightforward applications of the definition in 8.1. The transform of $t^{-1/2}$ requires the improper integral of e^{-t^2} over the interval $0 \le t < \infty$.

Table 8.1

$f(t)$	$\tilde{f}(s)$	$f(t)$	$\tilde{f}(s)$
t^n	$\dfrac{n!}{s^{n+1}}$	$\sinh at$	$\dfrac{a}{s^2 - a^2}$
e^{at}	$\dfrac{1}{s-a}$	$\cosh at$	$\dfrac{s}{s^2 - a^2}$
$\sin at$	$\dfrac{a}{s^2 + a^2}$	$t \sinh at$	$\dfrac{2as}{(s^2 - a^2)^2}$
$\cos at$	$\dfrac{s}{s^2 + a^2}$	$t \cosh at$	$\dfrac{s^2 + a^2}{(s^2 - a^2)^2}$
$t \sin at$	$\dfrac{2as}{(s^2 + a^2)^2}$	$\dfrac{1}{\sqrt{t}}$	$\sqrt{\dfrac{\pi}{s}}$
$t \cos at$	$\dfrac{s^2 - a^2}{(s^2 + a^2)^2}$		

Because the Laplace transform is an integral transform, F is unique for given f, but there exist many functions f having the same transform F. For example, the functions defined by

$$f(t) = t^2 \quad \text{and} \quad g(t) = \begin{cases} 0, & t = 1 \\ t^2, & t \neq 1, 2 \\ 0, & t = 2, \end{cases}$$

which are identical except for their values at $t = 1$ and $t = 2$, both have the same transform, $2/s^3$. What we are saying is that because the Laplace transform is not a one-to-one operation, the inverse transform $\mathcal{L}^{-1}\{F\}$ in 8.2 is not an inverse in the true sense of inverse. In Section 8.3 we derive a formula for calculating inverse transforms, and this formula always yields a continuous function f, if this is possible. In the event that it is not possible, the formula gives a piecewise-continuous function whose value is the average of right and left limits at discontinuities, namely, $[f(t+)+f(t-)]/2$. The importance of this formula is that it defines $f = \mathcal{L}^{-1}\{F\}$ in a unique way. Other functions that have the same transform F differ from f only in their values at isolated points; they cannot differ from f over an entire interval $a \leq t \leq b$. In compliance with this anticipated formula, we adopt the procedure in this section of always choosing a continuous function $\mathcal{L}^{-1}\{F\}$ for given F or, when this is not possible, a piecewise-continuous function.

The Laplace transform and its inverse are linear operators. Some of their simple properties are summarized next.

One of two shifting properties is

$$\mathcal{L}\{e^{at} f(t)\} = F(s - a), \tag{8.3a}$$

$$\mathcal{L}^{-1}\{F(s - a)\} = e^{at} f(t) \tag{8.3b}$$

(see Exercise 1). It states that multiplication by an exponential e^{at} in the t-domain is equivalent to a translation in the s-domain. For example, since $\mathcal{L}\{\cos 2t\} = s/(s^2+4)$, 8.3(a) implies that

$$\mathcal{L}\{e^{3t}\cos 2t\} = \frac{s-3}{(s-3)^2+4}.$$

The use of $\mathcal{L}$ and $\mathcal{L}^{-1}$ in 8.3 is not quite as described earlier. The Laplace transform and its inverse operate on functions, not on function values, as is suggested by 8.3. These equations are more properly stated in the form

$$\mathcal{L}\{e^{at}f\}(s) = F(s-a), \tag{8.3c}$$

$$\mathcal{L}^{-1}\{F(s-a)\}(t) = e^{at}f(t). \tag{8.3d}$$

We feel that the shifting property is more clearly conveyed for most readers by 8.3(a) and (b), and we apologize to readers who are offended by the notation. It is convenient to repeat this practice in describing other properties of the Laplace transform, but we shall attempt to minimize its use.

The second shifting property of the Laplace transform is

$$\mathcal{L}\{f(t-a)H(t-a)\} = e^{-as}F(s), \tag{8.4a}$$

$$\mathcal{L}^{-1}\{e^{-as}F(s)\} = f(t-a)H(t-a), \tag{8.4b}$$

where $H(t-a)$ is the Heaviside unit step function shown in Figure 8.1. It has value 0 when $t < a$ and value 1 when $t > a$.

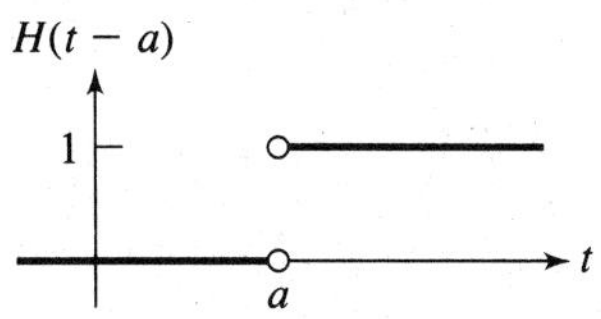

Figure 8.1

See Exercise 2 for a proof of properties 8.4. They imply that multiplication by an exponential e^{-as} in the s-domain is equivalent to a translation in the t-domain. Graphs of $f(t)$ and $f(t-a)H(t-a)$ are shown in Figure 8.2.

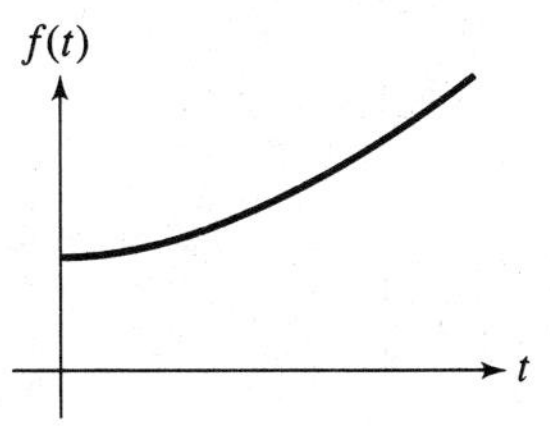

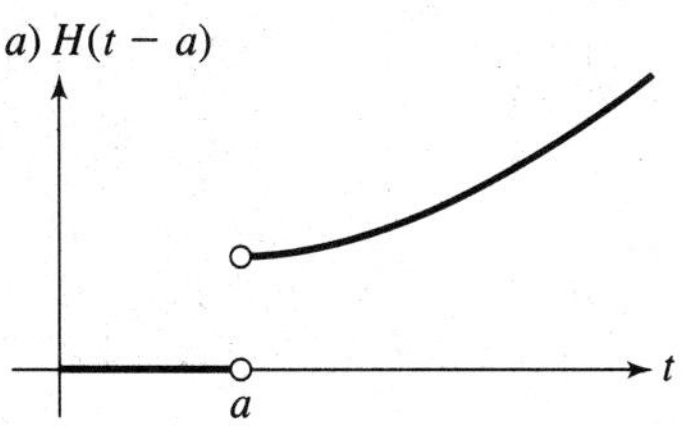

Figure 8.2

For example, because $\mathcal{L}^{-1}\{2/s^3\} = t^2$, it follows that

$$\mathcal{L}^{-1}\left\{\frac{2e^{-4s}}{s^3}\right\} = (t-4)^2 H(t-4).$$

This result is shown in Figure 8.3.

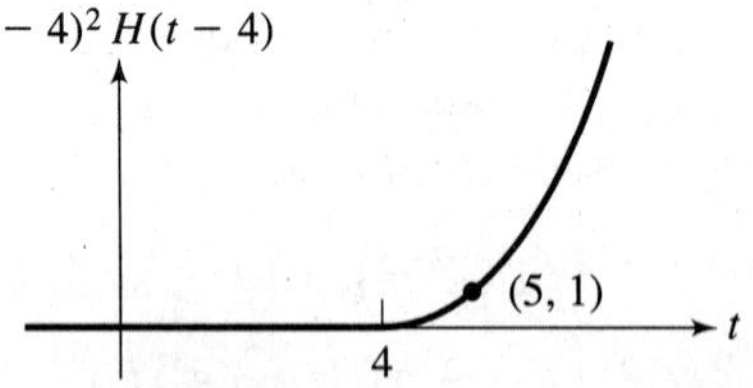

Figure 8.3

EXAMPLE 8.1 ....

Find the inverse transform for $F(s) = \dfrac{e^{-s} - e^{-2s}}{s}$.

Solution Because $\mathcal{L}^{-1}\{1/s\} = 1$, property 8.4(b) gives

$$f(t) = \mathcal{L}^{-1}\left\{\frac{e^{-s}}{s}\right\} - \mathcal{L}^{-1}\left\{\frac{e^{-2s}}{s}\right\} = H(t-1) - H(t-2).$$

This function is shown in Figure 8.4.

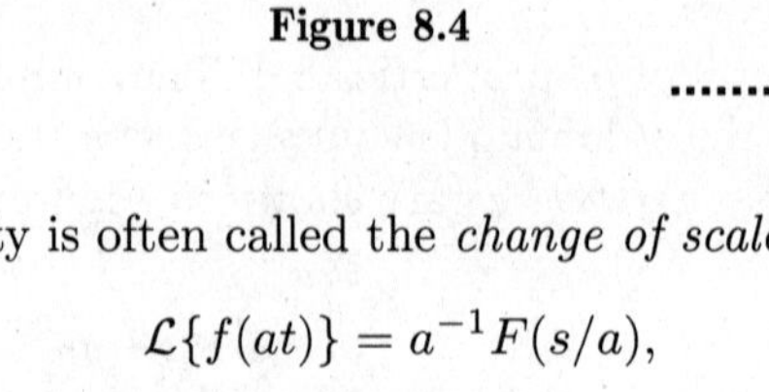

Figure 8.4

....

The following property is often called the *change of scale* property:

$$\mathcal{L}\{f(at)\} = a^{-1}F(s/a), \qquad \textbf{(8.5a)}$$

$$\mathcal{L}^{-1}\{F(as)\} = a^{-1}f(t/a) \qquad \textbf{(8.5b)}$$

(see Exercise 3). For instance, because $\mathcal{L}^{-1}\{1/(s^2+2)\} = (1/\sqrt{2})\sin\sqrt{2}\,t$, 8.5(b) implies that

$$\mathcal{L}^{-1}\left\{\frac{1}{4s^2+2}\right\} = \mathcal{L}^{-1}\left\{\frac{1}{(2s)^2+2}\right\} = \frac{1}{2}\left(\frac{1}{\sqrt{2}}\right)\sin\sqrt{2}\,(t/2) = \frac{1}{2\sqrt{2}}\sin\left(\frac{t}{\sqrt{2}}\right).$$

When a function f is periodic with period P, the improper integral in 8.1 may be replaced by an integral over $0 \leq t \leq P$,

$$F(s) = \frac{1}{1 - e^{-sP}} \int_0^P e^{-st} f(t)\, dt \tag{8.6}$$

(see Exercise 4). For example, the function in Figure 8.5 has period 2, and therefore

$$F(s) = \frac{1}{1 - e^{-2s}} \int_0^2 (1 - t)e^{-st}\, dt.$$

Integration now gives

$$F(s) = \frac{1}{1 - e^{-2s}} \left\{ \frac{(t-1)}{s} e^{-st} + \frac{1}{s^2} e^{-st} \right\}_0^2 = \frac{1 + e^{-2s}}{s(1 - e^{-2s})} - \frac{1}{s^2}.$$

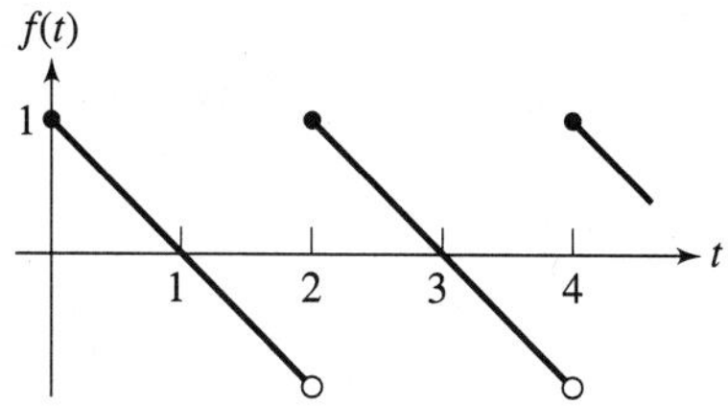

Figure 8.5

It is often necessary in applications to find the inverse transform of the product of two functions FG when inverse transforms f and g of F and G are known. We shall see shortly that the inverse of FG is what is called the convolution of f and g. The **convolution** of two functions f and g is defined as

$$f * g = \int_0^t f(u)g(t - u)\, du. \tag{8.7}$$

The following properties of convolutions are easily verified using the definition in 8.7:

$$f * g = g * f, \tag{8.8a}$$

$$f * (kg) = (kf) * g = k(f * g), \qquad k \text{ a constant} \tag{8.8b}$$

$$(f * g) * h = f * (g * h), \tag{8.8c}$$

$$f * (g + h) = f * g + f * h. \tag{8.8d}$$

The importance of convolutions lies in the following theorem.

■ **THEOREM 8.1** If f and g are $O(e^{\alpha t})$ and piecewise continuous on every finite interval $0 \leq t \leq T$, then

$$\mathcal{L}\{f * g\} = \mathcal{L}\{f\}\mathcal{L}\{g\}, \qquad s > \alpha. \tag{8.9a}$$

Proof If $F = \mathcal{L}\{f\}$ and $G = \mathcal{L}\{g\}$, then

$$F(s)G(s) = \int_0^\infty e^{-su} f(u)\,du \int_0^\infty e^{-s\tau} g(\tau)\,d\tau$$
$$= \int_0^\infty \int_0^\infty e^{-s(u+\tau)} f(u)g(\tau)\,d\tau\,du.$$

Suppose we change variables of integration in the inner integral with respect to τ by setting $t = u + \tau$. Then

$$F(s)G(s) = \int_0^\infty \int_u^\infty e^{-st} f(u)g(t-u)\,dt\,du$$
$$= \lim_{T\to\infty} \int_0^T \int_u^\infty e^{-st} f(u)g(t-u)\,dt\,du.$$

We would like to interchange orders of integration, but to do so requires that the inner integral converge uniformly with respect to u. To verify that this is indeed the case we note that because f and g are $O(e^{\alpha t})$ and piecewise continuous on every finite interval $0 \le t \le T$, there exists a constant M such that for all $t \ge 0$, $|f(t)| < Me^{\alpha t}$ and $|g(t)| < Me^{\alpha t}$. For each $u \ge 0$, we therefore have $|e^{-st} f(u)g(t-u)| < M^2 e^{-st} e^{\alpha u} e^{\alpha(t-u)} = M^2 e^{-t(s-\alpha)}$. Thus,

$$\left|\int_u^\infty e^{-st} f(u)g(t-u)\,dt\right| < M^2 \int_u^\infty e^{-t(s-\alpha)}\,dt = M^2 \left\{\frac{e^{-t(s-\alpha)}}{\alpha - s}\right\}_u^\infty$$
$$= \frac{M^2 e^{-u(s-\alpha)}}{s-\alpha} < \frac{M^2}{s-\alpha},$$

provided $s > \alpha$, and the improper integral is uniformly convergent with respect to u. The order of integration in the expression for $\tilde{f}(s)\tilde{g}(s)$ may therefore be interchanged (Figure 8.6), and we obtain

$$F(s)G(s) = \lim_{T\to\infty} \left[\int_0^T e^{-st} \int_0^t f(u)g(t-u)\,du\,dt \right.$$
$$\left. + \int_T^\infty e^{-st} \int_0^T f(u)g(t-u)\,du\,dt\right].$$

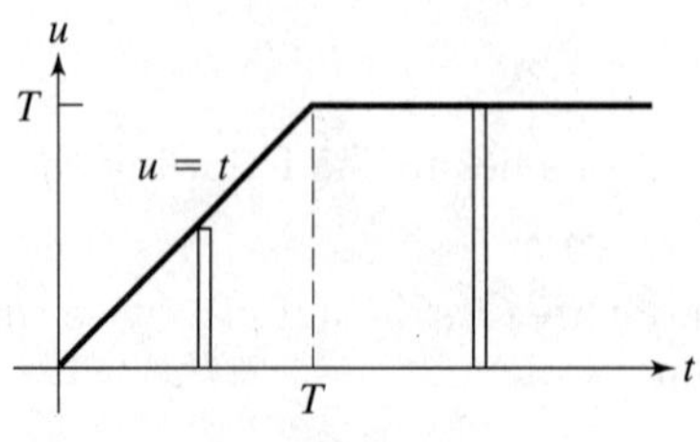

Figure 8.6

When a function f is periodic with period P, the improper integral in 8.1 may be replaced by an integral over $0 \le t \le P$,

$$F(s) = \frac{1}{1-e^{-sP}} \int_0^P e^{-st} f(t)\, dt \tag{8.6}$$

(see Exercise 4). For example, the function in Figure 8.5 has period 2, and therefore

$$F(s) = \frac{1}{1-e^{-2s}} \int_0^2 (1-t)e^{-st}\, dt.$$

Integration now gives

$$F(s) = \frac{1}{1-e^{-2s}} \left\{ \frac{(t-1)}{s} e^{-st} + \frac{1}{s^2} e^{-st} \right\}_0^2 = \frac{1+e^{-2s}}{s(1-e^{-2s})} - \frac{1}{s^2}.$$

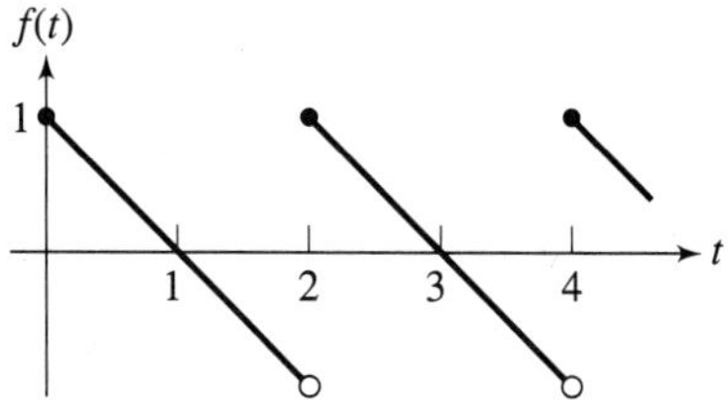

Figure 8.5

It is often necessary in applications to find the inverse transform of the product of two functions FG when inverse transforms f and g of F and G are known. We shall see shortly that the inverse of FG is what is called the convolution of f and g. The **convolution** of two functions f and g is defined as

$$f * g = \int_0^t f(u)g(t-u)\, du. \tag{8.7}$$

The following properties of convolutions are easily verified using the definition in 8.7:

$$f * g = g * f, \tag{8.8a}$$

$$f * (kg) = (kf) * g = k(f * g), \qquad k \text{ a constant} \tag{8.8b}$$

$$(f * g) * h = f * (g * h), \tag{8.8c}$$

$$f * (g + h) = f * g + f * h. \tag{8.8d}$$

The importance of convolutions lies in the following theorem.

■ **THEOREM 8.1** If f and g are $O(e^{\alpha t})$ and piecewise continuous on every finite interval $0 \le t \le T$, then

$$\mathcal{L}\{f * g\} = \mathcal{L}\{f\}\mathcal{L}\{g\}, \qquad s > \alpha. \tag{8.9a}$$

Proof If $F = \mathcal{L}\{f\}$ and $G = \mathcal{L}\{g\}$, then

$$\begin{aligned} F(s)G(s) &= \int_0^\infty e^{-su} f(u)\,du \int_0^\infty e^{-s\tau} g(\tau)\,d\tau \\ &= \int_0^\infty \int_0^\infty e^{-s(u+\tau)} f(u)g(\tau)\,d\tau\,du. \end{aligned}$$

Suppose we change variables of integration in the inner integral with respect to τ by setting $t = u + \tau$. Then

$$\begin{aligned} F(s)G(s) &= \int_0^\infty \int_u^\infty e^{-st} f(u)g(t-u)\,dt\,du \\ &= \lim_{T\to\infty} \int_0^T \int_u^\infty e^{-st} f(u)g(t-u)\,dt\,du. \end{aligned}$$

We would like to interchange orders of integration, but to do so requires that the inner integral converge uniformly with respect to u. To verify that this is indeed the case we note that because f and g are $O(e^{\alpha t})$ and piecewise continuous on every finite interval $0 \le t \le T$, there exists a constant M such that for all $t \ge 0$, $|f(t)| < Me^{\alpha t}$ and $|g(t)| < Me^{\alpha t}$. For each $u \ge 0$, we therefore have $|e^{-st} f(u)g(t-u)| < M^2 e^{-st} e^{\alpha u} e^{\alpha(t-u)} = M^2 e^{-t(s-\alpha)}$. Thus,

$$\begin{aligned} \left| \int_u^\infty e^{-st} f(u)g(t-u)\,dt \right| < M^2 \int_u^\infty e^{-t(s-\alpha)}\,dt &= M^2 \left\{ \frac{e^{-t(s-\alpha)}}{\alpha - s} \right\}_u^\infty \\ &= \frac{M^2 e^{-u(s-\alpha)}}{s-\alpha} < \frac{M^2}{s-\alpha}, \end{aligned}$$

provided $s > \alpha$, and the improper integral is uniformly convergent with respect to u. The order of integration in the expression for $\tilde{f}(s)\tilde{g}(s)$ may therefore be interchanged (Figure 8.6), and we obtain

$$\begin{aligned} F(s)G(s) = \lim_{T\to\infty} \Bigg[&\int_0^T e^{-st} \int_0^t f(u)g(t-u)\,du\,dt \\ &+ \int_T^\infty e^{-st} \int_0^T f(u)g(t-u)\,du\,dt \Bigg]. \end{aligned}$$

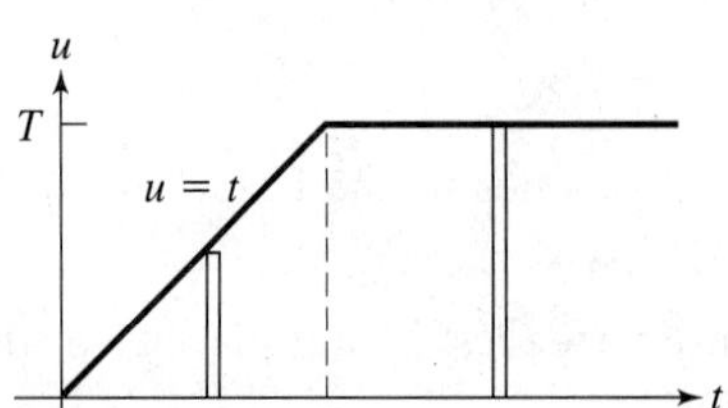

Figure 8.6

Because

$$\left|\int_T^\infty e^{-st}\int_0^T f(u)g(t-u)\,du\,dt\right| < \int_T^\infty\int_0^T M^2e^{-t(s-\alpha)}\,du\,dt$$

$$= M^2T\left\{\frac{e^{-t(s-\alpha)}}{\alpha-s}\right\}_T^\infty = \frac{M^2Te^{-T(s-\alpha)}}{s-\alpha},$$

provided $s > \alpha$, it follows that

$$\lim_{T\to\infty}\int_T^\infty e^{-st}\int_0^T f(u)g(t-u)\,du\,dt = 0.$$

Thus,

$$F(s)G(s) = \lim_{T\to\infty}\int_0^T e^{-st}\int_0^t f(u)g(t-u)\,du\,dt$$

$$= \lim_{T\to\infty}\int_0^T e^{-st}f*g\,dt = \mathcal{L}\{f*g\}.$$

More important in practice is the inverse of 8.9(a). ■

■ **COROLLARY** If $\mathcal{L}^{-1}\{F\} = f$ and $\mathcal{L}^{-1}\{G\} = g$, where f and g are $O(e^{\alpha t})$ and piecewise continuous on every finite interval, then

$$\mathcal{L}^{-1}\{FG\}(t) = \int_0^t f(u)g(t-u)\,du. \quad \textbf{(8.9b)}$$ ■

The following example illustrates this corollary.

EXAMPLE 8.2 ....

Find the inverse transform of $F(s) = \dfrac{2}{s^2(s^2+4)}$.

Solution Since $\mathcal{L}^{-1}\{2/(s^2+4)\} = \sin 2t$ and $\mathcal{L}^{-1}\{1/s^2\} = t$, convolution property 8.9(b) gives

$$\mathcal{L}^{-1}\left\{\frac{2}{s^2(s^2+4)}\right\} = \int_0^t u\sin 2(t-u)\,du = \left\{\frac{u}{2}\cos 2(t-u) + \frac{1}{4}\sin 2(t-u)\right\}_0^t$$

$$= \frac{t}{2} - \frac{1}{4}\sin 2t.$$

....

There are other algebraic properties of the Laplace transform and its inverse, but the ones discussed here suffice for our purposes. Finding inverse transforms is often a matter of finding the partial fraction decomposition of a rational function, together with the preceding properties and a set of tables, as illustrated in the following example.

EXAMPLE 8.3 ....

Find inverse Laplace transforms for the following functions.

(a) $F(s) = \dfrac{s^2 - 9s + 9}{s^3(s^2+9)}$ **(b)** $F(s) = \dfrac{e^{-s}}{s^2 - s}$ **(c)** $F(s) = \dfrac{1}{s^2(s^2-4)}$

Solution **(a)** The partial fraction decomposition of $F(s)$ is

$$F(s) = \frac{1}{s^3} - \frac{1}{s^2} + \frac{s}{s^2+9}.$$

Table 8.1 therefore gives $f(t) = \dfrac{t^2}{2} - t + \dfrac{1}{3}\sin 3t$.

(b) Partial fractions give $\dfrac{1}{s(s-1)} = \dfrac{1}{s-1} - \dfrac{1}{s}$, and therefore

$$\mathcal{L}^{-1}\left\{\frac{1}{s(s-1)}\right\} = e^t - 1.$$

Property 8.4(b) now gives $\mathcal{L}^{-1}\{F(s)\} = (e^{t-1} - 1)H(t-1)$.

(c) Partial fractions give $\dfrac{1}{s^2(s^2-4)} = \dfrac{1/4}{s^2-4} - \dfrac{1/4}{s^2}$, and therefore

$$\mathcal{L}^{-1}\left\{\frac{s+1}{s^2(s^2-4)}\right\} = \frac{1}{4}\left(\frac{1}{2}\sinh 2t\right) - \frac{1}{4}(t) = \frac{1}{8}\sinh 2t - \frac{t}{4}.$$

....

Exercises 8.1

1. **(a)** Verify shifting property 8.3.
   (b) Use 8.3(a) and Table 8.1 to calculate Laplace transforms for the following.
   (i) $f(t) = t^3 e^{-5t}$
   (ii) $f(t) = e^{-t}\cos 2t + e^{3t}\sin 2t$
   (iii) $f(t) = e^{at}\cosh 4t - e^{-at}\sinh 4t$
   (c) Use 8.3(b) and Table 8.1 to calculate inverse Laplace transforms for the following.
   (i) $F(s) = \dfrac{1}{s^2 - 2s + 5}$ **(ii)** $F(s) = \dfrac{1}{\sqrt{s+3}}$ **(iii)** $F(s) = \dfrac{s}{s^2 + 4s + 1}$
2. **(a)** Verify shifting property 8.4.
   (b) Use 8.4(a) and Table 8.1 to calculate Laplace transforms for the following.
   (i) $f(t) = \begin{cases} 0, & 0 < t < 3 \\ t-2, & t > 3 \end{cases}$

(ii) $f(t) = \begin{cases} 0, & 0 < t < a \\ 1, & t > a \end{cases}$

(iii) $f(t) = \begin{cases} 1, & 0 < t < a \\ 0, & t > a \end{cases}$

(iv) $f(t) = \begin{cases} 0, & 0 < t < a \\ 1, & a < t < b \\ 0, & t > b \end{cases}$

(c) Use 8.4(b) and Table 8.1 to calculate inverse Laplace transforms for the following.

(i) $F(s) = \dfrac{e^{-2s}}{s^2}$ (ii) $F(s) = \dfrac{e^{-3s}}{s^2+1}$ (iii) $F(s) = \dfrac{se^{-5s}}{s^2-2}$

3. (a) Verify change of scale property 8.5.

(b) Use 8.5(a) and Table 8.1 to calculate Laplace transforms for the following.

(i) $f(t) = 4t^2 + \sinh 2t$ (ii) $f(t) = e^{4t}\cos 4t$

(c) Use 8.5(b) and Table 8.1 to calculate inverse Laplace transforms for the following.

(i) $F(s) = \dfrac{s}{9s^2+2}$ (ii) $F(s) = \dfrac{1}{4s^2-6s-5}$

4. (a) Verify equation 8.6.

(b) Find Laplace transforms for the following functions.

(i) $f(t) = t, \quad 0 < t < a, \quad f(t+a) = f(t)$

(ii) $f(t) = \begin{cases} 1, & 0 < t < a \\ -1, & a < t < 2a, \end{cases} \qquad f(t+2a) = f(t)$

(iii) $f(t) = |\sin at|$

In Exercises 5–8 use convolutions to find inverse transforms for the function.

5. $F(s) = \dfrac{1}{s(s+1)}$

6. $F(s) = \dfrac{1}{(s^2+1)(s^2+4)}$

7. $F(s) = \dfrac{s}{(s+4)(s^2-2)}$

8. $F(s) = \dfrac{s}{(s^2-4)(s^2-9)}$

In Exercises 9–14 find the Laplace transform of the function.

9. $f(t) = \begin{cases} 2t, & 0 \le t \le 1 \\ t, & t > 1 \end{cases}$

10. $f(t) = \begin{cases} t^2, & 0 \le t \le 1 \\ 2t, & t > 1 \end{cases}$

11. $f(t) = \begin{cases} t, & 0 < t < a \\ 2a - t, & a < t < 2a, \end{cases} \qquad f(t+2a) = f(t)$

12. $f(t) = \begin{cases} 1, & 0 < t < a \\ 0, & a < t < 2a, \end{cases} \qquad f(t+2a) = f(t)$

13. $f(t) = \begin{cases} 0, & 0 < t < a \\ 1, & t > a \end{cases}$

14. $f(t) = \begin{cases} 0, & 0 < t < a \\ 1, & a < t < a+1 \\ 0, & t > a+1 \end{cases}$

Find the inverse Laplace transform in Exercises 15–24.

15. $F(s) = \dfrac{s}{s^2 - 3s + 2}$

16. $F(s) = \dfrac{4s+1}{(s^2+s)(4s^2-1)}$

17. $F(s) = \dfrac{e^{-3s}}{s+5}$

18. $F(s) = \dfrac{e^{-2s}}{s^2+3s+2}$

19. $F(s) = \dfrac{1}{s^3+1}$

20. $F(s) = \dfrac{5s-2}{3s^2+4s+8}$

21. $F(s) = \dfrac{e^{-s}(1-e^{-s})}{s(s^2+1)}$

22. $F(s) = \dfrac{s}{(s+1)^5}$

23. $F(s) = \dfrac{s^2+2s+3}{(s^2+2s+2)(s^2+2s+5)}$

24. $F(s) = \dfrac{s^2}{(s^2-4)^2}$

25. Verify that the Laplace transform of a function f that is piecewise continuous on every finite interval $0 \le t \le T$ and is $O(e^{\alpha t})$ exists for $s > \alpha$.

26. **(a)** Prove that when n is a nonnegative integer, t^n is $O(e^{\epsilon t})$ for every $\epsilon > 0$.
(b) Prove that when f is $O(e^{\alpha t})$, then $t^n f(t)$ is $O(e^{(\alpha+\epsilon)t})$ for every $\epsilon > 0$.

27. Let f and f' be $O(e^{\alpha t})$, f' be piecewise continuous on every finite interval $0 \le t \le T$, and f have only a finite number of finite discontinuities for $t \ge 0$. Verify the *initial value theorem*,

$$\lim_{s\to\infty} s\tilde{f}(s) = \lim_{t\to 0+} f(t).$$

Assume the result that $\lim_{s\to\infty} \tilde{f}(s) = 0$ for functions that are piecewise continuous and of exponential order.

8.2 Application of Laplace Transforms to Ordinary Differential Equations

Laplace transforms are a powerful technique for solving ordinary and partial differential equations. They replace differentiations with algebraic operations. The following theorem and its corollary simplify this process immensely.

■ **THEOREM 8.2** Suppose f is continuous with a piecewise-continuous first derivative on every finite interval $0 \le t \le T$. If f is $O(e^{\alpha t})$, then $\mathcal{L}\{f'\}$ exists for $s > \alpha$, and

$$\mathcal{L}\{f'\} = sF(s) - f(0). \tag{8.10a}$$

(A more exact representation of the left side of this equation is $\mathcal{L}\{f'\}(s)$.)

Proof If t_j, $j = 1, \ldots, n$ denote the discontinuities of f' in $0 \le t \le T$, then

$$\int_0^T e^{-st} f'(t)\,dt = \sum_{j=0}^{n} \int_{t_j}^{t_{j+1}} e^{-st} f'(t)\,dt,$$

where $t_0 = 0$ and $t_{n+1} = T$. Because f' is continuous on each subinterval, we may integrate by parts on these subintervals:

$$\int_0^T e^{-st} f'(t)\,dt = \sum_{j=0}^{n} \left[\{e^{-st} f(t)\}_{t_j}^{t_{j+1}} + s \int_{t_j}^{t_{j+1}} e^{-st} f(t)\,dt \right].$$

Because f is continuous, $f(t_j+) = f(t_j-)$, $j = 1, \ldots, n$, and therefore

$$\int_0^T e^{-st} f'(t)\,dt = -f(0) + e^{-sT} f(T) + s \int_0^T e^{-st} f(t)\,dt.$$

Thus,

$$\begin{aligned}
\mathcal{L}\{f'\} &= \int_0^\infty e^{-st} f'(t)\,dt = \lim_{T\to\infty} \int_0^T e^{-st} f'(t)\,dt \\
&= \lim_{T\to\infty} \left[-f(0) + e^{-sT} f(T) + s \int_0^T e^{-st} f(t)\,dt \right] \\
&= s\tilde{f}(s) - f(0) + \lim_{T\to\infty} e^{-sT} f(T),
\end{aligned}$$

provided the limit on the right exists. Because f is $O(e^{\alpha t})$, there exists M and $\overline{T}$ such that for $t > \overline{T}$, $|f(t)| < Me^{\alpha t}$. Thus, for $T > \overline{T}$,

$$e^{-sT}|f(T)| < e^{-sT} M e^{\alpha T} = M e^{(\alpha - s)T}$$

which approaches 0 as $T \to \infty$ (provided $s > \alpha$). Consequently,

$$\mathcal{L}\{f'\} = sF(s) - f(0). \qquad \blacksquare$$

This result is easily extended to second-order derivatives. The extension is stated in the following corollary and verified in Exercise 1. For extensions when f is only piecewise continuous, see Exercise 39.

■ **COROLLARY** Suppose f and f' are continuous and f'' is piecewise continuous on every finite interval $0 \le t \le T$. If f and f' are $O(e^{\alpha t})$, then $\mathcal{L}\{f''\}$ exists for $s > \alpha$, and

$$\mathcal{L}\{f''\} = s^2 F(s) - sf(0) - f'(0). \qquad \textbf{(8.10b)} \ \blacksquare$$

The following examples use these properties and at the same time indicate how Laplace transforms reduce ordinary differential equations to algebraic equations.

EXAMPLE 8.4

Solve the differential equation

$$y'' - 2y' + y = 2e^t, \qquad y(0) = y'(0) = 0.$$

Solution When we take Laplace transforms of both sides of the differential equation and use linearity of the operator,

$$\mathcal{L}\{y''\} - 2\mathcal{L}\{y'\} + \mathcal{L}\{y\} = 2\mathcal{L}\{e^t\}.$$

Properties 8.10(a) and (b) yield

$$[s^2Y(s) - sy(0) - y'(0)] - 2[sY(s) - y(0)] + Y(s) = \frac{2}{s-1}.$$

We now use the initial conditions $y(0) = y'(0) = 0$,

$$s^2Y(s) - 2sY(s) + Y(s) = \frac{2}{s-1},$$

and solve this equation for $Y(s)$:

$$Y(s) = \frac{2}{(s-1)^3}.$$

The required function $y(t)$ can now be obtained by taking the inverse transform of $Y(s)$:

$$\begin{aligned} y(t) &= \mathcal{L}^{-1}\left\{\frac{2}{(s-1)^3}\right\} = 2\mathcal{L}^{-1}\left\{\frac{1}{(s-1)^3}\right\} && \text{(by linearity)} \\ &= 2e^t\mathcal{L}^{-1}\left\{\frac{1}{s^3}\right\} && \text{(by 8.3(b))} \\ &= 2e^t\left(\frac{t^2}{2}\right) && \text{(from Table 8.1)} \\ &= t^2e^t. \end{aligned}$$

EXAMPLE 8.5

Solve the differential equation

$$y'' + 4y = 3\cos 2t, \qquad y(0) = 1, \qquad y'(0) = 0.$$

Solution When we take Laplace transforms of the differential equation and use the initial conditions,

$$[s^2Y - s(1) - 0] + 4Y = \frac{3s}{s^2+4}.$$

The solution of this equation for $Y(s)$ is

$$Y(s) = \frac{3s}{(s^2+4)^2} + \frac{s}{s^2+4},$$

and Table 8.1 gives

$$y(t) = 3\left(\frac{t}{4}\sin 2t\right) + \cos 2t.$$

Convolutions are particularly useful when solving ordinary differential equations that contain unspecified forcing functions and/or unspecified initial conditions.

EXAMPLE 8.6

Find the solution of the problem

$$y'' + 2y' - y = f(t), \qquad y(0) = A, \qquad y'(0) = B$$

for arbitrary constants A and B and an arbitrary function $f(t)$.

Solution When we take Laplace transforms,

$$[s^2Y - As - B] + 2[sY - A] - Y = F(s),$$

and solve for Y, the result is

$$Y(s) = \frac{F(s)}{s^2+2s-1} + \frac{As+B+2A}{s^2+2s-1}.$$

To find the inverse transform of this function, we first note that

$$\mathcal{L}^{-1}\left\{\frac{1}{s^2+2s-1}\right\} = \mathcal{L}^{-1}\left\{\frac{1}{(s+1)^2-2}\right\} = e^{-t}\mathcal{L}^{-1}\left\{\frac{1}{s^2-2}\right\}$$
$$= \frac{1}{\sqrt{2}}e^{-t}\sinh\sqrt{2}\,t.$$

Convolution property 8.9(b) on the first term of $Y(s)$ now yields

$$y(t) = \int_0^t f(u)\frac{1}{\sqrt{2}}e^{-(t-u)}\sinh\sqrt{2}(t-u)\,du + \mathcal{L}^{-1}\left\{\frac{A(s+1)+(B+A)}{(s+1)^2-2}\right\}$$
$$= \frac{1}{\sqrt{2}}\int_0^t f(u)e^{-(t-u)}\sinh\sqrt{2}(t-u)\,du + e^{-t}\mathcal{L}^{-1}\left\{\frac{As+(B+A)}{s^2-2}\right\}$$
$$= \frac{1}{\sqrt{2}}\int_0^t f(u)e^{-(t-u)}\sinh\sqrt{2}(t-u)\,du + e^{-t}\left(A\cosh\sqrt{2}\,t + \frac{B+A}{\sqrt{2}}\sinh\sqrt{2}\,t\right).$$

Exercises 8.2

1. Verify property 8.10(b).

In Exercises 2–20 use Laplace transforms to solve the differential equation.

2. $y'' + 2y' - y = e^t, \quad y(0) = 1, \quad y'(0) = 2$
3. $y'' + y = 2e^{-t}, \quad y(0) = y'(0) = 0$
4. $y'' + 2y' + y = t, \quad y(0) = 0, \quad y'(0) = 1$
5. $y'' - 2y' + y = t^2e^t, \quad y(0) = 1, \quad y'(0) = 0$
6. $y'' + y = t, \quad y(0) = 1, \quad y'(0) = -2$
7. $y'' + 2y' + 5y = e^{-t}\sin t, \quad y(0) = 0, \quad y'(0) = 1$
8. $y'' + 6y' + y = \sin 3t, \quad y(0) = 2, \quad y'(0) = 1$
9. $y'' + y' - 6y = t + \cos t, \quad y(0) = 1, \quad y'(0) = -2$
10. $y'' - 4y' + 5y = te^{-3t}, \quad y(0) = -1, \quad y'(0) = 2$
11. $y'' + 4y = f(t), \quad y(0) = 0, \quad y'(0) = 1$, where $f(t) = \begin{cases} 1, & 0 < t < 1 \\ 0, & t > 1 \end{cases}$
12. $y'' + 2y' - 4y = \cos^2 t, \quad y(0) = 0, \quad y'(0) = 0$
13. $y'' - 3y' + 2y = 8t^2 + 12e^{-t}, \quad y(0) = 0, \quad y'(0) = 2$
14. $y'' + 4y' - 2y = \sin 4t, \quad y(0) = 0, \quad y'(0) = 0$
15. $y'' + 8y' + 41y = e^{-2t}\sin t, \quad y(0) = 0, \quad y'(0) = 1$
16. $y'' + 2y' + y = f(t), \quad y(0) = 0, \quad y'(0) = 0$, where $f(t) = \begin{cases} t, & 0 < t < 1 \\ 0, & t > 1 \end{cases}$
17. $y'' + 9y = \cos 2t, \quad y(0) = 1, \quad y(\pi/2) = -1$
18. $y'' + 3y' - 4y = 2e^{-4t}, \quad y(0) = 1, \quad y(1) = 1$
19. $y'' + 2y' + 5y = e^{-t}\sin t, \quad y(0) = 0, \quad y(\pi/4) = 1$
20. $y'' - 4y' + 3y = f(t), \quad y(0) = 1, \quad y'(0) = 0$

In Exercises 21–26 use Laplace transforms to find a general solution of the differential equation.

21. $y'' - 2y' + 4y = t^2$
22. $y'' - 2y' + y = t^2e^t$
23. $y'' + y = f(t)$
24. $y'' + 2y' + 5y = e^{-t}\sin t$
25. $y'' + 4y' + y = t + 2$
26. $y'' - 4y = f(t)$
27. To find a general solution of $y'' + 9y = t\sin t$, replace $t\sin t$ by te^{ti}, solve the equation, and then take imaginary parts.
28. To find a general solution of $y'' - 2y' + 3y = t\cos 2t$, replace $t\cos 2t$ by te^{2ti}, solve the equation, and then take real parts.
29. Verify that

$$\mathcal{L}\{f'''\} = s^3F(s) - s^2f(0) - sf'(0) - f''(0)$$

when f, f', and f'' are continuous and f''' is piecewise continuous on every finite interval $0 \le t \le T$, and f, f', and f'' are $O(e^{\alpha t})$.

Use the result of Exercise 29 to solve each differential equation in Exercises 30 and 31.

30. $y''' - 3y'' + 3y' - y = t^2e^t, \quad y(0) = 1, \quad y'(0) = 0, \quad y''(0) = -2$

31. $y''' - 3y'' + 3y' - y = t^2e^t$

One end of a spring with constant k newtons per meter is attached to a mass of M kilograms and the other end is attached to a wall (see the following figure).

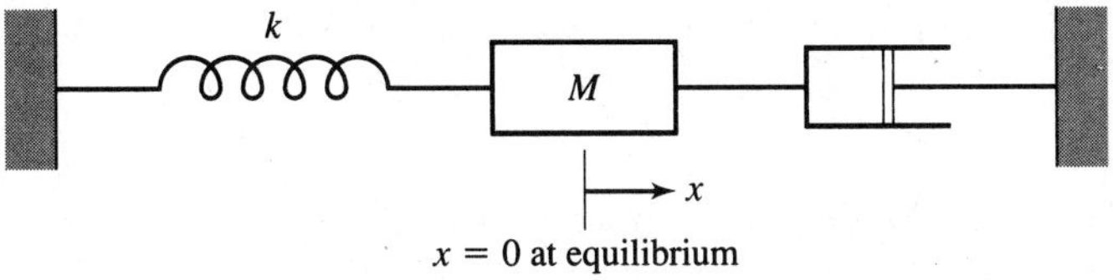

Attached to the mass is a dashpot — a device that provides (or represents) a resistive force on the mass directly proportional to the velocity of the mass. If all other forces are grouped into a function denoted by $f(t)$, the differential equation governing motion of the mass is

$$M\frac{d^2x}{dt^2} + \beta\frac{dx}{dt} + kx = f(t),$$

where $\beta > 0$ is a constant. The position of M when the spring is unstretched corresponds to $x = 0$. Accompanying the differential equation will be two initial conditions, $x(0) = A$ and $x'(0) = B$, representing the initial position and velocity of M. In Exercises 32–38, solve the differential equation with the given information.

32. $M = \frac{1}{5}, \quad \beta = 0, \quad k = 10, \quad f(t) = 0, \quad x(0) = -0.03, \quad x'(0) = 0$

33. $M = \frac{1}{5}, \quad \beta = \frac{3}{2}, \quad k = 10, \quad f(t) = 0, \quad x(0) = -0.03, \quad x'(0) = 0$

34. $M = \frac{1}{5}, \quad \beta = \frac{3}{2}, \quad k = 10, \quad f(t) = 4\sin 10t, \quad x(0) = 0, \quad x'(0) = 0$

35. $M = 2, \quad \beta = 0, \quad k = 16, \quad f(t) = 0, \quad x(0) = 0.1, \quad x'(0) = 0$

36. $M = \frac{1}{10}, \quad \beta = \frac{1}{20}, \quad k = 5, \quad f(t) = 0, \quad x(0) = -\frac{1}{20}, \quad x'(0) = 2$

37. $M = \frac{1}{10}, \quad \beta = 0, \quad k = 4000, \quad f(t) = 3\cos 200t, \quad x(0) = 0, \quad x'(0) = 10$

38. $M = 1, \quad \beta = 0, \quad k = 64, \quad f(t) = 2\sin 8t, \quad x(0) = 0, \quad x'(0) = 0$

39. **(a)** Let f be $O(e^{\alpha t})$ and be continuous for $t \geq 0$ except for a finite discontinuity at $t = t_0 > 0$, and let f' be piecewise continuous on every finite interval $0 \leq t \leq T$. Show that

$$\mathcal{L}\{f'\} = sF(s) - f(0) - e^{-st_0}[f(t_0+) - f(t_0-)].$$

(b) What is the result in (a) if $t_0 = 0$?

8.3 The Complex Inversion Integral

Finding the inverse Laplace transform in Sections 8.1 and 8.2 was a matter of organization and tables; we used properties 8.3(b), 8.4(b), and 8.5(b), and partial fractions to organize a given transform $F(s)$ into a form for which the inverse transform can be found in tables. In many applications, transform functions are so complicated, they cannot be found in tables; this is especially true in partial differential equations. What we need is a direct method for inverting the Laplace transform. In this section we use the theory of functions of a complex variable to derive such a formula.

We first note that the results in equations 8.3–8.10 remain valid when s is complex; the complex derivation may be somewhat different than its real counterpart, but each result is valid when s is complex.

The following theorem shows that Laplace transforms are analytic functions of the complex variable s.

■ **THEOREM 8.3** If f is $O(e^{\alpha t})$ and piecewise continuous on every finite interval $0 \le t \le T$, then the Laplace transform $F(s) = F(x + yi)$ of f is an analytic function of s in the half-plane $x > \alpha$.

Proof If the real and imaginary parts of $F(s)$ are denoted by $u(x, y)$ and $v(x, y)$, where

$$F(s) = u + vi = \int_0^\infty e^{-(x+yi)t} f(t)\,dt,$$

then

$$u(x,y) = \int_0^\infty e^{-xt}\cos yt\, f(t)\,dt, \qquad v(x,y) = \int_0^\infty -e^{-xt}\sin yt\, f(t)\,dt.$$

To verify analyticity of F, we show that $u(x, y)$ and $v(x, y)$ have continuous first partial derivatives that satisfy the Cauchy-Riemann equations when $x > \alpha$. Now,

$$\left\{\begin{array}{l} |e^{-xt}\cos yt\, f(t)| \\ |e^{-xt}\sin yt\, f(t)| \end{array}\right\} \le e^{-xt}|f(t)|,$$

and because f is $O(e^{\alpha t})$, there exist constants M and T such that for all $t > T$, $|f(t)| < Me^{\alpha t}$. Consequently, whenever $x > \alpha' > \alpha$ and $t > T$,

$$\left\{\begin{array}{l} |e^{-xt}\cos yt\, f(t)| \\ |e^{-xt}\sin yt\, f(t)| \end{array}\right\} < e^{-xt}Me^{\alpha t} \le Me^{(\alpha-\alpha')t},$$

and

$$\begin{aligned}\left\{\begin{array}{l} |u(x,y)| \\ |v(x,y)| \end{array}\right\} &< \int_0^T e^{-xt}|f(t)|\,dt + \int_T^\infty Me^{(\alpha-\alpha')t}dt \\ &\le \int_0^T e^{-\alpha' t}|f(t)|\,dt + M\left\{\frac{e^{(\alpha-\alpha')t}}{\alpha-\alpha'}\right\}_0^\infty \\ &= \int_0^T e^{-\alpha' t}|f(t)|\,dt + \frac{M}{\alpha'-\alpha}.\end{aligned}$$

Thus, the integrals representing u and v converge absolutely and uniformly with respect to x and y in the half-plane $x > \alpha' > \alpha$. Because f is piecewise continuous, u and v are continuous functions for $x > \alpha'$. Now,

$$\int_0^\infty \frac{\partial}{\partial x}\left[e^{-xt}\cos yt\, f(t)\right] dt = \int_0^\infty -te^{-xt}\cos yt\, f(t)\, dt$$

and

$$\int_0^\infty \frac{\partial}{\partial y}\left[-e^{-xt}\sin yt\, f(t)\right] dt = \int_0^\infty -te^{-xt}\cos yt\, f(t)\, dt.$$

Becuase $tf(t)$ is $O(e^{(\alpha+\epsilon)t})$ for any $\epsilon > 0$ (see Exercise 26 in Section 8.1) and piecewise continuous on every finite interval $0 < t < T$, a similar argument to the preceding one shows that this integral is absolutely and uniformly convergent with respect to x and y for $x > \alpha' > \alpha$. Because $\alpha' > \alpha$ is arbitrary, it follows that this integral converges to a continuous function, which is equal to both $\partial u/\partial x$ and $\partial v/\partial y$ for $x > \alpha$. We have shown then that the first of the Cauchy-Riemann equations $\partial u/\partial x = \partial v/\partial y$ is satisfied for $x > \alpha$. In a similar way, we can show that $\partial u/\partial y = -\partial v/\partial x$, and therefore $F(s)$ is analytic for $x > \alpha$. ∎

To obtain the complex inversion integral for $\mathcal{L}^{-1}\{F(s)\}$, we use the extension of Cauchy's integral formula contained in Theorem 4.18.

When a function f is $O(e^{\alpha t})$, we know that its transform $F(s)$ is analytic for $x > \alpha$ (Theorem 8.3). Suppose further that in some half-plane $x > \delta > 0$, there exist constants M, R, and $k > 0$ such that $|F(s)s^k| < M$ for $|s| > R$ (Figure 8.7). Then according to Theorem 4.18, if s is a point with $\operatorname{Re} s > \gamma > \delta$, $F(s)$ can be expressed in the form

$$F(s) = -\frac{1}{2\pi i}\lim_{\beta\to\infty}\int_{\gamma-\beta i}^{\gamma+\beta i}\frac{F(\zeta)}{\zeta - s}\, d\zeta.$$

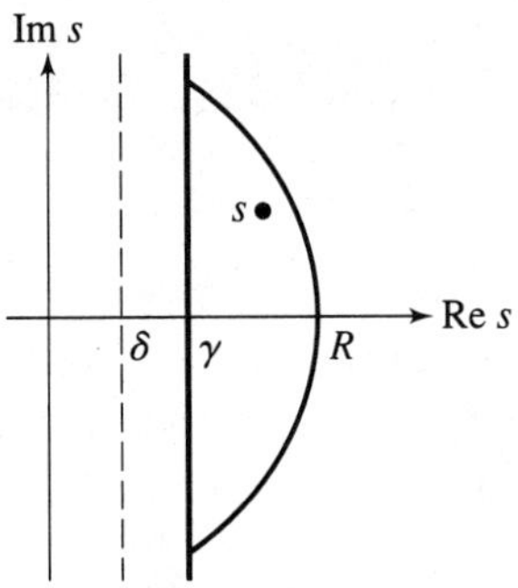

Figure 8.7

If we formally take inverse transforms of both sides of this equation and interchange the order of integration and $\mathcal{L}^{-1}$, we obtain

$$f(t) = -\frac{1}{2\pi i} \lim_{\beta\to\infty} \int_{\gamma-\beta i}^{\gamma+\beta i} -F(\zeta)\mathcal{L}^{-1}\left(\frac{1}{s-\zeta}\right) d\zeta = \frac{1}{2\pi i} \lim_{\beta\to\infty} \int_{\gamma-\beta i}^{\gamma+\beta i} e^{\zeta t} F(\zeta)\, d\zeta.$$

This expression,

$$f(t) = \frac{1}{2\pi i} \lim_{\beta\to\infty} \int_{\gamma-\beta i}^{\gamma+\beta i} e^{st} F(s)\, ds, \qquad \textbf{(8.11)}$$

is called the **complex inversion integral** for the Laplace transform. Although it is an integral in the complex plane along the line $x = \gamma$, it can be written as a complex combination of real improper integrals. But even for very simple functions $F(s)$, the integrations involved in this real form are usually very difficult (see Exercise 17). Fortunately, in Theorem 8.6 we prove that residues of $e^{st}F(s)$ may be used to evaluate the integral. First, however, we give conditions on a function F sufficient to guarantee that a function f exists whose transform is F and that f is given by this inversion integral.

■ **THEOREM 8.4** Let F be any function of the complex variable s that is analytic for all $s = x + yi$ in a half-plane $x > \alpha$ and such that $F(x)$ is real when $x > \alpha$. Suppose further that there exist constants $k > 1$, M, and r such that $|F(s)s^k| < M$ for $|s| > r$ in the half-plane $x > \alpha$. Then the inversion integral of F along any line $x = \gamma$ $(\gamma > \alpha)$ converges to a real-valued function f that is independent of γ and whose Laplace transform is F for $x > \alpha$. Furthermore, f is $O(e^{\alpha t})$, it is continuous, and $f(t) = 0$ for $t < 0$. ■

The conditions on F in this theorem are severe. They are not, for instance, satisfied by $F(s) = 1/s$, because for this function $k = 1$. By qualifying the function f instead of F, it is possible to relax conditions on the inversion integral formula.

■ **THEOREM 8.5** If F is the Laplace transform of any function f of $O(e^{\alpha t})$, which is piecewise smooth on every finite interval $0 \le t \le T$, then the inversion integral of F along any line $x = \gamma > \alpha$ exists and represents f. At any point of discontinuity of f, the inversion integral represents $[f(t+) + f(t-)]/2$, and if $t = 0$, it represents $f(0+)/2$. ■

We omit proofs of these results. One might argue that Theorem 8.5 is of little use in applications, because we do not know f; it is precisely f that is the unknown in the problem. For instance, suppose the Laplace transform is applied to an ordinary differential equation in $y(t)$, and some analysis is performed, leading to an expression for $Y(s)$. It now remains to apply the inverse transform to find $y(t)$. But how then can we use Theorem 8.5? What we do is ignore the conditions of the theorem and simply apply the inversion integral to $Y(s)$ to obtain a function $y(t)$, a function that we hope is both the inverse transform of $Y(s)$ and a solution of our differential equation. To verify that this is indeed the case, we can proceed in two ways. First, we can take the Laplace transform of $y(t)$; and if we obtain $Y(s)$, then there is no question that $y(t)$ is the inverse of $Y(s)$. Alternatively, we can set aside Laplace transforms completely and verify that $y(t)$ is a solution of the original differential equation with which we began.

As we already mentioned, the inversion integral is seldom used to find inverse transforms; it is circumvented with residues of the complex function $e^{st}F(s)$. The main argument of the method is contained in the following theorem, wherein F is assumed to satisfy conditions like those of Theorems 8.4 or 8.5.

THEOREM 8.6 Let F be a function for which the inversion integral along a line $x = \gamma$ represents the inverse function f, and let F be analytic except for isolated singularities s_n $(n = 1, \ldots)$ in the half-plane $x < \gamma$. Then the series of residues of $e^{st}F(s)$ at $s = s_n$ converges to f for each positive t,

$$f(t) = \text{sum of residues of } e^{st}F(s) \text{ at its singularities,}$$

provided a sequence C_n of contours can be found that satisfies the following properties:

1. C_n consists of the straight line $x = \gamma$ from $\gamma - \beta_n i$ to $\gamma + \beta_n i$ and some curve Γ_n beginning at $\gamma + \beta_n i$, ending at $\gamma - \beta_n i$, and lying in $x \le \gamma$ (Figure 8.8).
2. C_n encloses $s_1, s_2, \ldots, s_n$.
3. $\lim_{n\to\infty} \beta_n = \infty$.
4. $\lim_{n\to\infty} \int_{\Gamma_n} e^{st}F(s)\,ds = 0$.

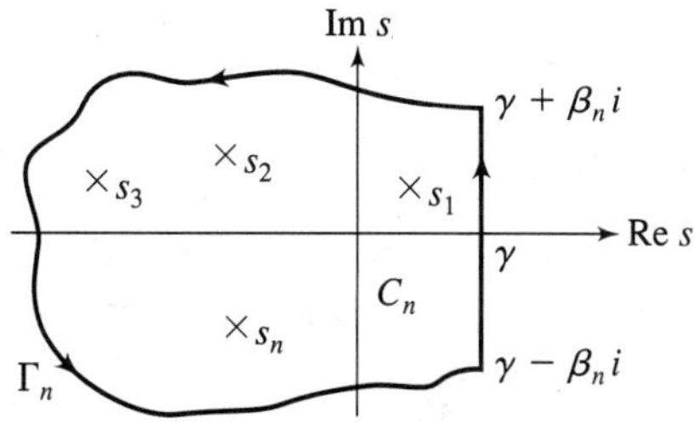

Figure 8.8

Proof Because $e^{st}F(s)$ is analytic in C_n except at $s_1, \ldots, s_n$, the residue theorem states that

$$\begin{pmatrix}\text{sum of residues of} \\ e^{st}F(s) \text{ at } s_1, \ldots, s_n\end{pmatrix} = \frac{1}{2\pi i}\oint_{C_n} e^{st}F(s)\,ds$$

$$= \frac{1}{2\pi i}\int_{\gamma-\beta_n i}^{\gamma+\beta_n i} e^{st}F(s)\,ds + \int_{\Gamma_n} e^{st}F(s)\,ds.$$

When we take limits on n (and use conditions 3 and 4 in the statement of the theorem),

$$\begin{pmatrix} \text{sum of residues of} \\ e^{st}F(s) \text{ at } s_1, \ldots, s_n, \ldots \end{pmatrix} = \frac{1}{2\pi i} \lim_{n\to\infty} \int_{\gamma-\beta_n i}^{\gamma+\beta_n i} e^{st}F(s)\,ds = f(t).$$

It is not essential, as condition 2 requires, that C_n contain precisely n of the singularities of $F(s)$. In fact, this could be very difficult to accomplish, depending on how the singularities are enumerated. What is essential is that as n increases, the C_n expand to enclose eventually all singularities of $F(s)$. ■

As a result of Theorem 8.6, finding the inverse transform of a function $F(s)$ is now a matter of calculating residues of the function $e^{st}F(s)$ at its singularities.

EXAMPLE 8.7 ....

Use Theorem 8.6 to find inverse transforms when $F(s)$ is equal to

(a) $\dfrac{1}{s^m}$, $m \geq 2$ an integer **(b)** $\dfrac{1}{s^2+9}$ **(c)** $\dfrac{s^2}{(s^2+1)^2}$

Solution **(a)** The function $F(s) = 1/s^m$ has a pole of order m at $s = 0$, as does $e^{st}F(s)$. According to equation (6.2), the residue there is

$$\lim_{s\to 0} \frac{1}{(m-1)!} \frac{d^{m-1}}{ds^{m-1}} \left[s^m e^{st}F(s)\right] = \frac{1}{(m-1)!} \lim_{s\to 0} \frac{d^{m-1}}{ds^{m-1}}(e^{st}) = \frac{t^{m-1}}{(m-1)!}.$$

The contours Γ_n in Figure 8.9 clearly satisfy conditions 1 and 3 in Theorem 8.6. Furthermore, on Γ_n,

$$\left|\frac{e^{st}}{s^m}\right| = \left|\frac{e^{(x+yi)t}}{n^m e^{m\theta i}}\right| = \frac{e^{xt}}{n^m} \leq \frac{e^t}{n^m}.$$

According to inequality 4.22, we may write

$$\left|\int_{\Gamma_n} \frac{e^{st}}{s^m}\,ds\right| < \frac{2\pi n e^t}{n^m} = \frac{2\pi e^t}{n^{m-1}}.$$

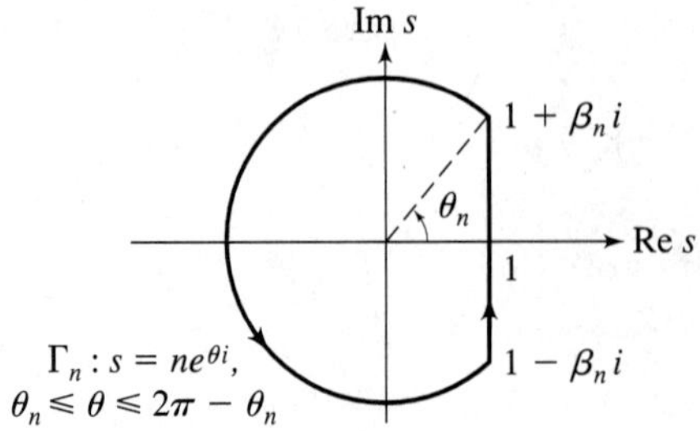

Figure 8.9

This expression approaches zero as $n \to \infty$. Consequently, the Γ_n satisfy condition 4 in Theorem 8.6, and by this theorem,

$$\mathcal{L}^{-1}\left\{\frac{1}{s^m}\right\} = \frac{t^{m-1}}{(m-1)!}.$$

(b) The function $F(s) = 1/(s^2+9)$ has poles of order 1 at $s = \pm 3i$, as does $e^{st}F(s)$. The residue at $3i$ is

$$\text{Res}[e^{st}F(s), 3i] = \lim_{s\to 3i} \frac{(s-3i)e^{st}}{(s+3i)(s-3i)} = \frac{e^{3ti}}{6i} = -\frac{i}{6}e^{3ti}.$$

Similarly, $\text{Res}[e^{st}F(s), -3i] = (i/6)e^{-3ti}$. On the contour Γ_n $(n \geq 4)$ in Figure 8.9,

$$\left|\frac{e^{st}}{s^2+9}\right| \leq \frac{|e^{(x+yi)t}|}{|s|^2-9} = \frac{e^{xt}}{n^2-9} \leq \frac{e^t}{n^2-9}.$$

Thus,

$$\left|\int_{\Gamma_n} \frac{e^{st}}{s^2+9}\,ds\right| < \frac{2\pi n e^t}{n^2-9},$$

and this expression approaches zero as $n \to \infty$. By Theorem 8.6, then,

$$\mathcal{L}^{-1}\left\{\frac{1}{s^2+9}\right\} = -\frac{i}{6}e^{3ti} + \frac{i}{6}e^{-3ti} = \frac{1}{3}\sin 3t.$$

(c) The function $F(s) = s^2/(s^2+1)^2$ has poles of order two at $s = \pm i$, as does $e^{st}F(s)$. The residue at i is

$$\begin{aligned}\text{Res}[e^{st}F(s), i] &= \lim_{s\to i} \frac{d}{ds}\left[\frac{(s-i)^2 e^{st} s^2}{(s+i)^2(s-i)^2}\right]\\ &= \lim_{s\to i}\left[\frac{(s+i)^2(2se^{st}+ts^2e^{st}) - s^2e^{st}(2)(s+i)}{(s+i)^4}\right]\\ &= \frac{1}{4}(t-i)e^{ti}.\end{aligned}$$

Similarly, $\text{Res}[e^{st}F(s), -i] = (1/4)(t+i)e^{-ti}$. On the contour Γ_n $(n \geq 2)$ in Figure 8.9,

$$\left|\frac{s^2e^{st}}{(s^2+1)^2}\right| \leq \frac{|s|^2e^{xt}}{(|s|^2-1)^2} \leq \frac{n^2e^t}{(n^2-1)^2}.$$

Thus,

$$\left|\int_{\Gamma_n} \frac{s^2e^{st}}{(s^2+1)^2}\,ds\right| < \frac{2\pi n^3 e^t}{(n^2-1)^2},$$

and this expression approaches zero as $n \to \infty$. By Theorem 8.6

$$\mathcal{L}^{-1}\left\{\frac{s^2}{(s^2+1)^2}\right\} = \frac{1}{4}(t-i)e^{ti} + \frac{1}{4}(t+i)e^{-ti} = \frac{t}{2}\cos t + \frac{1}{2}\sin t.$$

More complicated illustrations of Theorem 8.6 are contained in Example 8.8. This example is more typical of problems when Laplace transforms are used to solve partial differential equations.

EXAMPLE 8.8

Find inverse transforms for

(a) $F(s) = \dfrac{1}{s^3}(1 - \cosh sx) + \dfrac{\sinh s \sinh sx}{s^3 \cosh s}$ **(b)** $F(s) = \dfrac{\sinh \sqrt{s}\, x}{s \sinh \sqrt{s}}$

where contrary to previous notation we use $\sqrt{s}$ here to represent the branch of $s^{1/2}$ defined by

$$s^{1/2} = |s|^{1/2} e^{(1/2)\arg s}, \qquad 0 \le \arg s < 2\pi.$$

Solution **(a)** This transform has singularities at $s = 0$ and $s = (2n-1)\pi i/2$, n an integer (the zeros of $\cosh s$). The Laurent series of $F(s)$ about $s = 0$ can be found by expanding the hyperbolic functions in Maclaurin series,

$$\begin{aligned} F(s) &= \frac{1}{s^3}\left(1 - 1 - \frac{s^2x^2}{2!} - \frac{s^4x^4}{4!} - \cdots\right) \\ &\quad + \frac{1}{s^3}\left(\frac{s + \frac{s^3}{3!} + \frac{s^5}{5!} + \cdots}{1 + \frac{s^2}{2!} + \frac{s^4}{4!} + \cdots}\right)\left(sx + \frac{s^3x^3}{3!} + \cdots\right) \\ &= \left(-\frac{x^2}{2s} - \frac{x^4 s}{24} - \cdots\right) + \left(s - \frac{s^3}{3} + \cdots\right)\left(\frac{x}{s^2} + \frac{x^3}{6} + \cdots\right) \\ &= \frac{x}{2s}(2 - x) + \frac{xs}{24}(-x^3 + 4x^2 - 8) + \cdots. \end{aligned}$$

Consequently, $F(s)$ has a pole of order 1 at $s = 0$, as does $e^{st}F(s)$. Multiplication of this series by the Maclaurin series for e^{st} gives

$$\begin{aligned} e^{st}F(s) &= \left[1 + st + \frac{(st)^2}{2!} + \cdots\right]\left[\frac{x}{2s}(2 - x) + \frac{xs}{24}(-x^3 + 4x^2 - 8) + \cdots\right] \\ &= \frac{x}{2s}(2 - x) + \frac{xt}{2}(2 - x) + \cdots, \end{aligned}$$

and therefore the residue of $e^{st}F(s)$ at $s = 0$ is $x(2 - x)/2$.

Because the derivative of $\cosh s$ does not vanish at $s = (2n-1)\pi i/2$, these singularities are also poles of order 1, and the residues of $e^{st}F(s)$ at these poles are given by the limits

$$\begin{aligned}
&\lim_{s\to(2n-1)\pi i/2}\left[s-\frac{(2n-1)\pi i}{2}\right]e^{st}\left[\frac{1}{s^3}(1-\cosh sx)+\frac{\sinh s\sinh sx}{s^3\cosh s}\right]\\
&= \frac{e^{(2n-1)\pi ti/2}}{-(2n-1)^3\pi^3 i/8}\sinh\frac{(2n-1)\pi i}{2}\sinh\frac{(2n-1)\pi xi}{2}\\
&\quad\times\lim_{s\to(2n-1)\pi i/2}\frac{s-(2n-1)\pi i/2}{\cosh s}\\
&= \frac{8e^{(2n-1)\pi ti/2}}{(2n-1)^3\pi^3 i}\sin\frac{(2n-1)\pi}{2}\sin\frac{(2n-1)\pi x}{2}\\
&\quad\times\lim_{s\to(2n-1)\pi i/2}\frac{1}{\sinh s}\qquad\text{(using L'Hôpital's rule)}\\
&= \frac{8(-1)^{n+1}e^{(2n-1)\pi ti/2}}{(2n-1)^3\pi^3 i}\sin\frac{(2n-1)\pi x}{2}\frac{1}{\sinh\frac{(2n-1)\pi i}{2}}\\
&= -\frac{8e^{(2n-1)\pi ti/2}}{(2n-1)^3\pi^3}\sin\frac{(2n-1)\pi x}{2}.
\end{aligned}$$

The sum of the residues of $e^{st}F(s)$ at its singularities is therefore

$$f(x,t)=\frac{x}{2}(2-x)-\frac{8}{\pi^3}\sum_{n=-\infty}^{\infty}\frac{e^{(2n-1)\pi ti/2}}{(2n-1)^3}\sin\frac{(2n-1)\pi x}{2}.$$

To simplify this expression we separate it into two summations, one over positive n and the other over nonpositive n, and in the latter we set $m = 1-n$:

$$\begin{aligned}
f(x,t)&=\frac{x}{2}(2-x)-\frac{8}{\pi^3}\sum_{n=1}^{\infty}\frac{e^{(2n-1)\pi ti/2}}{(2n-1)^3}\sin\frac{(2n-1)\pi x}{2}\\
&\quad-\frac{8}{\pi^3}\sum_{n=-\infty}^{0}\frac{e^{(2n-1)\pi ti/2}}{(2n-1)^3}\sin\frac{(2n-1)\pi x}{2}\\
&=\frac{x}{2}(2-x)-\frac{8}{\pi^3}\sum_{n=1}^{\infty}\frac{e^{(2n-1)\pi ti/2}}{(2n-1)^3}\sin\frac{(2n-1)\pi x}{2}\\
&\quad-\frac{8}{\pi^3}\sum_{m=1}^{\infty}\frac{e^{[2(1-m)-1]\pi ti/2}}{[2(1-m)-1]^3}\sin\frac{[2(1-m)-1]\pi x}{2}.
\end{aligned}$$

If we now replace m by n in the second summation and combine it with the first, we have

$$f(x,t) = \frac{x}{2}(2-x) - \frac{8}{\pi^3}\sum_{n=1}^{\infty}\frac{e^{(2n-1)\pi ti/2}}{(2n-1)^3}\sin\frac{(2n-1)\pi x}{2}$$

$$-\frac{8}{\pi^3}\sum_{n=1}^{\infty}\frac{e^{-(2n-1)\pi ti/2}}{(2n-1)^3}\sin\frac{(2n-1)\pi x}{2}$$

$$= \frac{x}{2}(2-x) - \frac{8}{\pi^3}\sum_{n=1}^{\infty}\frac{e^{(2n-1)\pi ti/2}+e^{-(2n-1)\pi ti/2}}{(2n-1)^3}\sin\frac{(2n-1)\pi x}{2}$$

$$= \frac{x}{2}(2-x) - \frac{16}{\pi^3}\sum_{n=1}^{\infty}\frac{1}{(2n-1)^3}\cos\frac{(2n-1)\pi t}{2}\sin\frac{(2n-1)\pi x}{2}.$$

Verification that this function is the inverse of $F(s)$ still requires establishment of a sequence of contours satisfying the conditions of Theorem 8.6. We omit this part of the argument. Transforms of this type occur in vibration problems.

(b) The function $F(s)$ has singularities along its branch cut (the positive real axis) and also at the zeros of $\sinh\sqrt{s}$; that is, when $\sqrt{s} = n\pi i$, or $s = -n^2\pi^2$, $n \geq 0$ an integer. To determine the nature of the singularity at $s = 0$, we substitute $\sqrt{s}\,x$ and $\sqrt{s}$ into the Maclaurin series for $\sinh s$; the result is

$$F(s) = \frac{1}{s}\left[\frac{\sqrt{s}\,x + \frac{1}{3!}(\sqrt{s}\,x)^3 + \cdots}{\sqrt{s} + \frac{1}{3!}(\sqrt{s})^3 + \cdots}\right] = \frac{1}{s}\left[x + \frac{s}{6}(x^3 - x) + \cdots\right].$$

This must be the Laurent series for $F(s)$ about $s = 0$. It shows that, notwithstanding the fact that $\sqrt{s}$ has a branch point at $s = 0$ and a branch cut along the positive real axis, as do $\sinh\sqrt{s}\,x$ and $\sinh\sqrt{s}$, $F(s)$ has a pole of order 1 at $s = 0$. The following expansion shows that the residue of $e^{st}F(s)$ at this pole is x:

$$e^{st}F(s) = \left[1 + st + \frac{(st)^2}{2!} + \cdots\right]\left(\frac{1}{s}\right)\left[x + \frac{s}{6}(x^3 - x) + \cdots\right]$$

$$= \frac{1}{s}\left[x + \frac{s}{6}(6xt + x^3 - x) + \cdots\right].$$

Because the derivative of $\sinh\sqrt{s}$ does not vanish at the remaining singularities $s = -n^2\pi^2$ ($n > 0$), these are also poles of order 1, and the residues of $e^{st}F(s)$ at these poles are given by

$$\lim_{s\to -n^2\pi^2}(s + n^2\pi^2)e^{st}\frac{\sinh\sqrt{s}\,x}{s\,\sinh\sqrt{s}} = e^{-n^2\pi^2 t}\frac{\sinh n\pi xi}{-n^2\pi^2}\lim_{s\to -n^2\pi^2}\frac{s + n^2\pi^2}{\sinh\sqrt{s}}.$$

L'Hôpital's rule can be used to evaluate this limit, which, combined with the facts that $\sinh\theta i = i\sin\theta$ and $\cosh\theta i = \cos\theta$, gives for these residues

$$
\begin{aligned}
-\frac{i}{n^2\pi^2}e^{-n^2\pi^2 t}\sin n\pi x \lim_{s\to -n^2\pi^2}\frac{1}{\frac{1}{2\sqrt{s}}\cosh\sqrt{s}} &= -\frac{2i}{n^2\pi^2}e^{-n^2\pi^2 t}\sin n\pi x\frac{n\pi i}{\cosh n\pi i}\\
&= \frac{2}{n\pi}e^{-n^2\pi^2 t}\sin n\pi x\frac{1}{\cos n\pi}\\
&= \frac{2(-1)^n}{n\pi}e^{-n^2\pi^2 t}\sin n\pi x.
\end{aligned}
$$

Thus, the sum of the residues of $e^{st}F(s)$ at its singularities is

$$f(x,t) = x + \frac{2}{\pi}\sum_{n=1}^{\infty}\frac{(-1)^n}{n}e^{-n^2\pi^2 t}\sin n\pi x.$$

Once again we omit verification of existence of a sequence of contours satisfying Theorem 8.6. Transforms of this type arise in heat-conduction problems.

Exercises 8.3

In Exercises 1–16 use residues to find the inverse Laplace transform of the given function. In Exercises 1–5 verify the existence of contours satisfying the requirements of Theorem 8.6; in Exercises 6–16 neglect this verification.

1. $F(s) = \dfrac{s}{(s-1)^3}$

2. $F(s) = \dfrac{s}{(s^2+4)^2}$

3. $F(s) = \dfrac{1}{s^2(s+3)}$

4. $F(s) = \dfrac{s^2+2}{(s+1)^2(s-3)^3}$

5. $F(s) = \dfrac{s^2}{(s^2+1)(s^2+4)}$

6. $F(s) = \dfrac{s}{s^2-1}$

7. $F(s) = \dfrac{s^3}{(s^2-4)^3}$

8. $F(s) = \dfrac{1}{(s^2-2s+2)^2}$

9. $F(s) = \dfrac{s-1}{(s^2-2s+2)^2}$

10. $F(s) = \dfrac{s^2}{(s^2-2s+2)^2}$

11. $F(x,s) = \dfrac{1}{s}\left(x - \dfrac{\sinh\sqrt{s}\,x}{\sinh\sqrt{s}}\right)$, where $\sqrt{s} = |s|^{1/2}e^{(1/2)\arg s}$, $0 \le \arg s < 2\pi$

12. $F(x,u,s) = \dfrac{\sinh sx \sinh s(1-u)}{s\sinh s}$

13. $F(x,s) = \dfrac{2\sinh sx}{s^3\sinh s}(1-\cosh s) + \dfrac{2}{s^3}(\cosh sx - 1) + \dfrac{x}{s}(1-x)$

14. $F(x,s) = \dfrac{1}{s^3} + \dfrac{\cosh sx}{s^2\sinh s}$

15. $F(x,s) = \dfrac{\sinh sx}{(4s^2+\pi^2)\sinh s}$

16. $F(x,s) = \dfrac{\sinh sx}{(s^2+\pi^2)\sinh s}$

17. We have claimed that to use inversion integral 8.11 directly is usually impossible. Set up the complex combination of real improper integrals for 8.11 when $F(s) = 1/s^2$; that is, express 8.11 in the form

$$\mathcal{L}^{-1}\{1/s^2\} = I_1 + I_2 i$$

where I_1 and I_2 are real improper integrals. Use the line $\gamma = 1$.

8.4 Application of Laplace Transforms to Partial Differential Equations

In Section 8.2 we illustrated the effective use of Laplace transforms in solving ordinary differential equations. The transform replaces a differential equation in $y(t)$ with an algebraic equation in its transform $Y(s)$. It is then a matter of finding the inverse transform of $Y(s)$ either by partial fractions and tables (Section 8.1) or by residues (Section 8.3). Laplace transforms also provide a potent technique for solving partial differential equations. When the transform is applied to the variable t in a partial differential equation for a function $y(x,t)$, the result is an ordinary differential equation for the transform $Y(x,s)$. The ordinary differential equation is solved for $Y(x,s)$, and the function is inverted to yield $y(x,t)$. We illustrate this procedure with four physical examples. The first two examples are on unbounded spatial intervals; inverse transforms are found in tables. The last two examples are on bounded spatial intervals; inverse transforms are calculated with residues.

EXAMPLE 8.9 ....

A very, very long cylindrical rod is placed along the positive x-axis with one end at $x = 0$ (Figure 8.10). The rod is so long that any effects due to its right end may be neglected. Its sides are covered with perfect insulation so that no heat can enter or escape. At time $t = 0$, the temperature of the rod is 0°C throughout. Suddenly the left end of the rod has its temperature raised to h_0 and maintained at this temperature thereafter. The initial boundary-value problem describing temperature $h(x,t)$ at points in the rod is

$$\frac{\partial h}{\partial t} = k\frac{\partial^2 h}{\partial x^2}, \qquad x > 0, \quad t > 0, \tag{8.12a}$$

$$h(0,t) = h_0, \qquad t > 0, \tag{8.12b}$$

$$h(x,0) = 0, \qquad x > 0, \tag{8.12c}$$

where k is a constant called the *thermal diffusivity* of the material in the rod. Use Laplace transforms on variable t to find $h(x,t)$.

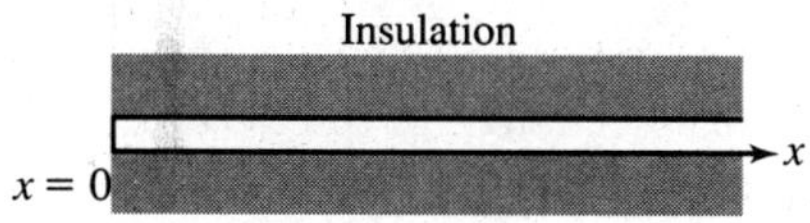

Figure 8.10

Solution When we apply the Laplace transform to the partial differential equation and use property 8.10(a),

$$sH(x,s) - h(x,0) = k\,\mathcal{L}\left\{\frac{\partial^2 h}{\partial x^2}\right\}.$$

Because the integration with respect to t in the Laplace transform and the differentiation with respect to x are independent, we interchange the order of operations on the right,

$$sH(x,s) = k\frac{\partial^2 H}{\partial x^2},$$

where we have also used initial condition 8.12(c). Because only derivatives with respect to x remain, we replace the partial derivative with an ordinary derivative,

$$sH = k\frac{d^2 H}{dx^2}, \qquad x > 0. \tag{8.13a}$$

When we take Laplace transforms of boundary condition 8.12(b), we obtain

$$H(0,s) = \frac{h_0}{s}, \tag{8.13b}$$

a boundary condition to accompany differential equation 8.13(a). The general solution of 8.13(a) is

$$H(x,s) = Ae^{\sqrt{s/k}\,x} + Be^{-\sqrt{s/k}\,x}.$$

Because $h(x,t)$ must remain bounded as x becomes infinite, so also must $H(x,s)$. We must therefore set $A = 0$, in which case 8.13(b) requires $B = h_0/s$. Thus,

$$H(x,s) = \frac{h_0}{s}e^{-\sqrt{s/k}\,x}. \tag{8.14}$$

The inverse transform of this function can be found in tables:

$$h(x,t) = h_0\,\mathcal{L}^{-1}\left\{\frac{e^{-\sqrt{s/k}\,x}}{s}\right\} = h_0\,\mathrm{erfc}\left(\frac{x}{2\sqrt{kt}}\right), \tag{8.15a}$$

where $\mathrm{erfc}(x)$ is the complementary error function

$$\mathrm{erfc}(x) = 1 - \mathrm{erf}(x) = \frac{2}{\sqrt{\pi}}\int_x^{\infty} e^{-u^2}\,du. \tag{8.15b}$$

Notice that for any $x > 0$ and any $t > 0$, temperature $h(x,t)$ is positive. A positive value indicates that the abrupt change in temperature at the end $x = 0$ from 0°C to h_0 is felt instantaneously at every point in the rod. We have shown a plot of $h(x,t)$ for various fixed values of t in Figure 8.11.

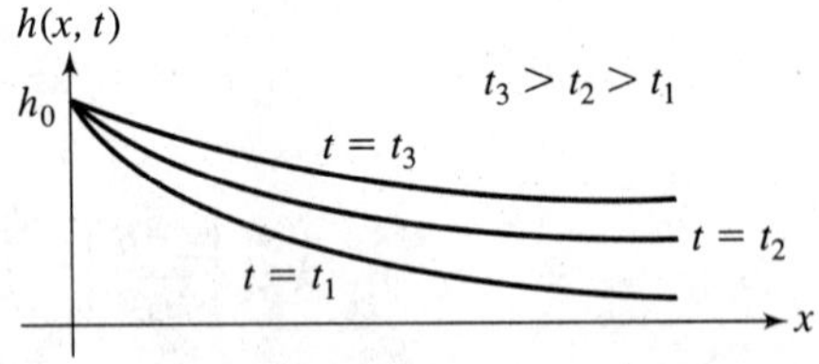

Figure 8.11

EXAMPLE 8.10

A very, very long taut string is supported from below so that it lies motionless on the positive x-axis. At time $t = 0$, the support is removed and gravity is permitted to act on the string. If the end $x = 0$ is fixed at the origin, the initial boundary-value problem describing displacements $y(x,t)$ of points in the string is

$$\frac{\partial^2 y}{\partial t^2} = c^2 \frac{\partial^2 y}{\partial x^2} + g, \qquad x > 0, \quad t > 0, \tag{8.16a}$$

$$y(0,t) = 0, \qquad t > 0, \tag{8.16b}$$

$$y(x,0) = 0, \qquad x > 0, \tag{8.16c}$$

$$y_t(x,0) = 0, \qquad x > 0, \tag{8.16d}$$

where $g = -9.81$ and $c > 0$ is a constant depending on the material and tension of the string. Initial condition 8.16(d) expresses the fact that the initial velocity of the string is zero. Use Laplace transforms to solve this problem.

Solution When we apply Laplace transforms to the partial differential equation and use property 8.10(b),

$$s^2 Y(x,s) - sy(x,0) - y_t(x,0) = c^2 \mathcal{L}\left\{\frac{\partial^2 y}{\partial x^2}\right\} + \frac{g}{s}.$$

We now use initial conditions 8.16(c) and (d) and interchange operations on the right:

$$s^2 Y = c^2 \frac{d^2 Y}{dx^2} + \frac{g}{s},$$

or

$$\frac{d^2 Y}{dx^2} - \frac{s^2}{c^2} Y = -\frac{g}{c^2 s}, \qquad x > 0. \tag{8.17a}$$

This ordinary differential equation is subject to the transform of 8.16(b),

$$Y(0, s) = 0. \tag{8.17b}$$

The general solution of 8.17(a) is

$$Y(x, s) = Ae^{sx/c} + Be^{-sx/c} + \frac{g}{s^3}.$$

For this function to remain bounded as $x \to \infty$, we must set $A = 0$, in which case 8.17(b) implies that $B = -g/s^3$. Thus,

$$Y(x, s) = \frac{g}{s^3} - \frac{g}{s^3}e^{-sx/c}. \tag{8.18}$$

Property 8.4(b) gives

$$y(x, t) = \frac{gt^2}{2} - \frac{g}{2}\left(t - \frac{x}{c}\right)^2 H\left(t - \frac{x}{c}\right) \tag{8.19}$$

where $H(t - x/c)$ is the Heaviside unit step function. What this says is that a point x in the string falls freely under gravity for $0 < t < x/c$, after which it falls with constant velocity gx/c (because for $t > x/c$, $y(x, t) = -(g/2)(-2xt/c + x^2/c^2)$). A picture of the string at any given time t_0 is shown in Figure 8.12. It is parabolic for $0 < x < ct_0$ and horizontal for $x > ct_0$. As t_0 increases, the parabolic portion lengthens and the horizontal section drops.

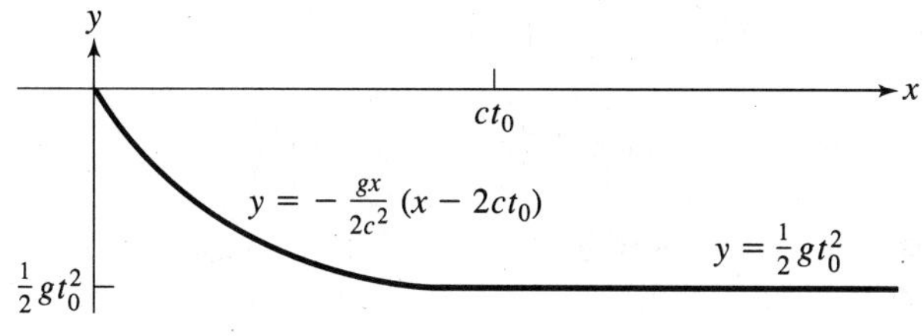

Figure 8.12

The next two examples are on bounded intervals; we use residues to find inverse transforms.

EXAMPLE 8.11

A cylindrical rod of length L has its ends at $x = 0$ and $x = L$ on the x-axis. Its sides are insulated so that no heat can enter or escape through them. At time $t = 0$, the left end ($x = 0$) has temperature 0°C and the right end ($x = L$) has temperature L degrees Celsius, and the temperature rises linearly between the ends. Suddenly at time $t = 0$, the temperature of the right end is reduced to 0°C, and both ends are held at temperature zero thereafter. The initial boundary-value problem describing temperature $h(x,t)$ at points in the rod is

$$\frac{\partial h}{\partial t} = k\frac{\partial^2 h}{\partial x^2}, \qquad 0 < x < L, \quad t > 0, \tag{8.20a}$$

$$h(0,t) = 0, \qquad t > 0, \tag{8.20b}$$

$$h(L,t) = 0, \qquad t > 0, \tag{8.20c}$$

$$h(x,0) = x, \qquad 0 < x < L, \tag{8.20d}$$

where k is a constant called the *thermal diffusivity* of the material in the rod. Use Laplace transforms to find $h(x,t)$.

Solution When we take Laplace transforms of 8.20(a), and use property 8.10(a),

$$sH(x,s) - x = k\frac{d^2 H}{dx^2},$$

or

$$\frac{d^2 H}{dx^2} - \frac{s}{k}H = -\frac{x}{k}, \qquad 0 < x < L. \tag{8.21a}$$

This ordinary differential equation is subject to the transforms of 8.20(b) and (c),

$$H(0,s) = 0, \tag{8.21b}$$

$$H(L,s) = 0. \tag{8.21c}$$

The general solution of 8.21(a) is

$$H(x,s) = A\cosh\sqrt{\frac{s}{k}}\,x + B\sinh\sqrt{\frac{s}{k}}\,x + \frac{x}{s}.$$

For Example 8.9 on an unbounded interval, we used exponentials in the solution of 8.13(a). On bounded intervals, hyperbolic functions are preferable. Boundary conditions 8.21(b) and (c) require

$$0 = A, \qquad 0 = A\cosh\sqrt{\frac{s}{k}}\,L + B\sinh\sqrt{\frac{s}{k}}\,L + \frac{L}{s}.$$

From these,

$$H(x,s) = \frac{1}{s}\left(x - \frac{L\sinh\sqrt{s/k}\,x}{\sinh\sqrt{s/k}\,L}\right). \tag{8.22}$$

Although $\sqrt{s/k}$ denotes the principal square root function, $H(x,s)$ in 8.22 is a solution of 8.21 for any branch of the root function.

It remains now to find the inverse transform of $H(x,s)$. We do this by calculating residues of $e^{st}H(x,s)$ at its singularities. To discover the nature of the singularity at $s=0$, we expand $H(x,s)$ in a Laurent series around $s=0$. Provided we use the same branch for the root function in numerator and denominator of 8.22, we may write that

$$H(x,s) = \frac{1}{s}\left\{x - \frac{L[\sqrt{s/k}\,x + (\sqrt{s/k}\,x)^3/3! + \cdots]}{\sqrt{s/k}\,L + (\sqrt{s/k}\,L)^3/3! + \cdots}\right\}$$
$$= \frac{1}{s}\left[x - \frac{x + sx^3/(6k) + \cdots}{1 + sL^2/(6k) + \cdots}\right]$$
$$= \frac{1}{s}\left[\frac{sx(L^2 - x^2)}{6k} + \cdots\right]$$
$$= \frac{x(L^2 - x^2)}{6k} + \text{terms in } s,\ s^2,\ \cdots.$$

Because this is the Laurent series of $H(x,s)$ valid in some annulus around $s = 0$, it follows that $H(x,s)$ has a removable singularity at $s = 0$.

To invert $H(x,s)$ with residues, the function cannot have a branch cut in the left half-plane $\operatorname{Im} s < 0$. Suppose we choose a branch of $\sqrt{s/k}$ in 8.22 with branch cut along the positive real axis. We continue to denote this branch by $\sqrt{s/k}$, notwithstanding the fact that it is no longer the principal branch. Singularities of $H(x,s)$ in the left half-plane then occur at the zeros of $\sinh\sqrt{s/k}\,L$ — that is, when $\sqrt{s/k}\,L = n\pi i$, or, $s = -n^2\pi^2k/L^2$, n a positive integer. Because the derivative of $\sinh\sqrt{s/k}\,L$ does not vanish at $s = -n^2\pi^2k/L^2$, this function has zeros of order 1 at $s = -n^2\pi^2k/L^2$. It follows that $H(x,s)$ has simple poles at these singularities, as does $e^{st}H(x,s)$. Residues of $e^{st}H(x,s)$ at these poles are

$$\lim_{s\to -n^2\pi^2k/L^2}\left(s + \frac{n^2\pi^2k}{L^2}\right)\frac{e^{st}}{s}\left(x - \frac{L\sinh\sqrt{s/k}\,x}{\sinh\sqrt{s/k}\,L}\right)$$
$$= -\frac{e^{-n^2\pi^2kt/L^2}}{-n^2\pi^2k/L^2}L\sinh\frac{n\pi xi}{L}\lim_{s\to -n^2\pi^2k/L^2}\frac{s + n^2\pi^2k/L^2}{\sinh\sqrt{s/k}\,L}.$$

L'Hôpital's rule, together with the facts that $\sinh\theta i = i\sin\theta$ and $\cosh\theta i = \cos\theta$, yields

$$\frac{iL^3}{n^2\pi^2k}e^{-n^2\pi^2kt/L^2}\sin\frac{n\pi x}{L}\lim_{s\to -n^2\pi^2k/L^2}\frac{1}{\frac{L}{2\sqrt{ks}}\cosh\sqrt{s/k}\,L}$$
$$= \frac{2iL^2}{n^2\pi^2k}e^{-n^2\pi^2kt/L^2}\sin\frac{n\pi x}{L}\frac{1}{\frac{L}{n\pi ki}\cosh n\pi i}$$
$$= \frac{2L}{n\pi}(-1)^{n+1}e^{-n^2\pi^2kt/L^2}\sin\frac{n\pi x}{L}.$$

We sum these residues to find the inverse transform of $H(x,s)$,

$$h(x,t) = \frac{2L}{\pi}\sum_{n=1}^{\infty}\frac{(-1)^{n+1}}{n}e^{-n^2\pi^2kt/L^2}\sin\frac{n\pi x}{L}. \tag{8.23}$$

....

Readers familiar with other methods for solving partial differential equations will be well aware that this solution can also be obtained by separation of variables. Exponentials enhance convergence for large values of t. For instance, suppose the rod is $\frac{1}{5}$ m in length and is made from stainless steel with thermal diffusivity $k = 3.87 \times 10^{-6}$ m^2/s. Consider finding the temperature at the midpoint $x = \frac{1}{10}$ m of the rod at the four times $t = 2$, 5, 30, and 100 min. Series 8.23 gives

$$h\left(\frac{1}{10}, 120\right) = \frac{2}{5\pi}\sum_{n=1}^{\infty}\frac{(-1)^{n+1}}{n}e^{-0.1145861n^2}\sin\frac{n\pi}{2} \approx 0.100°\text{C};$$

$$h\left(\frac{1}{10}, 300\right) = \frac{2}{5\pi}\sum_{n=1}^{\infty}\frac{(-1)^{n+1}}{n}e^{-0.28646526n^2}\sin\frac{n\pi}{2} \approx 0.092°\text{C};$$

$$h\left(\frac{1}{10}, 1800\right) = \frac{2}{5\pi}\sum_{n=1}^{\infty}\frac{(-1)^{n+1}}{n}e^{-1.7187915n^2}\sin\frac{n\pi}{2} \approx 0.023°\text{C};$$

$$h\left(\frac{1}{10}, 6000\right) = \frac{2}{5\pi}\sum_{n=1}^{\infty}\frac{(-1)^{n+1}}{n}e^{-5.7293052n^2}\sin\frac{n\pi}{2} \approx 0.0004°\text{C}.$$

To obtain these approximations we used four nonzero terms in the first series, three in the second, and one in each of the third and fourth.

Laplace transforms also yield a representation for temperature in the rod that is particularly valuable when t is small. This representation is not available through separation of variables. We write the transform 8.22 in the form

$$H(x,s) = \frac{x}{s} - \frac{L\sinh\sqrt{s/k}\,x}{s\sinh\sqrt{s/k}\,L} = \frac{x}{s} - \frac{L}{s}\frac{e^{\sqrt{s/k}\,x} - e^{-\sqrt{s/k}\,x}}{e^{\sqrt{s/k}\,L} - e^{-\sqrt{s/k}\,L}}$$

$$= \frac{x}{s} - \frac{L}{s}\frac{e^{-\sqrt{s/k}\,L}(e^{\sqrt{s/k}\,x} - e^{-\sqrt{s/k}\,x})}{1 - e^{-2\sqrt{s/k}\,L}}.$$

If we regard $1/(1 - e^{-2\sqrt{s/k}\,L})$ as the sum of an infinite geometric series with common ratio $e^{-2\sqrt{s/k}\,L}$, then

$$H(x,s) = \frac{x}{s} - \frac{L}{s}\left[e^{\sqrt{s/k}\,(x-L)} - e^{-\sqrt{s/k}\,(x+L)}\right]\sum_{n=0}^{\infty}e^{-2n\sqrt{s/k}\,L}$$

$$= \frac{x}{s} - L\sum_{n=0}^{\infty}\left[\frac{e^{-\sqrt{s/k}\,[(2n+1)L-x]}}{s} - \frac{e^{-\sqrt{s/k}\,[(2n+1)L+x]}}{s}\right].$$

Tables of Laplace transforms indicate that

$$\mathcal{L}^{-1}\left\{\frac{e^{-a\sqrt{s}}}{s}\right\} = \text{erfc}\left(\frac{a}{2\sqrt{t}}\right)$$

where erfc(x) is the complementary error function in 8.15(b). Hence $h(x,t)$ may be expressed as a series of complementary error functions,

$$h(x,t) = x - L\sum_{n=0}^{\infty}\left\{\operatorname{erfc}\left[\frac{(2n+1)L-x}{2\sqrt{kt}}\right] - \operatorname{erfc}\left[\frac{(2n+1)L+x}{2\sqrt{kt}}\right]\right\}$$

$$= x - L\sum_{n=0}^{\infty}\left\{\operatorname{erf}\left[\frac{(2n+1)L+x}{2\sqrt{kt}}\right] - \operatorname{erf}\left[\frac{(2n+1)L-x}{2\sqrt{kt}}\right]\right\},$$

where we have used the fact that erfc$(x) = 1 - \operatorname{erf}(x)$. This series converges rapidly for small values of t (as opposed to 8.23, which converges rapidly for large t). To see this consider temperature at the midpoint of the stainless steel rod at $t = 300$ s:

$$h\left(\frac{1}{10}, 300\right) = \frac{1}{10} - \frac{1}{5}\sum_{n=0}^{\infty}\left\{\operatorname{erf}\left[\frac{(2n+1)/5 + \frac{1}{10}}{2\sqrt{3.87\times 10^{-6}(300)}}\right] - \operatorname{erf}\left[\frac{(2n+1)/5 - \frac{1}{10}}{2\sqrt{3.87\times 10^{-6}(300)}}\right]\right\}.$$

For $n > 0$, all terms essentially vanish, and

$$h\left(\frac{1}{10}, 300\right) \approx \frac{1}{10} - \frac{1}{5}[\operatorname{erf}(4.402) - \operatorname{erf}(1.467)] = 0.092°\text{C}.$$

For $t = 1800$,

$$h\left(\frac{1}{10}, 1800\right) = \frac{1}{10} - \frac{1}{5}\sum_{n=0}^{\infty}\left\{\operatorname{erf}\left[\frac{(2n+1)/5 + \frac{1}{10}}{2\sqrt{3.87\times 10^{-6}(1800)}}\right] - \operatorname{erf}\left[\frac{(2n+1)/5 - \frac{1}{10}}{2\sqrt{3.87\times 10^{-6}(1800)}}\right]\right\}.$$

Once again, only the $n = 0$ term is required; it yields

$$h\left(\frac{1}{10}, 1800\right) \approx \frac{1}{10} - \frac{1}{5}[\operatorname{erf}(0.1797) - \operatorname{erf}(0.5991)] = 0.023°\text{C}.$$

Even for t as large as 6000, we need only the $n = 0$ and $n = 1$ terms, to give

$$h\left(\frac{1}{10}, 6000\right) = \frac{1}{10} - \frac{1}{5}\sum_{n=0}^{\infty}\left\{\operatorname{erf}\left[\frac{(2n+1)/5 + \frac{1}{10}}{2\sqrt{3.87\times 10^{-6}(6000)}}\right] - \operatorname{erf}\left[\frac{(2n+1)/5 - \frac{1}{10}}{2\sqrt{3.87\times 10^{-6}(6000)}}\right]\right\}$$

$$\approx \frac{1}{10} - \frac{1}{5}[\operatorname{erf}(0.9843) - \operatorname{erf}(0.3281) + \operatorname{erf}(2.297) - \operatorname{erf}(1.641)]$$

$$= 0.004°\text{C}.$$

EXAMPLE 8.12 ....

The ends of a taut string are fixed at $x = 0$ and $x = L$ on the x-axis. The string is initially at rest along the axis; then at time $t = 0$, it is allowed to drop under its own weight. The initial boundary-value problem describing displacements $y(x,t)$ of points in the string is

$$\frac{\partial^2 y}{\partial t^2} = c^2\frac{\partial^2 y}{\partial x^2} + g, \qquad 0 < x < L, \quad t > 0, \tag{8.24a}$$

$$y(0,t) = 0, \qquad t > 0, \tag{8.24b}$$

$$y(L,t) = 0, \qquad t > 0, \tag{8.24c}$$

$$y(x,0) = 0, \qquad 0 < x < L, \tag{8.24d}$$

$$y_t(x,0) = 0, \qquad 0 < x < L, \tag{8.24e}$$

where $g = -9.81$ and $c > 0$ is a constant that depends on the material and tension of the string. Initial condition 8.24(e) expresses the fact that the initial velocity of the string is zero. Use Laplace transforms to solve this problem.

Solution When we take Laplace transforms of 8.24(a) and use initial conditions 8.24(d) and (e),

$$s^2Y(x,s) = c^2\frac{\partial^2 Y}{\partial x^2} + \frac{g}{s},$$

or

$$\frac{d^2Y}{dx^2} - \frac{s^2}{c^2}Y = -\frac{g}{c^2 s}, \qquad 0 < x < L. \tag{8.25a}$$

This ordinary differential equation is subject to the transforms of 8.24(b) and (c),

$$Y(0,s) = 0, \tag{8.25b}$$

$$Y(L,s) = 0. \tag{8.25c}$$

The general solution of 8.25(a) is

$$Y(x,s) = A\cosh\frac{sx}{c} + B\sinh\frac{sx}{c} + \frac{g}{s^3}.$$

Boundary conditions 8.25(b) and (c) require

$$0 = A + \frac{g}{s^3}, \qquad 0 = A\cosh\frac{sL}{c} + B\sinh\frac{sL}{c} + \frac{g}{s^3}.$$

Thus,

$$Y(x,s) = -\frac{g}{s^3}\cosh\frac{sx}{c} + \frac{g}{s^3\sinh sL/c}\left(-1+\cosh\frac{sL}{c}\right)\sinh\frac{sx}{c} + \frac{g}{s^3}. \tag{8.26}$$

We invert this transform function by finding residues of $e^{st}Y(x,s)$ at its singularities. To determine the type of singularity at $s = 0$, we expand $Y(x,s)$ in a Laurent series around $s = 0$:

$$Y(x,s) = \frac{g}{s^3}\left[1 - \left(1 + \frac{s^2x^2}{2c^2} + \frac{s^4x^4}{24c^4} + \cdots\right)\right.$$

$$\left. + \left(\frac{s^2L^2}{2c^2} + \frac{s^4L^4}{24c^4} + \cdots\right)\left(\frac{\frac{sx}{c} + \frac{s^3x^3}{6c^3} + \cdots}{\frac{sL}{c} + \frac{s^3L^3}{6c^3} + \cdots}\right)\right]$$

$$= \frac{g}{s^3}\left[-\left(\frac{s^2x^2}{2c^2} + \frac{s^4x^4}{24c^4} + \cdots\right)\right.$$

$$\left. + \left(\frac{s^2L^2}{2c^2} + \frac{s^4L^4}{24c^4} + \cdots\right)\left(\frac{x}{L} + \frac{s^2x(x-L)}{6Lc^2} + \cdots\right)\right]$$

$$= \frac{gx(L-x)}{2c^2s} + \cdots$$

This result shows that $Y(x,s)$ has a simple pole at $s = 0$, and from the product

$$e^{st}Y(x,s) = \left(1 + st + \frac{s^2t^2}{2!} + \cdots\right)\left[\frac{gx(L-x)}{2c^2s} + \cdots\right]$$

$$= \frac{gx(L-x)}{2c^2s} + \cdots,$$

the residue of $e^{st}Y(x,s)$ at $s = 0$ is $gx(L-x)/(2c^2)$.

The remaining singularities of $Y(x,s)$ occur at the zeros of $\sinh(sL/c)$; that is, when $sL/c = n\pi i$, or $s = n\pi ci/L$, n an integer. Because the derivative of $\sinh(sL/c)$ does not vanish at $s = n\pi ci/L$, this function has zeros of order 1 at $s = n\pi ci/L$. When n is even, $s = n\pi ci/L$ is a simple zero of $-1 + \cosh(sL/c)$, and therefore these are removable singularities of $Y(x,s)$. When n is odd, $s = n\pi ci/L$ is not a zero of $-1 + \cosh(sL/c)$, and these are therefore simple poles of $Y(x,s)$. Residues of $e^{st}Y(x,s)$ at these poles are

$$\lim_{s \to n\pi ci/L}\left(s - \frac{n\pi ci}{L}\right)\frac{ge^{st}}{s^3}\left[1 - \cosh\frac{sx}{c} + \left(-1 + \cosh\frac{sL}{c}\right)\frac{\sinh(sx/c)}{\sinh(sL/c)}\right]$$

$$= \frac{g}{(n\pi ci/L)^3}e^{n\pi cti/L}(-1 + \cosh n\pi i)\sinh\frac{n\pi xi}{L}\lim_{s\to n\pi ci/L}\frac{s - n\pi ci/L}{\sinh(sL/c)}$$

$$= \frac{gL^3 i}{n^3\pi^3c^3}e^{n\pi cti/L}(-1 + \cos n\pi)i\sin\frac{n\pi x}{L}\lim_{s\to n\pi ci/L}\frac{1}{(L/c)\cosh(sL/c)}$$

$$= \frac{gL^3}{n^3\pi^3c^3}e^{n\pi cti/L}[1 + (-1)^{n+1}]\sin\frac{n\pi x}{L}\frac{1}{(L/c)\cosh n\pi i}$$

$$= -\frac{gL^2}{n^3\pi^3c^2}e^{n\pi cti/L}[1 + (-1)^{n+1}]\sin\frac{n\pi x}{L}.$$

Because n is odd, we may write that residues of $e^{st}Y(x,s)$ at the poles $s = (2n-1)\pi ci/L$ are

$$-\frac{2gL^2}{(2n-1)^3\pi^3c^2}e^{(2n-1)\pi cti/L}\sin\frac{(2n-1)\pi x}{L}.$$

Displacements of points in the string are given by

$$y(x,t) = \frac{gx(L-x)}{2c^2} + \sum_{n=-\infty}^{\infty}\frac{-2gL^2}{(2n-1)^3\pi^3c^2}e^{(2n-1)\pi cti/L}\sin\frac{(2n-1)\pi x}{L}.$$

We separate the summation into two, one over positive n and the other over non-positive n, and in the latter we set $m = 1-n$:

$$\begin{aligned}
y(x,t) &= \frac{gx(L-x)}{2c^2} - \frac{2gL^2}{\pi^3c^2}\sum_{n=1}^{\infty}\frac{1}{(2n-1)^3}e^{(2n-1)\pi cti/L}\sin\frac{(2n-1)\pi x}{L}\\
&\quad - \frac{2gL^2}{\pi^3c^2}\sum_{n=-\infty}^{0}\frac{1}{(2n-1)^3}e^{(2n-1)\pi cti/L}\sin\frac{(2n-1)\pi x}{L}\\
&= \frac{gx(L-x)}{2c^2} - \frac{2gL^2}{\pi^3c^2}\sum_{n=1}^{\infty}\frac{1}{(2n-1)^3}e^{(2n-1)\pi cti/L}\sin\frac{(2n-1)\pi x}{L}\\
&\quad - \frac{2gL^2}{\pi^3c^2}\sum_{m=1}^{\infty}\frac{1}{[2(1-m)-1]^3}e^{[2(1-m)-1]\pi cti/L}\sin\frac{[2(1-m)-1]\pi x}{L}.
\end{aligned}$$

If we now replace m by n in the second summation and combine it with the first,

$$\begin{aligned}
y(x,t) &= \frac{gx(L-x)}{2c^2} - \frac{2gL^2}{\pi^3c^2}\sum_{n=1}^{\infty}\frac{1}{(2n-1)^3}e^{(2n-1)\pi cti/L}\sin\frac{(2n-1)\pi x}{L}\\
&\quad - \frac{2gL^2}{\pi^3c^2}\sum_{n=1}^{\infty}\frac{1}{(2n-1)^3}e^{-(2n-1)\pi cti/L}\sin\frac{(2n-1)\pi x}{L}\\
&= \frac{gx(L-x)}{2c^2} - \frac{2gL^2}{\pi^3c^2}\sum_{n=1}^{\infty}\frac{e^{(2n-1)\pi cti/L}+e^{-(2n-1)\pi cti/L}}{(2n-1)^3}\sin\frac{(2n-1)\pi x}{L}\\
&= \frac{gx(L-x)}{2c^2} - \frac{4gL^2}{\pi^3c^2}\sum_{n=1}^{\infty}\frac{1}{(2n-1)^3}\cos\frac{(2n-1)\pi ct}{L}\sin\frac{(2n-1)\pi x}{L}.
\end{aligned}\tag{8.27}$$

The first term is the static position that the string would occupy were it slowly lowered under the force of gravity. The series represents oscillations about this position due to the fact that the string was dropped from a horizontal position.

The technique of separation of variables and eigenfunction expansions leads to the following solution of the problem stated in 8.24:

$$y(x,t) = \frac{2gL^2}{\pi^3 c^2} \sum_{n=1}^{\infty} \frac{[1+(-1)^{n+1}]}{n^3} \left(1 - \cos\frac{n\pi ct}{L}\right) \sin\frac{n\pi x}{L}.$$

We can see the advantage of Laplace transforms. They have rendered part of the series solution in closed form, namely, the term $gx(L-x)/(2c^2)$ in 8.27.

Exercises 8.4

In these exercises use Laplace transforms to solve the initial boundary-value problem.

1. A very, very long cylindrical rod is placed along the positive x-axis with one end at $x = 0$. Its curved sides are perfectly insulated so that no heat can enter or escape through them. At time $t = 0$, the temperature of the rod is 0°C throughout. For $t > 0$, heat is added at a constant rate to the left end. The initial boundary-value problem for temperature $h(x,t)$ in the rod is

$$\frac{\partial h}{\partial t} = k\frac{\partial^2 h}{\partial x^2}, \qquad x > 0, \quad t > 0,$$

$$h_x(0,t) = C, \qquad t > 0,$$

$$h(x,0) = 0, \qquad x > 0,$$

   where $k > 0$ and $C < 0$ are constants.

2. A very, very long cylindrical rod is placed along the positive x-axis with one end at $x = 0$. Its curved sides are perfectly insulated so that no heat can enter or escape through them. At time $t = 0$, the temperature of the rod is a constant $\overline{h}$ throughout. For $t > 0$, the left end has temperature h_0. The initial boundary-value problem for temperature $h(x,t)$ in the rod is

$$\frac{\partial h}{\partial t} = k\frac{\partial^2 h}{\partial x^2}, \qquad x > 0, \quad t > 0,$$

$$h(0,t) = h_0, \qquad t > 0,$$

$$h(x,0) = \overline{h}, \qquad x > 0,$$

   where $k > 0$ is a constant.

3. Use convolutions to express the solution to Exercise 1 in integral form when the boundary condition at $x = 0$ is $h_x(0,t) = f(t)$, $t > 0$.

4. Use convolutions to express the solution to Exercise 2 in integral form when the boundary condition at $x = 0$ is $h(0,t) = f(t)$, $t > 0$.

5. A very, very long string lies motionless along the positive x-axis. If the left end ($x = 0$) is subjected to vertical motion described by $f(t)$ for $t > 0$, subsequent displacements $y(x,t)$ of the string are described by the initial boundary-value problem

$$\begin{aligned}
\frac{\partial^2 y}{\partial t^2} &= c^2 \frac{\partial^2 y}{\partial x^2}, && x > 0, \quad t > 0,\\
y(0,t) &= f(t), && t > 0,\\
y(x,0) &= 0, && x > 0,\\
y_t(x,0) &= 0, && x > 0,
\end{aligned}$$

where $c > 0$ is a constant.

6. A very, very long string lies motionless along the positive x-axis. At time $t = 0$, the support is removed and gravity is permitted to act on the string. If the left end ($x = 0$) is subjected to periodic vertical motion described by $\sin \omega t$ for $t > 0$, subsequent displacements $y(x,t)$ of the string are described by the initial boundary-value problem

$$\begin{aligned}
\frac{\partial^2 y}{\partial t^2} &= c^2 \frac{\partial^2 y}{\partial x^2} + g, && x > 0, \quad t > 0,\\
y(0,t) &= \sin \omega t, && t > 0,\\
y(x,0) &= 0, && x > 0,\\
y_t(x,0) &= 0, && x > 0,
\end{aligned}$$

where $g = -9.81$ and $c > 0$ is a constant.

7. A cylindrical rod of length L has its ends at $x = 0$ and $x = L$ on the x-axis. Its curved sides are perfectly insulated so that no heat can enter or escape through them. At time $t = 0$, the temperature of the rod is given by $f(x) = \sin(m\pi x/L)$, $0 \le x \le L$, where $m > 0$ is an integer. For $t > 0$, both ends of the rod are held at temperature $0°$C. The initial boundary-value problem for temperature $h(x,t)$ in the rod is

$$\begin{aligned}
\frac{\partial h}{\partial t} &= k \frac{\partial^2 h}{\partial x^2}, && 0 < x < L, \quad t > 0,\\
h(0,t) &= 0, && t > 0,\\
h(L,t) &= 0, && t > 0,\\
h(x,0) &= \sin \frac{m\pi x}{L}, && 0 < x < L,
\end{aligned}$$

where $k > 0$ is a constant.

8. A cylindrical rod of length L has its ends at $x = 0$ and $x = L$ on the x-axis. Its curved sides are perfectly insulated so that no heat can enter or escape through them. At time $t = 0$, the temperature of the rod is given by $f(x) = x$, $0 \le x \le L$. For $t > 0$, both ends of the rod are insulated. The initial boundary-value problem for temperature $h(x,t)$ in the rod is

$$\begin{aligned}
\frac{\partial h}{\partial t} &= k \frac{\partial^2 h}{\partial x^2}, && 0 < x < L, \quad t > 0,\\
h_x(0,t) &= 0, && t > 0,\\
h_x(L,t) &= 0, && t > 0,\\
h(x,0) &= x, && 0 < x < L,
\end{aligned}$$

where $k > 0$ is a constant.

9. A cylindrical rod of length L has its ends at $x = 0$ and $x = L$ on the x-axis. Its curved sides are perfectly insulated so that no heat can enter or escape through them. At time $t = 0$, the temperature of the rod is 0°C throughout. For $t > 0$, its left end ($x = 0$) is kept at 0°C, and its right end ($x = L$) is kept at a constant h_L degrees Celsius. The initial boundary-value problem for temperature $h(x, t)$ in the rod is

$$\begin{aligned} \frac{\partial h}{\partial t} &= k\frac{\partial^2 h}{\partial x^2}, && 0 < x < L, \quad t > 0, \\ h(0,t) &= 0, && t > 0, \\ h(L,t) &= h_L, && t > 0, \\ h(x,0) &= 0, && 0 < x < L, \end{aligned}$$

where $k > 0$ is a constant.

10. A cylindrical rod of length L has its ends at $x = 0$ and $x = L$ on the x-axis. Its curved sides are perfectly insulated so that no heat can enter or escape through them. At time $t = 0$, the temperature of the rod is a constant h_0 degrees Celsius. For $t > 0$, its end $x = 0$ is insulated, and heat is added to the end $x = L$ at a constant rate. The initial boundary-value problem for temperature $h(x, t)$ in the rod is

$$\begin{aligned} \frac{\partial h}{\partial t} &= k\frac{\partial^2 h}{\partial x^2}, && 0 < x < L, \quad t > 0, \\ h_x(0,t) &= 0, && t > 0, \\ h_x(L,t) &= C, && t > 0, \\ h(x,0) &= h_0, && 0 < x < L, \end{aligned}$$

where $k > 0$ and $C > 0$ are constants.

11. A cylindrical rod of length L has its ends at $x = 0$ and $x = L$ on the x-axis. Its curved sides are perfectly insulated so that no heat can enter or escape through them. At time $t = 0$, the temperature of the rod is a constant 100°C. For $t > 0$, its end $x = 0$ is kept at temperature 0°, and end $x = L$ has temperature $100e^{-t}$. The initial boundary-value problem for temperature $h(x, t)$ in the rod is

$$\begin{aligned} \frac{\partial h}{\partial t} &= k\frac{\partial^2 h}{\partial x^2}, && 0 < x < L, \quad t > 0, \\ h(0,t) &= 0, && t > 0, \\ h(L,t) &= 100e^{-t}, && t > 0, \\ h(x,0) &= 100, && 0 < x < L, \end{aligned}$$

where $k > 0$ is a constant and $k \neq L^2/(n^2\pi^2)$ for any integer n.

12. A cylindrical rod of length L has its ends at $x = 0$ and $x = L$ on the x-axis. Its curved sides are perfectly insulated so that no heat can enter or escape through them. At time $t = 0$, the temperature of the rod is 0°C, and for $t > 0$ the ends of the rod continue to be held at 0°C. When heat generation at each point of the rod is described by the function $e^{-\alpha t}$, where $\alpha \neq n^2\pi^2/L^2$ is a positive constant (n an integer), the initial boundary-value problem for temperature $h(x,t)$ in the rod is

$$\begin{aligned} \frac{\partial h}{\partial t} &= k\frac{\partial^2 h}{\partial x^2} + e^{-\alpha t}, && 0 < x < L, \quad t > 0, \\ h(0,t) &= 0, && t > 0, \\ h(L,t) &= 0, && t > 0, \\ h(x,0) &= 0, && 0 < x < L, \end{aligned}$$

where $k > 0$ is a constant.

13. A taut string has its ends fixed at $x = 0$ and $x = L$ on the x-axis. If it is given an initial displacement at time $t = 0$ of $f(x) = kx(L - x)$, where $k > 0$ is a constant, and no initial velocity, the initial boundary-value problem for displacements $y(x,t)$ of points in the string is

$$\begin{aligned} \frac{\partial^2 y}{\partial t^2} &= c^2\frac{\partial^2 y}{\partial x^2}, && 0 < x < L, \quad t > 0, \\ y(0,t) &= 0, && t > 0, \\ y(L,t) &= 0, && t > 0, \\ y(x,0) &= f(x), && 0 < x < L, \\ y_t(x,0) &= 0, && x > 0, \end{aligned}$$

where $c > 0$ is a constant.

14. Repeat Exercise 13 if the initial displacement is zero and $f(x)$ is the initial velocity of the string.

15. A taut string has its left end fixed at $x = 0$ on the x-axis. Its right end is looped around a vertical support at $x = L$ and is free to move without friction. The string is initially at rest along the x-axis, and then at $t = 0$ it is allowed to drop under its own weight. The initial boundary-value problem for displacements $y(x,t)$ of points in the string is

$$\begin{aligned} \frac{\partial^2 y}{\partial t^2} &= c^2\frac{\partial^2 y}{\partial x^2} + g, && 0 < x < L, \quad t > 0, \\ y(0,t) &= 0, && t > 0, \\ y_x(L,t) &= 0, && t > 0, \\ y(x,0) &= 0, && 0 < x < L, \\ y_t(x,0) &= 0, && 0 < x < L, \end{aligned}$$

where $g = -9.81$ and $c > 0$ is a constant.

16. A taut string initially at rest along the x-axis has its ends fixed at $x = 0$ and $x = L$ on the x-axis. If a periodic external force $F_0 \sin \omega t$, $t \geq 0$, acts at every point on the string, the initial boundary-value problem for displacements $y(x,t)$ of points in the string is

$$\begin{aligned}
&\frac{\partial^2 y}{\partial t^2} = c^2 \frac{\partial^2 y}{\partial x^2} + \frac{F_0}{\rho} \sin \omega t, && 0 < x < L, \quad t > 0, \\
&y(0,t) = 0, && t > 0, \\
&y(L,t) = 0, && t > 0, \\
&y(x,0) = 0, && 0 < x < L, \\
&y_t(x,0) = 0, && 0 < x < L,
\end{aligned}$$

where $c > 0$ and ρ are constants. Assume $\omega \neq n\pi c/L$ for any positive integer n.

17. A taut string initially at rest along the x-axis has its end $x = 0$ fixed on the axis. Its end $x = L$ is subjected to vertical motion described by $A_0 \sin \omega t$, $t \geq 0$. The initial boundary-value problem for displacements $y(x,t)$ of points in the string is

$$\begin{aligned}
&\frac{\partial^2 y}{\partial t^2} = c^2 \frac{\partial^2 y}{\partial x^2}, && 0 < x < L, \quad t > 0, \\
&y(0,t) = 0, && t > 0, \\
&y(L,t) = A_0 \sin \omega t, && t > 0, \\
&y(x,0) = 0, && 0 < x < L, \\
&y_t(x,0) = 0, && 0 < x < L,
\end{aligned}$$

where $c > 0$ is a constant. Consider both the possibility that $\omega \neq n\pi c/L$ for any positive integer n and $\omega = n\pi c/L$ for some positive integer n.

18. A cylindrical rod of length L has its ends at $x = 0$ and $x = L$ on the x-axis. Its curved sides are perfectly insulated so that no heat can enter or escape through them. At time $t = 0$, the temperature of the rod is 0°C, and for $t > 0$, the ends $x = 0$ and $x = L$ of the rod are held at temperatures 100°C and 0°C, respectively. The initial boundary-value problem for temperature $h(x,t)$ in the rod is

$$\begin{aligned}
&\frac{\partial h}{\partial t} = k \frac{\partial^2 h}{\partial x^2}, && 0 < x < L, \quad t > 0, \\
&h(0,t) = 100, && t > 0, \\
&h(L,t) = 0, && t > 0, \\
&h(x,0) = 0, && 0 < x < L,
\end{aligned}$$

where $k > 0$ is a constant. Find two solutions, one in terms of error functions and the other in terms of time exponentials.

19. A cylindrical rod of length L has its ends at $x = 0$ and $x = L$ on the x-axis. Its curved sides are perfectly insulated so that no heat can enter or escape through them. At time $t = 0$, the temperature of the rod is 0°C, and for $t > 0$, its left end $x = 0$ continues to be kept at temperature 0°C. If heat is added to the end $x = L$ at a constant rate, the initial boundary-value problem for temperature $h(x,t)$ in the rod is

$$\frac{\partial h}{\partial t} = k\frac{\partial^2 h}{\partial x^2}, \qquad 0 < x < L, \quad t > 0,$$

$$h(0,t) = 0, \qquad t > 0,$$

$$h_x(L,t) = C, \qquad t > 0,$$

$$h(x,0) = 0, \qquad 0 < x < L,$$

where $k > 0$ and $C > 0$ are constants. Find two solutions, one in terms of error functions and the other in terms of time exponentials.

Appendix A

Answers to Even-Numbered Exercises

Section 1.1

2. $(a+c)+(b+d)\sqrt{3}$ **4.** $(ac+3bd)+(ad+bc)\sqrt{3}$

Section 1.2

2. $-1+6i$ **4.** $-2+11i$ **6.** $7+i$ **8.** $(1/13)-(5/13)i$ **10.** $5-3i$ **12.** $(-24/5)+(33/5)i$ **14.** $-4i$ **18.** $z=0,\ r$ **22.** $(-5\pm\sqrt{13})/2$ **24.** $-4,\ -4$ **26.** $-1\pm\sqrt{6}i$ **28.** $-[5/(2\sqrt{3})]\pm[\sqrt{12\sqrt{5}-25}/(2\sqrt{3})]i$ **30.** $\pm i,\ \pm\sqrt{3}i$ **32.** $3\pm i$ **34.** $\pm(5-2i)$ **36.** $\pm(\sqrt{\sqrt{2}-1}/\sqrt{2}+\sqrt{2\sqrt{2}+2}\,i/2)$ **38.** $\pm\sqrt{3}/2+(1\pm\sqrt{3}/2)i$ **40.** $(-1\pm 3/\sqrt{2\sqrt{10}-2})\pm\sqrt{\sqrt{10}-1}\,i/\sqrt{2}$ **42.** $[-1\pm\sqrt{\sqrt{65}-1}/(2\sqrt{2})]+[\frac{1}{2}\mp\sqrt{\sqrt{65}+1}/(2\sqrt{2})]i$

Section 1.3

2. $2[\cos(-\pi/2)+\sin(-\pi/2)\,i]$ **4.** $5[\cos(0.927)+\sin(0.927)\,i]$ **6.** $2[\cos(2\pi/3)+\sin(2\pi/3)\,i]$ **8.** $0.170+0.985i$ **10.** $4i$ **12.** 1 **14.** $(3/5)-(1/5)i$ **16.** $-1.66-0.654i$

Section 1.4

2. $\pm(5-2i)$ **4.** $1,\ -\frac{1}{2}\pm\sqrt{3}i/2$ **6.** $\pm(\sqrt{3}+i)$ **8.** $1.671-0.364i,\ -0.520+1.629i,\ -1.151-1.265i$ **10.** $\pm(0.607+1.367i),\ \pm(0.607-1.367i)$ **12.** $\cos(2k\pi/n)+\sin(2k\pi/n)i,\ k=0,\dots,n-1$ **14.** Circle with radius 3 and center $z=2-i$ and its interior **16.** Region between concentric circles with center $z=-3$ and radii 1 and 4, including the circles **18.** Ellipse with foci $z=\pm1$ and major axis of length 4 and its interior **20.** Region between, and including, two ellipses with foci $z=\pm1$, one with major axis of length 4, the other with major axis of length 6 **24.** $R^{m/n}[\cos(m\theta/n+2k\pi/n)+\sin(m\theta/n+2k\pi/n)i]$, where R and θ are modulus and principal argument of Z **26.** $\frac{3}{4}[\cos(2\pi/5)-1\pm\sin(2\pi/5)i]/\sin^2(\pi/5)$, $\frac{3}{4}[\cos(4\pi/5)-1\pm\sin(4\pi/5)i]/\sin^2(2\pi/5)$

Section 2.1

2.

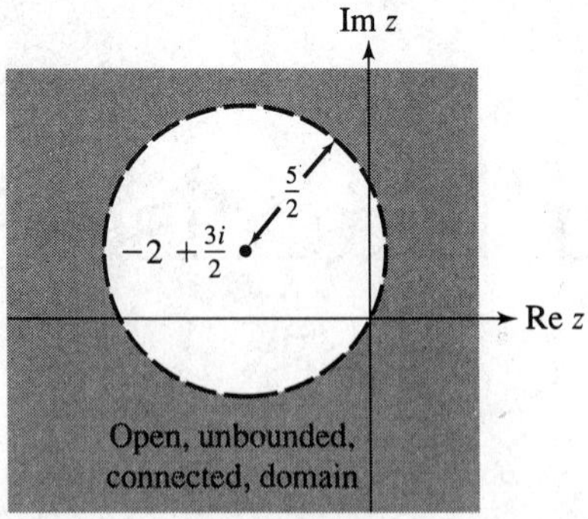

4.

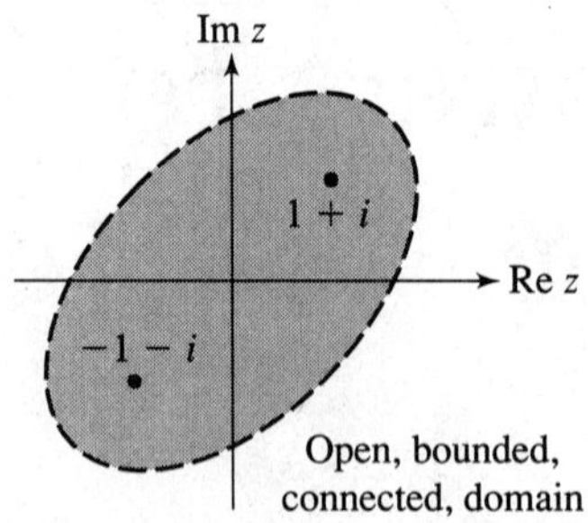

6.

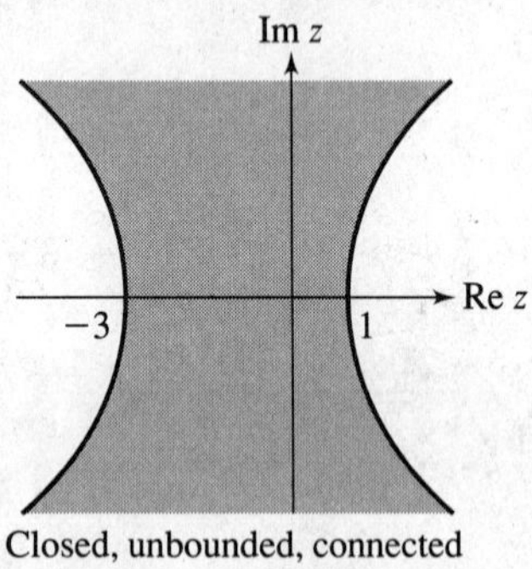

8.

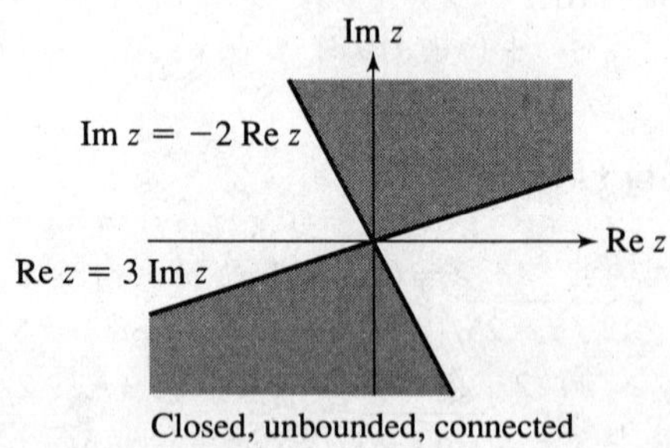

10.

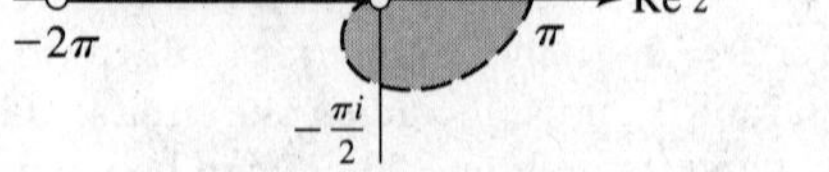

12. Yes **14.** No

Section 2.2

2. $[(3x+1)(1-x)-3y^2]/[(1-x)^2+y^2]$, $4y/[(1-x)^2+y^2]$ **4.** $\sqrt{x^2+y^2}(x^2+y^2+y)/[x^2+(y+1)^2]$, $-x\sqrt{x^2+y^2}/[x^2+(y+1)^2]$ **6.** $r^5(r\cos 4\theta-3\cos 5\theta)/(r^2-6r\cos\theta+9)$, $r^5(r\sin 4\theta-3\sin 5\theta)/(r^2-6r\cos\theta+9)$ **16.** $|w+1/6|=1/3$ **18.** Ellipses $u^2/(R^2+1)^2+v^2/(R^2-1)^2=1/R^2$ when $R\neq 1$; Real axis between $w=\pm 2$ when $R=1$ **20.** Real axis for which $|u|\geq 1/2$ **22.** No **24.** $1/z$ **26.** $z/(z+1)$ **28.** Real axis between $w=\pm 2a$

Section 2.3

2. i **4.** Does not exist **6.** 0 **8.** 0 **10.** $\mathcal{C}$ with $z = -3/2 \pm (\sqrt{23}/2)i$ deleted **12.** $\mathcal{C}$ **14.** $\mathcal{C}$ **16.** $\mathcal{C}$ **18.** $\mathcal{C}$ with $z = 0$ and negative real axis deleted **20.** Yes, if we define $f(0) = 0$

Section 2.4

2. $2z(1+z^3)^9(1+16z^3)$, entire **4.** $3z^2(1-z)^2(-5z^2+2z+2i-4iz)/(3z+2i)^2$, $z \neq -2i/3$ **6.** Does not exist **8.** Does not exist **10.** $(3x^2 - 3y^2 - 2) + 6xyi$, $\mathcal{C}$ **12.** Does not exist **14.** $\cos y \cosh x + \sin y \sinh x\, i$, $\mathcal{C}$ **16.** Not analytic **18.** $(\cos\theta - \sin\theta\, i)/r$ **22.** No, yes

Section 2.5

2. $\mathcal{C}$, $3x^2y - y^3 + 2y + C$ **4.** $\mathcal{C}$, $\cos x \sinh y + C$ **6.** $z \neq 0$, $-y/(x^2+y^2) + C$ **8.** (a) $z \neq 3i$, $6x/(9 + x^2 + y^2 - 6y) + C$ (b) $2z/(z-3i) + Ci$ **10.** $2xy + C$

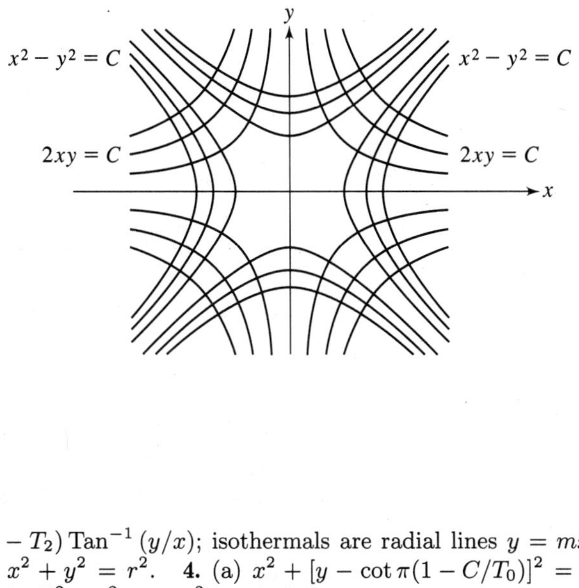

12. No

Section 2.6

2. $T_2 + (2/\pi)(T_1 - T_2)\,\mathrm{Tan}^{-1}(y/x)$; isothermals are radial lines $y = mx$; heat-flow lines are circular arcs $x^2 + y^2 = r^2$. **4.** (a) $x^2 + [y - \cot \pi(1 - C/T_0)]^2 = \csc^2 \pi(1 - C/T_0)$ (c) $[x + \coth(\pi C/T_0)]^2 + y^2 = \mathrm{csch}^2(\pi C/T_0)$

Section 2.7

2. (b) $[-q/(4\pi\epsilon_0)]\ln(x^2+y^2) + C$ (c) $[-q/(2\pi\epsilon_0)]\tan^{-1}(y/x) + D$

Section 2.8

2.

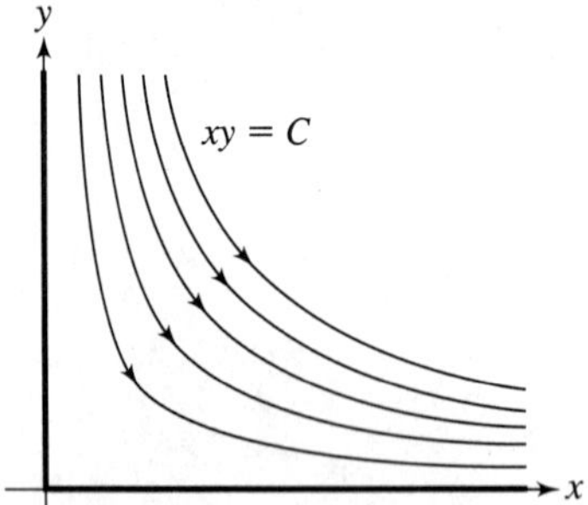

Stagnation point at $z = 0$

4.

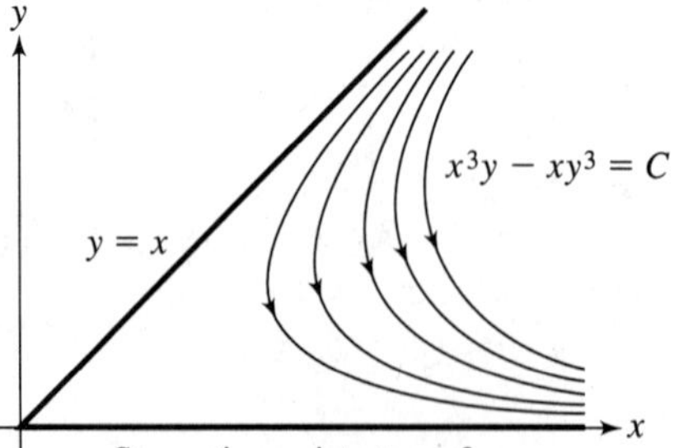

Stagnation point at $z = 0$

6. Uz

Section 3.1

2. Isolated singularity at $z = \frac{2}{3}$ **4.** Zeros of order 1 at $z = \pm 2i$ **6.** Zero of order 2 at $z = -1$ **8.** Zeros of order 1 at $z = \pm\sqrt{3}$; isolated singularities at $z = (-5 \pm \sqrt{21})/2$ **10.** Zero of order 3 at $z = 0$; isolated singularity at $z = -\frac{4}{3}$ **12.** No zeros or singularities **14.** Zero of order 2 at $z = 1$; isolated singularity at $z = -2$ **16.** Zeros of order 1 at $z = -1$, $(1 \pm \sqrt{3}i)/2$; isolated singularities at $z = 1$, $(-1 \pm \sqrt{3}i)/2$ **18.** Zeros of order 1 at $z = (-2 \pm \sqrt{2})i$ **20.** Zeros of orders 2 and 1 at $z = 0$ and $z = 2$; isolated singularities at $z = (-2/3 \pm 1/\sqrt{2}) \pm i/(3\sqrt{2})$ **22.** Zeros of order 1 at $z = \pm(0.676 + 0.978i)$, $\pm(0.676 - 0.978i)$; isolated singularity at $z = -5$

Section 3.2

2. $2e^{2\pi i/3}$ **4.** $5e^{i\text{Tan}^{-1}(4/3)}$ **6.** $5e^{[\text{Tan}^{-1}(4/3) - \pi]i}$ **8.** $4e^{9\pi i/10}$ **10.** $3e^{3\pi i/4}$ **12.** $(e^{-1}\cos 4) - (e^{-1}\sin 4)\,i$ **14.** $\cos 1 + (\sin 1)\,i$ **16.** $\ln 4 + (4n + 1)\pi i/2$ **18.** $(1/2)\ln 13 + [2n\pi - \text{Tan}^{-1}(3/2)]i$ **20.** $(2n + 1)\pi i$ **22.** $(1/2)\ln 5 + [2n\pi \pm \text{Cos}^{-1}(-3\sqrt{5}/10)]i$ **24.** $(\cos 1 - 4\sin 1) + (\sin 1 + 4\cos 1)i$ **26.** $e^{-3}\sin 1 - e^{-3}\cos 1i$ **28.** $w = e^{\pi z/a}$ **30.** Flow around a circle of radius a placed in uniform flow with speed V_0 at angle α with the positive x-axis

Section 3.3

2. $\cos 1 \cosh 1 + \sin 1 \sinh 1\, i$ **4.** $\sin 5.6 \cosh 4.1 + \cos 5.6 \sinh 4.1\, i$ **6.** $\cosh 1 \sinh 1\, i$ **8.** $-2\sin 1 + 3\sin 1\, i$ **10.** $\coth 1\, i$ **12.** $-0.0417 - 0.0906i$ **14.** $2n\pi + \ln(4 \pm \sqrt{15})\, i$ **16.** No solution **20.** $-\sin 1 \sinh 2 + \cos 1 \cosh 2\, i$ **22.** $2\coth 1 \operatorname{csch} 1$ **24.** $8\cos 8\, i$ **28.** The half-plane $\operatorname{Im} w \geq 0$ **30.** (a) Real z or $z = (2n + 1)\pi/2 + yi$ (b) $z = n\pi + yi$ **32.** (a) $\cos x \cosh y\,\hat{\imath} + \sin x \sinh y\,\hat{\jmath}$ (b) Lines $y = 0$ and $x = (2n + 1)\pi/2$ (c) Stagnation points are $(\pm\pi/2, 0)$.

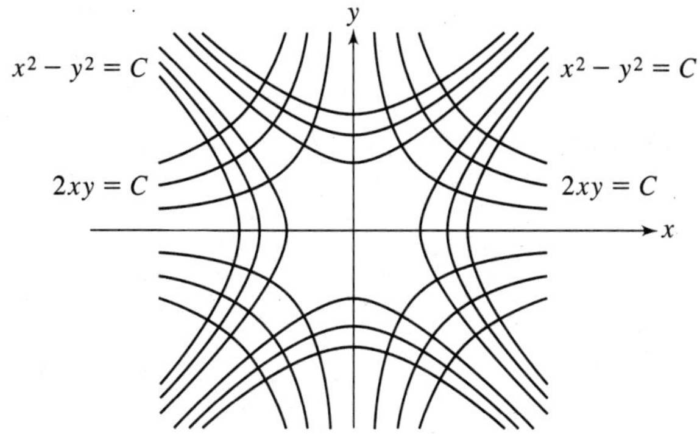

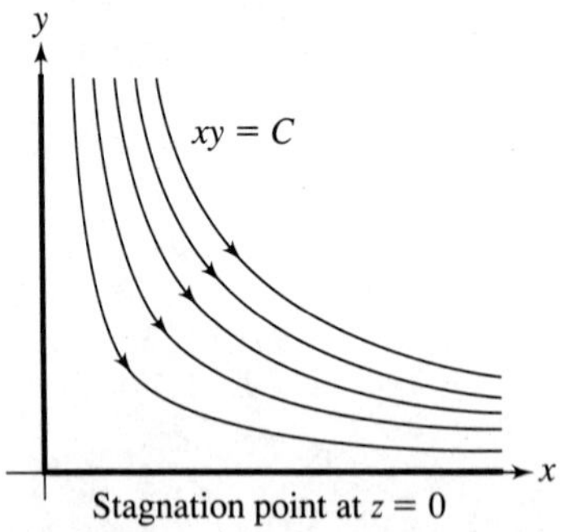

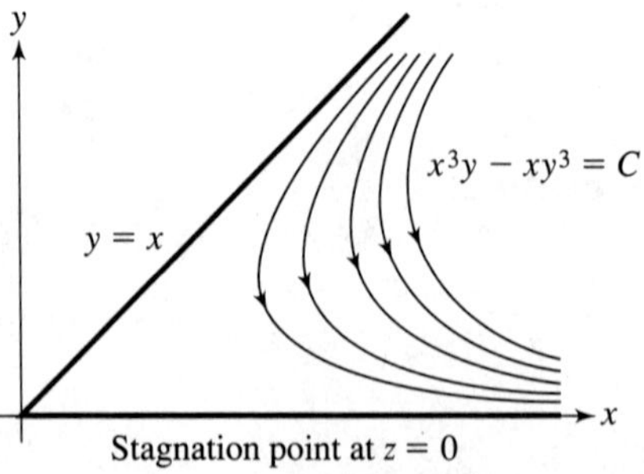

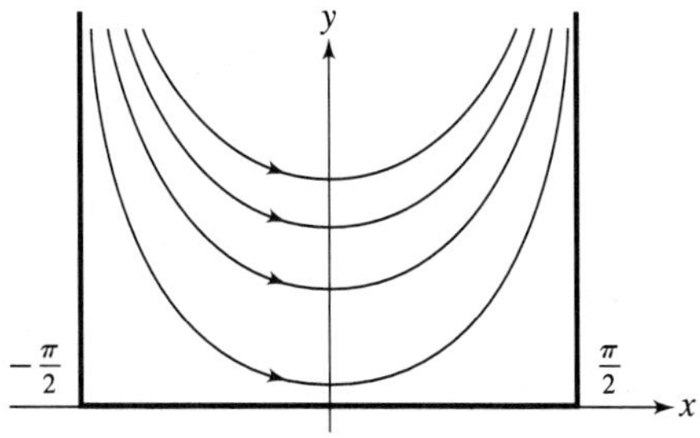

34. (b) $(-2/a)(\sin\phi\cosh\psi\,\hat{\imath}+\cos\phi\sinh\psi\,\hat{\jmath})/(\cosh 2\psi-\cos 2\phi)$, $(\sqrt{2}/a)/\sqrt{\cosh 2\psi-\cos 2\phi}$, clockwise

Section 3.4

2. $\cos 2\cosh 1-\sin 2\sinh 1\,i$ **4.** i **6.** $(\cos 2\sinh 1-\sin 2\cosh 1\,i)/(\sinh^2 1+\sin^2 2)$ **8.** $(\cos 5\cosh 4-\sin 5\sinh 4\,i)/(\sinh^2 4+\cos^2 5)$ **10.** $\ln(3\pm 2\sqrt{2})+(4n-1)\pi i/2$ **12.** $n\pi i$ **16.** $3\sec^2 3$ **18.** $-9\sec 5(1+2\tan^2 5)$ **22.** $i\sinh[\pi z/(2a)]$

Section 3.5

2. $\frac{1}{2}\ln 40-(\text{Tan}^{-1}3)\,i$ **4.** $\frac{1}{2}\ln 13+[\pi-\text{Tan}^{-1}\frac{3}{2}]\,i$ **6.** $e^2(\cos 3+\sin 3\,i)$ **8.** $(-2+\ln 4)+2n\pi i$ **10.** $2n\pi-i\ln(\sqrt{5}-2)$, $(2n+1)\pi-\ln(\sqrt{5}+2)\,i$; $2n\pi-\ln(4\pm\sqrt{15})\,i$; $(4n+1)\pi/2-i$ **14.** (a) $z=1$ (b) No zeros **16.** 1 **18.** (a) $[q/(4\pi\epsilon_0)]\ln\{[(x-x_1)^2+(y-y_1)^2]/[(x-x_0)^2+(y-y_0)^2]\}$ (b) Equipotentials are circles with centers on the x-axis; lines of force are circles passing through q and $-q$ with centers on the y-axis. **20.** (a) $|k|/\sqrt{x^2+y^2}$ (b) 0, 0 (c) 0, $\pm 2\rho k\pi$ **22.** Only if $-\pi<\text{Im}\,z\le\pi$ **24.** $z=-i$; the half-line $\text{Im}\,z=-1$, $\text{Re}\,z<0$ **26.** $z=\frac{3}{2}$; the half-line $\text{Im}\,z=0$, $\text{Re}\,z>\frac{3}{2}$ **28.** (a) $z=\pm 2$ **30.** $z=2i$; the half-line $\text{Im}\,z=2$, $\text{Re}\,z>0$ **32.** $z=2$; the half-line $\text{Re}\,z=2$, $\text{Im}\,z>0$ **34.** (a) $z=\pm i$ **38.** The same branch cuts; that part of the real z-axis satisfying $|\text{Re}\,z|>2$ **40.** Branch cuts for $f(z)$ are those parts of $x^2-y^2=4$ in the first and third quadrants. Branch cuts for $g(z)$ are the half-lines $\text{Re}\,z=-2$, $\text{Im}\,z<0$ and $\text{Re}\,z=2$, $\text{Im}\,z>0$.

Section 3.6

2. $e^{-3(4n+1)\pi/2}\cos(3\ln 2)+e^{-3(4n+1)\pi/2}\sin(3\ln 2)\,i$ **4.** $\pm[2^{1/4}\cos(\pi/8)+2^{1/4}\sin(\pi/8)]\,i$ **6.** $\pm\sqrt{3}+i$, $-2i$ **8.** $\pm[\cos(3\pi/8)+\sin(3\pi/8)\,i]$, $\pm[\cos(7\pi/8)+\sin(7\pi/8)\,i]$ **10.** $\sqrt{2}e^{(8n+1)\pi/4}\cos(\pi/4-\ln\sqrt{2})+\sqrt{2}e^{(8n+1)\pi/4}\sin(\pi/4-\ln\sqrt{2})i$ **12.** $2^{-4/3}(\sqrt{3}+i)/3$ **14.** No **16.** $0.325+0.0768i$ **18.** Not always **20.** $\cos(2\sqrt{2}n\pi)+\sin(2\sqrt{2}n\pi)\,i$ **24.** $z=\frac{2}{3}$; the half-line $\text{Im}\,z=0$, $\text{Re}\,z>\frac{2}{3}$ **26.** $z=\frac{2}{3}-i$; the half-line $\text{Im}\,z=-1$, $\text{Re}\,z<\frac{2}{3}$ **28.** $z=\pm 2$; that part of the real z-axis satisfying $|\text{Re}\,z|>2$ **30.** $z=-2\pm i$; the half-line $\text{Re}\,z=-2$, $|\text{Im}\,z|>1$ **32.** $z=-3$; that part of the real z-axis satisfying $\text{Re}\,z>-3$ **34.** $z=-\frac{2}{3}$; the half-line $\text{Re}\,z=-\frac{2}{3}$, $\text{Im}\,z<0$ **36.** $z=2-i$; the half-line $\text{Im}\,z=-1$, $\text{Re}\,z<2$ **38.** $z=\pm 2i$; the real axis and line segment joining $z=\pm 2i$ **40.** $z=\pm 3i$; parts of $y^2-x^2=9$ in second and fourth quadrants **42.** $z=-2\pm i$; the real axis and the line segment joining $z=-2\pm i$ **44.** (a) $\text{Re}\,z=0$, $|\text{Im}\,z|>a$ (b) $\text{Im}\,z=\pm a$, $\text{Re}\,z<0$ (c) Parts of $y^2-x^2=a^2$ in second and fourth quadrants (d) $\text{Re}\,z=0$, $|\text{Im}\,z|<a$ **46.** (a) The boundary of the domain $\text{Arg}\,z=0$ and $\text{Arg}\,z=\alpha$ (b) $(\pi/\alpha)r^{\pi/\alpha-1}\{\cos[(\pi/\alpha-1)\theta]\,\hat{\imath}-\sin[(\pi/\alpha-1)\theta]\,\hat{\jmath}\}$; $z=0$

(c) x-component positive, y-component negative (d) x-component positive, y-component negative (e) $(\pi/\alpha)r^{\pi/\alpha-1}[\sin(\alpha/2)\,\hat{\imath} - \cos(\alpha/2)\,\hat{\jmath}]$; yes **48.** (a) The rays $\operatorname{Arg} z = \pi$, $\operatorname{Arg} z = \pm\alpha$ (b) $[\pi/(\pi-\alpha)]r^{\alpha/(\pi-\alpha)}\{\cos[\alpha(\theta-\pi)/(\pi-\alpha)]\,\hat{\imath} - \sin[\alpha(\theta-\pi)/(\pi-\alpha)]\,\hat{\jmath}\}$; $z = 0$

Section 3.7

2. $(1-2n)\pi/2 - \frac{1}{2}\operatorname{Tan}^{-1} 2 - \frac{1}{4}\ln 5\, i$ **4.** $\ln(2+\sqrt{5}) + 2n\pi i$, $\ln(\sqrt{5}-2) + (2n+1)\pi i$ **6.** $\ln(\sqrt{3} \pm \sqrt{2}) \pm (4n+1)\pi i/2$ **8.** $-(4n+1)^2\pi^2/16$ **10.** $(4n+1)\pi/2$ **12.** $3/[\sqrt{1-9z^2}\operatorname{Sin}^{-1}(3z)]$ **14.** $3ie^{\operatorname{Sinh}^{-1}(3iz+6)}/\sqrt{37+36iz-9z^2}$ **16.** $-2\operatorname{Cos}^{-1}[\operatorname{Log}(3z)]/[z\sqrt{1-\operatorname{Log}^2(3z)}]$ **18.** $2n\pi i$; $\ln(3 \pm 2\sqrt{2}) + (4n-1)\pi i/2$; $(4n+1)\pi i/4$ **26.** $\operatorname{Im} z = 0$, $|\operatorname{Re} z| \le 1$ **28.** $\log[1/z + (1+1/z^2)^{1/2}]$, $\log[1/z + (1/z^2-1)^{1/2}]$, $(1/2)\log[(z+1)/(z-1)]$ **32.** $z = \pm i$; that part of the imaginary axis satisfying $|\operatorname{Im} z| > 1$

Section 4.1

2. $-1/z - z^2 + C$, $z \ne 0$ **4.** $(4z^3+2)^7/84 + C$ **6.** $-1/(z^2-3z+5)^2 + C$, $z \ne (3 \pm \sqrt{11}i)/2$ **8.** $-\frac{1}{8}\cos(4z^2) + C$ **10.** $\tan z + C$, $z \ne (2n+1)\pi/2$ **12.** $\frac{1}{3}\log_\phi(3z+4i) + C$, except at branch point $z = -4i/3$ and on branch cut **14.** $-\cos z + (1/3)\cos^3 z + C$ **16.** $\frac{4}{9}\log_\phi(z+2) + \frac{5}{9}\log_\psi(z-1) - \frac{1}{3}/(z-1) + C$, except at branch points $z = 1, -2$ and on branch cuts **18.** $-2\sqrt{1-z} + \frac{2}{3}(\sqrt{1-z})^3 + C$, except at branch point $z = 1$ and on branch cut $\operatorname{Im} z = 0$, $\operatorname{Re} z > 1$ **20.** $\cos(1/z) + C$, $z \ne 0$

Section 4.2

2. $x = 3\cos t$, $y = 3\sin t$, $0 \le t \le 2\pi$; $(-3\sin t, 3\cos t)$ **4.** $x = -t$, $y = t^2-2t+5$, $0 \le t \le 2$; $(-1, 2t-2)$ **6.** $x = -1+\sqrt{5}\cos t$, $y = 2+\sqrt{5}\sin t$, $0 \le t \le 2\pi$; $(-\sqrt{5}\sin t, \sqrt{5}\cos t)$ **8.** $x = (5t^2-t^3)/(t+1)$, $y = t$, $1 \le t \le 2$; $((10t+2t^2-2t^3)/(t+1)^2, 1)$ **10.** $x = x_1 + (x_2-x_1)t$, $y = y_1 + (y_2-y_1)t$, $0 \le t \le 1$; (x_2-x_1, y_2-y_1) **12.** $x = t$, $y = t^{5/3}$, $-1 \le t \le 1$ **14.** $\frac{1}{2}$ **16.** $-\frac{8}{3}$ **18.** 4π **20.** 0 **22.** $-\frac{4}{3}$ **24.** $\frac{8}{35}$ **26.** -8π **28.** 9 **30.** (a) 2π (b) 0 **32.** 2π **34.** -4π **36.** $2n\pi$

Section 4.3

2. $-2-i$ **4.** 0 **6.** $\cosh 2 - \cos 2$ **8.** $(3\ln 3 + \ln 2 - 2 - \pi/2) + (3\ln 3 + \ln 2 - 2 + \pi/2)i$

Section 4.4

10. $-2i/5$ **12.** $(2\cosh 2 - \sinh 2)/4 + (2\cos 2 - \sin 2)i/4$ **14.** $\frac{1}{2}\ln 5 - [\pi/2 + \operatorname{Tan}^{-1}(1/2)]i$ **16.** $4\sqrt{2}\sin(\pi/8)\, i$ **18.** $-32/3$ **20.** $-4\sqrt{3}i$

Section 4.5

2. 0 **4.** 0 **6.** 0 **8.** 0 **10.** $\pi(\sqrt{3}-i)/3$ **12.** Yes **14.** No **16.** Yes **18.** No **20.** No **22.** $(1+e^{-\pi})(1-i)/2$

Section 4.6

2. $-\pi^4/4$ **4.** 0 **6.** $-\sqrt{3}\pi/6$ **8.** (a) $-2\pi i/3$ (b) $2\pi i/3$ (c) 0 **10.** $(2\pi \ln 2)i$ **12.** $4\pi^2 - 2\pi(1+\ln 2)i$ **14.** Yes

Section 4.7

2. $f(z) = i$ **4.** $\frac{4}{7}$ **6.** Larger of $\cosh 2a$ and $\cosh 2b$ **8.** 0 **10.** $\frac{1}{10}$ **12.** Only if it is a constant function with $f(z) = k$ where $|k| < 1$

Section 4.8

6. 1 **8.** $(2/\pi)\,\mathrm{Tan}^{-1}(x/y)$ **10.** $\{\pi x + (y/2)\ln\{[(1-x)^2+y^2]/[(1+x)^2+y^2]\} - x\,\mathrm{Tan}^{-1}[(1-x)/y] - x\,\mathrm{Tan}^{-1}[(1+x)/y]\}/[\pi(x^2+y^2)]$ **12.** $u(0,\theta)$
16. (b) No (c) $u_2(x,y)$

Section 5.1

2. Does not exist **4.** $(\sinh 1)i$ **6.** 0 **8.** $-\pi + (\ln 2)i$ **10.** 1 **14.** (a) 0 (b) $-\pi i$ (c) Does not exist

Section 5.2

2. Converges **4.** Converges **6.** Diverges **8.** Converges **10.** Converges **12.** Converges **14.** Converges **16.** Diverges **18.** Converges **20** Converges **22.** $\mathrm{Re}\, z \le 0$

Section 5.3

2. $|z-2| < 1$ **4.** $|z| < 1$ **6.** $|z+4+i| < 1$ **8.** $|z-4| < 1/2$ **10.** $|z-4| < 2^{-1/4}$ **12.** $|z-2+i| < e^{-2}$ **14.** $|z+1| < 3^{1/3}$ **16.** $|z+1| < 4$ **18.** $|z-2| < 1$ **20.** $|z| < 1$ **22.** $-(z+1)^2/[1-i(z+1)]$, $|z+1| < 1$ **24.** $(1+i)(z-i)^3/[(1-i)-(1+i)(z-i)^3]$, $|z-i| < 1$ **26.** $2/(2+i-z)^3$, $|z-1-i| < 1$ **28.** $(1+z)/(1-z)^3$, $|z| < 1$
30. $(3z-z^2)/(1-z)^3$, $|z| < 1$ **32.** $|z| < \infty$ **34.** $\displaystyle\sum_{n=2}^{\infty} \frac{z^n}{2^{n-1}}$, $|z| < 2$
36. $\displaystyle\sum_{n=0}^{\infty} \frac{(-1)^n}{(1+i)^{n+1}}(z-i)^n$, $|z-i| < \sqrt{2}$ **38.** $\displaystyle\sum_{n=1}^{\infty} n3^{n-1}z^n$, $|z| < 1/3$ **42.** $|z| < \sqrt{2}$
44. Diverges for all $|z| = 1$ **46.** Converges absolutely for all $|z| = 1$ **48.** Diverges for all $|z| = 1/3$

Section 5.4

2. $\displaystyle\sum_{n=0}^{\infty} \frac{3(-1)^n}{2^{n+1}}(z-1)^n$, $|z-1| < 2$ **4.** $\displaystyle\sum_{n=0}^{\infty} \frac{1}{n!}z^n$, $|z| < \infty$ **6.** $\displaystyle\sum_{n=0}^{\infty} \frac{(-1)^n}{(2n)!}z^{2n}$, $|z| < \infty$
8. $\displaystyle\sum_{n=0}^{\infty} \frac{1}{(2n)!}z^{2n}$, $|z| < \infty$ **10.** $\displaystyle\sum_{n=1}^{\infty} \frac{(-1)^{n+1}2^{2n-1}}{(2n)!}z^{2n}$, $|z| < \infty$
12. $\displaystyle\sum_{n=0}^{\infty} \frac{(-1)^n}{\sqrt{2}(2n)!}(z-\pi/4)^{2n} + \sum_{n=0}^{\infty} \frac{(-1)^n}{\sqrt{2}(2n+1)!}(z-\pi/4)^{2n+1}$, $|z| < \infty$

14. $\sum_{n=0}^{\infty}(-1)^n\left[1+\frac{2^2}{2!}+\frac{2^4}{4!}+\cdots+\frac{2^{2n}}{(2n)!}\right]z^{2n}$, $|z|<1$

16. $\sum_{n=0}^{\infty}\frac{(-1)^n}{(1+i)^{n+1}}(z-i)^n$, $|z-i|<\sqrt{2}$

18. $\frac{1+i}{2+i}+\sum_{n=1}^{\infty}\frac{(-1)^{n+1}}{(2+i)^{n+1}}(z-1-i)^n$, $|z-1-i|<\sqrt{5}$

20. $\sum_{n=0}^{\infty}\frac{e^{1+i}}{n!}(z-1-i)^n$, $|z-1-i|<\infty$

22. $\frac{\pi i}{2}+\sum_{n=1}^{\infty}\frac{(-1)^{n-1}}{ni^n}(z-i)^n$, $|z-i|<1$

24. $z=0$ of order 3 **26.** $z=n\pi i$ of order 2 **28.** $z=2$ of order 1

30. $\sum_{n=0}^{\infty}\frac{3(-1)^n(1-3^{2n})}{4(2n+1)!}z^{2n+1}$, $|z|<\infty$

32. $\frac{1+i}{\sqrt{2}}+\frac{1-i}{2\sqrt{2}}z+\sum_{n=2}^{\infty}\frac{(-1)^{n+1}(2n-3)!e^{(1-2n)\pi i/4}}{2^{2n-1}(n-1)!\,n!}z^n$, $|z|<1$

36. $\frac{\ln 2}{2}+\frac{3\pi i}{4}+\sum_{n=1}^{\infty}\frac{-e^{n\pi i/4}}{n2^{n/2}}(z+1-i)^n$, $\sqrt{2}$ **38.** No **44.** $|z|>0$, $z=1/(n\pi)$, yes

Section 5.5

2. $\sum_{n=-1}^{\infty}\frac{(-1)^{n+1}}{2^{n+2}}z^n$ **4.** $\sum_{n=0}^{\infty}\frac{(-1)^n}{(2n+1)!\,z^{2n+1}}$

6. $\frac{1/2}{z-2i}+\sum_{n=0}^{\infty}\frac{i^{n-1}}{2^{2n+3}}(z-2i)^n$, $0<|z-2i|<4$ **8.** $\sum_{n=-1}^{\infty}\frac{1}{e(n+1)!}(z+1)^n$, $|z+1|>0$

10. $\sum_{n=1}^{\infty}\frac{(-1)^{n+1}}{2^n(z-1)^n}+\sum_{n=0}^{\infty}-(z-1)^n$ **12.** (b) 8 **14.** $z=0$, essential **16.** $z=0$, pole of order 4 **18.** None **20.** $z=n\pi$, poles of order 1 **22.** $z=\pm i$, $[(2n+1)\pi/2-1]i$, poles of order 1 **24.** $z=n\pi$, $(6n\pm 1)\pi/3$, poles of order 1 **26.** No **28.** (a) Isolated singularity (b) Not a singularity (c) Not a singularity (d) Singularity, but not isolated

Section 6.1

2. 0 **4.** $-\pi^4/4$ **6.** 0 **8.** $-\sqrt{3}\pi/6$ **10.** 0 **12.** $(2\pi\ln 2)i$ **14.** $4\pi^2-2\pi(1+\ln 2)i$ **16.** $\pi(e^2-1)i/4$ **18.** 0 **20.** 0 **22.** $2\pi i$ **24.** $2\pi i$ **26.** $-4\pi i$

Section 6.2

2. $2\pi/\sqrt{5}$ **4.** $2\pi/\sqrt{11}$ **6.** $\pi/\sqrt{5}$ **8.** 0.676 **10.** 0.463 **12.** $\pi/\sqrt{15}$ **14.** $\sqrt{2}\pi/4$ **16.** $2\pi/3^{11/6}$ **18.** $-(\pi\sin 1)/e$ **24.** $\pi/2$ **28.** $\pi(\cos 3-\sin 3)/(2e^3)$ **30.** $(\pi/b)e^{-b}\cos a$

Section 6.3

2. $\pi/2$ **4.** $\pi/\sqrt{2}$ **8.** $(\pi/\sqrt{3})\sin(2/\sqrt{3})$ **10.** $\pi e^{-2\sqrt{3}}(\sqrt{3}\sin 2-\cos 2)/24+(\pi\cos 4)/24$ **12.** $\sqrt{2}\pi(2\ln a-\pi/2)/(8a^3)$ **14.** $\pi/2$ **16.** (a) $2\pi e^{a\pi i/2}$ (d) $(\pi/2)\sec(a\pi/2)$ **18.** $\sqrt{2}\pi/4$

Section 7.1

2. (a) $e < |w| < e^2, -\pi/2 < \operatorname{Arg} w < \pi/2$ (b) $1/e < |w| < e, 0 < \operatorname{Arg} w < 2$ (c) $\operatorname{Im} w > 0$ (d) $|w| > 1$ (e) $|w| > 1$, $\operatorname{Im} w > 0$ **6.** (a) $|w| < 1$, $0 < \operatorname{Arg} w < \pi$ (b) $\operatorname{Im} w > 0$ **10.** (a) $\operatorname{Im} z = 2$, $\operatorname{Re} z < 0$ (b) $0 < \operatorname{Im} w < \pi$

Section 7.2

2. $w = [(1+2i)z - 2i]/(iz+2)$ **4.** $w = [-(18+9i)z - 25 + 27i]/[(14+i)z + 1 + 25i]$ **6.** $w = (z - 5 + 4i)/(3z - 3)$ **8.** $w = [(5-12i)z - 40 - 11i]/(z + 4 + 3i)$ **10.** $w = -[(1+\sqrt{2})iz + 2 + \sqrt{2}]/(z + \sqrt{2}i)$ **12.** $w = (z+1)/(1-z)$ **16.** $w = (R-z)/(R+z)$ **20.** (b) (i) Inverses (ii) Not inverses (iii) Not inverses **24.** (a) $w = (az+bR)/(cz+dR)$ (b) $w = (az + bR - az_0)/(cz + dR - cz_0)$

Section 7.3

14. $w = \sin^2[\pi z/(2a)]$ **16.** $w = -i\sinh(\pi z/2)$ **18.** $w = e^{\pi i z/2}$ **20.** (b) $w = z - 1/(2a) - [i/(2a)]\sqrt{4az-1}$ **22.** $w = z^2 + a^4/z^2$

Section 7.4

8. 10 **10.** $3\sqrt{\sinh^2 2 + \sin^2 3}$ **12.** No

Section 7.5

2. $(2a/\pi)\operatorname{Tan}^{-1}(u/v)$ **4.** $-a + (2a/\pi)\tan^{-1}[2cv/(u^2+v^2-c^2)]$
6. $(1/\pi)\{(a-b)\operatorname{Tan}^{-1}[(c-u)/v] + (b-a)\operatorname{Tan}^{-1}[(c+u)/v] + 2a\operatorname{Tan}^{-1}(u/v)\}$
8. $1/2 + (1/\pi)\operatorname{Tan}^{-1}[(x^2-y^2)/(2xy)]$
10. $(1/\pi)\operatorname{Tan}^{-1}\{[1+\sin(\pi y/2a)\cosh(\pi x/2a)]/[\cos(\pi y/2a)\sinh(\pi x/2a)]\}$
$-(1/\pi)\operatorname{Tan}^{-1}\{[1-\sin(\pi y/2a)\cosh(\pi x/2a)]/[\cos(\pi y/2a)\sinh(\pi x/2a)]\}$
12. $\frac{3}{2} - (1/\pi)\operatorname{Tan}^{-1}[(x^3-3xy^2)/(3x^2y-y^3)]$
14. $b/2 + (1/\pi)\{a\operatorname{Tan}^{-1}(\cot\pi x) + a\operatorname{Tan}^{-1}[(1-e^{-\pi y}\cos\pi x)/(e^{-\pi y}\sin\pi x)]$
$-b\operatorname{Tan}^{-1}[(1+e^{-\pi y}\cos\pi x)/(e^{-\pi y}\sin\pi x)]\}$

Section 7.6

6. $z = (a/\pi)(w + \log_{-\pi/2} w + 1)$

Section 7.7

2. $T_0 + (T_0/\pi)\{\operatorname{Tan}^{-1}[(1+x)/y] - \operatorname{Tan}^{-1}[(1-x)/y]\}$
4. $(2/\pi)\operatorname{Tan}^{-1}[\cos(\pi x/2)/\sinh(\pi y/2)]$
6. $(T_0/\pi)\{\operatorname{Tan}^{-1}[(1+x^2-y^2)/(2xy)] - 2\operatorname{Tan}^{-1}[(x^2-y^2)/(2xy)]$
$-\operatorname{Tan}^{-1}[(1-x^2+y^2)/(2xy)]\}$
8. $(T_0/\pi)\tan^{-1}[\sin(\pi y/a)/\sinh(\pi x/a)]$
10. $(T_1+T_2)/2 + [(T_2-T_1)/\pi]\operatorname{Tan}^{-1}[2Rx/(R^2-x^2-y^2)]$
12. $(2T_0/\pi)\operatorname{Tan}^{-1}\{[R^4-(x^2+y^2)^2]/(4R^2xy)\}$
14. $(2T_0/\pi)\operatorname{Sin}^{-1}\{[\sqrt{(x+1)^2+y^2} - \sqrt{(x-1)^2+y^2}]/2\}$
20. (b) $100 - (100/\pi)\tan^{-1}\{4y(x^2+y^2-1)/[(x^2+y^2)^2 - 2x^2 - 6y^2 + 1]\}$

Section 7.8

4. $V_0 + (V_\alpha - V_0)\theta/\alpha$ **6.** $(V_0/\pi)\tan^{-1}\{(R^2 - x^2 - y^2)/[(x-R)^2 + (y-R)^2 - R^2]\}$, $V_0/4$
8. $(V_0/\pi)\tan^{-1}[\sin(\pi y/a)/\sinh(\pi x/a)]$ **10.** $V_1 + 2a(V_0 - V_1)y/(x^2+y^2)$

Section 7.9

2. (a) Stream curves are $r = [(C/U)\csc(\pi\theta/\alpha)]^{\alpha/\pi}$, C and U constants
(b) $(U\pi/\alpha)r^{\pi/\alpha-1}(-\cos\alpha\,\hat{\mathbf{i}} - \sin\alpha\,\hat{\mathbf{j}}]$ (c) $(U\pi/\alpha)r^{\pi/\alpha-1}[\sin(\alpha/2)\,\hat{\mathbf{i}} - \cos(\alpha/2)\,\hat{\mathbf{j}}]$, yes
(d) y-component negative; x-component sometimes negative and sometimes positive
(e) Yes **4.** (a) $c^2 = (a^2-b^2)/4$, $R = (a+b)/2$ (b) $[U(a+b)/2][(z+\sqrt{z^2+b^2-a^2})/(a+b) + (z-\sqrt{z^2+b^2-a^2})/(a-b)]$ **6.** (a) $U[z - 1/(2a) - i\sqrt{4az-1}/(2a)]$ **8.** $\psi(x,y) = U\cos(\pi x/a)\sinh(\pi y/a)$; $\mathbf{v} = (U\pi/a)[\cos(\pi x/a)\cosh(\pi y/a)\,\hat{\mathbf{i}} + \sin(\pi x/a)\sinh(\pi y/a)\,\hat{\mathbf{j}}]$;
$|\mathbf{v}| = (U\pi/a)\sqrt{\cos^2(\pi x/a)\cosh^2(\pi y/a) + \sin^2(\pi x/a)\sinh^2(\pi y/a)}$
10. Lemniscates $r^2 = -(Ua^3\sin 2\theta)/(2C)$, C a constant
12. $x = (d/\pi)\{1 + e^{\pi u/(\rho U d)}\cos[\pi C/(\rho U d)] + \pi u/(\rho U d)\}$,
$y = (d/\pi)\{e^{\pi u/(\rho U d)}\sin[\pi C/(\rho U d)] + \pi C/(\rho U d)\}$

Section 7.10

2. $k\log\{\sinh[\pi z/(2a)]\}$ **4.** (a) $[Ua/(2\pi h)][\log(z-h) - \log(z+h)]$ (b) $-Ua/(\pi z)$;
$x^2 + [y - Ua/(2\pi C)]^2 = U^2a^2/(4\pi^2C^2)$; $[x + Ua/(2\pi C)]^2 + y^2 = U^2a^2/(4\pi^2C^2)$
6. $k\log(z+a) + k\log(z-a)$; smallest at $(0,0)$; largest at $(0,\pm a)$
8. $k\log(z + a^2/z - c - a^2/c)$ **12.** $k\log\{\sinh[\pi z/(2a)]\}$

Section 8.1

2. (b) (i) $e^{-3s}(s+1)/s^2$ (ii) $(1/s)e^{-as}$ (iii) $(1-e^{-as})/s$ (iv) $(e^{-as} - e^{-bs})/s$ (c) (i) 0, $0 < t < 2$; $t-2$, $t > 2$ (ii) 0, $0 < t < 3$; $\sin(t-3)$, $t > 3$ (iii) 0, $0 < t < 5$; $\cosh\sqrt{2}(t-5)$, $t > 5$ **4.** (b) (i) $[1-e^{-as}(1+as)]/[s^2(1-e^{-as})]$ (ii) $(1-e^{-as})/[s(1+e^{-as})]$ (iii) $a(1+e^{-\pi s/a})/[(s^2+a^2)(1-e^{-\pi s/a})]$ **6.** $(2\sin t - \sin 2t)/6$ **8.** $(\cosh 3t - \cosh 2t)/5$
10. $2(1-e^{-s})/s^3 + (1/s)e^{-s}$ **12.** $1/[s(1+e^{-as})]$ **14.** $e^{-as}(1-e^{-s})/s$ **16.** $e^{-t} + 2\sinh(t/2) - 1$ **18.** 0, $t < 2$; $e^{-(t-2)} - e^{-2(t-2)}$, $t > 2$ **20.** $\frac{1}{3}e^{-2t/3}[5\cos(2\sqrt{5}t/3) - (8/\sqrt{5})\sin(2\sqrt{5}t/3)]$
22. $t^3e^{-t}(4-t)/24$ **24.** $(2t\cosh 2t + \sinh 2t)/4$

Section 8.2

2. $\frac{1}{2}e^t + \frac{1}{2}e^{-t}(\cosh\sqrt{2}t + 2\sqrt{2}\sinh\sqrt{2}t)$
4. $2(1+t)e^{-t} + t - 2$ **6.** $\cos t - 3\sin t + t$
8. $-\frac{9}{194}\cos 3t - \frac{2}{97}\sin 3t + \frac{1}{194}e^{-3t}[397\cosh 2\sqrt{2}t + (1397/2\sqrt{2})\sinh 2\sqrt{2}t]$
10. $\frac{1}{338}[(5+13t)e^{-3t} + e^{2t}(1364\sin t - 343\cos t)]$
12. $-\frac{1}{8} + (\sin 2t - 2\cos 2t)/40 + \frac{1}{40}e^{-t}(7\cosh\sqrt{5}t + \sqrt{5}\sinh\sqrt{5}t)$
14. $-(8\cos 4t + 9\sin 4t)/290 + [2/(145\sqrt{6})]e^{-2t}(2\sqrt{6}\cosh\sqrt{6}t + 13\sinh\sqrt{6}t)$
16. $(t+2)e^{-t} + t - 2 + (2 - t - e^{1-t})H(t-1)$
18. $-(2t/5)e^{-4t} + [(5e^5 - 5e^4 - 2)/(5e^5 - 5)]e^{-4t} + [(5e^4 - 3)/(5e^5 - 5)]e^t$
20. $\frac{3}{2}e^t - \frac{1}{2}e^{3t} + \frac{1}{2}\int_0^t f(u)[e^{3(t-u)} - e^{t-u}]\,du$
22. $(C_1 + C_2t)e^t + (t^4/12)e^t$ **24.** $e^{-t}(C_1\cos 2t + C_2\sin 2t) + \frac{1}{3}e^{-t}\sin t$
26. $C_1\cosh 2t + C_2\sinh 2t + \frac{1}{2}\int_0^t f(u)\sinh 2(t-u)\,du$
28. $e^t(C_1\cos\sqrt{2}t + C_2\sin\sqrt{2}t) - (t/17)\cos 2t - (4t/17)\sin 2t - \frac{62}{289}\cos 2t - \frac{44}{289}\sin 2t$
30. $e^t(t^5 - 30t^2 - 60t + 60)/60$ **32.** $-0.03\cos 5\sqrt{2}t$

34. $-(12\cos 10t + 8\sin 10t)/65 + \frac{4}{65}e^{-15t/4}[3\cos(5\sqrt{23}t/4) + (25/\sqrt{23})\sin(5\sqrt{23}t/4)]$
36. $\frac{1}{20}e^{-t/4}[(159/\sqrt{799})\sin(\sqrt{799}t/4) - \cos(\sqrt{799}t/4)]$
38. $(\sin 8t - 8t\cos 8t)/64$

Section 8.3

2. $(t/4)\sin 2t$ **4.** $[(1+8t+88t^2)e^{3t} - (1+12t)e^{-t}]/256$ **6.** $\cosh t$ **8.** $e^t(\sin t - t\cos t)/2$

10. $(t+1)e^t\sin t$ **12.** $\dfrac{2}{\pi}\sum_{n=1}^{\infty}\dfrac{1}{n}\sin n\pi t\sin n\pi u\sin n\pi x$

14. $t^2 + \dfrac{x^2}{2} - \dfrac{1}{6} + \dfrac{2}{\pi^2}\sum_{n=1}^{\infty}\dfrac{(-1)^{n+1}}{n^2}\cos n\pi t\cos n\pi x$

16. $\dfrac{1}{2\pi^2}[\sin\pi t(\sin\pi x - 2\pi x\cos\pi x) - 2\pi t\cos\pi t\sin\pi x] + \dfrac{2}{\pi^2}\sum_{n=2}^{\infty}\dfrac{(-1)^n}{n^2-1}\sin n\pi t\sin n\pi x$

Section 8.4

2. $h_0 + (h_0 - \overline{h})\,\mathrm{erfc}[x/(2\sqrt{kt})]$

4. $\overline{h} - \overline{h}\,\mathrm{erfc}[x/(2\sqrt{kt})] + \dfrac{2}{\sqrt{\pi}}\displaystyle\int_{x/(2\sqrt{kt})}^{\infty} f\left(t - \dfrac{x^2}{4kv^2}\right)e^{-v^2}\,dv$

6. $gt^2/2 + [\sin\omega(t - x/c) - (g/2)(t - x/c)^2]H(t - x/c)$

8. $\dfrac{L}{2} - \dfrac{4L}{\pi^2}\sum_{n=1}^{\infty}\dfrac{1}{(2n-1)^2}e^{-(2n-1)^2\pi^2kt/L^2}\cos\dfrac{(2n-1)\pi x}{L}$

10. $h_0 + \dfrac{C(3x^2 - L^2)}{6L} + \dfrac{Ckt}{L} + \dfrac{2LC}{\pi^2}\sum_{n=1}^{\infty}\dfrac{(-1)^{n+1}}{n^2}e^{-n^2\pi^2kt/L^2}\cos\dfrac{n\pi x}{L}$

12. $\dfrac{e^{-\alpha t}}{\alpha}\left[-1 + \dfrac{\cos\sqrt{\frac{\alpha}{k}}\left(\frac{L}{2} - x\right)}{\cos\sqrt{\frac{\alpha}{k}}\frac{L}{2}}\right]$

$+\dfrac{4L^2}{\pi}\sum_{n=1}^{\infty}\dfrac{e^{-(2n-1)^2\pi^2kt/L^2}}{(2n-1)[\alpha L^2 - (2n-1)^2\pi^2k]}\sin\dfrac{(2n-1)\pi x}{L}$

14. $\dfrac{8kL^3}{\pi^4 c}\sum_{n=1}^{\infty}\dfrac{1}{(2n-1)^4}\sin\dfrac{(2n-1)\pi ct}{L}\sin\dfrac{(2n-1)\pi x}{L}$

16. $\dfrac{F_0}{\rho\omega^2\sin(\omega L/c)}\left[\sin\dfrac{\omega x}{c} - \sin\dfrac{\omega L}{c} + \sin\dfrac{\omega(L-x)}{c}\right]\sin\omega t$

$+\dfrac{4F_0\omega L^3}{\rho c\pi^2}\sum_{n=1}^{\infty}\dfrac{1}{(2n-1)^2[\omega^2L^2 - (2n-1)^2\pi^2c^2]}\sin\dfrac{(2n-1)\pi ct}{L}\sin\dfrac{(2n-1)\pi x}{L}$

18. $100\sum_{n=0}^{\infty}\left\{\mathrm{erf}\left[\dfrac{2(n+1)L - x}{2\sqrt{kt}}\right] - \mathrm{erf}\left(\dfrac{2nL + x}{2\sqrt{kt}}\right)\right\}$,

$100\left(1 - \dfrac{x}{L}\right) - \dfrac{200}{\pi}\sum_{n=1}^{\infty}\dfrac{1}{n}e^{-n^2\pi^2kt/L^2}\sin\dfrac{n\pi x}{L}$

34. $-(12\cos 10t + 8\sin 10t)/65 + \frac{4}{65}e^{-15t/4}[3\cos(5\sqrt{23}t/4) + (25/\sqrt{23})\sin(5\sqrt{23}t/4)]$
36. $\frac{1}{20}e^{-t/4}[(159/\sqrt{799})\sin(\sqrt{799}t/4) - \cos(\sqrt{799}t/4)]$
38. $(\sin 8t - 8t\cos 8t)/64$

Section 8.3

2. $(t/4)\sin 2t$ **4.** $[(1+8t+88t^2)e^{3t} - (1+12t)e^{-t}]/256$ **6.** $\cosh t$ **8.** $e^t(\sin t - t\cos t)/2$

10. $(t+1)e^t \sin t$ **12.** $\displaystyle \frac{2}{\pi}\sum_{n=1}^{\infty}\frac{1}{n}\sin n\pi t \sin n\pi u \sin n\pi x$

14. $\displaystyle t^2 + \frac{x^2}{2} - \frac{1}{6} + \frac{2}{\pi^2}\sum_{n=1}^{\infty}\frac{(-1)^{n+1}}{n^2}\cos n\pi t \cos n\pi x$

16. $\displaystyle \frac{1}{2\pi^2}[\sin \pi t(\sin \pi x - 2\pi x\cos \pi x) - 2\pi t\cos \pi t \sin \pi x] + \frac{2}{\pi^2}\sum_{n=2}^{\infty}\frac{(-1)^n}{n^2-1}\sin n\pi t \sin n\pi x$

Section 8.4

2. $h_0 + (h_0 - \overline{h})\,\mathrm{erfc}[x/(2\sqrt{kt})]$

4. $\displaystyle \overline{h} - \overline{h}\,\mathrm{erfc}[x/(2\sqrt{kt})] + \frac{2}{\sqrt{\pi}}\int_{x/(2\sqrt{kt})}^{\infty} f\left(t - \frac{x^2}{4kv^2}\right)e^{-v^2}\,dv$

6. $gt^2/2 + [\sin\omega(t - x/c) - (g/2)(t - x/c)^2]H(t - x/c)$

8. $\displaystyle \frac{L}{2} - \frac{4L}{\pi^2}\sum_{n=1}^{\infty}\frac{1}{(2n-1)^2}e^{-(2n-1)^2\pi^2kt/L^2}\cos\frac{(2n-1)\pi x}{L}$

10. $\displaystyle h_0 + \frac{C(3x^2 - L^2)}{6L} + \frac{Ckt}{L} + \frac{2LC}{\pi^2}\sum_{n=1}^{\infty}\frac{(-1)^{n+1}}{n^2}e^{-n^2\pi^2kt/L^2}\cos\frac{n\pi x}{L}$

12. $\displaystyle \frac{e^{-\alpha t}}{\alpha}\left[-1 + \frac{\cos\sqrt{\frac{\alpha}{k}}\left(\frac{L}{2} - x\right)}{\cos\sqrt{\frac{\alpha}{k}}\frac{L}{2}}\right]$

$\displaystyle + \frac{4L^2}{\pi}\sum_{n=1}^{\infty}\frac{e^{-(2n-1)^2\pi^2kt/L^2}}{(2n-1)[\alpha L^2 - (2n-1)^2\pi^2k]}\sin\frac{(2n-1)\pi x}{L}$

14. $\displaystyle \frac{8kL^3}{\pi^4c}\sum_{n=1}^{\infty}\frac{1}{(2n-1)^4}\sin\frac{(2n-1)\pi ct}{L}\sin\frac{(2n-1)\pi x}{L}$

16. $\displaystyle \frac{F_0}{\rho\omega^2\sin(\omega L/c)}\left[\sin\frac{\omega x}{c} - \sin\frac{\omega L}{c} + \sin\frac{\omega(L-x)}{c}\right]\sin\omega t$

$\displaystyle + \frac{4F_0\omega L^3}{\rho c\pi^2}\sum_{n=1}^{\infty}\frac{1}{(2n-1)^2[\omega^2L^2 - (2n-1)^2\pi^2c^2]}\sin\frac{(2n-1)\pi ct}{L}\sin\frac{(2n-1)\pi x}{L}$

18. $\displaystyle 100\sum_{n=0}^{\infty}\left\{\mathrm{erf}\left[\frac{2(n+1)L - x}{2\sqrt{kt}}\right] - \mathrm{erf}\left(\frac{2nL + x}{2\sqrt{kt}}\right)\right\},$

$\displaystyle 100\left(1 - \frac{x}{L}\right) - \frac{200}{\pi}\sum_{n=1}^{\infty}\frac{1}{n}e^{-n^2\pi^2kt/L^2}\sin\frac{n\pi x}{L}$

Index